Information Technology for Management

B. MUTHUKUMARAN

Deputy General Manager
HTC Global
Chennai

Oxford University Press is a department of the University of Oxford.
It furthers the University's objective of excellence in research, scholarship,
and education by publishing worldwide. Oxford is a registered trademark of
Oxford University Press in the UK and in certain other countries

Published in India by
Oxford University Press
YMCA Library Building, 1 Jai Singh Road, New Delhi 110001, India

© Oxford University Press 2010

The moral rights of the author have been asserted

First published in 2010
Second impression 2012

All rights reserved. No part of this publication may be reproduced, stored in
a retrieval system, or transmitted, in any form or by any means, without the
prior permission in writing of Oxford University Press, or as expressly permitted
by law, by licence, or under terms agreed with the appropriate reprographics
rights organization. Enquiries concerning reproduction outside the scope of the
above should be sent to the Rights Department, Oxford University Press, at the
address above

You must not circulate this book in any other form
and you must impose this same condition on any acquirer

ISBN-13: 978-0-19-806414-5
ISBN-10: 0-19-806414-4

Typeset in Baskerville
by Anvi Composers, New Delhi 110 063
Printed in India by Tara Art Printers (P) Ltd, Noida

Third-Party website addresses mentioned in this book are
provided by Oxford University Press in good faith and for information only.
Oxford University Press disclaims any responsibility for the material contained therein.

Preface

The term information technology or IT is commonly used to refer to an entire industry. In actuality, information technology is the use of computers and software to manage information. In some companies, this is referred to as management information system (MIS) or simply as information system (IS). The IT department of a large company is responsible for storing, protecting, processing, transmitting, and retrieving information.

Today, IT involves more than just computer literacy; it also takes into account how computers work and how they can further be used not just for information processing but also for communications and problem-solving tasks. Information technology management is concerned with exploring and understanding information technology as a corporate resource that determines both strategic and operational capabilities of the firm in designing and developing products and services.

In the past decade, major industries have seen an operational change in terms of their structure, regulation, technology, and services. The growth in technology has not only brought a change in the way businessmen saw IT, but has also helped IT in gaining wide recognition in business management. It is quite evident that many primary business processes would no longer function without the contributions made by information systems. Information technology departments and their chief information officers (CIOs) are often confronted with a number of challenges, which force them to reconsider past information management strategies and solutions. Traditional networks are getting replaced by wireless, optical, satellite, and other advanced networks. Information technology departments are under pressure to maintain quality, functionality, and transparency in their services. The effciency of a company's IT utilization is central to its success. Therefore, to adapt and contribute to such changes effectively, communication and network engineers must acquire a solid foundation of the fundamentals of IT.

Information technology has defiantly made a dramatic impact in the business world. From a simple installation of hardware to database management, IT is required for carrying out various business activities. The study of information technology is inevitable in acquiring the knowledge of design, development, installation, and implementation of all types of computer information systems and networks.

This book addresses all the above areas and provides insights into the important terminology, concepts, theories and analytic techniques, used in the field of information technology management. This book, therefore, provides the much-needed technical and management knowledge and skills to effectively integrate information and communication technologies, and business processes in support of the organizational strategic goals.

About the Book

IT for Management is a comprehensive textbook, specially designed for business management students. The fundamentals explained in the book are well supported by examples, exhibits, and cases. Apart from a basic course on IT management, this book can also be used for courses on IT infrastructure management, IT strategy, enterprise governance of IT, information systems management, etc.

The book discusses information technology management in the context of the recent trends and covers major topics such as IT infrastructure, IT production tools, and the Internet. It also elucidates on the hardware and software requirements of computing. A detailed discussion on the application of IT in other fields, such as health, content, supply chain, and customer relationship management, aims to disseminate academic knowledge and would be particularly relevant to practitioners in the field.

Pedagogical Features

Each chapter of the book begins with learning objectives, which give an introduction to the various topics discussed in the chapter. These are followed by a brief introduction to the topic which gives way to the various concepts and theories, which are explained through exhibits, examples and figures. Each chapter includes a summary and a list of key terms and their definitions.

The review questions at the end of each chapter will test the grasp on the various concepts discussed in each chapter and will also guide the readers for interviews/discussions based on technical knowledge. Each chapter concludes with projects (both individual as well as group) for students to ponder upon.

The highlight of the book is a run-through business case study. It is a unique concept that introduces a fresh problem in various chapters, based on a common case. This would help students to examine every possible IT problem that could emerge in a business firm.

The book is written in a simple, lucid, and easy to understand language, to ensure that the students are able to grasp the concepts and apply them effectively.

Other key features of the book can be listed as follows:
- The book explores production tools, various protocols which run the infrastructure system, and management of the business systems.
- It includes an exclusive section on importance of IT tools in content management.
- The book contains a detailed text on MIS, its types, and application in various businesses.

Coverage and Structure

The book consists of 21 chapters, divided across five parts: IT Infrastructure, IT Production Tools, Internet and Network Protocols, IT Management, and IT Applications-Business Systems.

Part I, *IT Infrastructure* begins with a detailed introduction on information technology and the related infrastructures. It covers Chapters 1 to 7 namely, Information Technology, Computing Infrastructure: Hardware, Computing Infrastructure: Software, Networking Infrastructure, Cabling Infrastructure, Wireless Infrastructure, and Storage Infrastructure.

Part II, *IT Production Tools* covers the basics of security infrastructure and various IT production tools. Chapters 8 to 12 in this part are Security Infrastructure, Office Tools, Data Management Tools, and Web Tools.

Part III, *Internet and Network Protocols* elucidates the Internet and network protocols. It consists of chapter 12, which deals with Network Management Tools and Chapter 13, which deals with Protocols and Global Connectivity.

Part IV, *IT Management* discusses management of IT through Chapters 14 to 18. Chapters are on E-business Highway—Business Automation Platform, Infrastructure Management, Security Management, Information Management, and Audit respectively.

Part V, *IT Applications—Business Systems* consisting of Chapters 19 to 21 delineates applications of IT in business systems such as MIS, customer relationship management, E-business, supply chain management, etc. Chapters are on Governance, Connected World and E-commerce, and Information System and Business Systems.

Acknowledgements

I would like to pay a sincere tribute to all those people who took out time from their busy schedules to critically read through all or part of the manuscript. I would like to specially thank Mr Arunachalam. V. S., for his valuable comments. Not only did he highlight inconsistencies and areas of disagreement, but also improvised my presentation style. I sincerely thank the Storage Consultant at Gemini Communications Ltd., for his inputs on the storage section. I also thank Mr Ramkumar R., and the CEO, Gemini Communications Ltd., for their constant support throughout the book.

The project has benefitted from all the reviews from the experts in this field. I thank the editorial team of Oxford University Press for their support in the making of this book. Their contributions throughout the publishing process, from inception of the initial idea to final publication, have been invaluable. Without their patience and support, this book would not have existed.

I express my gratitude towards my wife and children for their constant encouragement and understanding, and especially to my daughter for having gone through the manuscript along with me. Any errors and unclear sections are my responsibility.

I also thank all those, who have been responsible for my growth as a teacher, a writer, and a businessman.

B. MUTHUKUMARAN

Brief Contents

Preface   *iii*

PART I   IT INFRASTRUCTURE

1. Information Technology   3
2. Computing Infrastructure: Hardware   46
3. Computing Infrastructure: Software   84
4. Networking Infrastructure   121
5. Cabling Infrastructure   163
6. Wireless Infrastructure   197
7. Storage Infrastructure   217

PART II   IT PRODUCTION TOOLS

8. Security Infrastructure   257
9. Office Tools   301
10. Data Management Tools   340
11. Web Tools   370

PART III   INTERNET AND NETWORK PROTOCOLS

12. Network Management Tools   389
13. Protocols and Global Connectivity   407

PART IV   IT MANAGEMENT

14. E-business Highway—Business Automation Platform   429
15. Infrastructure Management   453
16. Security Management   479
17. Information Management   515
18. Audit   536

PART V   IT APPLICATIONS—BUSINESS SYSTEMS

19. Governance   569
20. Connected World and E-commerce   599
21. Information Systems and Business Systems   628

Index   *678*

Detailed Contents

Preface   *iii*

PART I   IT INFRASTRUCTURE

1. Information Technology   **3**
   1.1 Introduction   3
     1.1.1 Functional Roles and Responsibilities   5
   1.2 Business Value of IT   7
   1.3 Role of Computers in Modern Business   8
   1.4 Infrastructure Management   8
   1.5 Elements of IT Infrastructure   9
     1.5.1 Computing Infrastructure   11
     1.5.2 Networking Infrastructure   11
     1.5.3 Storage Infrastructure   14
     1.5.4 Security Infrastructure   15
   1.6 Internetworks   16
     1.6.1 Local Area Network (LAN)   16
     1.6.2 Metropolitan Area Network (MAN)   17
     1.6.3 Wide Area Network (WAN)   17
     1.6.4 Wireless Network   18
     1.6.5 Ad Hoc Networks   18
     1.6.6 Mobile Networks   20
     1.6.7 Optical Networks   20
   1.7 IT Systems   21
     1.7.1 Management Information Systems   23
     1.7.2 Expert Systems   23
     1.7.3 Geographic Information Systems   23
     1.7.4 Health Information Systems   25
   1.8 Management   26
     1.8.1 Service Management   26
     1.8.2 Data Management   26
     1.8.3 Disaster Management   27
     1.8.4 Remote Infrastructure Management   28
     1.8.5 Measures and Metrics   29
   1.9 Standards, Audit, and Governance   31
     1.9.1 Standards   32
     1.9.2 Audit   33
     1.9.3 Governance   33
   1.10 Current Trends   35
     1.10.1 Data Centre   35
     1.10.2 Green Computing   35
     1.10.3 Grid Computing   36
     1.10.4 Virtualization   36
     1.10.5 Server Consolidation   37
     1.10.6 Storage Consolidation   38
     1.10.7 IT Practices in Banks   39

2. Computing Infrastructure: Hardware   **46**
   2.1 Introduction   46
     2.1.1 Elements of a Computer   48
     2.1.2 A Brief History   48
     2.1.3 Processor Philosophies   49
     2.1.4 Processor Address   50
     2.1.5 Processor Registers   50
     2.1.6 Memory   51
   2.2 Hardware   51
     2.2.1 Intel   52
     2.2.2 Advanced Micro Devices   54
     2.2.3 Servers   55
     2.2.4 Blade Servers and Enclosures   57
     2.2.5 Hard Disk Drives   58
     2.2.6 Desktop   59
   2.3 Input-Output   59
     2.3.1 Touch Screens   61
     2.3.2 Input Devices   61
   2.4 Computing Infrastructure Deployments   62
   2.5 Data Centre   62
   2.6 Virtualization   65
     2.6.1 Virtual Machines   65
     2.6.2 Templates   65
     2.6.3 Virtual Image   66
     2.6.4 Hypervisors   67
     2.6.5 Xen Hypervisor   68

    2.6.6 Types of Virtualization 69
    2.6.7 Hyper-V 70
    2.6.8 Virtualization Standards 70
    2.6.9 Server Virtualization 71
    2.6.10 Single Point Failures 71
2.7 Server Farm 72
2.8 Cluster Computing 72
    2.8.1 High Availability 73
2.9 Grid Computing 74
2.10 Cloud Computing 75
    2.10.1 Cloud Platforms 78
    2.10.2 Prominent Cloud Platforms 79

3. Computing Infrastructure: Software   84

3.1 Software Industry 84
    3.1.1 Software 86
    3.1.2 Types of Software 86
3.2 System Software 87
    3.2.1 Assemblers 87
    3.2.2 Interpreters 88
    3.2.3 Compiler 89
    3.2.4 Linker 89
    3.2.5 Loader 90
3.3 Operating System 90
    3.3.1 Types of Operating Systems 91
    3.3.2 Real-time Operating System 91
    3.3.3 Multi-user and Single-user Operating Systems 91
    3.3.4 Multitasking–Single-tasking Operating Systems 92
    3.3.5 Distributed Operating System 92
    3.3.6 Embedded System 92
3.4 Kernel 93
    3.4.1 System Calls 94
    3.4.2 Memory Management 95
    3.4.3 Process Management 95
    3.4.4 Thread Management 97
    3.4.5 Process Scheduler 98
    3.4.6 Interprocess Communication 98
    3.4.7 Interrupt Handling 99
    3.4.8 Device Drivers 100
3.5 Windows Operating System 100
    3.5.1 Windows Registry 101
    3.5.2 Microsoft Windows 2000 102
    3.5.3 Microsoft Windows XP 102
    3.5.4 Microsoft Windows Vista 103
    3.5.5 Microsoft Windows Server 2008 103
    3.5.6 Windows Script 103
3.6 Open Source Movement 104
    3.6.1 Open Source License 107
3.7 Linux Operating System 108
    3.7.1 Debian Linux 110
    3.7.2 OpenSolaris 111
3.8 Real-time Operating System (RTOS) 111
3.9 Programming Languages 113
    3.9.1 Low-level Programming Language 113
    3.9.2 High-level Programming Language 114

4. Networking Infrastructure   121

4.1 Introduction 121
4.2 Protocols—The Communication Enabler 123
4.3 Open System Interconnection 123
    4.3.1 N-layer Service 125
    4.3.2 OSI Layers 125
4.4 TCP/IP Suite 126
    4.4.1 Ethernet 128
4.5 LAN and WAN 130
4.6 Communication Medium 133
    4.6.1 ISP 134
    4.6.2 Digital Subscriber Line (DSL) Network 135
    4.6.3 Leased Lines 136
    4.6.4 Integrated Services Digital Network 137
    4.6.5 Terrestrial Microwave 137
    4.6.6 Very Small Aperture Terminal 138
4.7 Racks 138
4.8 Networking Elements 138
    4.8.1 Network Interface Card 139
    4.8.2 Overview of Routers 139
    4.8.3 Switches 145
    4.8.4 Availability 155
    4.8.5 Business Benefits 156
    4.8.6 Application Accelerator 156
    4.8.7 Modems 157

Contents

        4.8.8 Access Points 159
4.9 Network Resilience 159

5. Cabling Infrastructure 163
5.1 Introduction 163
5.2 Ethernet—The Common Denominator 166
    5.2.1 10 Gigabit Ethernet 167
5.3 Banking on Structured Cabling 167
    5.3.1 Structured Cabling as a Solution 168
    5.3.2 Advantages of Structured Cable 169
    5.3.3 Design of Structured Cabling 170
    5.3.4 Need for Cabling Standards 170
    5.3.5 Roles of Management 171
    5.3.6 Cable Solutions in the Market 172
5.4 Technology Drivers 172
5.5 Classification of Cables 172
    5.5.1 Twisted-pair Cable 173
    5.5.2 Unshielded Twisted-pair 174
    5.5.3 Shielded Twisted Pair (STP) 176
5.6 Fibre-optic Cable (FOC) 177
    5.6.1 Optical Fibre Media 178
    5.6.2 Strengths of Fibre-optic Cables 182
    5.6.3 Packing Density 185
5.7 Cables in Buildings 185
    5.7.1 Horizontal Cabling 186
    5.7.2 Backbone Cabling 187
    5.7.3 Pathways 187
    5.7.4 Glides 187
    5.7.5 Connectors 188
    5.7.6 Wiring Patterns 190
    5.7.7 Wall Plates 190
5.8 Testing and Certification 191

6. Wireless Infrastructure 197
6.1 Introduction 197
    6.1.1 Radio Waves 198
    6.1.2 ISM Band 200
    6.1.3 Antenna 200
    6.1.4 Spread Spectrum 200
6.2 Wireless Networks and Connectivity 201
    6.2.1 Overview of Standards 202
    6.2.2 Types of Wireless Deployments 204
6.3 Wireless Local Area Network 204
6.4 Wi-Mesh 207
6.5 WiMAX 208
    6.5.1 Fixed WiMAX 210
    6.5.2 Mobile WiMAX 210
    6.5.3 Wireless Personal Area Network 211
    6.5.4 Business Benefits 211
    6.5.5 Wireless System 212
6.6 Bluetooth 212
    6.6.1 Bluetooth Connections 213
    6.6.2 Benefits of Bluetooth 214

7. Storage Infrastructure 217
7.1 A Quick History of Data Storage 217
7.2 Storage I/O Basics 220
7.3 Storage Media 222
    7.3.1 Types of Storage Media 222
    7.3.2 Holographic Storage 226
7.4 Storage Protocols 227
    7.4.1 Internet SCSI 228
    7.4.2 Fibre Channel 229
    7.4.3 Types of Fibre Channel 229
7.5 Storage Models 230
    7.5.1 Direct Attached Storage 230
    7.5.2 Network Attached Storage 232
    7.5.3 Storage Area Network (SAN) 234
    7.5.4 Fibre Channel Topologies 235
    7.5.5 Data Backup 237
7.6 RAID 237
    7.6.1 RAID Defined 238
    7.6.2 Need for RAID 239
    7.6.3 Types of RAID 240
    7.6.4 RAID 0 240
    7.6.5 RAID 1 241
    7.6.6 RAID 2 242
    7.6.7 RAID 3 242
    7.6.8 RAID 4 243
    7.6.9 RAID 5 243
    7.6.10 RAID Controller 245
7.7 Backup Solutions 246
    7.7.1 Tape-based Solutions 246
    7.7.2 Virtual tape library 248
7.8 Continuous Data Protection 248

PART II   IT PRODUCTION TOOLS

8. **Security Infrastructure** 257
   - 8.1 Introduction 257
   - 8.2 Vulnerability—Gaps in Computing System 261
     - 8.2.1 Plugging Vulnerability—Management Game 262
     - 8.2.2 Common Vulnerabilities and Exposures 263
     - 8.2.3 Adaptive Management for Changing Threats 264
   - 8.3 Penetration Testing 266
   - 8.4 Patch Management 266
   - 8.5 Integrated Network Security 266
   - 8.6 Types of Attack Vectors 267
     - 8.6.1 Computer Worms 267
     - 8.6.2 Denial of Service and Distributed Denial of Service 268
     - 8.6.3 SPAM 268
     - 8.6.4 BOTS 269
     - 8.6.5 Insider Threat 269
     - 8.6.6 Social Engineering 272
     - 8.6.7 Countermeasures 273
   - 8.7 Firewall 273
     - 8.7.1 Firewall Defined 275
     - 8.7.2 Connection Table 277
     - 8.7.3 Stateful Firewall 277
     - 8.7.4 Packet Filtering Firewall 277
     - 8.7.5 Application Proxy Firewall 277
     - 8.7.6 Software Firewall 278
     - 8.7.7 Firewall Deployments 278
     - 8.7.8 Demilitarized Zone 278
     - 8.7.9 UTM 278
     - 8.7.10 Challenges 280
   - 8.8 Intrusion detection and Prevention Systems 280
     - 8.8.1 IDS Defined 281
     - 8.8.2 Types of IDS 281
   - 8.9 Network Access Control 281
   - 8.10 Endpoint Security 282
   - 8.11 Security Surveillance 283
     - 8.11.1 Surveillance Defined 283
     - 8.11.2 Surveillance Cameras 284
     - 8.11.3 IP—CCTV 285
     - 8.11.4 Video Walls 286
     - 8.11.5 Camera Deployments 286
   - 8.12 Biometric Systems 286
     - 8.12.1 Biometric Access 288
   - 8.13 RFID 288
     - 8.13.1 Associations and Forums 288
   - 8.14 Contact-less Card-based Systems 289
     - 8.14.1 RFID System 289
     - 8.14.2 RFID Components 291
     - 8.14.3 RFID inlays 292
     - 8.14.4 Passive Tags 293
     - 8.14.5 Active Tags 293
     - 8.14.6 RFID Readers 294
     - 8.14.7 RFID Deployments 295
     - 8.14.8 Business Benefits 296

9. **Office Tools** 301
   - 9.1 Introduction 301
   - 9.2 Microsoft and Office Automation 302
   - 9.3 Office Productivity 303
     - 9.3.1 MS Office Suite 303
     - 9.3.2 MS Word 304
     - 9.3.3 MS Excel 313
     - 9.3.4 MS PowerPoint 324
   - 9.4 Lotus Development Corporation 330
     - 9.4.1 Open Office 330
   - 9.5 Collaboration Tools 331
     - 9.5.1 Essentials of E-mail 332
     - 9.5.2 MS Outlook 334
     - 9.5.3 Evolution 336
     - 9.5.4 Thunderbird 336

10. **Data Management Tools** 340
    - 10.1 Data Management 340
    - 10.2 Database 341
      - 10.2.1 Database Management System 342
      - 10.2.2 Data Models 343
      - 10.2.3 Microsoft Access 347
      - 10.2.4 DBMS and RDBMS 349
    - 10.3 Transaction Processing Systems 352
      - 10.3.1 Flat Transactions 355
    - 10.4 Data Warehousing 357
    - 10.5 Business Intelligence and Data Mining 358
      - 10.5.1 Data Mining Tools 359

xii Contents

    10.6 Content Management 361
        10.6.1 Need for Content Management 362
        10.6.2 Portals 362
        10.6.3 Open Source Content Management System 364
        10.6.4 Commercial Tools 365
        10.6.5 Microsoft Content Management Server 367

11. Web Tools 370
    11.1 Web Servers and Browsers 370
        11.1.1 Deceptive Website 371
        11.1.2 Web Servers 371
        11.1.3 Web Browsers 373
        11.1.4 Internet Explorer 373
        11.1.5 FireFox 374
    11.2 Web Authoring Tools 375
        11.2.1 SGML 375
        11.2.2 Hypertext Markup Language 376
        11.2.3 Cookies 379
        11.2.4 XML 380
        11.2.5 Front Page Editor 381
        11.2.6 Dreamweaver 381
    11.3 Web System 382

PART III INTERNET AND NETWORK PROTOCOLS

12. Network Management Tools 389
    12.1 Basics of IT Management 389
        12.1.1 Protocols 390
        12.1.2 Simple Network Management Protocol (SNMP) 390
        12.1.3 RMON 394
    12.2 Network Management 394
    12.3 Remote Management Tools 396
        12.3.1 Dashboards 397
        12.3.1 Dashboards 397
        12.3.3 CA-Unicenter 398
        12.3.4 HP OpenView 398
    12.4 Open Source—IT Management 399
        12.4.1 Nagios 401
        12.4.2 Cacti 402
        12.4.3 Big Brother 403

13. Protocols and Global Connectivity 407
    13.1 Internet 407
        13.1.1 ARPANET and DARPANET 409
        13.1.2 Current Systems 410
    13.2 Protocols 411
        13.2.1 Seven Layer OSI Stack 411
        13.2.2 TCP/IP Stack 415
        13.2.3 Internet Protocol 416
    13.3 IP Addressing Mechanism 416
        13.3.1 IP V4-Address System 417
        13.3.2 Limitations of IPV4 417
        13.3.3 IPV6 Address system 418
        13.3.4 Transmission Control Protocol 419
        13.3.5 User Datagram Protocol 419
        13.3.6 Multi-protocol Label Switching 420
    13.4 Telephone System 420
    13.5 Voice Over Internet 421

PART IV IT MANAGEMENT

14. E-business Highway—Business Automation Platform 429
    14.1 Intranet 429
        14.1.1 Purpose of Intranet 431
        14.1.2 Intranet Benefits 434
        14.1.3 Weblog 436
        14.1.4 Return on Investment 437
    14.2 Extranet 438
    14.3 Benefits of Intranets and Extranets 438
        14.3.1 Tangible Benefits 438
        14.3.2 Intangible Benefits 439
    14.4 Internet Services 439
    14.5 World Wide Web 440
        14.5.1 Telnet 440
        14.5.2 FTP 441

14.5.3 Search Engine and
Data Delivery 441
14.5.4 Web Services 443
14.5.5 Surfing 446
14.6 The Global Village 446
14.7 Bandwidth 447
14.7.1 Cost of Bandwidth 448
14.7.2 Internet Service Providers 448
14.7.3 Types of Bandwidth 449

15. Infrastructure Management 453
15.1 Introduction 453
15.2 IT Function 456
15.2.1 IT Management 456
15.3 Overview of ITIL Framework 457
15.3.1 BS15000 and ISO20000 459
15.4 Management and Measurements 459
15.4.1 IT Service as a Pproduct 461
15.4.2 Service Life Cycle 462
15.4.3 Interaction Management 464
15.4.4 Request and Incident
Management 464
15.4.5 Problem Management 464
15.4.6 Change Management 465
15.4.7 Service Centre Organization 466
15.4.8 Immediate Response Model 466
15.4.9 Managed Response Model 467
15.4.10 Service Desk/Help Desk 467
15.4.11 Levels of Support 467
15.5 Information Technology
Measurements 468
15.5.1 Measurement Metrics 468
15.5.2 Uptime and Downtime 468
15.5.3 Availability 469
15.5.4 Bit Error Rate 469
15.5.5 Mean Time between Failure 469
15.5.6 Mean Time to
Repair MTTR 469
15.6 Outage Management 470
15.6.1 System and Network
Outage 470
15.7 Service Level Agreements 471
15.7.1 Components of SLA 472
15.8 Outsourcing 473
15.8.1 Types of Outsourcing 473
15.8.2 Technology Services
Outsourcing 474

15.8.3 Business Process
Outsourcing 474
15.8.4 Outsourcing Trends 475

16. Security Management 479
16.1 Introduction to Security 480
16.1.1 CIA Triangle 480
16.2 Two Views of Security 481
16.2.1 Sources of Threats 482
16.3 Security Management Controls 483
16.4 Physical Security 484
16.4.1 Locks and Physical
Security 486
16.4.2 Asset Tagging and Asset Life
Cycle Management 486
16.5 Access Control 487
16.5.1 Access Control Principles and
Objectives 488
16.5.2 Access Control Mechanisms 489
16.5.3 Logical Access Control 490
16.5.4 Physical Access Control 491
16.5.5 Passwords/Keys/Tokens 491
16.5.6 Biometric Identification 492
16.6 System Security 494
16.6.1 Hardening Systems 494
16.7 Password Management 494
16.7.1 Password Guidelines 495
16.7.2 Secure Password 495
16.7.3 Age and Length of
Password 495
16.7.4 Password Best Practices 496
16.8 Communication Security 497
16.9 Information Security 497
16.10 Risk Management and Business
Continuity Planning 498
16.10.1 Risk Analysis 498
16.10.2 Risk Analysis and
Assessment 498
16.10.3 Business Continuity
Planning 499
16.10.4 Business Continuity in
Distributed Environments 501
16.11 Security Standards and Assurance 502
16.11.1 ISO 27001 502
16.11.2 OCTAVE 503
16.12 Information Infrastructure 503
16.12.1 Information Warfare 504

16.12.2 Electronic Warfare   505
16.13 Information Operations and Cyber Weapons   506
   16.13.1 Cyber Weapons   507
   16.13.2 Virtual Cyber Weapons   508
   16.13.3 Cyber Warrior   508
   16.13.4 Information Exploitation   508
16.14 Operations and Information Dominance   509
   16.14.1 Defensive Information Warfare   510
   16.14.2 Defensive Measures   510

17. Information Management   515
17.1 Information Architecture   515
   17.1.1 Information Architecture Components   518
   17.1.2 Challenges of Organizing Information   519
   17.1.3 Information Maps   520
17.2 Information Life Cycle Management   520
   17.2.1 Unstructured and Structured Information   521
   17.2.2 Importance of Life Cycle Management   521
   17.2.3 Data Classification   521
17.3 Information Economics   522
   17.3.1 Copyright   523
17.4 Data Quality Problem   525
   17.4.1 Data Quality—Why is it Important?   526
   17.4.2 Manifestation of Data Quality Problems   529
   17.4.3 Improved Data Quality Benefits   530
   17.4.4 Data Quality Assurance   531
   17.4.5 Assessing Data Quality   531
   17.4.6 Data Quality Metrics   532

18. Audit   536
18.1 Introduction   536
18.2 Systems   538
18.3 Audits   541
   18.3.1 Information Audit   543
   18.3.2 Audit Teams   545
   18.3.3 Audit Schedule   546
   18.3.4 Audit Plan   547
   18.3.5 Audit Preparation   548
   18.3.6 Audit Procedures   549
   18.3.7 Internal Audit   549
   18.3.8 Audit Findings and Conclusions   551
   18.3.9 Audit Reports   551
   18.3.10 Working Papers   552
18.4 Controls   552
   18.4.1 Internal Controls   554
   18.4.2 IT Controls   556
   18.4.3 General Controls   556
   18.4.4 Application Controls   556
   18.4.5 Management Controls   557
   18.4.6 IT Controls Practices   557
   18.4.7 Nature of Controls   558
   18.4.8 Assessing System Reliability   559
   18.4.9 Controls and Classifications   559
18.5 Knowledge Audit and Evaluation   560

PART V   IT APPLICATIONS—BUSINESS SYSTEMS

19. Governance   569
19.1 Introduction   569
19.2 Governance   570
19.3 Corporate Governance   571
19.4 IT Governance   573
19.5 Operational Risk and Governance   576
19.6 Organizational Framework—Value Creation   577
19.7 Internet Governance   581
19.8 Governance of Internal IT Processes   582
   19.8.1 Modern Governance of IT   582
   19.8.2 COSO   583
   19.8.3 COBIT   584
   19.8.4 SAC   586
   19.8.5 IT Control Dependencies   587

	19.8.6	Benefits of IT Governance 587	
19.9	E-governance Framework 588		
	19.9.1	Definition of E-governance 589	
	19.9.2	E-governance Initiatives of India 589	
	19.9.3	State-wide Area Network 591	
	19.9.4	State Data Centres (SDCs) 592	
	19.9.5	E-governance PPP Projects 592	
	19.9.6	E-governance BOOT Projects 593	
	19.9.7	Benefits of E-governance 594	

20. Connected World and E-commerce 599

- 20.1 Websites and E-business 599
- 20.2 E-business 600
- 20.3 E-commerce 601
- 20.4 Business on the Net 604
  - 20.4.1 Software Development 605
  - 20.4.2 Payment Systems 608
  - 20.4.3 Value Creation 608
  - 20.4.4 Types of E-commerce 610
- 20.5 Digital Markets 611
  - 20.5.1 Infomediary 612
  - 20.5.2 E-auctions 613
  - 20.5.3 Agents 614
  - 20.5.4 Digital Supermarkets 616
  - 20.5.5 E-commerce and E-procurement 617
  - 20.5.6 E-commerce and E-contracting 618
  - 20.5.7 E-retailing 620
  - 20.5.8 E-commerce and E-marketing 620
- 20.6 Electronic Data Interchange 622
  - 20.6.1 Need for EDI 623
  - 20.6.2 EDI Standards 623
- 20.7 Electronic Data Security 624

21. Information Systems and Business Systems 628

- 21.1 Information Systems 629
  - 21.1.1 Process 631
  - 21.1.2 Process Reengineering 633
  - 21.1.3 Business Process Reengineering 633
  - 21.1.4 Classical Business 634
  - 21.1.5 Enterprise Application Integration 635
  - 21.1.6 Data-level Integration 637
  - 21.1.7 Function-level Integration 637
  - 21.1.8 Brokered Integration 637
- 21.2 Workflow Automation 638
  - 21.2.1 Workflow Management System
  - 21.2.2 Value Chain 640
- 21.3 Strategic Information Systems 641
  - 21.3.1 Cycle Time—Lead Time Measurements 642
- 21.4 Enterprise Resource Planning 643
  - 21.4.1 Need for ERP Systems 645
  - 21.4.2 Evolution of ERP Systems 646
  - 21.4.3 Materials Requirement Planning 646
  - 21.4.4 Types of Manufacturing 648
  - 21.4.5 Order Processing 649
  - 21.4.6 Current ERP Systems 650
  - 21.4.7 ERP Implementation—Integration 651
  - 21.4.8 Popular ERP Systems 652
  - 21.4.9 Business Benefits 657
- 21.5 Change Management 658
- 21.6 Customer Relationship Management (CRM) 659
  - 21.6.1 Need for CRM Systems 661
  - 21.6.2 Evolution of CRM systems 662
  - 21.6.3 Process Flow 663
  - 21.6.4 Popular CRM Systems 663
  - 21.6.5 Business Benefits 665
- 21.7 Supply Chain and Integrated Supply Chain Management 665
  - 21.7.1 Supply Chain 666
  - 21.7.2 Supply Chain Systems 666
  - 21.7.3 Integrated Supply Chain Management 668
  - 21.7.4 Bullwhip Effect 670
  - 21.7.5 Logistics 671
  - 21.7.6 Supply Chain Process Visibility 671
- 21.8 Business Analytics and Knowledge Management 672
- 21.9 Management Information Systems 673
- 21.10 Geographic Information System 673

Index 678

PART I

IT Infrastructure

- **Chapter 1**　Information Technology
- **Chapter 2**　Computing Infrastructure: Hardware
- **Chapter 3**　Computing Infrastructure: Software
- **Chapter 4**　Networking Infrastructure
- **Chapter 5**　Cabling Infrastructure
- **Chapter 6**　Wireless Infrastructure
- **Chapter 7**　Storage Infrastructure

CHAPTER 1

Information Technology

A market is never saturated with a good product.
But it is very quickly saturated with a bad one.
–Henry Ford

Learning Objectives

After reading this chapter, you should be able to understand:

- the current scenario of IT in business
- the various elements of IT infrastructure
- the relevance of sizing infrastructure for business needs
- and appreciate the current IT trends and their business impacts
- the need for IT audits and IT governance

1.1 Introduction

The expansion of the Internet has been called the most revolutionary development in the history of human communications. It is ubiquitous and is changing politics, economics, and social relations. The borderless nature of the Internet produces particular needs for global institutions and has opened the door for innovative approaches.

Information technology (IT) provides a significant advantage in a corporation's ability to compete in the growing Internet marketplace. Corporations have adopted technology to increase productivity, reduce costs, drive revenues, offer new capabilities to customers and suppliers, and improve competitive positioning to respond to real-time requirements.

The on-demand, instant action–reaction business environment today requires real-time responsiveness to change, whether it is to meet new demands by customers, changes in the supply chain, or unexpected competitive moves. In order to be able to respond quickly, enterprises must provide their employees with immediate access to accurate and updated information. This greater dependence on information translates into greater dependence on the effectiveness of IT infrastructure as a whole.

Infrastructure is a concept that depends on context. To a city planner (domain/context), infrastructure is transportation and communications systems, water and power lines, and public institutions including schools, post offices, etc. The term has diverse meanings in different fields, but is perhaps most widely understood to refer to roads, airports, and utilities. These various elements may collectively be termed as civil infrastructure, municipal infrastructure, or simply public works, although they may be developed and operated as private-sector or government enterprises.

In other applications, infrastructure may refer to information technology, informal and formal channels of communication, software development tools, political and or social networks. To the chief executive officer (CEO) of an organization, infrastructure might be facilities, security, logistics, power, waste disposal, and large chunks of information technology. A flexible and robust IT infrastructure also plays an important competitive role by enabling employee productivity and globalization, allowing pervasive and secure business communications anywhere and anytime, and finally, by managing operational complexity and providing greater utilization of resource assets.

Over the past decade, as information and computer-based systems have become larger and more complex, the importance of and reliance on IT systems have grown substantially. Information technology has all the hallmarks of an infrastructural technology. Infrastructural technologies offer far more value when shared than when used in isolation. The value of infrastructure sharing has emerged as a business model to increase the bottom line. In fact, its mix of characteristics guarantees particularly rapid communalization. Information technology is, first of all, a transport mechanism. It carries digital information just as railroads carry goods and power grids carry electricity. And like any transport mechanism, it is far more valuable when shared than when used in isolation.

In recent years, companies have worked hard to reduce the cost of IT infrastructure data centre, networks, databases, and software tools that support businesses. These efforts to consolidate; standardize; and streamline assets, technologies, and processes have delivered major savings. Yet, even the most effective cost-cutting program eventually hits a wall, the complexity of the infrastructure itself.

Large consulting organizations, such as the Gartner Group, perform regular studies on the state of the technical executive. In a 2004 report, surveyed chief information officers (CIOs) agreed that the ability to communicate effectively, strategic thinking, and planning and understanding business processes are critical skills for the CIO position. The switch in information technology's value from a function to control costs and increase productivity to a vehicle to generate revenue, is permanent. Nostalgic yearnings for IT spending levels of the past is a misuse of important time; the business model that once supported lower levels of IT spending no longer exists for many industries and will be less relevant in the future [Stevan 2001].

In the present century, a company's success will be even more driven by the extent to which it can target its products to specialized customer needs. Thus, over time, many companies will evolve to become customer-driven businesses. Market mapping is, therefore, going to be a key organizational capability. While new technologies will be developed to help companies understand customers' needs and to identify their best customers, successful companies will have to do more. They will have to start thinking in reverse, i.e., finding out what their customers want and respond immediately to those needs. They will also have to look at the world through 'new lenses' and develop a learning relationship with each customer. This relationship will be an ongoing connection, which becomes smarter as the company and customer interact with each other. This form of intimacy between a company and a customer can only be built up over time but, if successful, will yield a substantial competitive advantage.

1.1.1 Functional Roles and Responsibilities

Throughout this book, we refer to the chief technology officer as CTO. The CTO's primary responsibility is to contribute to the strategic direction of the company by identifying the role that specific technologies will play in its future growth. As Sun Tzu said, strategy without tactics is the slowest route to victory and tactics without strategy is the noise before defeat. A CTO is expected to provide a technical strategy that seamlessly segues into the corporate business strategy.

Throughout this book, we refer to the chief information officer as CIO. Just as the chief financial officer (CFO) is a senior person responsible for devising the capital structure best suited to an enterprise's business needs, the CIO is the senior person responsible for making sure that the enterprise's IT infrastructure best supports its business needs. The CIO is the senior-level liaison between the business and technical sides of an enterprise. He/she is the person who helps define and translate business goals and strategies into a system's performance requirements and oversees a portfolio of IT development projects to deliver systems that meet these requirements. The CIO leads the application of IT to internal processes and services. This is a title that is used as a form of shorthand to infer the member of the executive team who has responsibility for all IT functions. There are other members in the IT team who are responsible for carrying out the treadmill of IT functions.

Value delivery meshes with strategic planning and the implementation of an enterprise architecture to support the business mission. The CIO needs tools and methods to identify the most productive investments and then to communicate the value to stakeholders at all levels of the organization.

According to the IT Governance Institute (ITGI), there are four critical questions for value delivery. These are as follows.

1. Are we doing the right things?
2. Are we getting the benefits?
3. Are we getting them done well?
4. Are we doing them the right way?

Many CIOs have had the nagging feeling that the traditional IT risk analysis starts out with what the analyst thinks is the answer. Then the details, weighings, and issues are aligned to show the applications, infrastructure components, and projects that score the highest risk. A comprehensive understanding of risk measures will aid the CIO to take the right decisions.

Four key business requirements (also referred to as Four A framework) have been identified for the success of an organization.

Availability Systems must be up and running. Recovery from failure should be rapid, based on the firm's business requirements.

Access Systems should be sufficiently secure to prevent loss and destruction of data but flexible enough to enable employees to do their job.

Accuracy Information must be timely, complete, and correct when presented to both internal and external users.

Agility Ability to change IT systems to meet new business requirements with requisite speed and reasonable cost.

1.2 Business Value of IT

Information technology can have a significant impact on the quality of services and solutions and the performance of a company. Efficiently and effectively managed IT investments that meet business and mission needs can create new value in revenue generation, build important competitive advantages and barriers to entry, improve productivity and performance, and decrease costs. Similarly, poorly aligned and unmanaged IT investments can sink a company. Concurrent to cutbacks in IT spending and a short-term focus, management within companies is demanding an increase in IT productivity, expanding IT's role from internally focused to customer facing and making IT more relevant to business strategy.

Misalignment between information technology and the strategic intent, inability to establish a common IT architecture, and a highly redundant and undocumented as-is architecture will result in high operations and maintenance costs. Web services and services-oriented development of applications (SODA) will continue to make the business and IT relationship more critical as IT continues to become increasingly more integral to business processes.

Business value is just one output of the collection of processes through which businesses today try to maximize the age-old equation of *profit equals revenue minus expenses*. Most businesses today rely on information technology to realize some of their business value. How do we measure the value of information technology? It is a question that is on everyone's mind, from business managers to board rooms.

The spectacular growth of IT has enormous potential for improving the performance of organizations. However, the huge investment made in IT puts increasing pressure on the management to justify the outlay by quantifying the business value of IT. In today's fast moving competitive business environment, companies increasingly demand that IT investments demonstrate business value through measurable results.

For many companies, the link between business technology investment and business performance remains elusive.

There is a general global consensus that developments in information technology in recent years have contributed to the emergence of what has been described as the information economy characterized by significant productivity benefits at the macro level. However, there is a critical need to understand the complex relationships between IT investments and business value at more micro levels, as underscored by policy makers, practitioners, and researchers.

Implementing an innovative approach to determine the business value of information technology is an enormous task, but the payoff for the IT organization and the enterprise as a whole is worth the effort.

1.3 Role of Computers in Modern Business

It was not many years ago that the personal computer (PC) was first introduced. The benefits of PCs were obvious but limited because each computer was an island unto itself. Over the years, a great deal of effort has gone into figuring out how to interconnect computer systems. Computers are used to process, store, and exchange information in digital form. Computers have today become the backbone of business processes. The act of computing is built over computing software which runs on top of bare metal computing hardware. A variety of computing and processing software are available for various kinds of applications. Different computer applications generate different kinds of data. It is easy to imagine that the requirements for a network linking a bank's automatic teller machines (ATMs) to its computers are different from those of a network of computers that control air traffic or a car manufacturer's assembly line.

1.4 Infrastructure Management

Management has been defined as assembling the resources to achieve a mutually agreed upon objective. This reflects the two typical management structures American companies now employ: command and control, or collaboration. Based on the classic military structure, this style was popular for most of American corporate history. But now this style has lost popularity. While some environments still operate under this style, many corporations are revisiting their commitment to such a rigid method of management.

The Five C Model, proposed by Gartner, addresses the performance areas that CEOs, CFOs, and the senior leadership team members expect their CIOs to address. The five C's—clarity, context, competence, commitment, and competition—represent more than a clever memory device. They represent key areas on enterprise balanced scorecards—performance management systems that include forward-looking, predictive metrics in addition to financials.

Collaborative management is a more modern way of handling the art of management. It ensures that all levels of the corporate ladder are actively involved in the execution of a business.

1.5 Elements of IT Infrastructure

The merging of computers and communications has had a profound influence on the way computer systems are organized. Many companies worldwide have invested on substantial number of computers. For example, a company may have separate computers to monitor production, keep track of inventories, and do the payroll. Initially, each of these computers may have worked in isolation from the others, but at some point, management may have decided to connect them in order to be able to extract and correlate information about the entire company.

In a business climate that punishes the inefficient and the slow-moving, enterprises are under pressure to manage their information and infrastructure assets more effectively and efficiently than ever. The information and infrastructure management framework is no longer an adjunct support structure; it is the essential foundation for corporate performance. A company's success in managing its information assets is a function of infrastructure, process, people, and culture, all working in concert. Shrinking business cycles have put many staid, slow-moving organizations into acceleration. Time to market, which was once measured in years, is now measured in weeks. In the intensely competitive, web-fuelled marketplace, today's window could close into tomorrow's missed opportunity. Infrastructure technologies play an important part in bringing markets closer, aiding economic development, aggregating demand, and reducing transaction costs. They also make reduction of communication travel and other costs feasible, consequently creating new forms of corporate activity.

The IT infrastructure is characterized by the 7S model (Figure 1.1). The 7S model identifies the essential IT components to ensure a good IT service.

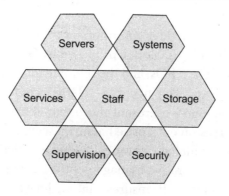

FIGURE 1.1 7S Model

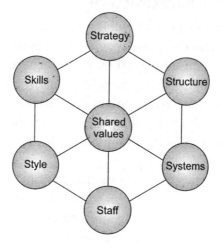

FIGURE 1.2 McKinsey 7S Model

The 7S model includes servers, systems, storage, security, supervision, services, and staff. All the seven components work together to achieve the overall business benefit. This is in line with the 7S model attributed to McKinsey.

The 7S model popularized by McKinsey & Company (Figure 1.2) characterizes a business using a set of seven distinct attributes: strategy, structure (organization and geography), systems (in the sense of both processes and IT systems), staff, style (for example, dynamic versus bureaucratic), skills, and lastly unifying shared values (such as respect for the environment and shareholder value).

1.5.1 Computing Infrastructure

Computing infrastructure is a set of technologies. Computing infrastructure is built on a growing infrastructure of rack-mounted data servers and parallel computing clusters attached to large redundant array of inexpensive disks (RAID) storage systems. Computing infrastructure also encompasses all the associated infrastructure that are required to exploit the available computing power with a good return on investment (ROI). Due to the tremendous strides made by computing technology, machines with enormous computing power are now available. It is well known that today even an entry-level machine has more computing power, in terms of storage, memory, central processing unit (CPU), and so on, than the powerful machines that existed a decade ago. It is, therefore, essential to harness the potential of computing infrastructure and increase the return on investment through suitable optimization of business processes.

1.5.2 Networking Infrastructure

Networking infrastructure is made up of a good number of gadgets built on standards. The interworking of networking infrastructure provides the essential superhighway for the movement of data traffic. A brief overview of some of the essential networking elements are listed below.

Structured Cabling

In the old days, copper wires were the only means of transmitting information. Technically, they are known as unshielded twisted pair (UTP) in comparison to their counterpart which are called shielded twisted pair (STP). The wire pairs are called twisted because they are physically twisted. They do not have a shield, and therefore a high frequency part of the signal can leak out.

A structured cabling system provides a universal platform upon which an overall information system's strategy is built. With a flexible cabling infrastructure, a structured cabling system can support multiple voice, data, video, and multimedia systems regardless of their manufacturer. A well-designed cabling plant may include several independent cabling solutions of different media types, installed at each workstation to support multiple system performance requirements.

Before an enterprise chooses to go for cabling, it needs to understand the purpose for cabling. It is important to have a foresight of business growth as structured cabling will be a long-term commitment. While selecting a structured cabling system, organizations must consider the applications that may be required to be added in future. Once installed, it is expected to last for many years because unlike the active components it cannot be changed at will. Size of the enterprise, the expansion growth rate expected, transmission speeds, and capacities needed are the key issues.

Structured cabling offers consistency and flexibility, provides support for multi-vendor equipment, simplifies troubleshooting, and provides support for future applications.

Selection of structured cabling components as per their designed performance levels is a complex task. For this, the system integrator or cable installer has to consider key issues such as size of the enterprise, expansion growth rate expected, physical spread, and transmission speeds and capacities needed. It is recommended that one uses modular components that are pre-terminated and tested at the factory prior to shipment, where just connecting the components makes the link go up and running. Continuous growth in the structured cabling industry is directly related to infrastructure investments.

Routers

Routers are network gadgets which direct data packets following Internet protocol (IP) rules from one network to another. Routers are more complex devices. Routers determine the most efficient path to different elements in the Internet complex routing algorithms. Identified path information are stored in routing tables available within the router. Routing tables are lookup tables similar to telephone directories. Routers deal with large routing tables for the global Internet and find appropriate routing addresses on demand.

Router equipment can be classified as core routers and edge routers. Core routers reside in the core of the network. Edge routers have more diversified functions as they reside on the edge of the network and deal with network traffic. Detailed discussion on the various aspects of routers and the associated protocol are discussed later in the book.

Switches

Networks are commonly used to interconnect computers or other devices. Each network generally includes two or more computers, often referred to as nodes or stations, which are coupled together through selected media and various other network devices for relaying, transmitting, repeating, translating, filtering, etc. A network system is a communication system that links two or more computers and peripheral devices, and allows users to access resources on other computers and exchange messages with other users.

Typically, the switch is a computer comprising a collection of components interconnected by a backplane of wires. Switches split large networks into smaller segments, decreasing the number of users sharing the same resources. They allow different nodes of a network to communicate directly with one another in a smooth and efficient manner.

A majority of switches became digital in the 1980s. Vendors and operators saw that, like computers, they could be programmed, making it possible to introduce intelligent services.

Load Balancing and Load Balancers

The world of server load balancing (and network-based load balancing in general) is filled with confusing jargon and inconsistent terminology. Because of the relative youth and the fierce competition of the server load balancing industry, vendors have come up with their own sets of terminology, which makes it difficult to compare one product and technology to another. A load balancer is a device that distributes load among several machines. Load balancing is the process by which inbound IP traffic can be distributed across multiple servers. Load balancing enhances the performance of the servers, leads to their optimal utilization and ensures that no single server is overwhelmed. Server load balancing (SLB) is defined as a process and technology that distributes site traffic among several servers using a network-based device. This device intercepts traffic destined for a site and redirects that traffic to various servers. The load balancing process is completely transparent to the end-user. Load balancers are an integral part of the corporate network which is primarily used for load balancing.

A load balancer performs the following functions.

- Intercepts network-based traffic (such as web traffic) destined for a site.
- Splits the traffic into individual requests and decides which servers receive individual requests.
- Maintains a watch on the available servers, ensuring that they are responding to traffic. If they are not, they are taken out of rotation.
- Provides redundancy by employing more than one unit in a fail-over scenario.
- Offers content-aware distribution by reading uniform resource locator (URL) intercepting cookies, and extensible markup language (XML) parsing.

Load balancing is particularly important for busy networks where it is difficult to predict the number of requests that will be issued to a server. Load balancing allows the service to continue even in the face of server downtime due to server failure or server maintenance. Load balancers are an integral part of today's web infrastructure. They are also complex and under-documented pieces of hardware. The ability for a load balancer to peer into the hypertext transfer protocol (HTTP) headers of incoming connections was once an advanced feature, but now is fairly commonplace. A business overview of load balancers and load balancing is provided in the later chapters.

Direct server return (DSR) is a method of bypassing the load balancer on the outbound connection. This can increase the performance of the load balancer by significantly reducing the amount of traffic running through the device and its packet-rewriting processes.

1.5.3 Storage Infrastructure

Deploying IT, aligning it with the organization's business goals, and creating a scalable architecture is a fundamental yet significant challenge facing today's CIOs. A successful IT backbone is a judicious blend of applications and hardware that create an infrastructure to be used by thousands of users. If we look at the key components of an IT infrastructure, storage is the vital link that secures the data and an enterprise's digital assets reside on a plethora of storage topologies and devices. Information technology infrastructures are stressed because of an avalanche of information, most of it coming from completely new sources and in completely new forms such as image, voice, and video. Since IT budgets have become largely flat, the toughest challenge faced by CIOs is to architect and manage the right storage infrastructure at a reduced budget, which address the data and information growth. Additionally, there are challenges in managing and protecting the data efficiently. Considering the need for content of various forms to be managed effectively, storage has become a big issue because of which organizations need to look at solutions like enterprise content management (ECM) and archiving.

The explosion of data and its management is driving the storage market to glory. New data-intensive applications and access to data from across geographically distributed sites are the key storage driver (refer to Chapter 7 on Storage Infrastructure for more information).

Standards

The Storage Networking Industry Association (SNIA) launched the Storage Management Initiative (SMI) in mid-2002 to create and encourage the universal adoption of an open interface for managing storage networks. The SMI's main aim is to deliver open storage network management interface technology in the form of an SMI Specification (SMI-S). Storage Management Initiative Specification is based on the Common Information Model (CIM) and Web-Based Enterprise Management (WBEM) standards developed by the Distributed Management Task Force (DMTF). Common Information Model is an open standard prescribed for storage vendors by the SNIA to bring uniformity to the storage industry. Storage vendors are beginning to adopt CIM by developing their products according to the standard.

With storage systems getting more complex by the day, it is a challenge for enterprises to manage them. Enterprises need to utilize their storage infrastructure more efficiently by improving storage utilization and performance.

Storage Software

Storage software helps bring down the cost of labour by automating processes that required manual intervention, reducing the dependence on specialists for keeping the corporate data centres up and running. It has become important for enterprises to allocate storage as per the priority of the data, and then replicate it as per the monetary tag attached to it, on the basis of data priority.

1.5.4 Security Infrastructure

Electronic security measures in today's corporate world are limited to anti-virus measures and firewalls (Figure 1.3). But the rise in security incidents and attacks on well-known websites have led to increased awareness of various aspects of information and infrastructure security.

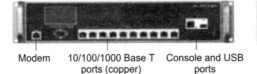
Modem   10/100/1000 Base T ports (copper)   Console and USB ports

Redundant hot swappable power supplies

FIGURE 1.3 Firewall

The integration of physical, logical, network security, and security standards in order to provide a reliable security framework is viewed as security infrastructure. The prime driver for enterprise security is (Internet) connectivity. The other driver for security is globalization. The third driver for increased security awareness is the security regulators. The emphasis on security arises from various aspects of change in the business environment. With business models evolving and competition on the rise, there is a need for greater emphasis on the information and physical security of IT infrastructure. Some of the common security appliances include firewall, intruder detection systems, and intruder prevention systems from different vendors and different makes.

Security can be best achieved by ensuring multiple layers of security and not depending on a single measure. This principle is very evident here.

The controls for physical and environmental security are defined in the following three areas.

1. Security of the premise
2. Security of the equipment
3. Secure behaviour

1.6 Internetworks

One way to categorize the different types of computer network designs is by their scope or scale. For historical reasons, the networking industry refers to nearly every type of design as some kind of area network.

The networking protocol used in most modern computer networks is called Ethernet. Networks connected through Ethernet are called Ethernet networks. Information in an Ethernet network is exchanged in a packet format. The packet provides grouping of the information for transmission that includes the header, the data, and the trailer.

1.6.1 Local Area Network (LAN)

Interconnection allows users to exchange information (data) with other network members. It also allows resource sharing of expensive equipment such as file servers and high-quality graphics printers, or access to more powerful computers for tasks too complicated for the local computer to process. The network commonly used to accomplish this interconnection is called a local area network (LAN). A LAN connects network devices over a relatively short distance. Today, local area networking is a shared access technology. This means that all of the devices attached to the LAN share a communication medium. In addition to operating in a limited space, LANs are also typically owned, controlled, and managed by a single person or organization. Various aspects of LAN are dealt in Chapter 4.

1.6.2 Metropolitan Area Network (MAN)

A metropolitan area network (MAN) is a class of network which serves a role similar to an Internet service provider (ISP), but for corporate users with large LANs. Metropolitan area networks connect multiple geographically-nearby LANs to one another at high speeds. Its geographic scope falls between a wide area network (WAN) and LAN. A MAN typically covers an area of between 5 km and 50 km diameter. The MAN, its communications links and equipment are generally owned by either a consortium of users or by a single network provider who sells the service to the users. Due to the emergence of new services, the requirements for MANs have increased and diversified.

1.6.3 Wide Area Network (WAN)

A WAN is a geographically dispersed collection of LANs. A WAN spans a large geographic area such as a state, province, or country. The world's most popular WAN is the Internet. A network device called router connects LANs to a WAN. Wide area networks generally utilize different and more expensive networking equipment than LANs do. Wide area networks are composed of long-haul networks. Long-haul networks provide transmission services over long distances, typically over hundreds of kilometres.

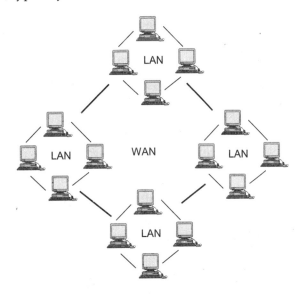

FIGURE 1.4 Wide Area Network

Wide area networks are used to connect LANs and other types of networks together, so that users and computers in one location can communicate with users and computers in other locations. Wide area networks (Figure 1.4) are typically utilized by organizations operating out of multiple office locations that require a secure, flexible, and cost-effective means for their employees to communicate and share information across a central computer network.

A major factor impacting WAN design and performance is a requirement that they lease communications circuits from telephone companies or other communications carriers. Transmission rates are typically 2 Mbps, 34 Mbps, 45 Mbps, 155 Mbps, 625 Mbps (or sometimes considerably more).

> Indian Institute of Information Technology, Bangalore (IIT-B) is building a wireless local area network (WLAN) in its campus to give its students and faculty the ability to perform better research and collaboration in an anytime-anywhere environment. Going unwired has been a dream of IIIT-B from the very beginning. It moved to a new campus in 2003, which is an Intel-approved wireless local loop (WLL) site. Intel is funding the entire wireless initiative of the institute as part of its vision to spread mobile computing with Centrino technology. Wireless computing is accessible across the campus.

1.6.4 Wireless Network

Wireless is the new buzzword among networking and software vendors. The wireless communication revolution is bringing fundamental changes to data networking, telecommunication, and is making integrated networks a reality. The wireless phenomenon is reshaping enterprise connectivity worldwide. By freeing the user from the cord, personal communications networks, WLANs, mobile radio networks, and cellular systems harbour the promise of fully distributed mobile computing and communications, anytime, anywhere (refer to Figures 1.5 and 1.6).

1.6.5 Ad Hoc Networks

Since the inception of wireless networking there have been two types of wireless networks: the infrastructure network, including some LANs and the ad hoc network. Ad hoc networks are created on the fly and for one-time or temporary use. Ad hoc networks are generally closed in so that they do not connect to the Internet and are typically created between participants. They are self organizing, self healing, distributed networks which most often employ wireless transmission techniques.

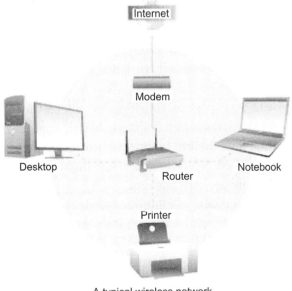

FIGURE 1.5 Wireless Deployment

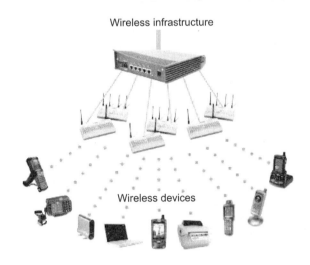

FIGURE 1.6 Current Wireless Network

A mobile ad hoc network (MANET) is an autonomous collection of mobile users that communicate over relatively bandwidth-constrained wireless links.

1.6.6 Mobile Networks

Mobile computing represents a new paradigm that aims to provide continuous network connectivity to users, regardless of their location. A wide spectrum of portable, personalized computing devices ranging from laptop computers to hand-held personal digital assistants, are currently being used. Their explosive growth has sparked considerable interest in providing continuous network coverage to such mobile hosts (MHs), regardless of their location. Such hosts cannot depend on traditional forms of network connectivity and routing because their location, and hence the route to reach them, cannot be deduced from their network address.

To facilitate continuous network coverage for mobile hosts, a static network augmented with mobile support stations (MSSs) have emerged that are each capable of directly communicating with MHs within a limited geographical area cell, usually via a low-bandwidth wireless medium. The requirement of continuous coverage has opened up the market for wireless services. Wireless data services represent a significant revenue growth opportunity to carriers worldwide.

1.6.7 Optical Networks

Forecasting demand for telecommunications services and equipment has been a challenge with respect to optical networks. Telecommunication networks based on optical communication have been constantly evolving for the last decade following the changing industry landscape as shaped by the market conditions, technological innovations, and regulatory decisions. As networks face increasing bandwidth demand, network providers have started moving towards a different solution based on the principles of optics. The interconnection networks based on optical principles are called optical networks. The recent surge in bandwidth demand, driven by fast-growing video-on-demand (VOD) services and emerging applications, such as network gaming, peer-to-peer downloading, etc. has revitalized the optical communication industry.

Current optical networks are expected to support the increasing network load by employing advanced transmission wavelength division multiplexing (WDM), switching optical switches and cross-connects (XC) and routing technologies. After more than 20 years of active research, passive optical network (PON) based broadband optical access systems are finally seeing widescale deployments in Asia and North America.

Optical networks began with WDM, which arose to provide additional capacity on existing fibres. Optical networks are high-capacity telecommunications networks based on optical technologies and components that provide routing, grooming, and restoration at the wavelength level as well as wavelength-based services. The cornerstone of an optical network is the advanced optical technologies that perform the necessary all-optical functions.

An all-optical network (AON) is a network that uses light wave communication exclusively within the network. More precisely, in an AON all network-to-network interfaces are based on optical transmission, all user-to-network interfaces use optical transmission on the network side of the interface, and all switching and routing within AON network nodes is performed optically.

Wavelength division multiplexing and dense wave length division multiplexing (DWDM) have emerged to augment the capabilities of optical networks. They increase the capacity of optical networks by increasing the number of wavelengths, or colours of light that can be transmitted down a fibre optic path.

The evolution to the optical layer in telecommunications networks will occur in stages in different markets because of the traffic types and capacity demands.

Optical networks, based on the emergence of the optical layer in transport networks, provide higher capacity and reduced costs for new applications such as the Internet, video and multimedia interaction, and advanced digital services.

One of the great revenue-producing aspects of optical networks is the ability to resell bandwidth rather than fibre. By maximizing the capacity available on fibre, service providers can improve revenue by selling wavelengths, regardless of the data rate required.

Optical networks can be divided into multiple groups based on their functionalities and architecture. They are discussed below.

Passive optical network A passive optical network (PON) is a point-to-multipoint architecture for delivering last-mile connectivity without any active components in the distribution network. It is a single, shared optical fibre that uses inexpensive optical splitters to divide the single fibre into separate strands feeding individual subscribers. Passive optical networks are passive because other than at the central office (CO) and subscriber endpoints, there are no active electronics within the access network.

Ethernet passive optical network Ethernet passive optical network (EPON) is a passive optical network based on the Ethernet standard. It allows the users to utilize the economies-of-scale of Ethernet and provides simple, easy-to-manage connectivity to Ethernet-based IP equipment, both at the customer premises and at the central office.

Gigabit passive optical network Gigabit-capable passive optical network (GPON) is currently one of the fastest access technologies to attract market interest. It has received a lot of attention since the International Telecommunication Union (ITU) introduced the ITU-TG.984 recommendation in 2003. A GPON solution is an integral part of a full service broadband architecture, which is designed to meet the needs of fixed mobile convergence (FMC) and next generation networks (NGN) across residential and enterprise service offerings. A key characteristic is the 2.5 Gbps downstream data rate and the 1.25 Gbps upstream data rate. Gigabit passive optical network operates in a very similar fashion to gigabit Ethernet passive optical network (GEPON) when supporting Ethernet as its primary transport protocol.

1.7 IT Systems

Information is one of the most important resources for managers. It adds to the knowledge a person has about an entity of interest. For information to be useful to managers, it must possess certain attributes, which include accuracy, timeliness, relevance, and completeness.

The successful development and implementation of business information systems requires an integrated approach which includes the seamless design of both the business processes and the information systems supporting the business processes. Therefore, several frameworks and modelling methods have been developed for an integrated modelling of the entire enterprise with respect to both organizational and information systems aspects.

Information systems and processes are very important parts of our due diligence assessment of a company—yet the jargon is often more difficult to understand than many foreign languages. An information system is a set of interrelated components working together to provide useful information as needed by problem solvers and decision-makers. The five major components of an information system include hardware, software, people, data, and procedures. In other words, information systems are the software and hardware systems with a set of formal procedures that support, data-intensive applications for human consumption.

Information systems are classified into groups based on functionalities. The major types of information systems that serve the needs of different levels of managers in an organization includes transaction processing, office automation, management information, decision support, and executive support. Some of the prominent information system groups are management information systems (MIS), expert systems (ES), geographical information systems (GIS), health information systems (HIS), hospital management system (HMS), etc.

1.7.1 Management Information Systems

Management information system (MIS) is the term given to the discipline focused on the integration of computer systems to meet the aims and objectives of an organization. The components of an MIS include a hardware which is used for input/output process and storage of data, software used to process data and also to instruct the hardware component, database which is the location in the system where all the organization data will be automated, and procedures which are a set of documents that explain the structure of that MIS.

Management information systems are generally used by medium and larger-scale organizations. They help the manager to access relevant, accurate, up-to-date information.

1.7.2 Expert Systems

Expert systems are computer programs that are derived from a branch of computer science research called artificial intelligence (AI). Often, the term expert systems is reserved for programs whose knowledge base contains the knowledge used by human experts, in contrast to the knowledge gathered from textbooks or non-experts. Every expert system consists of two principal parts: the knowledge base and the reasoning or inference engine.

One of the most powerful attributes of expert systems is the ability to reason and explain the reasoning. Since the system remembers its logical chain of reasoning, a user may ask for an explanation of a recommendation and the system will display the factors it considered in providing a particular recommendation.

1.7.3 Geographic Information Systems

Geographic information systems (GIS) have become an increasingly important means for analysing and understanding geography. Geographic information system concepts and technologies help us collect and organize geographical data and understand their spatial relationships. A geographic information system can be viewed as a data-management system that permits access to and manipulation of spatial data and visual portrayal of data and analyses results. Geographic information system concepts and technologies arise from a wide variety of fields and GIS has become a generic term referring to all automated systems used primarily for the management of maps and geographic data. The development of GIS has relied on innovations made in many disciplines including geography, civil engineering, photogrammetry, remote sensing, surveying, geodesy, statistics, computer science, operations research, demography, and many other branches of engineering and natural and social sciences.

Various definitions have been offered that reinforce the major dimensions of GIS. Elements of a GIS include data and information technology (ie., computers, software, and networks) to support it. Spatial data include any data that has a geographic location.

A data model is a set of rules to identify and symbolize features of the real world, called entities, into digitally and logically represented spatial objects consisting of the attributes and geometry.

There are two basic categories of data involved, namely

1. Spatial data
2. Attribute data

Spatial data include the locations of features, such as the latitude/longitude of dams, gauging stations, etc. Spatial data are often represented as objects, such as points, lines, and polygons, which are used to represent the differing types of features. The spatial data are characterized as having a vector structure composed of features represented as points, lines, and polygons. Other GIS spatial data are handled as images or rasters, having simple row and column formats. Geographic information system functions for spatial data capture include the numerous technologies for data capture as well as the many ways for conversion of source data into GIS-compatible formats.

Attribute data include numerical and character-type data that characterize the resource. Geographic data are characterized by a series of attribute and behavioural values that define their spatial (location), graphical, textual, and numeric dimensions. Attribute data are handled in relational database software composed of records and fields, and the power of the relational model is applied to these data. Geographic information system databases incorporate two distinct branches, the spatial database and the associated attribute database.

1.7.4 Health Information Systems

Hospitals are the key institutions in providing relief against sickness and disease. Effectiveness of a health institution—hospital or nursing home—depends on its goals and objectives, its strategic location, soundness of its operations, and efficiency of its management systems. Hospital information system (HIS) is one of the most promising applications of information technology in the health care sector. The aim of HIS is to use a network of computers to collect, process, and retrieve patient care and administrative information from various departments for all hospital activities to satisfy the functional requirement of the users. It also helps as a decision support system for the hospital authorities for developing comprehensive health care policies.

The next system that most hospitals are looking to deploy is picture archiving and communication system (PACS). It is a filmless and computerized method of communicating and storing medical image data such as computed radiographic, digital radiographic, computed tomographic, ultrasound, fluoroscopic, magnetic resonance, and other special X-ray images. This is a system that is used to capture, store, distribute, and display medical images. Electronic images and reports are transmitted digitally via PACS, eliminating the need to manually file, retrieve, or transport film jackets. A PACS consists of image and data acquisition, storage, display stations integrated with various digital networks.

A PACS comprises four principal components of imaging, namely computed tomography (CT) and magnetic resonance imaging (MRI), a network for the transmission of patient information, workstations for interpreting and reviewing images, and long-term and short-term archives for retrieving images and reports. It has the ability to deliver timely access to images, interpretations, and related data. Typically, a PACS network consists of a central server which stores a database containing the images. This server is connected to one or more clients via a LAN or a WAN which provides and/or utilizes the images. Client workstations can use local peripherals for scanning image films into the system, printing image films from the system and interactive display of digital images. Picture archiving and communication systems workstations offer means of manipulating the images. The medical images are stored in an independent format. The most common format for image storage is digital imaging and communications in medicine (DICOM). For example, Samsung Medical Center, an 1100-bed general teaching hospital, started a four-phase PACS implementation plan in 1994. The medical centre had over 4,000 outpatient clinic visits per day and performed about 340,000 examinations per year. The PACS in Samsung provides support for primary and clinical diagnosis, conference, slide making, generation of teaching materials, and printing hard copies for referring physicians.

Health care in India is just beginning to realize the importance of integrated, configurable systems. Hospitals in India are in various stages of implementing HIS. The driver is return on investment (ROI), and in the long run, benefits are realized from the bottom line and quality of care delivered to patients.

1.8 Management

Strong IT leadership and a relationship of trust between business and IT executives are prerequisites to successfully exploit technology. A clear understanding of IT enables better management. Improving information management practices is a key focus for many organizations, across both the public and private sectors. Effective information management is not an easy job. There are many systems built on different technologies with different mandates and different protocols integration requirement, a huge range of business needs to meet, and complex organizational (and cultural) issues to address. Information Technology Infrastructure Library (ITIL) provides a management framework to address the management needs of information technology.

1.8.1 Service Management

A service offering is a defined entry in the enterprise service catalogue. It is a measurable and specific offering of the IT organization to external clients. It should be seen as a logical application programming interface (API) of the service provider; everything behind it (in theory) may be opaque to the service consumer. Service offerings are of two major types: orderable service and hosting service.

1.8.2 Data Management

Data is the lifeblood of an organization and a valuable enterprise asset. It provides the foundation on which critical and everyday business decisions are based. It is, therefore, essential that decision-makers can access and depend on quality data to operate confidently in a high-performance environment.

Data management has become increasingly important as businesses face compliance consequent to modern legislation, such as Basel II and the Sarbanes–Oxley Act, which regulate how organizations must deal with particular types of data. The data management dilemma that all organizations around the world face today are the regulatory requirements that cause data growth and longer retention periods. Data management experts stress that data life cycle management (DLM) is not simply a product, but a comprehensive approach to managing organizational data, involving procedures and practices as well as applications.

Data management is essentially the process of managing data as a resource that is valuable to an organization or business. Key elements of data management includes specification of data formats (metadata), data access protocols (transport), and data transformation rules (mapping). A good data management solution would provide all these capabilities in an easy-to-use single product package. The data management platform coupled with its process and methodology provides auditing, tracking, and controlling mechanisms to manage the data effectively. A detailed discussion on data life cycle is presented in the later chapters.

1.8.3 Disaster Management

Disasters come in many forms. Natural disasters kill one million people around the world each decade, and leave millions more homeless. Natural disasters may include earthquakes, floods and flash floods, landslides and mud flows, wildland fires, winter storms, and others. Technological disasters include house and building fires, hazardous materials, terrorism, and nuclear power plant emergencies.

The mission of an effective disaster communications strategy is to provide timely and accurate information in all the following four phases of emergency management.

1. *Mitigation*, which is to promote implementation of strategies, technologies, and actions that will reduce the loss of lives, business, and property.
2. *Preparedness*, which is to communicate preparedness messages that encourage and educate business in anticipation of disaster events.
3. *Response*, which is to provide to the appropriate notification, warning, evacuation, and situation reports on an ongoing disaster.
4. *Recovery*, which is to provide business units affected by a disaster with information to recover from the business data loss and the time taken to bring the business functions to operational readiness.

1.8.4 Remote Infrastructure Management

Over the past few years, the infrastructure outsourcing industry has witnessed substantive shifts. Key drivers behind these shifts include enterprise customers that seek to enhance service and performance levels while exploring innovative delivery models to reduce costs, technology that has improved infrastructure efficiency, and management and maturing offshore capabilities. Businesses today face a considerable challenge to effectively optimize their IT infrastructure and related operations and deliver ever-improving service levels to meet and exceed the expectations of their business users without compromising on quality and security. More and more companies are turning towards infrastructure management service (IMS) as the answer to this need. Infrastructure management as a service has resulted in the genesis of remote infrastructure management services (RIMS).

Remote infrastructure management services comprise day-to-day management of IT infrastructure needs of an organization from a remote location. Remote infrastructure management services consist of remote monitoring and managing the infrastructure components and taking proactive steps and remedial actions across the IT landscape. Remote infrastructure management (RIM) capabilities are bounded with service level agreements (SLA) with penalties on downtimes. These value propositions appeal to two distinct demands of enterprise customers: cost reduction and optimization that leads to transformation.

Networking infrastructure has emerged as a key differentiator to drive RIMS. Since infrastructure management is a critical issue and even minutes of downtime can cripple a client's business, the network has emerged as a crucial element of an IT vendor's business strategy. The entire network must, therefore, be resilient. Business resilience refers to the operational and technological readiness that allows IT service providers to operate their networks efficiently. Business-resilient networks help businesses respond quickly to opportunities and react appropriately to unplanned events.

The study conducted by McKinsey & Company highlights that the $524 billion infrastructure management services (IMS) industry— that manages an enterprise's core IT systems, including hardware, software, connectivity and people could become as important as business process outsourcing (BPO) industries that have dominated the rise of offshoring in the last decade.

Increasingly, enterprises worldwide have been waking up to the challenges involved in ensuring the availability and the predictability of their networks and devices. The rigorous processes and operations combined with the critical nature of business makes it imperative for the vendors to specialize in RIMS service delivery. A detailed discussion on remote infrastructure management is presented in the later chapters.

1.8.5 Measures and Metrics

It is arguably the number one rule of business—if one cannot measure it, one cannot improve it. Or as many a CIO has had to learn the hard way, if one cannot measure it, one cannot communicate its value. Applying manufacturing-style metrics to the global delivery of IT application development and maintenance does more than streamline and reduce cost. It provides organizations with a clear path to IT productivity to help them achieve higher levels of performance.

> In April 2008, Forrester published the report 'The Five Essential Metrics for Managing IT'. This has become one of the IT industry's most popular documents in terms of readership and reader feedback. Principal Analyst Craig Symons lays out five IT metrics that are extremely relevant to IT's business stakeholders: investment alignment to business strategy, business value of IT investments, IT budget balance, service-level excellence, and operational excellence.

Chief information officers frequently ask what IT should measure and report to business executives. The key to success is choosing a small number of metrics that are relevant to the business and have the maximum impact on business outcomes. A beginning step in the development of an IT performance management program is the identification of information technology's role as an enabler of both the strategic and operational requirements of the business. As understood and promoted by the ITIL framework and information technology service management (ITSM) in general, the primary accomplishment of IT should be the effective alignment of services with the current and future needs of the business and its customers.

Information technology metrics are hardly new but the understanding of how to learn from and leverage them is steadily evolving, and their importance is growing as corporations weave IT into virtually all of their activities. Information technology measures must have context to have meaning; the appropriate context is derived by deciding which investments in measurement will actually provide business value. Effective service level agreements (SLAs) are extremely important to assure effective operations. Information technology projects, deployments, operations and maintenance, and other IT exercises are measured through the planned and executed SLAs. The metrics used to measure and manage performance to SLA commitments are the heart of a successful agreement and are a critical long-term success factor. Lack of experience in the use and implementation of performance metrics causes problems for many organizations as they attempt to formulate their SLA strategies and select and set the metrics needed to support those strategies. Among the most salient differences between performance measurement and integrated performance management is the ability to develop and apply measures that determine the causal factors of performance.

Now, yet another category of metrics is gaining attention, namely metrics that aim to help IT in better justifying itself at a business level. This means fewer bits and bytes and more dollars and cents, which in turn means calculating fully costed ROIs and being able to associate measurable gains in revenues or market share.

Leading organizations apply metrics and use them to continually boost the quality of their output, increase the predictability of that output and improve their overall efficiency. The proactive use of metrics is increasingly important as organizations adhere to industry standards for measuring quality such as Six Sigma and Capability Maturity Model for Software. The importance of metrics will only increase as IT departments strive to organize and manage themselves as in-house service providers.

Return on Investment

The return on investment(ROI) for corporate information technology investments has been the subject of considerable research in the last decade. When capital to invest is scarce, new e-business and IT projects must show a good ROI in order to be funded. One conceptual definition of ROI is that it is a project's net output (cost savings and/or new revenue that results from a project less the total project costs), divided by the project's total inputs (total costs), and expressed as a percentage. The inputs are all of the project costs such as hardware, software, programmers' time, external consultants, and training. Therefore, if a project has an ROI of 100 per cent, from this definition the cash benefits out of the project will be twice as great as the original investment.

Return on investment was defined in the introduction as

$$\text{ROI} = \frac{\text{Project outputs} - \text{Project inputs}}{\text{Project inputs}} \times 100\% \qquad (1.1)$$

where the project outputs are all of the benefits of the project quantified in terms of cost savings and revenue generation, and the project inputs are all of the costs of the project.

Return on investment is an important component of the IT investment decisions made in many large companies. Full life cycle ROI analysis translates into better information to make better decisions, which in turn should impact the returns for the total corporate IT portfolio of investments.

The method of calculating ROI for an e-business or IT project is in principle no different from the method for calculating ROI for a new manufacturing plant, marketing plan, or research and development project. However, e-business and IT projects can be incredibly complex, so that estimates and generalities that are good enough for a manufacturing project can potentially destroy an IT project if any element goes wrong. Building the ROI model on sound assumptions and developing a risk-management strategy can, therefore, significantly impact the actual ROI realized for IT projects.

Profitability Index

Profitability index identifies the relationship of investment to payoff of a proposed project. It is the ratio of the present value of a project's cash flows to the initial investment. Profitability index is also known as profit investment ratio, abbreviated to PI and value investment ratio (VIR). Profitability index is a good tool for ranking projects because it allows you to clearly identify the amount of value created per unit of investment.

$$\text{Profitability index} = \frac{\text{Net present value}}{\text{Investment}} \quad (1.2)$$

1.9 Standards, Audit, and Governance

Technology leadership is as important for the managing executives of a large Fortune 500 company as it is for the new entrepreneur. Whether you are a technology person who has learned the business side, or a business person who has learned the technology side, a technology map gets generated as a brain map. A technology map is a mental map of how technology works and how the different components of the systems in a company fit together. The map gives the reader a good understanding of how technology interacts and fits together. The map demonstrates the layering effect of IT.

The IT system, so understood, needs to be audited based on some common criteria called best practices or standards periodically, in order to find the dynamic interplay of IT systems. A clear understanding of national and international standards are required to capture the big picture in order to fine-tune the governance mechanisms.

1.9.1 Standards

There is an old joke in the technology industry about standards—'the wonderful thing about standards is that there are so many to choose from'. A standard, literally, is an approved way of accomplishing a technical goal that is published by a standards body such as the Institute of Electrical and Electronics Engineers (IEEE), International Organization for Standardization (ISO), the World Wide Web Consortium (W3C), and several others. Old or young, a standards organization is usually needed to broker negotiations for shared standards so that businesses can capitalize on interrelationships.

International Committee for Information Technology Standards (INCITS) is a US-based standardization organization in the field of information and communications technologies (ICT), encompassing storage, processing, transfer, display, management, organization, and retrieval of information. As such, INCITS also serves as American National Standards Institute (ANSI) Technical Advisory Group (TAG) for ISO/IEC Joint Technical Committee 1 (JTC1). Joint Technical Committee 1 is responsible for international standardization in the field of information technology.

Competing standards are often not interoperable. One cannot have a network card talking (transferring data) at gigabit Ethernet speeds while the hub port is talking (transferring data) at fast Ethernet speeds (100). They must match, or the network does not work. Information technology products are useless unless they interface with other IT products. So IT products must comply to standards providing the essential compliance and interoperability. Hence, the products and services sold by IT vendors have a relationship with the standards published by standards bodies.

Although a de facto standard is not a standards body but a standard, it is relevant to explain this term here. If the market commonly has adopted the implementation of a specific company, or a standard specified by an unofficial standards body, then this standard is called a de facto standard. When this kind of a standard exists, it can be very difficult for competitors to enter the market with different directions, including those companies that implement an official standard on the subject. Microsoft is an example of a company whose products are de facto standards. Participation in the standards processes can assist the formulation of company strategy, marketing plans, finance, and development.

The most important benefits of standards are as follows.

- They ensure that equipment from different suppliers can interwork. This benefits the user by enabling competitive procurement.

- They enable equipment supplied by a manufacturer to be used for different applications and in different regions. This increases the market size and reduces costs due to economies of scale. This benefits both the suppliers and the consumers.
- They provide the means to deal with a changeable environment. Today, businesses must do more with fewer resources in changeable conditions and environments. It is difficult to keep up to date, to understand the complex technologies, to be a specialist in all system parts, and to respond effectively to changes, because of the speed at which change takes place.

1.9.2 Audit

In most companies, key operational processes are managed by information technology systems. An IT organization, with well-defined internal controls, enables companies to identify and manage their IT-related risks. Ability to manage and contain such risks is critical to ensuring compliance with regulations and mandates such as Sarbanes–Oxley Act (SOX), Gramm–Leach–Bliley Act (GLBA), and Health Insurance Portability and Accountability Act (HIPAA). Moreover, companies leveraging outsourced services that impact their own control environment rely on SAS 70 service auditor reports to gain an understanding of the IT processes of their service providers.

Auditing, in general, is formally described as 'the independent examination of records and other information in order to form an opinion on the integrity of a system of controls and recommend control improvements to limit risks'. Information technology auditing is a branch of general auditing concerned with governance (control) of information and communications technologies (computers). Information technology auditors primarily study computer systems and networks from the point of view of examining the effectiveness of their technical and procedural controls to minimize risks. All audits are performed in relation to certain risks identified by the auditor which he/she believes are important. The specific controls that are actually embedded in or associated with IT systems and processes are then assessed to determine whether they adequately address the risks. Most organizations regularly test the internal controls within their IT organization to ensure secure and continuous operation of their entire information system's infrastructure.

1.9.3 Governance

Governance essentially means systems of control. Information security managers develop, implement, and operate information security control systems for ICT governance. Information technology governance is a framework for the leadership, organizational structures and business processes, standards and compliance to these standards, which ensure that the organization's IT supports and enables the achievement of its strategies and objectives. It is a critical component of corporate governance. Essentially, governance addresses proper management of organizations. Information technology governance takes these concepts one step lower and applies them to the IT group. Information technology governance is the key to integrating people, processes, technology, and information necessary to achieve business goals according to International Business Machines (IBM).

Perhaps the best definition can be found in the executive summary of control objectives for information and related technology (COBIT), which identifies IT governance as a structure of relationships and processes to direct and control the enterprise in order to achieve the enterprise's goals by adding value while balancing risk versus return over IT and its processes. The COBIT framework is comprised of 34 high-level control objectives and 318 detailed control objectives that have been designed to help businesses maintain effective control over IT.

The ISO/IEC 38500:2008 provides guiding principles for directors of organizations on the effective, efficient, and acceptable use of IT. It relates to the information and communication services used by an organization. These standards provide a framework for the governance of ICT, comprising a model, principles, and a vocabulary. They also reference other standards and relate to methodologies used for project management and control. This governance framework for IT complements a number of standards, frameworks, and methodologies. It also relates to record keeping, fiduciary duties, and privacy regulatory and legislative requirements and the organization's internal policies covering fraud control, whistle blowing and corporate social responsibility(CSR).

Weill and Ross [2004] recommend the 10 principles of governance as listed below.

1. Actively design governance
2. Know when to redesign
3. Involve senior managers
4. Make choices
5. Clarify the exception-handling process
6. Provide the right incentives

7. Assign ownership and accountability for IT governance
8. Design governance at multiple organizational levels
9. Provide transparency and education
10. Implement common mechanisms across the key assets.

1.10 Current Trends

Aligning information technology with business strategy is essential to meet and beat competition. Information technology is dynamic and changes over time.

1.10.1 Data Centre

The modern corporation runs on data. Data centres house thousands of servers that power applications, provide information, and automate a range of processes. The data centre is a physical place that houses computers, computer networks, critical computing systems, data storage, including backup power supplies, air conditioning, and security applications. Wikipedia defines data centre as a facility used to house computer systems and associated components such as telecommunications and storage systems. It generally includes redundant or backup power supplies, redundant data communications connections, environmental controls (for example, air conditioning, fire suppression), and security devices. According to Gartner forecasts, the total data centre capacity in India will be almost 5.1 square million feet by 2012 growing at a compound annual growth rate (CAGR) of 31 per cent. Captive data centres will grow at a CAGR of 29 per cent and hosted data centres will grow at 33 per cent to reach 2.571 and 2.573 million square feet respectively.

1.10.2 Green Computing

Energy is a major concern especially in developing countries like India where there is a power crisis. Data centre operational costs have also been heavily impacted by the rising costs of energy. Data centres today consume a substantial percentage of generated power for any country. Especially, the cost of energy needed to power and cool the data centre facilities, computer rooms, and data centres.

> A National Action Plan on Climate Change (NAPCC) was released by Prime Minister Manmohan Singh. The central ministries were directed to present a detailed blueprint to the Prime Minister's Council on Climate Change on the specific implementation plans they would adopt to mitigate the greenhouse emissions.

Stacking servers ever closer together in rows and rows of racks and making their processors work faster and harder has recreated many of the old cooling problems. Today, the industry finds that for every kilowatt of power it uses to drive a server, another kilowatt is needed to cool it. Businesses are looking to minimize energy waste and reduce the carbon footprint of computing resources.

> Mckinsey's cost curve studies of potential ways to reduce carbon emissions show that the incremental improvement of today's technology and energy consumption patterns cannot have a significant effect and new low carbon technologies will have to be developed [Enkvist et al. 2007].

Green IT efficiently utilizes computing resources reducing carbon footprint based on reduction of energy requirements and decrease in total energy usage.

1.10.3 Grid Computing

Computer facilities are expensive to build and maintain, so being able to share cycles across the globe represents a more effective way to utilize IT resources. Grid computing enables the creation of a single IT infrastructure that can be shared by multiple business processes. The basic idea of grid computing is to create a grid computing infrastructure to harness the power of remote high-end computers, databases, and other computing resources owned by various people across the globe through the Net. Grids are intrinsically distributed and heterogeneous but must be viewed by the user as a virtual environment with uniform access to resources. Grid computing extends the web services concept by providing task and resource management functions across heterogeneous computing environments [Franc 2003, Lucio 2005]. Compute grids can be deployed at a local level or encompass computing facilities around the nation.

1.10.4 Virtualization

Virtualization is one of the IT buzzwords being thrown around the industry more and more lately. Virtualization technology is a way of making a physical computer function as if it were two or more computers. Each non-physical or virtualized computer is provided with the same basic architecture as that of a generic physical computer. There is an increasing recognition of virtualization as a tool that helps in better utilization of server hardware, reduces floor-space requirements, lowers power and cooling costs, and improves productivity of personnel. Virtualization today is such an exciting topic because people realize that through virtualization they can consolidate and get more out of their infrastructure.

Virtualization provides the benefits necessary to give IT organizations the ability to save costs on hardware and increase the efficiency of server deployments, provisioning, and management. Virtualization also enables physical hardware independence, which gives IT the flexibility and freedom of not being locked into a single vendor's hardware solution. Virtualization allows an operator to control a guest operating system's use of CPU, memory, storage, and other resources, so that each guest receives only the resources that it needs. This distribution eliminates the danger of a single runaway process consuming all the available memory or CPU. It also helps IT staff to satisfy service-level requirements for specific applications.

1.10.5 Server Consolidation

Consolidation allows companies to improve overall business processing through the following three primary IT objectives.

1. A higher and more consistent level of service
2. Greater efficiency and control over operations
3. The flexibility to respond to constantly changing business requirements

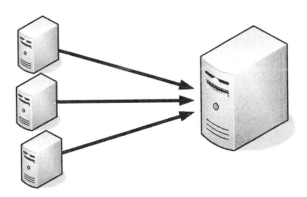

FIGURE 1.7 Server Consolidation

Server consolidation (Figure 1.7) is becoming an increasingly popular technique to manage and utilize systems. Server consolidation is an approach to efficient usage of computer server resources in order to reduce the total number of servers or server locations that an organization requires. The practice developed in response to the problem of server sprawl, a situation in which multiple, under-utilized servers take up more space and consume more resources than can be justified by their workload. Although consolidation can substantially increase the efficient use of server resources, it may also result in complex configurations of data, applications, and servers that can be confusing for the average user to contend with.

There are a number of steps an organization could take after successfully executing a server consolidation strategy. One of them is to orchestrate workloads to make maximum use of physical server capacity, reduce power consumption, reduce administrative and operational expenses, and create an environment which can assure that each and every workload can meet service-level objectives. Orchestration is a process of identifying the need for a virtual spin for a virtual server. Information technology coordinates resource allocation among different infrastructure components such as switch provisioning, Internet protocol (IP) addressing, domain name servers (DNS), and active directory services (ADS).

1.10.6 Storage Consolidation

During the past few years, customers have focused on day-to-day business and and have created storage islands interns of dedicated storage. As these storage islands have expanded to more significant capacity levels, total storage utilization across the enterprise and the inability to share data between applications have become inhibitors to further growth.

The proliferation of data throughout every facet of industry, combined with an increasing demand for secure access from a variety of devices, make data storage a critical component of every business strategy and organization. As the cost of storage management and maintenance continues to skyrocket, the business is under increasing pressure to streamline storage management and maximize investments in the current business practices.

Consolidating the storage systems, platforms, and applications can help the organization ease manageability and improve capacity utilization factors. Storage consolidation, also called storage convergence, is a method of centralizing data storage among multiple servers. The objective of storage consolidation is to facilitate data backup and archiving for all subscribers in an enterprise, while minimizing the time required to access and store data. By creating large storage pools accessible by many servers, IT asset utilization is increased and operating costs are reduced.

Storage consolidation provides a common platform to allocate and manage growing demands for data storage and helps reduce administrative effort by establishing a single management console. The formation of centralized storage pools that can be managed as a whole and shared across applications and servers improves storage utilization and efficiency.

1.10.7 IT Practices in Banks

Core banking solutions (CBS) have been around in banking circles for quite some time now. Today, the situation is such that an increasing number of banks are recognizing the need for a strategic transformation of their day-to-day operations. Core banking solutions implementation is the first step towards transformation and automation of all of a bank's business processes. Core banking solutions have become a part of basic hygiene for a bank's operation. Today, if a bank does not have a CBS in place, the ecosystem solutions, such as anywhere banking, Internet banking, etc. become a huge challenge for operations. The CBS solution forms the primary IT environment for any bank today. This has become an integral part of business rather than being a support function. Thereby IT, and in turn CBS, becomes a business driver rather than a back-office operation.

By integrating all the varied and disparate legacy systems and self-contained services, a CBS provides a bank with a solid, flexible, and scalable foundation on which other systems can be easily built and harmonized into a robust scalable banking solution. The focus of many banks is now on business agility as a key differentiator in their attempt to gain a competitive edge. With business agility, banks achieve improved operational efficiency, enhanced customer service experience, a proactive approach towards risk, and most importantly, a drastic reduction in operating costs.

Meanwhile, the largest opportunity at the moment is for the Regional Rural Bank (RRB). According to a mandate by the Reserve bank of India (RBI), all commercial and RRBs are supposed to implement CBS by March 2010. As per RBI guidelines, the CBS in RRBs should be geared towards better management control and monitoring, wider range of services offered and enhanced level of customer satisfaction. Adoption of CBS would lead to uniformity in work environment, more informed decision-making, centralized processing and better MIS and reporting and improved regulatory compliance.

The centralized funds management system (CFMS) initiative provides for a centralized viewing of balance positions of the account holders across different accounts maintained at various locations of the RBI. The electronic clearing service (ECS) and electronic funds transfer (EFT) are also being enhanced in terms of security by means of implementation of public key infrastructure (PKI) and digital signatures using the facilities offered by the certifying authority as per the guidelines of the RBI. Modernization of clearing and settlement through MICR-based cheque clearing, popularizing electronic clearing services and integration of RBI-EFT scheme with funds transfer schemes of banks, introduction of centralized funds management system are significant milestones.

Summary

In this chapter we discussed the basic IT components, networks, and the types of resources networks can share. Computer networks are groups of computers connected together by some type of media (the physical connection among the devices on a network) to allow them to communicate and share information. Servers are large, powerful computers that provide services to clients. Clients are smaller desktop computers that users use to access network services. The physical connection between the computers on a network is referred to as the media. Client/server architectural network is an arrangement used on LAN that makes use of distributed intelligence to treat both the server and the individual workstations as intelligent, programmable devices, thus exploiting the full computing power of each. The client and server machines work together to accomplish the processing of the application being used. Local area networks are small networks usually contained in one office or building. They have high speed, low error rates, and they are inexpensive. Metropolitan area networks are larger networks that consist of individual LANs to interconnect large campus-type environments such as organizations spread over a city. Wide area networks can cover an entire organization's enterprise network.

Information technology managers have to understand IT more effectively to make it dance to their tunes. The chapter outlined the basic need for IT infrastructure and the various components that go with it. An overview of current trends in the field of IT has been discussed to enable the reader to get a feel of it. The topics discussed include data centre, grid computing, cloud computing, virtualization, server consolidation, and storage consolidation. This chapter is also the foundation chapter for the successive chapters.

Key Terms

Cloud computing It is the term used for anything that involves delivering hosted services over the Internet.

Data centre It is a well-equipped computer facility designed for continuous use by several users.

Ethernet passive optical network It is an optical network built on Ethernet technology.

Gigabit passive optical network It is a network built on optical technology.

Grid computing It harnesses unused processing cycles of all computers in a network.

Local area network It is an internal network of computers.

Metropolitan area network It is a network spread over a large geographic area.

Passive optical network It is a network built on optical technology.

Virtualization It is a technique for hiding physical characteristics of computing resources.

Wide area network It is a network of networks spread over geography.

REVIEW QUESTIONS

1.1 What is IT infrastructure and what is it composed of?

1.2 Discuss the four key business requirements for the success of a business organization.

1.3 Discuss the business value of information technology.

1.4 Briefly discuss the various elements of networking infrastructure.

1.5 Justify the statement 'structured cabling is a capital investment'.

1.6 Briefly discuss the various aspects of a networking switch and its types.

1.7 How is optical network different from wired network? Differentiate between PON, EPON, and GPON.

1.8 Briefly explain the concept of remote infrastructure management identifying the business value of RIM.

1.9 Differentiate IT audit and IT governance.

1.10 Discuss and differentiate grid computing and cloud computing and identify their business values.

Projects

1.11 Visit an organization which has implemented virtualization of IT assets. Study the business impact of virtualization of infrastructure in the organization outlining the pre- and post-business benefits including ROI spread over a period of three years.

1.12 Prepare a short business report on green initiatives, need for going green and its business impact on bottom line.

1.13 Discuss the technical and management plans which enabled ABC Unlimited to carry out the migration with no detectable disruptions. Note: AS/400 is a mid range server. Refer to the business case given below

Run-through Business Case

From his office on the sixth floor in New York, ABC Unlimited's state-of-the-art global headquarters, CIO Nayak watched the traffic loom away. ABC Unlimited is perhaps not a familiar name for the average Indian customer. It is a retail chain establishment with multiple outlets spread across various sectors of the USA and the UK. As Nayak sipped his morning coffee, he glanced at the internal memo from his boss, ABC Unlimited's CEO, David. The subject read 'Costs must come down'. The message from David was succinct, clear, and immediate. 'Our product offerings are clear and our balance sheet is strong.' David indicated that in order to sustain profitability 'We must lower our costs and deliver a premium service. The challenge to cut costs is never-ending, and our task is to work together to get them down year over year.' The message was crystal clear and echoed across the organization. Like many of his competitors, David hoped to move his company up the value chain. At ABC Unlimited, the plan was to accomplish this goal by providing clients with innovative, integrated services that would impact and even redefine their core way of doing business. David was clear that in order to survive he would have to change the way the company approached its customers. To do so, he wanted to differentiate the service offerings with unique propositions that would tie the customers with ABC Unlimited. In order to do this, he wanted to remove the silos of information spread across the organization and centralize it.

New York based ABC Unlimited was founded in 1991 by an Indian named Pooja as a venture capital attempt with an initial investment of about 1 million dollars. (Venture capital can be defined as equity or equity-linked investments in young, privately held companies, where the investor is a financial intermediary and is typically active as a director, advisor, or even manager of the firm). In 1994, one of Hong Kong's largest trading company acquired a 45 per cent share in the company. As a retail group, ABC Unlimited developed most of its systems in-house. John, who started his career in IT in 1998 right after his graduation, joined ABC Unlimited as a programmer and wrote the first integrated online billing system for ABC Unlimited called 'centralview', a program which was later sold to other retail outlets. John and his team of systems developers were responsible for writing the code for most of ABC's proprietary applications including accounting systems, personnel systems, logistics, and supply chain systems apart from a host of internal applications. The department started with less than 20 developers and by late 2003 had grown into a team of 120 members.

ABC Unlimited had earlier taken a decision to decentralize the operations across all the outlets. Each outlet had hardware and a copy of the appropriate software running in their hardware to take care of the business. However, of late the organization found it very difficult to get a consolidated view of the business due to multiple data sources and multiple reports. Based on the demand from the developers and answerable to the board of directors, David wanted to have a comprehensive consolidated view of the reports on demand and had sent an internal note to Nayak on the possibilities of centralizing the data. He also added that the centralization would considerably reduce the overheads on IT infrastructure and help the organization overcome the IT resource crunch.

Nayak was supposed to meet David with a plan of action on the cost benefit advantage apart from a plan to set up a data centre. The company needed to aggregate their data at a data centre to a new headquarters building several miles away. As identified, the IT costs at the new location including human resource (HR) cost, were slightly lesser than the current pay levels spread across the various outlets. Nayak believed that the move to the new location will also have a direct impact on the HR management issues. Apart from the location-specific competitive advantages, on the whole the new location had an extensive fibre-optic network infrastructure in place, with a fast growing cohort of IT professionals. Several large multinationals such as IBM, Cisco, and Dell have established their presence in the selected location. This, Nayak believed would result in rapid support delivery. He also understood that the new location was well secured from a seismic zone chart. He also understood that the new location was an emerging IT hub with a good residential support at an affordable cost.

Apart from the challenge of moving headquarters personnel, there was a need to move the existing servers and IT infrastructure which supported the warehouse operations and retail sales. While the warehouse only functioned six days a week, allowing one day to move an AS/400 and bring up new ISP links, many of the stores were open 24 hours a day and any significant downtime would quickly find its way to the bottom line. From the data processing viewpoint, the move included a large uninterruptible power supply (UPS), an AS/400, multiple application servers, and directory servers and with PCs. The network consisted of approximately 75 stores connected through ISPs to three routers at the data centre running a common LAN shared by headquarters (HQ) staff, the data centre, and the warehouse.

John, by now the head of IT operations, was asked to evaluate the configuration and dependency options of the IT infrastructure and identify a suitable systems migration plan. Needless to say, the prospect of shutting everything down on a Friday night, running backups on every server, dismantling all data servers and HQ employee systems, reconfiguring all warehouse systems which were not moving, and having everything up and running for the

	ABC Unlimited				
Particulars	**Year 2000**	**Year 2002**	**Year 2004**	**Year 2006**	**Year 2008**
Assets					
Cash and cash equivalents	49	95	249	539	953
Marketable securities	29	59	129	359	569
Loaned securities	37	76	169	335	359
Finance receivables net	109	462	861	1698	2968
Other receivables net	28	94	195	299	495
Inventories	79	386	756	1269	3496
Net property	98	248	445	699	1245
Defered income tax	104	126	327	692	1022
Other asset	58	88	166	226	333
Total assets	**591**	**1634**	**3297**	**6116**	**11503**
Liabilities and stockholder's equity					
Payables	95	292	393	596	1056
Accured liabilities and deferred revenues	35	183	377	825	1265
Debt	109	349	589	1615	2566
Liabilities of discontinued/held for sale operation	24	89	28	35	60
Total liabilities	**263**	**913**	**1387**	**3071**	**4947**
Stockholder's equity					
Capital stock (including in excess of par value of stock)	146	368	479	1096	1596
Accumulated and other income/(Loss)	137	285	1182	936	3395
Treasury stock	0	0	0	359	496
Retained earnings/(accumulated deficit)	45	68	249	654	1069
Total stockholder's equity	**328**	**721**	**1910**	**3045**	**6556**
Total liabilities and stockholder's equity	**591**	**1634**	**3297**	**6116**	**11503**

Balance Sheet (Figure in USD)(In Millions)

Statement of Income

ABC Unlimited

Particulars	Year 2000	Year 2002	Year 2004	Year 2006	Year 2008
Sales and revenues	987	2985	6045	10284	15875
Total revenues	**987**	**2985**	**6045**	**10284**	**15875**
Cost and expenses					
Selling, administrative and other expenses	429	1487	2945	5955	7744
IT expenses	189	809	995	1200	1395
Interest expense	98	178	245	497	529
Total expense	**716**	**2474**	**4185**	**7652**	**9668**
Income/(loss) before income taxes	271	511	1860	2632	6207
Provision for income tax	89	158	429	542	743
Net income/(loss)	**182**	**353**	**1431**	**2090**	**5464**
Profit after tax %	**18.44%**	**11.83%**	**23.67%**	**20.32%**	**34.42%**
Contribution of expense to sales (%)	**72.54%**	**82.88%**	**69.23%**	**74.41%**	**60.90%**

Statement of Income (Figure in USD)(In Millions)

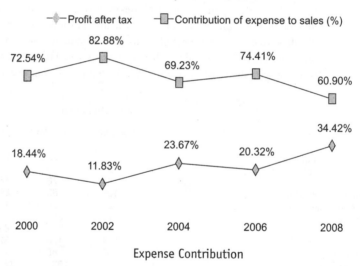

Expense Contribution

warehouse to begin operations at 6:00 a.m. on Monday caused the MIS manager to pass many sleepless nights.

Rather than even attempting to pick everything up and hope that it would all work when deposited in the new location, ABC Unlimited, worked up a network and data migration plan which would allow the move to occur in phases, allowing the MIS staff to concentrate on one critical factor at a time and minimize the danger of excessive downtime at any time during the move.

The data centre move took place on schedule and with no detectable disruption of retail support services.

REFERENCES

Baschab, John and Jon Piot 2007, *The Executive's Guide to Information Technology*, John Wiley & Sons.

Becker, Wendy M. and Vanessa M. Freeman 2006, 'Going from global trends to corporate strategy', *The Mckinsey Quarterly*.

Berman, Fran 2003, *Grid Computing: Making the Global Infrastructure a Reality*, John Wiley & Sons.

Enkvist, Per-Anders, Tomas Nauclér, and Jerker Rosander 2007, 'A cost curve for green house gas reduction', *The Mckinsey Quarterly*.

Enkvist, Per-Anders, Tomas Nauclér, and Jeremy M. Oppenheim 2008, 'Business strategies for climate change', *The Mckinsey Quarterly*.

Farrell, Diana and Jaana K. Remes 2008, 'How the world should invest in energy efficiency', *The Mckinsey Quarterly*.

Ford, Jerry Lee 2002, *Absolute Beginner's Guide to Personal Firewalls*, Que.

Grandinetti, Lucio 2005, *Grid Computing: The New Frontier of High Performance Computing*, Elsevier.

Harris, Michael D., David E. Herron, and Stasia Iwanicki 2008, *The Business Value of IT: Managing Risks, Optimizing Performance, and Measuring Results*, CRC Press.

Holtsnider, Bill, and Brian D. Jaffe 2008, *IT Manager's Handbook*, Morgan Kaufmann.

Kaplan, James M., Rishi Roy, and Rajesh Srinivasaraghavan 2008, 'Meeting the demand for data storage', *The McKinsey Quarterly*.

Krzysztof, Iniewski, Carl McCrosky, and Daniel Minoli 2008, *Network Infrastructure and Architecture*, Wiley Interscience.

Power, Richard 2000, *Tangled Web: Tales of Digital Crime from the Shadows of Cyberspace*, Que.

Reynders, Deon, Steve Mackay, and Edwin Wright 2005, *Practical Industrial Data Communications*, Elsevier.

Rhoads, C. J. 2008, *The Entrepreneur's Guide to Managing Information Technology*, Praeger.

Sivalingam, Krishna M. and Suresh Subramaniam 2005, *Emerging Optical Network Technologies: Architectures, Protocols and Performance*, Springer.

Stevan, Howard 2001, *Case Studies: New Tactics for Active Containment of IT Costs*, Gartner Symposium.

Weill, Peter and Jeanne W. Ross 2004, *Ten Principles of IT Governance*, http://hbswk.hbs.edu/archive/4241.html, download, accessed on 16 January 2009.

William, Forrest, James M. Kaplan, and Noah Kindler 2008, 'Data centers: How to cut carbon emissions and costs', McKinsey on Business Technology Winter.

CHAPTER 2
Computing Infrastructure: Hardware

Competition is a lot like cod liver oil. First it makes you sick. Then it makes you better.
–American Micro Devices

Learning Objectives

After reading this chapter, you should be able to understand:
- the basic elements which make up computing infrastructure
- of the need for operating systems
- the differences between various types of operating systems

2.1 Introduction

The focus of this chapter is on computing infrastructure. Computing infrastructure forms the backbone of any computational activity. Information technology (IT) is a generic name provided to computing infrastructure as a whole. Infact, IT is more than a collection of related components. The processes represent a fundamental value chain, and therefore require systems support as much as any other business process. Information technology should be architectured, implemented, and operated as a cohesive set of related systems.

Infrastructure runs the applications that process transactions, handles the customer data that yield market insights, and supports the analytical tools that help executives and managers make and communicate the decisions shaping complex organizations. In fact, infrastructure has helped in much of the corporate growth and rising productivity of recent years.

Infrastructure is generally a set of interconnected structural elements that provide the operational framework for an organization, a group, or a nation. In general, infrastructure defines the operational base for an organization or system. Still underlying the more general uses is the concept that infrastructure provides organizing structure and support for the system or organization it serves. Infrastructure systems include both the physical assets and the control systems and software required to operate, manage, and monitor the systems. Information technology infrastructures usually include a myriad collection of servers and storage and application platforms coupled with heterogenous technologies. In addition, data and applications often span across distributed or clustered servers and storage built on different technologies. Supporting and protecting these heterogeneous technologies and platforms works out to be a complex issue.

The ISO 9001:2000 definition of infrastructure is the system of facilities, equipment, and services needed for the operation of an organization.

Thanks to management gurus and management visionaries, technology now meshes tightly with operations in ways that was not possible a decade ago. Virtually all organizations—small, medium, and large—across the world depend to some degree on technology. Software development and systems administration are skills sought worldwide as gateways to greater economic security. And worldwide investment in IT continues apace, into hundreds of billions and trillions of dollars.

As IT's power and presence have expanded, companies have come to view IT as a resource ever more critical to their success. The whole IT infrastructure, on which business applications run and which ultimately provides users with a modern office environment, needs to be operated and maintained. The infrastructure itself covers personal computers (PCs), data centres, networks, and basic office tools such as e-mail and wordprocessing. Users, likewise, need to be supported in using the infrastructure. In order to create a strategic advantage, the IT systems need to be aligned and realigned with changing business focus. To add to this dynamic complexity, IT as a technology changes frequently. Not only does the hardware and software scene change relentlessly, but also ideas about the actual management of the IT function are being continuously modified, updated, and changed. There are few professions which require as much continuous updating as that of the IT executive. Thus, keeping abreast of what is going on is really a major task.

2.1.1 Elements of a Computer

As a start, let us consider the word 'computer'. It is an old word that has changed its meaning several times in the last few hundred years. Coming, originally, from Latin, by the mid-1600s it meant 'someone who computes'. It remained associated with human activity until about the middle of this century when it became applied to 'a programmable electronic device that can store, retrieve, and process data' as *Webster's Dictionary* defines it.

A computer is an electrical machine that can both process and store information. The information may consist of numbers, words, or both. Thus, the computer performs a wide variety of services controlled by inputting commands initiated by the users. The most commonly used inputting methods are keyboard and pointing device (for example, a mouse).

A computer system is composed of hardware and software. Hardware consists of printed circuits, central processing unit (CPU), memory chips, storage devices, connection ports, keyboards, printers, scanners, and monitors.

Software consists of digital bits downloaded onto the storage devices. All pieces of hardware connected to the main unit, which houses the CPU, are called peripherals. Software is sold in packages that are designed to perform different tasks commanded by the user(s) of the computer. One piece of software is called the operating system. This piece of software is crucial to the operation of the computer, as it acts as an interpreter between the machine (actually the machine language) and the wide variety of software that are designed to perform specific tasks.

2.1.2 A Brief History

The first of the modern computers can be considered to be divided into two different classes, depending on how they transferred information around inside the machine. The idea for the stored program computer originated, as stated earlier, from the work done on the Electronic Numerical Integrator and Computer (ENIAC) project in the USA. This was followed by other versions, which were either an upgrade of the basic system or an equivalent system.

Although magnetic recording had been invented in 1898, it was not until 1947 that magnetic memory for computers received attention. The typical magnetic tape in the 1950s was 2,400 feet long and 1/2 inch wide. Later, magnetic tape replaced punched cards for shelf storage of data. Punched cards remained the primary input medium. Tape drives became the primary auxiliary memory for computer systems. The ENIAC-on-a-Chip was later attempted. As a part of evolution the increasing number of transistors on a single chip gave the possibility of building a CPU on a chip. Networking started in the 1970s.

Computers first began to be widely used at home (therefore were given the name personal computers or PCs) in the late 1970s with the introduction of the Apple II by Apple Computers. Earlier brands existed before the Apple II, but were not adopted on a large scale. The Apple II was the first personal computer to be supported by over 500 software packages written specifically for its operating system. The market for personal computers was further expanded with the introduction of the IBM PC in 1981 operated by Microsoft's Disk Operating System (DOS) and later on, in 1984, by the Apple Macintosh operated by its own graphic-user interface (GUI) operating system, which was incompatible with all other operating systems. By the mid 1990s, 40 per cent of the households in developed countries owned at least one personal computer.

2.1.3 Processor Philosophies

There are two basic types of processor design philosophies: reduced instruction set computers (RISC) and complex instruction set computers (CISC). In the 1970s and the early 1980s, processors predominantly followed the CISC designs. The current trend is to use the RISC philosophy.

The processor acts as the controller of all actions or services provided by the system. It can be thought of as executing the following cycle forever.

1. Fetch an instruction from the memory
2. Decode the instruction
3. Execute the instruction

This process is often referred to as the fetch-decode-execute cycle, or simply the execution cycle. Fetching an instruction from the main memory involves placing the appropriate address on the address bus and activating the memory-read signal on the control bus to indicate to the memory unit that an instruction should be read from that location. The memory unit requires time to read the instruction at the addressed location. This time is called the access time. Identifying the instruction that has been fetched from the memory is called decoding. The system clock provides a timing signal to synchronize the operations of the system. The clock period is defined as the length of time taken by one clock cycle (Figure 2.1).

$$\text{Clock period} = \frac{1}{\text{Clock frequency}} \tag{2.1}$$

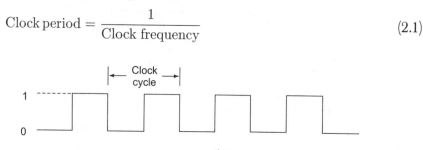

FIGURE 2.1 Clock Cycle of a Computer System

2.1.4 Processor Address

One of the characteristics that shapes the architecture of a processor is the number of addresses used in its instructions. In three-address machines, instructions carry all three addresses explicitly. Most current processors use three addresses. In two-address machines, one address doubles as a source and destination. The Pentium is an example processor that uses two addresses. In the early machines, a special set of registers was used to provide one of the input operands as well as to receive the result of the operation. Because of this, these registers are called the accumulators. In zero-address machines, the locations of both operands are assumed to be at a default location. These machines use the stack as the source of the input operands and the result goes back into the stack. Complex instruction set computer processors support a large number of addressing modes compared to reduced instruction set computer processors.

2.1.5 Processor Registers

Processors have registers to hold data, instructions, and state information. The registers are divided into general-purpose or special-purpose registers. Special-purpose registers can be further divided into those that are accessible to the user programs and those reserved for the system use. Reduced instruction set computer processors typically have a large number of registers. Some processors maintain a few special-purpose registers.

For example, the Pentium uses a couple of registers to implement the processor stack. The Pentium has ten 32-bit and six 16-bit registers. These registers are grouped into general, control, and segment registers. The general registers are further divided into data, pointer, and index registers.

> **Partnership for Advanced Computing in Europe**
>
> PRACE, the Partnership for Advanced Computing in Europe was formed in 2007 as an initiative of 14 and subsequently 18 European countries with the goal to implement the ESFRI vision of a world-leading, persistent, pan-European high-end computing infrastructure. This infrastructure is to be managed as a single European legal entity. It will comprise several world-class supercomputer centres offering a range of architectures to meet the needs of the different scientific and industrial domains and applications. PRACE is undertaking all legal, administrative, and technical work to establish the infrastructure to start its operation in 2010.
>
> *Source*: http://www.prace-project.eu/documents/Prace Hres.pdf

2.1.6 Memory

The memory unit is implemented using different types of memory chips available at different speeds from different manufacturing technologies, and of different sizes. The two basic types of memory are the read-only memory (ROM) and read/write memory. A volatile memory requires power to retain its contents. A non-volatile memory can retain its values even in the absence of power.

Read-only memory allows only read operations to be performed. The ROMs are non-volatile and are generally factory-programmed. Other types of ROM include programmable ROM(PROM) and erasable PROM(EPROM). Contents of an EPROM can be erased by exposing it to ultraviolet light for a few minutes. Read/write memory is commonly referred to as random access memory (RAM). Read/write memory can be divided into static and dynamic RAM. Static random access memory (SRAM) retains the data, once written. The other type of RAM is dynamic random access memory (DRAM). It is a complex memory device that uses a tiny capacitor to store a bit. The commonly available RAMs include SDRAM, DDR SDRAM in the commercial market.

2.2 Hardware

Honey, I shrunk the PC! It sounds like the movie *Honey, I Shrunk the Kids*. From desktops to laptops and netbooks and now pocket PCs, the world of computers is surely on a shrinking spree. Electronics major Sony recently took thinking small to new heights by launching a laptop that fits into one's pocket. In comparison to 10 years ago, the processor scene has become drastically different. Initially client/server computing centred around extremely large expensive devices called mainframes. Mainframes could provide access to multiple simultaneous users by running multiple user operating systems and were made available to business and academics. While in the period 1980–90, the proprietary processors and in particular the vector processors were the driving forces of the supercomputers of that period, today that role has been taken over by common off-the-shelf processors.

In fact, there are only two companies left that produce vector systems, while all other systems that are offered are based on RISC/EPIC CPUs or x86-like ones. The RISC processor scene has shrunk significantly in the last few years. The Alpha and PA-RISC processors have disappeared in favour of the Itanium processor product line and, interestingly, the million instructions per second (MIPS) processor disappeared some years ago and now reappears in the SiCortex systems.

The hardware consists of the processor, hard drives, video cards, sound cards, and more. Each processor has built into it a language that only it understands, plus each manufacturer creates a different language for its processor. For instance, an Intel x86 processor uses a different internal language than, say, a Motorola 68000 processor. Some of the platforms include:

- x86 (Intel [386, 486, Pentium, Pentium II, Pentium III, Celeron])
- AMD (K6-2, Athlon, or others equivalent to the Intel line)
- Alpha (Was DEC, now Compaq)
- Power PC, also known as PPC (Motorola/IBM Power PC)
- M68k (Motorola 68000 series)
- Sparc (Sun Microsystems SPARCstation)

2.2.1 Intel

Intel Corporation is a manufacturer of microprocessors, chipsets and motherboards, microprocessor peripherals, microcomputers and supercomputers, and semiconductors, including flash memory devices, best-known for its Intel Inside advertising slogan designed to push its Pentium processors for personal computers.

Intel has been in the business of introducing cutting-edge technology and introducing increased speeds in its new chips. Intel makes a wide range of computing products. These include video conferencing products, networking products, and a wide array of embedded devices. The mainstay of Intel's business is its microprocessors or chips. The most current of these is the Pentium line of microprocessors.

> **Know Your Firm—Brief Overview of Intel**
>
> Intel Corporation was the creation of a couple of engineers who had left Fairchild Semiconductor in 1968 in an entrepreneurial desire to develop large-scale integration technology for silicon-based chips. Intel pushes the boundaries of innovation so that its work can make people's lives more exciting, fulfilling, and manageable. Intel is a model of good technology branding, and positioning, and had it not already had a strong position, crisis management may not have been enough to save the day. The company really survived and prospered because of this, and has shown how a power positioning approach can solve the problems of consumer technophobia, with its now famous Intel Inside campaign. The Intel position has always been based on authenticity, quality, and performance, supported strongly by consistent global campaigns.
>
> - 1968: Intel is founded
> - 1971: An Intel engineer invents the microprocessor, the 4004 Microchip
> - 1972: The 8008 microprocessor is Intel's first chip to be actively marketed
> - 1981: IBM decides to put an Intel microprocessor into its first PC
> - 1985: The Intel386 is released with 275,000 transistors and 5 million instructions per second capacity
> - 1993: The Pentium processor debuts with 3.1 million transistors and 90 million instrusctions per second capacity
> - 1997: The Pentium II is released with 7.5 million transistors
> - 1998: Intel and Polaroid announce plans to produce a digital camera using Intel technology
>
> *Source*: www.intel.com

Intel introduced microprocessors way back in 1969. Their first 4-bit microprocessor was the 4004. This was followed by the 8080 and 8085 microprocessors. The first processor in the IA family was the 8086 processor, introduced in 1979. Intel introduced its first 32-bit processor, the 80386, in 1985. It has a 32-bit data bus and 32-bit address bus. The Intel 80486 was introduced in 1989. This is an improved version of the 80386. While maintaining the same address and data buses, it combined the coprocessor functions for performing floating-point arithmetic. The 80486 processor has added more parallel execution capability to instruction decode and execution units to achieve a scalar execution rate of one instruction per clock. The first Pentium was introduced in 1993. The Pentium is similar to the 80486 but uses a 64-bit-wide data bus. The Pentium II processor added multimedia (MMX) instructions to the Pentium architecture. The Pentium III processor introduced streaming SIMD extensions (SSE), cache prefetch instructions, and memory fences and the single-instruction multiple-data (SIMD) architecture for concurrent execution of multiple floating-point operations. Pentium 4 enhanced these features. The Itanium uses a 64-bit address bus to provide substantially large address space. The Itanium 2 is a representative of Intel's IA-64 64-bit processor family and as such the second generation. The first Itanium processor came out in 2001, but has not spread widely, primarily because the Itanium 2 would follow quickly with projected performance levels up to twice that of the first Itanium. The first Itanium 2 implementation ran at 0.8–1.0 GHz and has been followed quickly by a technology shrink (code named Madison) to the present-day Itanium 2 with clock frequencies in the range 1.3–1.66 GHz. The processor core is almost unaltered with respect to the Madison processor but it is now built with 90 nm feature size instead of 130 nm and two cores are put onto a chip working at a clock frequency of 1.66 GHz at maximum. The present processor (code named Montecito), is a dual core processor like most other processors are these days. Because of the technology shrink, the power requirements are lower than for the Madison processor, slightly over 100W, even if there are two processor cores on the chip.

Although the Intel Xeon processors are not applied in integrated parallel systems these days, they play a major role in the cluster community as the majority of compute nodes in Beowulf clusters are of this type. As of early 2006, Intel introduced an enhanced micro-architecture for the IA-32 instruction set architecture called the core architecture. The server version with the code named Woodcrest was the first implementation of this new micro-architecture. The Woodcrest processor has two processor cores like all high-end processors now do. In November 2007, Intel introduced a quad-core processor, code named Clovertown. In fact, it features two Woodcrest processors on one chip which share the 1333 MHz frontside bus to the memory.

2.2.2 Advanced Micro Devices

Advanced Micro Devices (AMD) is an innovative technology company dedicated to collaborating with customers and technology partners to ignite the next generation of computing and graphics solutions at work, home, and play. Advanced Micro Devices is the second-largest global supplier of microprocessors based on the x86 architecture after Intel Corporation, and the third-largest supplier of graphics processing units, after Intel and Nvidia. Advanced Micro Devices has a proud history of bringing innovations to the marketplace that benefit end-users and customers alike. Over the course of AMD's four decades in business, silicon and software have become the steel and plastic of the worldwide digital economy. Technology companies have become global pacesetters, making technical advances at a prodigious rate always driving the industry to deliver more and more, faster and faster.

> **Super Computer on a Chip**
>
> In early 2005, IBM, Fujitsu, and Sony announced the Cell chip. This joint project has produced a supercomputer on a chip designed with graphics and AV processing in mind. Sony will use this the core to their PS3 game machine. IBM has created a demo stand-alone workstation using the Cell chip with an outstanding computing benchmark of 16 Teraflops. The Cell chip is configured as many smaller CPUs networked together on one substrate. No doubt we will learn more about this exciting new device as it goes into production.

The Opteron dual core processor is currently the most common part of the AMD Opteron family. It is manufactured in 90 nm feature size (65 nm coming up shortly) and available since the end of 2005. It is a clone with respect to Intel's x86 Instruction Set Architecture, and it is quite popular for use in clusters.

Building on 10 years of AMD Athlon processor innovation, the new 45 nm AMD Athlon II X2 250 processor gives mainstream consumers exceptional performance, efficiency, and value. With the latest addition of AMD Phenom II X2 550 Black Edition processor, the first-ever dual-core AMD Phenom II CPU to the AMD Phenom II processor family, users can now experience the power of AMD platform technology, code name Dragon, with dual-, triple-, and quad-core configurations.

2.2.3 Servers

A server is primarily a program that runs on a machine, providing a particular and specific service to other machines connected to the machine on which it is found. A server is a machine with a specific configuration and specific set of programs that offer different types of service, which accept request from clients to do certain tasks. Server programs generally receive requests from client programs, execute database retrieval and updates, manage data integrity, and dispatch responses to client requests. Sometimes, server programs execute common or complex business logic.

A web server is a machine that hosts websites and allows Internet users (clients) to access these websites. Network servers typically are configured with additional processing, memory, and storage capacity to handle the load of servicing clients.

Know Your Firm—Brief Overview of Dell

Dell was founded by Michael Dell on a simple concept of selling computer systems directly to customers. Dell Inc., a multinational technology corporation, develops, manufactures, sells, and supports personal computers and other computer-related products. The company currently sells personal computers, servers, data storage devices, network switches, software, and computer peripherals. Dell also sells HDTVs, cameras, printers, MP3 players, and other electronics built by other manufacturers. Intel provides computer chips for many different market segments. A market segment is usually defined, for example, by age, gender, or geographical position. Intel identifies its market segments by product use, for example, notebooks, desktops, servers. Some products are for the business market, for example, desktop computers and laptops for companies. Other products are for personal use, for example, notebook computers for students.

- 1984: 19-year old Michael Dell registers Dell Computer Corporation with $1,000 in start-up capital.
- 1987: Dell establishes its first international subsidiary in the UK.
- 1988: Dell raises 30 million in its initial public offering, bringing the market capitalization of the company to 85 million.
- 1991: Converting its entire product line to the highest-performing Intel 486 microprocessors, Dell demonstrates its commitment to rapidly delivering the latest technology to its customers.
- 1992: Dell achieves more than 2 billion in sales, which represents a remarkable 127 per cent increase.
- 1993: 'Liquidity, profitability, and growth' become a company mantra, signifying its shift from a focus on growth alone to more balanced priorities.
- 1996: Dell challenges the traditional market for premium-priced servers based on proprietary technology with its introduction of its PowerEdge server line.
- 2004: Annual revenue of Dell Inc. exceeds 41 billion. The company ranks No. 6 on *Fortune* magazine's Global 'Most Admired' list.

Source: www.dell.com

2.2.4 Blade Servers and Enclosures

A server is a device running a service that shares the load among other services. A server typically refers to a hypertext transfer protocol (HTTP) server, although other or even multiple services would also be relevant. A server has an Internet protocol (IP) address and usually a transmission control protocol (TCP)/user datagram protocol (UDP) port associated with it and does not have to be publicly addressable.

Blades are still relatively new, their adoption still early. About three years ago, when blades first hit the market, compute servers sat in racks. Taking up physical space in and around those racks were switches, power units, fans, storage units, and all the other devices that make up a general major server. The blade approach, by contrast, effectively packages an entire compute server 1, 2, or 4 processors, memory, storage, network controllers, and operating system into a single box.

Blade systems are unique systems in the server market, as they are intelligent chassis that house multiple server motherboards and provide shared power, cooling, management, and networking to all of them. These systems yield many benefits—small, energy-efficient server blades, a significant reduction in cabling, much simpler configuration, and the ease of management of an all-in-one design. Blade systems pack 10 to 24 servers into as little as a 9 rack unit (U) package consuming, on average, fewer than 3 kW. They can support these configurations with as few as two fans and one high-efficiency power supply. Blade systems also provide significant configuration flexibility, offering a choice among myriad servers, input/output (I/O) options, and other internal components. The chassis can accommodate a mix of x86 (Intel or AMD CPUs) and UNIX RISC servers, storage blades, workstations, and PC blades as well as multiple I/O connections per blade. Most server blades have a variety of connectivity options on the motherboard, including peripheral component interconnect express (PCIe) slots, mezzanine and secure digital (SD) slots, and universal serial bus (USB) ports. In the case of the Sun blade system, PCIe is extended to the back of the chassis to allow for hot-swapping the individual I/O modules per blade without disconnecting the server blades.

Blade servers are stripped down computer servers with a modular design optimized to minimize the use of physical space. Blades come with management software that automates the initial setup, provisioning, and reprovisioning of the multiple blades within a blade system.

The big thing in blade computing is the engineering technologies encompassed by the blade enclosure. This is what makes it all possible. A blade enclosure, which can hold multiple blade servers, provides services such as power, cooling, networking, various interconnects, and management though different blade providers have differing principles around what to include in the blade itself. It is the job of the blade enclosure to provide infrastructure support such as communications, networking, and various interconnects for all blade units housed within it and for blade-enclosure-to-blade-enclosure interoperability. The cabling that connects all of these components together is already done within the enclosure. So at the very least, blades consolidate the space required for multiple compute servers. This tends to be through the utilization of fibre optic networking technologies as they deliver the greatest performance, least excess thermal energy production, and the greatest immunity to noise, electromagnetic interference (EMI), and environmental aberrations.

Together, blades and the blade enclosure form the blade system. Because everything is integrated in a blade system, the management software provides a single point of access for systems administration.

Certain vendors have begun to offer unique blade system designs for customers that require much greater server density than even a normal blade chassis can deliver. Silicon Graphics International (SGI), Verari Systems, and other specialty blade makers are now building cabinet-sized blade enclosures for maximum density, efficiency, and homogeneity. International Business Machine's iDataPlex is an example of one traditional vendor offering a specialty solution that is very different in design from its mainline blade system, BladeCenter. Hewlett Packard (HP) unveiled the new BladeSystem Matrix, a cloud infrastructure in a box that brings the economics, scalability, and response times of the cloud to different applications. Matrix is newly packaged and is built on the foundation of HP Systems Insight Manager (SIM), with a heaping helping of associated services such as rapid-deployment software (HP's RDP), Microsoft Active Directory, server virtualization (VMware, XenServer, or Microsoft Hyper-V), and hardware in the form of the HP BladeSystem c-Class blade chassis and HP StorageWork's EVA Fibre Channel storage framework.

2.2.5 Hard Disk Drives

The hard disk has had an immense impact on the evolution of IT systems. Two different types of hard disk drive (HDD) have emerged: one is the so-called small computer system interface (SCSI) HDD and the other is the advanced technology attachment/integrated drive electronics (ATA/IDE) drive. The drives are similar in many ways. The SCSI drive is aimed at enterprise data centres where top-notch performance was required. The ATA drive is aimed at the PC market where less performance is acceptable. Because of the different target markets, the common perception is that SCSI drives are the right choice for high-end applications and ATA drives are for home use and light business. More storage for less is the trend and it will likely continue.

2.2.6 Desktop

A desktop computer is a personal computer that is designed to fit conveniently on top of a typical office desk. The desktop computer was traditionally assumed to be a fairly specific layout. This type of computer would contain most of the computer components, such as the CPU, the hard disk, and the RAM, within a single case that would sit horizontally on a desktop. A monitor would then sit on top of the desktop computer case, and an external mouse and keyboard would be attached to it.

2.3 Input-Output

Information display devices play a major role in the transition of information. The global move toward information societies is causing a sharp increase in the demand for information display devices [Jiun Haw et al. 2008]. Display systems are used to produce and reproduce colour images. Typically, the formation of colours can be described by the following four-stage process.

1. The light source, either artificial or natural
2. Light object interaction, such as reflection, absorption, and transmission
3. Stimulation of the eyes
4. Recognition by the brain

A display is an interface containing information which stimulates human vision. Information may be in the form of pictures, animation, movies, and articles. One can say that the functions of a display are to produce or reproduce colours and images. There are a lot of electronic displays that use an electronic signal to create images on a panel and stimulate the human eye. Typically, they can be classified as emissive and non-emissive. Emissive displays emit light from each pixel which constitutes an image on the panel. In contrast, non-emissive displays modulate light, by means of absorption, reflection, refraction, and scattering, to display colours and images. For a non-emissive display, a light source is needed. Hence, these can be classified into transmissive and reflective displays. One of the most successful display technologies for home entertainment is the cathode ray tube (CRT). These displays were first invented in 1847 by German physicist Ferdinand Braun and became commercially available in 1922. Cathode ray tubes work by shining an electron beam against a phosphor surface which causes the phosphor to emit visible light. Deflecting the beam with a magnetic field enables the beam to be precisely controlled to sweep across the CRT screen to paint a complete picture line by line.

Flat panel displays (FPDs) are present in our daily lives—mobile phones, notebooks, monitors, televisions (TVs), traffic signals, and electronic signage are a few examples. Several FPD technologies, such as liquid crystal displays (LCDs), plasma display panels (PDPs), light-emitting diodes (LEDs), organic light-emitting devices (OLEDs), and field emission displays (FEDs), have been developed. They coexist because each technology has its own unique properties and applications. An LCD is a non-emissive display in which the liquid crystal molecules in each pixel work as an independent light switch. Thin-film transistors (TFTs) are widely used as electronic switches to turn the pixels of active matrix LCDs. The following three types of LCD have been developed.

1. Transmissive
2. Reflective
3. Transflective

A transmissive LCD uses a backlight for illuminating the LCD panel which results in high brightness and high contrast ratio. Transmissive microdisplays are commonly used in projection displays such as data projectors. Reflective LCDs are classified into direct-view and projection displays. Liquid crystal displays work by passing light from a cold cathode fluorescent tube through a vertical polarizing filter which is then rotated by the individual liquid crystal cells that make up the display. Liquid crystal displays have been a major competitor for plasma and while plasmas provide better colour gamut, LCDs have surpassed plasmas in their ability to display very high-resolution images. With technological improvements in LCD technology filtering down to consumer monitors, including 5 ms or lower response times, displays smaller than 24 inches with 1,920 x 1,080 native resolution, 1,000:1 or greater contrast ratios, many monitors now offer features similar to what is seen on LCD high definition televisions (HDTVs). Additionally, several monitor manufacturers, including Acer, HP, LG, and Samsung, are switching from the 16:10 aspect ratio to 16:9, so that you can watch movies or TV shows without wasting screen space for black bars or distorting the picture.

Plasma displays work by exciting xenon and neon gas contained in tiny cells that are sandwiched between two layers of glass and electrodes. The application of a charge to an individual cell causes the gas to ionize into plasma which then excites phosphors to emit light. Colour is achieved as a triad of sub-pixels. Each cell is coated with a particular phosphor that results in the emission of either red, green, or blue light. The brightness of the colour sub-pixel is achieved by varying the pulses of current flowing through the cell. The blending of the emitted light from the triad results in the overall colour of the pixel.

2.3.1 Touch Screens

While research in touch-screen technology, user interfaces, and applications has been active for over a decade, touch technologies saw a dramatic increase in popularity with the advent of Apple's iPhone. The most widespread use of touch screens has been in automated teller machines (ATMs). They are popular in public displays because they lack any form of mechanical part that could potentially suffer wear and tear over the course of millions of uses.

2.3.2 Input Devices

There is a number of ways to structure the vast field of user interface devices. There are a plethora of input devices according to categories of interface hardware. These are

- Buttons, keys, and keyboards
- Mice, joysticks
- Pen and touch input
- Biosignal sensors

Technically, a standard typewriter keyboard is nothing more than a collection of pushbuttons. Keyboards have been a (if not the) standard input device for computer. Keyboards may be reduced to only contain the most important keys and do so in an ergonomic fashion. The keycaps may contain small displays to facilitate assigning special functions to them.

Since its invention in 1967 by Douglas Engelbart, the operating principle of the computer mouse has not changed much, with the biggest addition to the original design possibly being the scroll wheel. A computer mouse senses a position set by the user's hand. Mice, joysticks, and steering wheels are also available equipped with force feedback.

Pen-based interaction with a screen dates back to the 1950s, which is even further back than the invention of the computer mouse. Advanced pen input devices such as graphics tablets and the screens of most tablet PCs operate through radio frequency position detection and even employ active electronics in the stylus. In contrast, simple resistive touch panels use two slightly separated layers of transparent, conductive, but resistive plastic on top of the display screen. The touch with a stylus brings the two layers into contact below the stylus' tip. Voltage measurements can determine the position where the contact took place. These panels can also be operated by a bare finger, which is more natural, but less precise.

2.4 Computing Infrastructure Deployments

Computing infrastructure deployments are just a common name used to describe a grouping of servers that you used within the local area network (LAN) or wide area network (WAN), usually tied to a high-speed backbone if designed properly. There are a variety of such deployments which are gaining prominence. Some of them are discussed in the following sections.

2.5 Data Centre

A data centre (Figure 2.2) is designed to suit business requirements. Data centre design is one of the most complex and important undertakings for any company. Since data centres represent a significant portion of the IT budget, every design decision has an impact on the budget. Designing a cost-effective data centre is greatly dependent on the mission of the data centre. Organizations today face data centre silos, apart from space and power constraints.

The basic guidelines for setting up a data centre includes the following.

Flexibility

Technology evolves over time and what is relevant today may not be relevant tomorrow. Emergence of destructive technology will generate newer models. However, it is a good guess that there will be some major changes. Making sure that the design is flexible and easily upgradable is critical to a successful long-term design.

Modularity

Data centres are highly complex and modular design allows creation of highly complex systems from smaller, more manageable building blocks.

FIGURE 2.2 Data Centre

A data centre can occupy one room of a building, one or more floors, or an entire building. Large data centres are some of the most significant energy consumers in an organization's IT infrastructure. Data centres are comprised of a high-speed, high-demand networking communication systems capable of handing the traffic for storage area networks (SANs), network attached storage (NAS), file/application/web server farms, and other components located in the controlled environment. There are literally hundreds of hardware and software combinations in the data centre, all with their own set of management tools, resulting in dozens of disconnected silos and creating a management nightmare.

In general, data centres can be divided into the following three interrelated components.

1. Enterprise: The central data processing facility for an enterprise is its computing network.
2. Internet: A facility that provides data and Internet services for other companies.
3. Storage area network (SAN): A network of interconnected storage devices and data servers usually located within an enterprise data centre or as an off-site facility offering leased storage space.

As data centres run an array of computing, storage, and networks infrastructure, the system creates management challenges and strains resources. The requirement comes to automate them to run smartly and efficiently.

Many cloud service data centres today may be termed mega data centres, having on the order of tens of thousands or more servers drawing tens of megawatts of power at peak. An area of rapid innovation in the industry is the design and deployment of micro data centres, having on the order of thousands of servers drawing power peaking in hundreds of kilowatt. Today, micro data centres are used primarily as nodes in content distribution networks. By infrastructure, we mean facilities dedicated to consistent power delivery and to evacuating heat. In some sense, infrastructure is the overhead of cloud services data centres. Driving the price of the infrastructure to these high levels is the requirement for delivering consistent power. Decreasing the power draw of each server clearly has the largest impact on the power cost of a data centre, and it would additionally benefit infrastructure cost by decreasing the need for infrastructure equipment. Those improvements are most likely to come from hardware innovation, including use of high-efficiency power supplies and voltage regulation modules.

The capital cost of networking gear for data centres is a significant fraction of the cost of networking, and is concentrated primarily in switches, routers, and load balancers. The remaining networking costs are concentrated in wide area networking.

- Peering, where traffic is handed off to the Internet service provider (ISP) that deliver packets to end-users
- The inter-data centre links carrying traffic between geographically distributed data centres
- Regional facilities (backhaul, metro-area connectivity, co-location space) needed to reach WAN interconnection sites

Infrastructure costs are high, largely because each data centre is designed so that it will never fail. These costs can be dramatically reduced by stripping out layers of redundancy inside each data centre, such as the generators and uninterruptible power supply (UPS), and instead using other data centres to mask a data centre failure. The challenge is creating the systems software and conducting the networking research needed to support this type of redundancy between data centres.

2.6 Virtualization

The origins of virtualization go back to the 1960s, when IBM developed the technology so that its customers could make better use of its mainframes. Yet, it lingered in obscurity until VMware applied it to the commodity computers in today's data centres. Virtualization has made it possible to decouple the functionality of a system as it is captured by the software stack (operating system, middleware, application, configuration, and data) from the physical computational resources on which it executes. Virtualization is a concept that allows a computer's resources to be divided or shared by multiple environments simultaneously. These environments can interoperate or be totally unaware of each other. A single environment may or may not be aware that it is running in a virtual environment. Environments are most commonly known as virtual machines (VMs).

2.6.1 Virtual Machines

A VM provides a virtualized abstraction of a physical machine. Software running on a host supporting VM deployment is typically called a virtual machine monitor. Virtual machines will almost always house an installation of an operating system commonly referred to as guest operating system. The guest operating system running in the virtual machine sees a consistent, normalized set of hardware regardless of what the actual physical hardware components are in the host server. Thus, virtualization allows multiple operating system instances to run concurrently on a single computer. It dramatically improves the efficiency and availability of resources and applications in the organization. Internal resources are underutilized under the old one-server, one-application model and IT administrators spend too much time managing servers rather than innovating.

2.6.2 Templates

A major component of virtualized computing systems is template image management. It is important to have a well-defined methodology which documents how templates are created, managed, and deployed. What makes templates so powerful is that they simplify the deployment of new virtual machines by reducing most of the work of image creation to a simple file copy operation. The idea is to make a copy of an existing template, either manually or by an automated process, which provides a bootable, usable virtual machine that only needs minimal configuration changes and possibly have specialized applications installed.

Templates, which are used to refer to a disk image that is intended to be cloned to enable fast VM deployment, may actually consist of one or more disk images for a single virtual machine. Templates can also consist of one or more disk images for multiple virtual machines. Template disk images are generally not compatible across different virtualization platforms by default. Because of this fact, it must be known ahead of time for which virtualization platform the new template will be created. For each virtual machine defined by the template, define the number of virtual disks that will be created and attached to each virtual machine. In most simple cases, this will be a single virtual machine with a single virtual disk. For each virtual disk defined in the template, define the maximum storage size of each disk and how the virtual disk is to be partitioned.

A very important part of planning a new template is to determine the number of virtual network adapters that will be used. Most templates will have at least one virtual network adapter, although it is not absolutely required and some use cases may not need virtual network adapters. Some virtualization platforms support more than one virtual network adapter type and this usually affects which network adapter driver is used within the template's guest operating system. It is very important to document the network adapter typesetting of each adapter.

2.6.3 Virtual Image

The virtualization layer is designed to normalize the hardware, which means that the virtual images installed on top of the virtualization layer are shielded from the details of the underlying physical server hardware. Because of the normalization that can be achieved in a virtualization environment, virtual server images do not need to take physical hardware differences into account. A VM representation (VM image) is composed of a full image of a VM RAM, disk images, and configuration files. Thus, a VM can be paused, its state serialized, and later resumed at a different time and in a different location, thus decoupling image preparation from its deployment and enabling migration.

Consequently, virtualization helps simplify image management of various installations.

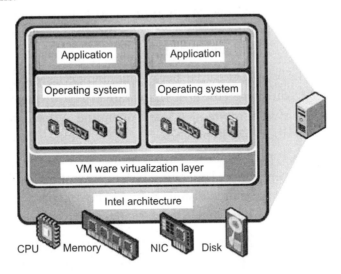

FIGURE 2.3 Virtualization

2.6.4 Hypervisors

Virtualization (Figure 2.3) is a way of making a physical computer function as if it were two or more computers. Virtualization does it by software partitioning that allows the resource of a single physical computer to be divided into multiple partitions handled by a special software program called hypervisor, that controls access to a computer's processor and memory. It allows servers to be split into several virtual machines, each of which can run its own operating system and application. The hypervisor is the virtualization layer that allows multiple operating systems to run on top of the same set of physical hardware. The hypervisor is a software that sits on the host hardware between the hardware and the virtual machines. The term virtual machine or hardware virtual machine refers to the hardware abstraction layer provided by the virtualization software, which allows each operating system to think that it is directly accessing the hardware. The virtualization software that provides this illusion is typically referred to as 'hypervisor'. It is also called virtual machine monitor (VMM). It is a computer hardware platform virtualization software that allows multiple operating systems to run on a host computer concurrently. A hypervisor provides the most efficient use of resources in a virtualization platform because it does not rely upon a host operating system and can minimize the resource overhead. This is called abstraction. Abstraction, as it relates to virtualization, is the representation of a set of common hardware devices that are entirely software driven. There are two types of hypervisors, namely Type I and Type II.

1. Type I (or native, bare-metal) hypervisors are software systems that run directly on the host's hardware as a hardware control and guest operating system monitor. A guest operating system, thus, runs on another level above the hypervisor.
2. Type II (or hosted) hypervisors are software applications running within a conventional operating system environment. Considering the hypervisor layer being a distinct software layer, guest operating systems, thus, run at the third level above the hardware.

A hypervisor is responsible for supporting this abstraction by intercepting and emulating certain instructions issued by the guest machines. It provides an interface allowing a client to start, pause, serialize, and shut down multiple guests. Each guest believes that it has the resources of an entire machine under its control, but beneath its feet, the hypervisor transparently ensures that physical resources are securely partitioned between different guests. The hypervisor manages all hardware structures, such as the memory management unit (MMU), I/O devices, direct memory access (DMA) controllers and the like, and presents a virtualized abstraction of those resources to each guest. Two main approaches have emerged: emulation and binary patching, used by VMware, and paravirtualization, pioneered on the x86 architecture by Xen and modelled after techniques used on IBM mainframes.

2.6.5 Xen Hypervisor

The Xen hypervisor is widely regarded as an open source industry standard for virtualization, because many of the major enterprise systems and software vendors, such as AMD, Intel, IBM, HP, Sun, Fujitsu, NEC, Dell, Unisys, Red Hat, Novell, and VA Linux, amongst others, have contributed to the open source Xen code base, and because the open source Xen code base is packaged and is delivered to market in commercial products by several vendors, including Novell's SLES 10 and in the forthcoming Red Hat RHEL 5 and Sun Solaris 10 updates. Xen also shares key architectural features with Microsoft's Windows Hypervisor, as a plug-in to the forthcoming Windows

Server Longhorn. XenSource itself packages the Xen hypervisor, with proprietary management tools and enhanced features to virtualize Windows virtual machines, in its XenServer product family. Paravirtualization, together with the complementary innovations of processor extensions for virtualization, eliminates the need for binary patching, and offer opportunity for near-native performance of virtualized guests that scales with Moore's law. It is for this reason maybe that Microsoft has seized on the Xen architecture as its role model for the forthcoming Windows Hypervisor.

On deployment, a VM needs to be made aware of its deployment context—its IP address may need to be assigned, or it may have to be pointed at site services. This is particularly important when VMs make up more complex constructs such as a virtual cluster—the cluster nodes may need to be configured to network with each other, share storage and other resources, or recognize a head node. Contemporary virtualization software often provides mechanisms to checkpoint-restart VMs. The checkpointing protocol coordinates activities among all parties involved in a checkpointing operation, while the disk image manipulation protocol handles disk accesses and disk structure manipulation during checkpointing. Unlike data checkpointing, the VM checkpointing typically involves saving the states of hardware devices, memory, and disk image of a VM to physical disks. This operation is known to be a time-consuming process due to potentially large memory and/or disk image size of virtual machines. Consequently, the turnaround time of applications running in a VM would increase significantly due to the delay in VM checkpointing.

2.6.6 Types of Virtualization

The two kinds of server or machine virtualization are discussed below.

Full Virtualization Full virtualization emulates the entire hardware platform of a guest computer. That can be effective for running an otherwise incompatible operating system like Windows on a Linux server. Here you have the advantage of same hardware as the Xen host, but the disadvantage is that it works slower.

Paravirtualization Paravirtualization uses a customized kernel, compatible with the host's kernel and hypervisor, that speaks compatibly and much more directly to your host's hardware. It is much lighter in weight, allows memory to be reallocated among guest domains so a server can run far more guest domains, and provides a really noticeable performance to any guest operation that has to talk to the disk.

2.6.7 Hyper-V

With the release of Windows Server 2008, Microsoft has included a built-in virtualization solution, Hyper-V. Hyper-V is a role of Windows Server 2008 that lets administrators create multiple virtual machines. Hyper-V is Microsoft's efficient hypervisor that enables operating-system virtualization in a server environment. Hyper-V is a core technology pillar of Microsoft's virtualization strategy. It is an installable feature of Windows Server 2008 and is available as a no-cost download as Hyper-V Server. Hyper-V takes the concept of virtualization to the mainstream IT environment by including it in the operating system.

Hyper-V provides two important features that enable business continuity: live backup and quick migration. Live backup uses Microsoft Volume Shadow Services functionality to make a backup of the entire system without incurring any downtime, as well as provide a backup of the VM at a known good point in time. When a backup request comes from the host, Hyper-V is notified, and all the VMs running on that host are placed into a state where they can be backed up without affecting current activity; they can then be restored later.

Quick migration is the ability to move a VM from one host to another in a cluster using Microsoft Failover Cluster functionality. During a quick migration, the user saves the state of the VM, moves storage connectivity from the source host to the target host, and then restores the state of the VM.

2.6.8 Virtualization Standards

Virtualization management initiative (VMAN) includes a set of specifications that address the management life cycle of a virtual environment. The open virtualization format (OVF) specification describes an open, secure, portable, efficient, and extensible format for the packaging and distribution of software to be run in virtual machines. All metadata about the package and its contents is stored in the OVF descriptor. This is an extensible markup language (XML) document for encoding information, such as product details, virtual hardware requirements, and licensing. The OVF does not require any specific disk format to be used, but to comply with this specification the disk format shall be given by a uniform resource identifier (URI) which identifies an unencumbered specification on how to interpret the disk format.

The virtual machine disk format (VMDF) specification describes and documents the virtual machine environment and how it is stored. The virtual machine disk format specification is critical to how virtual environments are provisioned, manipulated, patched, updated, scanned, and backed up.

2.6.9 Server Virtualization

Server virtualization technology is a way of making a physical computer function as if it were two or more computers, where each non-physical or virtualized computer is provided with the same basic architecture as that of a generic physical computer. This technology enables the flexible construction of virtual servers with utmost no hardware limitations and consequently reduces the total cost of ownership and makes it easier to use virtual servers in the changing business environment [Yoshihiko 2008]. For example, Essel Propack, part of the Rs 12,000 crore Essel Group is a specialty packaging company manufacturing laminated and seamless or extruded plastic tubes, and medical devices. Essel Propack functions in 13 countries, selling 4.5 billion tubes. Managing physical servers across 21 locations in 13 countries world wide for mailing, communication, and collaboration applications was proving to be quite a task for Essel Propack. Essel consolidated all servers at a data centre in India to reduce on operational costs and improve manageability. Essel Propack was able to remove 18 of 27 physical servers and is now using five blade servers to run a virtualized infrastructure, comprising about 21 virtual machines [Chidambaram 2006].

2.6.10 Single Point Failures

Virtualization and cloud computing can be used quite successfully to improve the resilience of an IT environment because they provide the means to recover quickly from component or system malfunctions using failover, and to back up essential applications and data.

Virtual machines can be migrated from one physical server to another in a live migration; virtual machine images can be restarted in a different location to provide for disaster recovery. Downtime exposure grows with the number of virtual machines on any single physical server. Easy provisioning of virtual machines also keeps virtualized and cloud environments in perpetual motion.

That very flexibility can introduce single points of failure when too many of the virtual machines that support a particular service or application are concentrated on the same physical server. Close attention must be paid to how many virtual machines reside on a physical server, and which virtual machines are placed on which physical server(s). Even though failover is provided for, the cause of the failure may be propagated to the secondary virtual machine. That allows the same failure to repeat along with the same threats of downtime and data corruption.

2.7 Server Farm

A server farm is a group of computers acting as servers and housed together in a single location. Server farms are increasingly being used instead of or in addition to mainframe computers by large enterprises. Server farms are typically co-located with the network switches and/or routers which enable communication between the different parts of the cluster and the users of the cluster.

2.8 Cluster Computing

Cluster computing originated within a few years of the inauguration of the modern electronic stored-program digital computer. SAGE was a cluster system built for North American Aerospace Defense Command (NORAD) under Air Force contract by IBM in the 1950s based on the Massachusetts Institute of Technology (MIT) Whirlwind computer architecture. Using vacuum tube and core memory technologies, SAGE consisted of a number of separate stand-alone systems cooperating to manage early warning detection of hostile airborne intrusion of the North American continent. Early commercial applications of clusters employed paired loosely coupled computers with one computer performing user jobs, while the other managed various input/output devices.

Clustering is a powerful concept and technique for deriving extended capabilities from existing classes of components. In nature, clustering is a fundamental mechanism for creating complexity and diversity through the aggregation and synthesis of simple basic elements. In the most general terms, a cluster is any ensemble of independently operational elements integrated by some medium for coordinated and cooperative behaviour. Computer clusters are ensembles of independently operational computers integrated by means of an interconnection network and supporting user-accessible software for organizing and controlling concurrent computing tasks that may cooperate on a common application program or workload. There are many kinds of computer clusters, ranging from among the world's largest computer's to collections of throwaway PCs.

The adoption of clusters, collections of workstations/PCs connected by a local network, has virtually exploded since the introduction of the first Beowulf cluster in 1994. The attraction lies in the (potentially) low cost of both hardware and software and the control that builders/users have over their system. The interest for clusters can be seen, for instance, from the Institute of Electrical and Electronics Engineers (IEEE) task force on cluster computing (TFCC) which reviews on a regular basis the current status of cluster computing.

Commodity clusters are local ensembles of computing nodes that are commercially available systems employed for mainstream data-processing markets. The interconnection network used to integrate the compute nodes of a commodity cluster is dedicated to the cluster system and is also commercially available from its manufacturer. Commodity clusters employ software, which is also available to the general community. The number of vendors that sell cluster configurations has become so large that it is not possible to include all their products in this section.

Beowulf-class systems are commodity clusters that exploit the attributes derived from mass-market manufacturing and distribution of consumer-grade digital electronic components. A number of different microprocessor families have been used successfully in Beowulfs such as Intel X86 family (80386 and above), AMD binary compatible counterparts, and the Compaq Alpha 64-bit architecture. The nodes of Beowulfs are either uniprocessor or symmetric multiprocessors (SMPs) of a few processors. At the other end of the cluster spectrum are the constellations which are clusters of large SMP nodes scaled such that the number of processors per node is greater than the number of such nodes making up the entire system. The first Beowulf-class PC cluster was developed at the NASA Goddard Space Flight Centre in 1994 using early releases of the Linux operating system and PVM running on 16 Intel 100 MHz 80486-based personal computers connected by dual 10 Mbps Ethernet LANs.

2.8.1 High Availability

Clustering is a means of providing high availability. Clustering is a group of machines acting as a single entity to provide resources and services to the network. In time of failure, a failover will occur to a system in that group that will maintain availability of those resources to the network. Clustering can allow for failover to other systems and it can also allow for load balancing between systems. Load balancing is using a device, which can be a server or an appliance, to balance the load of traffic across multiple servers waiting to receive that traffic. The device sends incoming traffic based on an algorithm to the most under-used machine or spreads the traffic out evenly among all machines that are on at the time. If Node A in a cluster senses a problem with Node B (Node B is down), then Node A comes online. This is done with heartbeat traffic, which is a way for Node A to know that Node B is no longer available and it must come online to take over the traffic. With load balancing, a single device sends traffic to any available node in the load balanced group of nodes.

Load balancing uses heartbeat traffic as well but, in this case, when a node comes offline, the load is recalculated among the remaining nodes in the group. Failover is the act of another server in the cluster group taking. Failback is the capability of the failed server to come back online and take the load back. Failing back is the process of going back to the original primary node that originally failed. Failback can be immediate or one can set a policy to allow timing to be put in place to have the failback occur in off-hours. Active/passive is defined as a cluster group where one server is handling the entire load and, in case of failure and disaster, a passive node is standing by waiting for failover. Active/active clustering is when you want all servers in the cluster group to service clients and still be able to take up the load of a failed server in case of disaster.

2.9 Grid Computing

The term 'grid' is analogous to the electrical power grid, which is the basic infrastructure that enables electricity to be distributed and made available as a utility to anyone who needs it. The grid is an emerging infrastructure that will fundamentally change the way we think about and use computing. The word grid is used by analogy with the electric power grid, which provides pervasive access to electricity and has had a dramatic impact on human capabilities and society. A computational grid is a hardware and software infrastructure that provides dependable, consistent, pervasive, and inexpensive access to high-end computational capabilities. The grid couples disparate and distributed heterogeneous software and hardware resources to provide a uniform computing environment for scientists and engineers to solve data and computation-intensive problems. A grid is considered a decentralized system, a spanning system containing multiple administrative domains, that provides a non-trivial quality of service where both the set of users and the total set of resources can (and do) vary dynamically and continuously. It handles large numbers of hardware and software systems to perform functions and computations on large volumes of data [Foster et al. 1999].

Grids can offer various services such as the following.

Compute services: CPU cycles by pooling computational power

Data services: Collaborative sharing of data generated by people, processes, and devices such as sensors and scientific instruments

Application services: Access to remote software services/libraries and license management

Interaction services: E-learning, virtual tables, group communication, and gaming

Knowledge services: Data mining and knowledge acquisition, processing, and management

Many of these services are provided and managed by self-interested parties and no single organization has control over the entire grid.

A data grid is typically responsible for housing the access provisions to data across multiple organizations. The open-source Globus project (http://www.globus.org) aims to create a software infrastructure that facilitates the development of grids and grid computing. The Globus Alliance is a community of organizations and individuals developing fundamental technologies behind the 'grid', which lets people share computing power, databases, instruments, and other online tools securely across corporate, institutional, and geographic boundaries without sacrificing local autonomy. The Globus Toolkit is an open-source software toolkit used for building grid systems and applications. Many ongoing research projects have already adopted the Globus Toolkit, a collection of various software services and program libraries, that helps programmers easily create grid applications. Grid infrastructure will provide us with the ability to dynamically link together resources as an ensemble to support the execution of large-scale, resource-intensive, and distributed applications.

The successful implementation of the grid computing infrastructure will certainly have far-reaching implications for the business, scientific, and individual computing users. The grid promises to fundamentally change the way we think about and use computing. This infrastructure will connect multiple regional and national computational grids, creating a universal source of pervasive and dependable computing power that supports dramatically new classes of applications.

The long-term vision for grid computing is to make computing available as a true utility, with consumers not having to worry about generation, distribution, or management of computing resources.

2.10 Cloud Computing

Cloud computing hype centres largely around the outsourcing of IT needs to cloud services available over the Internet. According to Gartner, the future of corporate IT is in private clouds, flexible computing networks modelled after public providers, such as Google and Amazon, yet built and managed internally for each business' users. While this trend is expected to accelerate, Gartner predicts it will also become standard for large companies to build their own highly automated private cloud networks in which all resources can be managed from a single point and assigned to applications or services as needed.

Cloud computing is the most recent successor to grid computing, utility computing, virtualization, and clustering. The traditional method of using computers and, indeed, any other technology is currently in the midst of a drastic transformation. The business web has evolved from a static destination where businesses merely publish and search for information, into a set of dynamic virtual workplaces where employees create and share information in a flexible, continuously evolving way. This flexibility provided by the Internet has accelerated the birth of cloud computing. Cloud computing overlaps those concepts but has its own meaning—the ability to connect to software and data on the Internet (the cloud) instead of a local hard drive or a local network. Cloud computing describes a system where users can connect to a vast network of computing resources, data, and servers that reside somewhere 'out there', usually on the Internet. Cloud computing can give on-demand access to supercomputer-level power, even from a thin client or mobile device such as a smart phone or laptop. There is not yet a standard definition for cloud computing, but a good working definition is to say that clouds provide on-demand resources or services over the Internet.

The foundation of cloud computing is virtualization. Cloud computing is redefining IT operations by eliminating routine infrastructure deployment, configuration, and maintenance. Forrester defines cloud computing as a pool of abstracted, highly scalable, and managed compute infrastructure capable of hosting end-customer applications and billed by consumption.

Cloud computing refers to services that sit in the cloud on a third-party server and individuals or businesses can tap and use as per their needs [Naone 2008]. The architecture behind cloud computing is a massive network of cloud servers interconnected as if in a grid running in parallel, sometimes using the technique of virtualization to maximize compute power per server. A cloud computing platform dynamically provisions, configures, reconfigures, and deprovisions servers as needed [Buyya et al. 2008]. Cloud computing packages computing resources such as processing power, storage, connectivity, etc. as a service and delivering the same to the consumer in a scale-free, cost efficient and timely manner over the Web. The key feature of cloud computing is that both the software and the information held in it live on centrally located servers rather than on an end-user's computer. With cloud computing, excess computing capacity of an organization can be put to use and be profitably sold to consumers. This transformation of computing and IT infrastructure into a utility, which could be available to all, is the basis of cloud computing.

The two key advantages of this model are ease of use and cost-effectiveness. Though there remain questions on aspects such as security and vendor lock-in, the benefits this model offers are many. Cloud computing enables users to access applications from the Internet and use it when they need, for as long as they need it. Cloud computing offers almost unlimited computing power and collaboration at a massive scale for enterprises of all sizes. Cloud computing is particularly valuable to small and medium businesses, where effective and affordable IT tools are critical to helping them become more productive without spending lots of money on in-house resources and technical equipment.

Microsoft Steams into Services Era with Azure

Claiming to set the stage for the next 50 years of computing, Microsoft has unveiled a cloud operating system and a complementary slate of developer resources that will become the core of its services platform and provide an online delivery option for all its current software. The Azure Service Platform, which includes the cloud operating system called Windows Azure, defines the scalable back-end that will support the services portion of Microsoft's software-plus-services strategy, which it has been laying out in bits and pieces over the past three years. The platform is the coming to life of the well-known Internet Services Disruption memo that Ray Ozzie penned three years ago before becoming Microsoft's chief software architect. The memo laid out how Microsoft needed to embrace software plus services to remain relevant [Chappell 2009]. The Azure platform, which has at its core a highly tuned operating system two years in the making, is no less than Microsoft's largest bet yet to achieve that relevancy as it moves into a market already active with cloud environments from Amazon, Google, IBM, and others.

Source: www.computerworld.com.au, accessed in November 2008

Cloud computing offers the integrated deployment of software, hosts, storage, and networking to shared virtualized resource pools. An essential component of any cloud computing environment is servers equipped with processors that deliver a balance of performance, I/O capabilities, low energy consumption, and hardware-assisted virtualization. The growth of virtualization has been critical to the success of cloud computing, allowing one computer to act as if it were another—or many others. Server virtualization lets clouds support more applications than traditional computing grids, hosting various kinds of middleware on virtual machines throughout the cloud.

Information technology organizations can choose to deploy applications on public, private, or hybrid clouds, each of which has its trade-offs. The terms public, private, and hybrid do not dictate location. While public clouds are typically 'out there' on the Internet and private clouds are typically located on premises, a private cloud might be hosted at a colocation facility as well. Public clouds are run by third parties, and applications from different customers are likely to be mixed together on the cloud's servers, storage systems, and networks. Public clouds are most often hosted away from customer premises, and they provide a way to reduce customer risk and cost by providing a flexible, even temporary extension to enterprise infrastructure. Portions of a public cloud can be carved out for the exclusive use of a single client, creating a virtual private data centre. Private clouds are built for the exclusive use of one client, providing the utmost control over data, security, and quality of service. Private clouds can be built and managed by a company's own IT organization or by a cloud provider. Hybrid clouds combine both public and private cloud models. Hybrid clouds introduce the complexity of determining how to distribute applications across both a public and private cloud [Rochwerger et al. 2009].

2.10.1 Cloud Platforms

Cloud computing is reminiscent of the application service provider and database-as-a-service (DaaS) paradigms. One of the most important part of that shift is the advent of cloud platforms. As its name suggests, this kind of platform lets developers write applications that run in the cloud, or use services provided from the cloud, or both. In practice, cloud computing platforms, like those offered by Amazon Web Services, AT&T's Synaptic Hosting, AppNexus, GoGrid, Rackspace Cloud Hosting, and to an extent, the HP/Yahoo/ Intel Cloud Computing Testbed, and the IBM/Google cloud initiative, work differently.

An application platform comprises the following three parts.

A foundation Nearly every application uses some platform software on the machine it runs on. This typically includes various support functions, such as standard libraries and storage, and a base operating system.

A group of infrastructure services In a modern distributed environment, applications frequently use basic services provided on other computers. It is common to provide remote storage, for example, integration services, an identity service, and more.

A set of application services As more and more applications become service-oriented, the functions they offer become accessible to new applications. Even though these applications exist primarily to provide services to end-users, this also makes them a part of the application platform.

2.10.2 Prominent Cloud Platforms

Some prominent cloud platforms are discussed below.

Eucalyptus

Eucalyptus is an open source software framework developed by the University of California, Santa Barbara, for cloud computing that implements what is commonly referred to as infrastructure as a service (IaaS); systems that give users the ability to run and control entire virtual machine instances deployed across a variety physical resources. Eucalyptus is implemented using commonly available Linux tools and basic web service technologies making it easy to install and maintain.

Nimbus

The University of Chicago Science Cloud, code named Nimbus, provides compute capability in the form of Xen virtual machines that are deployed on physical nodes of the University of Chicago TeraPort cluster using the Nimbus software. The Nimbus cloud is available to all members of scientific community wanting to run in the cloud.

Cumulus Project

The Cumulus project is an ongoing cloud computing project at the recently established Steinbuch Centre for Computing (SCC) at the Karlsruhe Institute of Technology (KIT). It intends to provide virtual machines, virtual applications, and virtual computing platforms for scientific computing applications. The Cumulus project currently is running on high-performance HP and IBM blade servers with Linux and the Xen hypervisor. Cumulus aims to build a test bed and infrastructure in the context mentioned above and mainly integrates already existing technology.

Amazon Elastic Compute Cloud (EC2)

It provides a virtual computing environment that enables a user to run Linux-based applications. The user can either create a new Amazon Machine Image (AMI) containing the applications, libraries, data, and associated configuration settings, or select from a library of globally available AMIs. The user then needs to upload the created or selected AMIs to Amazon Simple Storage Service (S3), before he/she can start, stop, and monitor instances of the uploaded AMIs. Amazon EC2 charges the user for the time when the instance is alive, while Amazon S3 charges for any data transfer. Amazon's Simple Queue Service (SQS) provides a straightforward way for applications to exchange messages via queues in the cloud. These simplifications let Amazon make SQS more scalable, but they also mean that developers must use SQS differently from an on-premises message queuing technology.

Google App Engine

It allows a user to run web applications written using the Python programming language. Other than supporting the Python standard library, Google App Engine also supports application programming interfaces (APIs) for the datastore, Google Accounts, URL fetch, image manipulation, and email services. Google App Engine also provides a Web-based administration console for the user to easily manage his/her running web applications.

Windows Azure

It is essentially a dynamic virtualization layer over Windows Server 2008—to be offered as a service. In choosing this approach to cloud platform architecture, Microsoft retains a good measure of backward compatibility, the same languages and development tools are used and most .NET framework medium trust applications will migrate to Azure with minimal or no code change.

Unlike in a typical computing environment, virtualization, coupled with grid computing, enables the disaggregation of operating systems, middleware, data stores and application software from the limitations of physical machines and the LAN. This new world is colliding with traditional vendor licensing practices, producing software compliance nightmares for both licensees and licensors, and often resulting in irrational license fees.

SUMMARY

This chapter highlighted the various elements of computing infrastructure with emphasis on hardware and implementations. A computer system can be thought of as consisting of three main components: a processor, a memory unit, and I/O devices. We started our discussion with a brief history of the Intel architecture. This architecture encompasses the X86 family of processors. All these processors, including the Pentium, belong to the CISC category. Even though the Pentium is a 32-bit processor, it maintains backward compatibility to the earlier 16-bit processors. Protected-mode architecture is the native mode for the Pentium processor. The real mode is provided to mimic the 16-bit 8086 memory architecture. In both modes, the Pentium supports segmented memory architecture. In the protected mode, it also supports paging to facilitate implementation of virtual memory. It is important for an assembly language programmer to understand the segmented memory organization supported by the Pentium.

Displays will no doubt become more ubiquitous, possessing higher resolution, and capable of both 2D and 3D. They will become more portable and configurable, capable of touch interaction, and will contain imbedded sensors, such as cameras, so that they are both an input and output device. Computers and displays are embedding themselves into every device and surface, and will continue to do so as technology becomes cheaper and more powerful so as to enable wholly new classes of applications. Within a decade, one can envision that flat screen displays will not only become ubiquitous, but will be embedded into every surface interacting with human race.

Creation and application of the most powerful computer systems (supercomputers) is an indicator of the scientific and technological potential and is among the top priorities of the developed countries. Cloud computing, as the term is popularly used, refers to the delivery of a range of IT capabilities (such as infrastructure applications) as an externally sourced service.

Key Terms

ActiveX control It is a Microsoft software module that enables another program to add functionality by calling ready-made components that blend in and appear as normal parts of the program.

Central processing unit (CPU) It is the main processing chip of a computer. The part of the computer or computer system which performs core processing functions.

Cluster It is a collection of application server instances with identical configuration and application deployment. Clusters enforce homogeneity between member instances so that a cluster of application server instances can appear and function as a single instance. With appropriate front-end load balancing, any instance in an application server cluster can serve client requests.

Thick client Thick clients, also called heavy clients, are full-featured computers that are connected to a network. Unlike thin clients, which lack hard drives and other features, thick clients are functional whether they are connected to a network or not.

REVIEW QUESTIONS

2.1 What are the prominent hardware platforms applicable to computing infrastructure?

2.2 Compare the current competing processor chips available in the market.

2.3 Differentiate clusters, grid, and cloud computing.

Projects

2.1 Visit the nearest IT firm and understand the IT computing infrastructure as implemented by it. Prepare a suitable report debating for and against the implemented IT computing infrastructure.

2.2 Prepare a short report on cloud computing infrastructure.

2.3 Prepare an advisory report to ABC unlimited to opt for blade servers or avoid blade servers with a suitable projection on ROI.

2.4 The office of a small independent realty company has four real estate agents, an executive secretary, and two assistant secretaries. They agreed to keep all common documents in a central location to which everyone has access. They also want to have a centralized computing infrastructure to cater to the graphic intensive CAD applications. The company has employed a sizable CAD operators for design purposes. They need a place to store their individual listings so that only the agent who has the listing should have access to them. The executive secretary has considerable network administration knowledge, but everyone else does not. It is expected that in the near future, the office will employ 500 more agents and three assistant secretaries. The company plans to expand the IT computing infrastructure with scaling as an option.

Advice the company on server and desktop infrastructure.

REFERENCES

Ball, S.R. 2002, *Embedded Microprocessor Systems: Real World Design*, Third Edition, Newnes.

Buyya, R., C. S. Yeo, and S. Venugopal 2008, 'Market-oriented cloud computing: vision, hype, and reality for delivering IT services as computing utilities,' proceedings of the 10th IEEE International Conference on High Performance Computing and Communications.

Carlow, G. D. 1984, 'Architecture of the Space Shuttle Primary Avionics Software System,' CACM, v 27, no. 9.

Chappell, David 2009, 'Introducing the Azure services Platform: An early look at Windows Azure, .NET services, SQL services, and live services,' Microsoft White Paper, March 2009.

Chidambaram, Varsha 2006, 'Propack virtualizes for speed', http://www.cio.in/case-study/propack-virtualizes-speed.

Dandamudi, Sivarama P. 2005, *Guide to RISC Processors for Programmers and Engineers*, Springer.

El-Rewini, Hesham and Mostafa Abd-El-Barr 2005, *Advanced Computer Architecture And Parallel Processing*, Wiley Inter Science.

Fawcett, Neil 1996, 'Happy Birthday, Micro!' *Computer Weekly*, 17 October 1996.

Foster, I. and C. Kesselman (eds) 1999, *The Grid: Blueprint for a New Computing Infrastructure*, Morgan Kaufmann.

Grandinetti, Lucio 2005, *Grid Computing: The New Frontier of High Performance Computing*, Elsevier.

Iniewski, Krzysztof, Carl Mccrosky, and Daniel Minoli 2008, *Network Infrastructure and Architecture*, Wiley Inter Science.

Kelbley, John, Mike Sterling, and Allen Stewart, 2009, *Windows Server 2008 Hyper-V: Insider's Guide to Microsoft's Hypervisor*, John Wiley & Sons.

Kirpatrick, David, 1997, ' Intel's Amazing Profit Machine, *Fortune*, 17 February 1997.

Lee Jiun-Haw, David N. Liu, and Shin-TsonWu 2008, *Introduction to Flat Panel Displays*, Wiley.

Lytras, Miltiadis 2007, *Open Source for Knowledge and Learning Management: Strategies Beyond Tools*, Idea Group Inc.

Marshall, David, Wade A. Reynolds, and Dave McCrory 2006, *Advanced Server Virtualization*, Auerbach.

Marshall, David, Stephen S. Beaver, and Jason W. McCarty 2009, *VMware ESX Essentials in the Virtual Data Center*, CRC Press.

Naone, E. 2008, 'Computer in the cloud', Technology Review, Retrieved on 2 July 2009, from http://www.technologyreview.com/ Infotech/19397/?a=f.

Oguchi, Yoshihiko 2008, 'Server virtualization technology and its latest trends', *FUJITSU Scientific Technical Journal*.

Rochwerger, B. et al. 2009, 'Reservoir model and architecture for open federated cloud computing', *IBM Journal of Research and Development*.

Severance, Charles 2009, *Using Google App Engine*, OReilly.

Wells, April J. 2008, *Grid Application System Design*, Auerbach.

Xiao Yang 2007, *Security in Distributed Grid, Mobile, and Pervasive Computing*.

CHAPTER

3 Computing Infrastructure: Software

The man who makes no mistakes does not usually make anything.
–Edward John Phelps

Learning Objectives

After reading this chapter, you should be able to understand:

- the concept of software
- the classification of software
- the types of system software
- the meaning of operating system software
- application software
- the functions of operating software
- software tools
- software system utilities

3.1 Software Industry

Software industry is an ambiguous concept. The industry is young and is characterized by rapid technological development. The industry has been shaped by numerous expansions and technological paradigm shifts during its less than fifty year history. Thousands of companies have grown and disappeared. Rapid growth in the demand for high-quality software and increased investment in software projects show that software development is one of the key markets worldwide. A fast changing market demands software products with ever more functionality, higher reliability, and higher performance. It has been claimed by some business analysts that hardware manufacturing will be of no great commercial consequence. The profit lies in programming—lead the world in the development of systems software. But it is now clear that in such a rapidly changing world, early access to new hardware designs gives the software industry an important marketing lead. The first software products to exploit some new hardware facility have a clear leadership in the marketplace. The neglect of the hardware side of the computing industry has never delivered any long-term advantage. Understanding basic principles and appreciating their application by modern technology within a range of current products is the central aim of this text. Many excellent machines became commercial flops because of their sub-standard software. These well-rehearsed public examples can be added to and confirmed by thousands of private disasters which all underline the need to pursue hardware and software developments in concert. We now recognize that despite their technical superiority, computer systems can fail to win acceptance for many reasons, such as a poorly thought out user interface, a lack of applications software, or an inappropriate choice of operating system.

According to Lehman's laws of software evolution, software systems must be continually adapted, or they become progressively less satisfactory to use in their environment. Therefore, we often need to extend the functionalities of an operational system by adding new features or removing defects discovered during usage of the software system. Software that lacks modifiability is sometimes re-architected to increase the modifiability. Typically, re-architecting does not necessarily add user value, because it does not provide new functionalities.

Today, only few companies that operated in the 1950s and 1960s have survived and most of the contemporary industry leaders were founded less than thirty years ago. Software business models used today can be distinguished from several perspectives. For example, depending on whether the software is sold as a product or service, structure of the sales channel, and income sources.

3.1.1 Software

To make a computer functional, it requires computational components, namely hardware and software for its proper functioning. Software, implies computer instructions or data. Anything that can be stored electronically is data for computer.

A software is a program or set of instructions which is required to use the computer. A software system usually consists of a number of separate programs, configuration files which are used to set up these programs, system documentation which describes the structure of the system, and user documentation which explains how to use the system. Software engineering is an engineering discipline that is concerned with all aspects of software production from the early stages of system specification to maintaining the system after it has gone into use.

3.1.2 Types of Software

At a macroscopic level, software is classified based on the way it is used. Software that is required to control the working of hardware and aid in effective execution of a general user's applications is called system software. This software performs a variety of functions such as file editing, storage management, resource accounting, I/O management, database management, etc. System software can be further categorized into the following three types.

1. System management software (operating systems, DBMSS, operating environments)
2. System development software (language translators, application generators, CASE tools)
3. System software utilities

System software helps an application programmer in abstracting away from hardware, memory, and other internal complexities of a computer. Operating systems software is used for energizing the hardware and preparing the hardware to understand requests. As such, the operating system is a generic system without which the system hardware is unusable.

Software that is required for general and special purpose applications, such as database management, word processing, accounting, etc., are called application software.

Application software can be further classified into the following two types.

1. General purpose application software (database management packages, word processors, spreadsheets, etc.).
2. Special purpose application software (accounting, inventory, production management, etc.)

Application software are built over operating systems to solve specific problems. There are a variety of application software available in the market. Open source software has perhaps shaped most of the Web server software, which has become new market for software products between personal computers and corporate mainframes. A software combination known with acronym LAMP (Linux, Apache, MySQL, and PHP/Perl) has been the first choice for many system integrators during the recent years.

3.2 System Software

One category of software which assist in the mechanics of software development is system software. Assembler, linker/loader, and compiler belong to the realm of system software.

Computers have a small fixed repertoire of instructions, known as the machine instruction set. Every manufacturer, such as IBM, Intel, or Sun, designs and produces computers containing a central processing unit (CPU) which has its own native instruction set. This typically includes between 100 and 200 instructions.

The input to a translation program is expressed in a source language. A source language might be assembly language or any high-level language. The output of a translator is a target language. Target language is often the machine language of some computer which is then able to execute the algorithm.

3.2.1 Assemblers

Assembly language programming is referred to as low-level programming because each assembly language instruction performs a much lower-level task compared to an instruction in a high-level language. Assembly language instructions are native to the processor used in the system. Programming in the assembly language also requires knowledge about system internal details such as the processor architecture, memory organization, and so on. Typically, there is a one-to-one correspondence between the assembly language and machine language instructions. Assembly language is directly influenced by the instruction set and architecture of the processor. Before high-level language compilers were widely available in the market, programmers used assembler. A program consists of a sequence of statements. Each statement is written on one line. Each statement consists of an optional (line) number, an optional label, and an instruction. A statement is either a compiler instruction, a memory instruction, or a processor instruction. An assembler is used to describe instructions of a processor. Assemblers are programs that translate programs from assembly language to machine language. The assembly language code must be processed by a program in order to generate the machine language code. Assembler is the program that translates assembly language code into the machine language. Netwide Assembler (NASM), Microsoft Assembler (MASM), and Borland Turbo Assembler (TASM) are some of the popular assemblers for the Pentium processors.

There are two types of assemblers based on how many passes through the source are needed to produce the executable program. One-pass assemblers go through the source code once and assume that all symbols will be defined before any instruction that references them. Two-pass assemblers (and multi-pass assemblers) create a table with all unresolved symbols in the first pass, then use the second pass to resolve these addresses. Assemblers translate assembly language code (source program) into machine language code (object program). After assembling, a linker program is used to convert an object program into an executable program.

Programming in the assembly language is a tedious and error-prone process. The natural preference of a programmer is to program in some high-level language.

3.2.2 Interpreters

Interpreters offer an alternative way of running high level language (HLL) programs. Instructions of a high-level language are coded in many statements. Instead of first translating all the HLL instructions into machine codes and creating an executable program, the interpreter reads the HLL instructions one at a time and carries out their orders using its own library of routines. At the time of their execution, they are converted statement by statement into machine code by using a system software called interpreters. In this way, the executable code is not generated from the source code, but is contained within the interpreter. The advantages offered by the interpreting mode of dealing with HLL programs are rapid start-up and the apparent removal of the complexities of compiling and linking.

3.2.3 Compiler

Programs called compilers translate the HLL instructions into machine code binary, capable of being executed directly on the computer. Compilers have become critical components of computer systems, often determining how efficiently the hardware will be exploited by the programmers. To compile is to transform a program written in a high-level programming language from source code into object code. The compiler derives its name from the way it works, looking at the entire piece of source code and collecting and reorganizing the instructions. Thus, a compiler differs from an interpreter, which analyses and executes each line of source code in succession, without looking at the entire program. Programmers write programs in a form called source code. Source code must go through several steps before it becomes an executable program. The first step is to pass the source code through a compiler, which translates the high-level language instructions into object code. It is often the same as or similar to a computer's machine language. Compilers do not translate and execute the instructions at the same time. They translate the entire program (source code) into machine code (object code). The final step in producing an executable program is to pass the object code through a linker (or assemblers, binders, loaders).

A language that functions as an interpreter takes the text of the program and translates it at the time of execution into commands, the computer can understand. A compiled program, on the other hand, has already had the program text translated into executable code before it is run, usually including some extra code needed to carry out necessary functions of input, output, and calculations. As such, an interpreted program usually runs more slowly, but has the advantage of being easier to modify and re-run without the delay of first re-compiling. A compiled program will ordinarily run faster, but may use more memory than an equivalent interpreted program.

3.2.4 Linker

The program that links several programs is called the linker. The linker produces a link pile which contains the binary codes for all compound modules. The linker also produces a link map which contains the address information about the linked files. The linker combines modules and gives real values to all symbolic addresses, thereby producing machine code. Using linker, the object code is converted into executable code. Compilers are widely used in translating codes of high-level languages (for example, COBOL, FORTRAN, PASCAL, Turbo/Quick BASIC, Turbo/Microsoft C, etc.)

3.2.5 Loader

A loader is a program that places programs into the main memory and prepares them for execution. The loader's target language is machine language, its source language is nearly machine language. Loading is ultimately bound with the storage management function of operating systems and is usually performed after assembly or compilation. The period of executions of user's program is called execution time.

3.3 Operating System

The computer is useless unless it is provided with the essential software that makes it ready for use. An operating system (OS) is the most essential system software that manages the operations of a computer. It is a software program that enables the computer hardware to communicate and operate with the computer software. An OS solves several problems arising from hardware variation. In the past, the OSs were computer-dependent. An OS running on one machine could not be run on another different model machine.

When a manufacturer introduces a new computer or new model of computer, then an operating system has to be developed to operate that machine. The application programs often would not work on the new computer because the applications were designed to work with a specific operating system. Operating systems perform basic tasks such as recognizing input from the keyboard, sending output to the display screen, keeping track of files and directories on the disk, and controlling peripheral devices such as disk drives and printers.

Multiprogramming is used to share a single CPU with two or more programs simultaneously. Using this technique, an OS keeps a CPU busy. Multiprogramming allows a processor to handle either multiple batch jobs at a time or multiple interactive jobs shared among multiple users. Multiprocessing refers to the use of two or more CPUs to perform a coordinated task simultaneously. Multitasking refers to an ability of OS to execute two or more tasks concurrently.

Virtual memory gives the operating system the responsibility for managing the memory hierarchy, in a manner invisible to the applications programmer, enabling the use of objects code or data larger than the available physical memory without any explicit management activities on the programmer's part. This is done by keeping objects on disk, and bringing necessary parts of them into memory as needed. The most usual approach is that of demand paging in which the address space is available to the programmer. The virtual address space is divided into units of a fixed size (perhaps 4 or 8 KB), which is both the atom of allocatable memory and the unit of transfer between disk and memory.

Operating systems host the several applications that run on a computer and handle the operations of computer hardware. Users and application programs access the services offered by the operating system, by means of system calls and application programming interfaces. Users interact with OS through command line interfaces (CLIs) or graphical user interfaces (GUIs).

3.3.1 Types of Operating Systems

An operating system is divided into multiple types based on its design and the way it is deployed. For example, based on the user interface it is divided into single-user and multi-user interface.

3.3.2 Real-time Operating System

Real-time operating system (RTOS) is a multitasking operating system that aims at executing real-time applications. Real-time operating systems often use specialized scheduling algorithms so that they can achieve a deterministic nature of behaviour. The main objective of RTOS is their quick and predictable response to events. They either have an event-driven or a time-sharing design. An event-driven system switches between tasks based of their priorities, while time-sharing operating systems switch tasks based on clock interrupt (refer the section on RTOS in the chapter).

3.3.3 Multi-user and Single-user Operating Systems

Multi-user and single-user operating systems are discussed below.

Single-user operating systems This variant of operating systems is used for computers having a single terminal.

Multi-user operating systems These operating systems are used for those computers (micro to mainframe) which have many terminals. There are several techniques used in multi-user operating systems for enabling many users to concurrently share the single or multiple CPU. A multi-user operating system allows for multiple users to use the same computer at the same time and/or different times. An operating system is capable of supporting and utilizing more than one computer processor. It is capable of allowing multiple software processes to run at the same time.

The operating system of this type allows multiple users to access a computer system concurrently. Time-sharing system can be classified as multi-user systems as they enable a multiple-user access to a computer through the sharing of time. Single-user operating systems, as opposed to a multi-user operating system, are usable by a single user at a time. Being able to have multiple accounts on a Windows operating system does not make it a multi-user system. Rather, only the network administrator is the real user. But for a Unix-like operating system, it is possible for two users to login at a time and this capability of the operating system makes it a multi-user operating system.

3.3.4 Multitasking–Single-tasking Operating Systems

When a single program is allowed to run at a time, the system is grouped under a single-tasking system. In case the operating system allows the execution of multipletasks at one time, it is classified as a multitasking operating system. Multitasking can be of two types namely, pre-emptive or cooperative. In pre-emptive multitasking, the OS slices the CPU time and dedicates one slot to each of the programs. Unix-like operating systems, such as Solaris and Linux, support pre-emptive multitasking. Cooperative multitasking is achieved by relying on each process to give time to the other processes in a defined manner. Microsoft Windows prior to Windows 95 used to support cooperative multitasking.

3.3.5 Distributed Operating System

An operating system that manages a group of independent computers and makes them appear to be a single computer is known as a distributed operating system. The development of networked computers that could be linked and communicate with each other, gave rise to distributed computing. Distributed computations are carried out on more than one machine. When computers in a group work in cooperation, they make a distributed system.

3.3.6 Embedded System

The operating systems designed for being used in embedded computer systems are known as embedded operating systems. They are designed to operate on small machines like PDAs with less autonomy. They are able to operate with a limited number of resources. They are very compact and extremely efficient by design. Windows CE, FreeBSD, and Minix 3 are some examples of embedded operating systems.

3.4 Kernel

The most important collection of system programs comprises the operating system. The OS provides a software platform on top of which other programs, called application programs, can run. It makes a computer ready for use by a process called booting and controls and co-ordinates the overall operations of the computer system. An operating system has two main parts: the kernel and the system programs. The kernel allocates machine resources, including memory, disk space, and CPU cycles, to all other programs that run on the computer. The system programs perform higher-level housekeeping tasks, often acting as servers in a client/server relationship. A portion of the OS in the main memory includes the kernel, or nucleus, which contains the most frequently used functions in the OS and, at a given time, other portions of the OS currently in use. The remainder of the main memory contains user programs and data. The OS masks the details of the hardware from the programmer and provides the programmer with a convenient interface for using the system.

The kernel is the low-level software that schedules tasks, allocates storage, and handles the interfaces to peripheral hardware. Typical components of a kernel are interrupt handlers to service interrupt requests, a scheduler to share processor time among multiple processes, a memory management system to manage process address spaces, and system services such as networking and inter-process communication. The hardware consists of the processor, hard drives, video cards, sound cards, and more. It controls a computer by providing a single standard way for applications to access hardware devices. Each processor has built into it a language that only it understands, and each manufacturer creates a different language for its processor. All interactions between the hardware and any programs must be negotiated through the kernel.

The monitor controls the sequence of events, and hence must always be in the main memory and available for execution. The portion which is present in the main memory is referred to as the resident monitor. The rest of the utilities and common functions that are loaded as subroutines to the user program at the beginning of any job that requires them are the other part of the monitor. The monitor reads in jobs one at a time from the input device. As it is read in, the current job is placed in the user program area, and control is passed to this job.

Most processors support at least two modes of execution. The reason for using two modes should be clear. It is necessary to protect the OS and key operating system tables, such as process control blocks, from interference by user programs. In the kernel mode, the software has complete control of the processor and all its instructions, registers, and memory. The less-privileged mode is often referred to as the user mode, because user programs typically would execute in this mode. The more-privileged mode is referred to as the system mode, control mode, or kernel mode. Considerations of memory-protection and privileged instructions lead to the concept of modes of operation. A user program executes in a user mode, in which certain areas of memory are protected from the user's use and in which certain instructions may not be executed. The monitor executes in a system mode, or what has come to be called the kernel mode, in which privileged instructions may be executed and in which protected areas of the memory may be accessed.

Most operating systems, until recently, featured a large monolithic kernel. Most of what is thought of as OS functionality is provided in these large kernels, including scheduling, file system, networking, device drivers, memory management, and more. Typically, a monolithic kernel is implemented as a single process, with all elements sharing the same address space. A micro-kernel architecture assigns only a few essential functions to the kernel, including address spaces, inter-process communication (IPC), and basic scheduling. Other OS services are provided by processes, sometimes called servers, that run in user mode and are treated like any other application by the micro-kernel. This approach decouples kernel and server development.

3.4.1 System Calls

Programs interact with the kernel through system calls and special functions with well-known names. System calls provide a layer between the hardware and user-space processes. They provide an abstracted hardware interface for user-space. System calls ensure system security and stability.

System calls are each assigned a syscall number. This is a unique number that is used to reference a specific system call. When a user-space process executes a system call, the syscall number delineates which syscall was executed; the process does not refer to the syscall by name. The system call handler is the function system_call() available in the kernel. The kernel keeps track of all registered system calls in the system call table.

The system call is the means by which a process requests a specific kernel service. There are several hundred system calls, which can be roughly grouped into six categories: filesystem, process, scheduling, inter-process communication, socket (networking), and miscellaneous.

3.4.2 Memory Management

Virtual memory is a facility that allows programs to address memory from a logical point of view, without regard to the amount of main memory physically available. The OS has the following five principal storage management responsibilities.

Process isolation The OS must prevent independent processes from interfering with each other's memory, both data and instructions.

Automatic allocation and management It ensures that programs be dynamically allocated across the memory hierarchy as required. Allocation should be transparent to the programmer.

Support of modular programming It ensures that programmers are able to define program modules, and to create, destroy, and alter the size of modules dynamically.

Protection and access control It takes care of sharing of memory at any level of the memory hierarchy and create the potential for one program to address the memory space of another.

Long-term storage Many application programs require means for storing information for extended periods of time, after the computer has been powered down.

Operating systems meet these requirements with virtual memory and file system facilities. The file system implements a long-term store, with information stored in named objects called files.

3.4.3 Process Management

The concept of process is fundamental to the structure of operating systems. A process may be thought of as an entity that consists of a number of elements. Two essential elements of a process are program code and a set of data associated with that code. Each process at any time, in one of a number of execution states, including ready, running, and blocked. The operating system keeps track of these execution states and manages the movement of processes among the states. For this purpose, the operating system maintains elaborate data structures describing each process.

The process can be identified by a number of elements, including the following.

- Identifier
- State

- Priority
- Program counter
- Memory pointers
- Context data
- I/O status information
- Accounting information

The data mentioned above is available in a process control block. The process control block is the key tool that enables the OS to support multiple processes and to provide for multi-processing. When a process is interrupted, the current values of the program counter and the processor registers (context data) are saved in the appropriate fields of the corresponding process control block, and the state of the process is changed to some other value, such as blocked or ready.

When the OS creates a process at the explicit request of another process, the action is referred to as process spawning. When one process spawns another, the former is referred to as the parent process, and the spawned process is referred to as the child process. A process that is not in the main memory is not immediately available for execution, whether or not it is awaiting an event. A process starts with the fork() system call, which creates a new process by duplicating an existing one. The process that calls fork() is the parent, whereas the new process is the child. The child differs from the parent only in its process identification value (PID), its parent's PID, and certain resources and statistics, such as pending signals, which are not inherited. The exec() function calls are used to create a new address space and load a new program into it. The process exits via the exit() system call. Typically, process destruction occurs when the process calls the exit() system call, either explicitly when it is ready to terminate or implicitly on return from the main sub-routine of any program. The exit function terminates the process and frees the resources used. A parent process can inquire about the status of a terminated child via the wait()system call, which enables a process to wait for the termination of a specific process. When a process exits, it is placed into a special zombie state that is used to represent terminated processes and subsequently removed. The kernel stores the list of processes in a circular doubly linked list called the task list. The system identifies processes by a unique process identification value. The state field of the process describes the current condition of the process.

3.4.4 Thread Management

Threads of execution, often shortened to threads, are the objects of activity within the process. There are two broad categories of thread implementation: user-level threads (ULTs) and kernel-level threads (KLTs). In user-level threads, all of the work of thread management is done by the application and the kernel is not aware of the existence of threads. In kernel-level threads all of the work of thread management is done by the kernel. There is no thread management code in the application-level, simply an application programming interface (API) to the kernel thread facility. Windows is an example of this approach. Each thread includes a unique program counter, process stack, and a set of processor registers. Processes can communicate with one another but remain fully protected from one another, just as the kernel is protected from all processes. As with processes, the key states for a thread are running, ready, and blocked.

There are four basic thread operations associated with a change in thread state transition management.

- Spawn
- Block
- Unblock
- Finish

All of the threads of a process share the same address space and other resources, such as open files. Any alteration of a resource by one thread affects the environment of the other threads in the same process. It is therefore necessary to synchronize the activities of the various threads so that they do not interfere with each other or corrupt data structures.

Therefore, all modern operating systems support the subdivision of processes into multiple threads of execution. Threads run independently, like processes, and no thread knows what other threads are running or where they are in the program unless they synchronize explicitly. The key difference between threads and processes is that the threads within a process share all the data of the process.

Each thread has its own instruction pointer (a register pointing to the place in the program where it is running) and stack (a region of memory that holds subroutine return addresses and local variables for subroutines), but otherwise a thread shares its memory.

Even the stack memory of each thread is accessible to the other threads, though when they are programmed properly, they do not step on each other's stacks. Threads within a process that run independently but share memory, have the obvious benefit of being able to share work quickly, because each thread has access to the same memory as the other threads in the same process. The operating system can view multiple threads as multiple processes that have essentially the same permissions to regions of memory.

Deadlock occurs when at least two tasks wait for each other, and none of the two will resume until the other task proceeds. A race condition occurs when multiple tasks read from and write to the same memory without proper synchronization. The race may finish correctly sometimes and therefore complete without errors, and at other times it may finish incorrectly. Race conditions are less catastrophic than deadlocks.

3.4.5 Process Scheduler

A key responsibility of the OS is to manage the various resources available to it and to schedule their use by the various active processes. The process scheduler (shortened as scheduler) is the component of the kernel that selects which process to run next. The scheduler can be viewed as the subsystem of the kernel that divides the finite resource of processor time between the runnable processes on a system. The scheduler is the basis of a multitasking operating system such as Linux. By deciding what process can run, the scheduler is responsible for best utilizing the system CPU resources.

The scheduler, which controls the mechanism of how processes share the CPU, is part of process management. A common strategy is to give each process in the queue some time in turn; this is referred to as the round-robin technique. In effect, the round-robin technique employs a circular queue. Another strategy is to assign priority levels to the various processes, with the scheduler selecting processes in priority order.

3.4.6 Interprocess Communication

The kernel is the big chunk of executable code in charge of handling interprocess communication. It communicates with the basic computer hardware like the microprocessor, memory, and device controllers. Processes communicate with each other and with the kernel to coordinate their activities. Linux supports a number of Inter-Process Communication (IPC) mechanisms. For example, signals and pipes are IPC mechanisms. Signals are one of the oldest inter-process communication methods. A signal could be generated by a keyboard interrupt or any other interrupt.

Named pipes allow two unrelated processes to communicate with each other. Named pipes are identified by their access points, which are basically in a file kept on the file system. Because named pipes have the path name of a file associated with them, it is possible for unrelated processes to communicate with each other. A named pipe supports blocked read and write operations by default: if a process opens the file for reading, it is blocked until another process opens the file for writing, and vice versa.

3.4.7 Interrupt Handling

Interrupts allow hardware to communicate with the processor. With interrupts, the processor can be engaged in executing other instructions while an I/O operation is in progress. An interrupt is physically produced by electronic signals originating from hardware devices and directed into input pins on an interrupt controller. Different devices are referred through unique interrupts by means of a unique value associated with each interrupt. The interrupt values are often called interrupt request (IRQ) lines. This enables the operating system to differentiate between interrupts and to know which hardware device caused which interrupt.

Nearly all architectures, including all systems that Linux supports, provide the concept of interrupts. When hardware wants to communicate with the system, it issues an interrupt that asynchronously interrupts the kernel. Interrupts are identified by a number. The kernel uses the number to execute a specific interrupt handler to process and respond to the interrupt. The kernel is in charge of creating and destroying processes and handling their connection to the outside world. For the user program, an interrupt suspends the normal sequence of execution. When the interrupt processing is completed, execution resumes. Thus, the user program does not have to contain any special code to accommodate interrupts; the processor and the OS are responsible for suspending the user program and then resuming it at the same point.

Kernel responds to a specific interrupt via the interrupt handler. The role of the interrupt handler depends entirely on the device and its reasons for issuing the interrupt. The interrupt's journey in the kernel begins at a pre-defined entry point, just as system calls enter the kernel through a pre-defined exception handler. This pre-defined point is generally set up by the kernel. To accommodate interrupts, an interrupt stage is added to the instruction cycle. In the interrupt stage, the processor checks to see if any interrupts have occurred, indicated by the presence of an interrupt signal. If no interrupts are pending, the processor proceeds with the next instruction of the current program. If an interrupt is pending, the processor suspends execution of the current program and executes an interrupt-handler routine. The interrupt-handler routine is generally part of the OS.

3.4.8 Device Drivers

Device drivers take on a special role in the Linux kernel. The kernel takes care of translating the requests into the form the particular device speaks. The user interface available with the kernel makes it possible for the individual to interact with the computer and launch programs, apart from controlling the computer.

They are closed software blocks that make a particular piece of hardware respond to a well-defined internal programming interface. The device drivers present in the kernel allow the kernel to talk to the various devices, such as hard drives and modems, which are connected to the computer. Each hardware device speaks its own language, and the operating system must be capable of interacting with it. They hide the details of the working of devices. User activities are performed by means of a set of standardized calls that are independent of the specific driver; mapping those calls to device-specific operations that act on real hardware is then the role of the device driver.

Shared resources require protection from concurrent access because if multiple threads of execution access and manipulate the data at the same time, the threads may overwrite each other's changes or access data while it is in an inconsistent state.

3.5 Windows Operating System

Windows is a popular operating system. It is developed by Microsoft Corporation of USA. This is why it is referred to as Microsoft Windows. The evolution of Windows was often uncertain and precarious. Its success was symbiotic with advancements in processor speed and memory capacity, and Microsoft relied heavily on third-party software to bridge the gap between concept and consumer.

On 10 November 1983, at the Plaza Hotel in New York City, Microsoft Corporation formally announced Microsoft Windows, a next-generation operating system that would provide a graphical user interface (GUI) and a multitasking environment for IBM computers. When referring to an operating system, Windows, or Win, is an operating environment created by Microsoft that provides an interface known as graphical user interface for IBM-compatible computers.

Like other operating systems, Windows is also a collection of programs that manage and co-ordinate the overall operations of a computer system. The user interacts with the computer by using input pointing devices, such as mouse, trackball, light pen, etc. in addition to the keyboard.

3.5.1 Windows Registry

Microsoft describes the registry as a central hierarchical database used in Microsoft Windows. It is required to store information that is necessary to configure the system for one or more users, applications, and hardware devices. The registry maintains a great deal of information about the configuration of the system services to run, when to run them, how the user likes their desktop configured, etc. The registry also maintains information about hardware devices added to the system, applications that were installed on the system, as well as information about how the user has configured.

Windows stores configuration data in the registry. The registry is a hierarchical database, which you can describe as a central repository for configuration data. A hierarchical database has characteristics that make it ideally suited for storing configuration data. Globally Unique Identifiers are better known as GUIDs. They are numbers that uniquely identify objects, including computers, program components, devices, and so on. These objects often have names, but their GUIDs remain unique even if two objects have the same name or their names change. An example of a real GUID is 645FF040-5081-101B-9F08-00AA002F954E, which represents the recycle bin object that you see on the desktop. Keys are so similar to folders that they have the same naming rules. Nesting one or more keys within another key is permitted as long as the names are unique within each key. Each key contains one or more values. A value's name is similar to a file's name.

A value's type is similar to a file's extension, which indicates its type. A value's data is similar to the file's actual contents. Every key contains at least one value, and that's the default value. The registry actually stores all values as binary values. The registry API identifies each type of value by a number, which programmers refer to as a constant. Physically, Windows organizes the registry in hives, each of which is in a binary file called a hive file. For each hive file, Windows creates additional supporting files that contain backup copies of each hive's data. These backups allow the operating system to repair the hive during the installation and boot processes if something goes terribly wrong.

3.5.2 Microsoft Windows 2000

Windows 2000, released in Professional, Server, Advanced Server, and Datacenter Server flavours, began its development in late 1996 with a projected release date sometime in 1997. This, however, was a long shot because there were so many features that needed to be added, removed, or fixed to Windows NT, such as improved directory services, plug-and-play support, and FAT32 support, just to name a few. The first Windows NT 5.0 beta was released in 1997.

3.5.3 Microsoft Windows XP

Plans for Windows XP, then code named Neptune (for the home release) and code named Odyssey (for the business release), started in early 1999. It was decided that this version of Windows would be the one that would end the days of the Win9x kernel. In 2001, Microsoft released Windows XP, code named Whistler. Windows XP is designed more for users who may not be familiar with all of Windows features and has several new abilities to make the Windows experience more easy for those users. The merging of the Windows NT/2000 and Windows 95/98/Me lines was finally achieved with Windows XP. The Windows eXPerience operating system is available as home and professional edition and are similarly suitable for the use on standalone computers. Windows XP includes various new features not found in previous versions of Microsoft Windows. Windows XP is available in a number of versions:

- Windows XP Home Edition, for home desktops and laptops
- Windows XP Professional, for business and power users
- Windows XP Media Center Edition (MCE), released in November 2002, for desktops and notebooks with an emphasis on home entertainment
- Windows XP Embedded, for embedded systems
- Windows XP Professional x64 Edition, released on 25 April 2005, for home and workstation systems utilizing 64-bit processors
- Windows XP 64-bit Edition, is a version for Intel's Itanium line of processors.

3.5.4 Microsoft Windows Vista

Microsoft Windows Vista is an upgrade to Microsoft Windows XP and Windows 2000 users that was released to the public on January 30, 2007. Windows Vista contains a dramatic new look for users used to previous versions of Microsoft Windows that has been designed to help create an overall better experience.

3.5.5 Microsoft Windows Server 2008

Windows Server 2008, released on February 27, 2008, was originally known as Windows Server, code named Longhorn. Windows 2008 includes an enhanced operating system kernel NT6.1 kernel. It is more manageable, reliable, and scalable than previous versions of windows operating system. Internal changes were made to how threads are allocated CPU cycles, so threads now get a fairer share of the processor. Other changes were made in how I/O completion is handled. Kernel Transaction Manager allows file and registry changes to be placed in a transaction and automically committed or rolled back as one unit.

> **Registry Virtualization**
>
> Beginning with the Windows Vista operating system, registry virtualization is supported, allowing registry write operations with global implications (that is, that affect the entire system) to be written to a specific location based on the user that installed that software application. This mechanism is transparent to applications, as well as to users, but can be extremely important to a forensic analyst.

3.5.6 Windows Script

Scripts themselves are text files that permit a user to edit with a simple text editor such as notepad. Windows script is a program written in an interpreted language with access to OS components through the common object model (COM).

Microsoft has split its scripting system into two parts: one called the script host takes care of managing the script's component objects, and the other called the script engine interprets the actual script language itself. Windows Script Host lets the user write scripts that manipulate files, process data, change operating system settings, install and uninstall software, send e-mail, and so on. Windows XP and Windows 2000 come with two script language interpreters: VBScript and JScript. VBScript is the dialect used in Windows Script Host, and it can also be used in Web browsers and servers. JScript is a programming language modeled after Netscape's JavaScript language. JScript is also used for Windows scripting. Windows Script files are still plain-text files, but they're structured with text markup called tags. Windows Script Host comes in two flavors: a windowed version named Wscript, and a command-line version named Cscript. Either version can run any script.

In Microsoft's parlance, the various outside software components are packaged up as objects. Objects are self-contained program modules that perform tasks for other programs through a set of well-defined programming links. Objects are usually meant to represent some real-world object or concept, such as a file or a computer user's account, and the programming links provide a way for other programs. Objects are little program packages that manipulate and communicate information. They are a software representation of something tangible, such as a file, a folder, a network connection, an e-mail message, or an Excel document. Objects have properties and methods. Properties are data values that describe the attributes of the thing the object represents. Methods are actions— program subroutines. Objects need a mechanism through which they can exchange property and method information with a program or a script. Because each programming language has a unique way of storing and transferring data, objects and programs must use some agreed-upon, common way of exchanging data. Microsoft uses what it calls the common object model.

To help maintain and write complex sets of scripts, Microsoft has developed a file format that allows several separate scripts to be stored in one text file, and even lets the user use several languages in one script. The new file format is called a Windows Script File, and the extension used is .wsf. The readers are advised to refer to advanced books to understand windows scripts in detail.

3.6 Open Source Movement

Over the last few years, the concept of openness has been spreading its wings far and wide in many guises. According to Stallman, software should be free from restrictions against copying or modification in order to make better and efficient computer programs. With his famous 1983 manifesto that declared the beginnings of the GNU project, he started a movement to create and distribute softwares that conveyed his philosophy. Incidentally, the name GNU is a recursive acronym which actually stands for 'GNU is not Unix'. Much of it started with the popularity of the free and open source software (FOSS) movement. The FOSS collaboration and learning model has attracted considerable attention, mostly because of its mere existence and the way it works contradicts existing theories and counteracts common business practices.

The term 'open source' as used by the open source initiative (OSI) is defined using the open source definition, which lists a number of rights a license has to grant in order to constitute an open source license. These include free redistribution, inclusion of source code, permission to allow for derived works which can be redistributed under the same license, and integrity of author's source code.

A software is defined as free if the user has the freedom to run the program, for any purpose, to study how the program works and adapt it to his or her needs, to redistribute copies and improve the program, and release these improvements to the public. Access to the source code is a necessary precondition. In this definition, open source and free software are largely interchangeable. Due to its globally distributed developer force and the possibility to collaborate on a large scale, FOSS software projects enjoy extremely rapid code evolution and highest software quality. With products such as Linux, Apache, Perl, KDE, and Gnome desktop, to name a few, FOSS development is also highly successful in an economic sense.

Though dating back to Richard M. Stallman's days at Massachussetts Institute of Techonology (MIT), it gathered popularity only on the arrival of the Linux kernel from Linus Torvalds in the early 1990s. Since then, the movement has never looked back. The term 'open source' refers to software in which the source code is freely available for others to view, amend, and adapt. Typically it is created and maintained by a team of developers that crosses institutional and national boundaries.

Open source is, primarily, a software development methodology. As is the norm with any other methodology or practice, open source has its unique ethos, rituals, and codes. Open source mandates a level of intense collaboration. Such engagement between developers and users leads to an improvement in the quality of the code.

Free and open source software is strongly based on the community metaphor of the bazaar development model described by Eric S. Raymond, which brings together a number of people from around the world to work on a common system. This necessarily demands explicit efforts to reduce the learning curve for others, and transparency in shared data structures. Open standards, where the complete description is freely accessible to everyone, and where the standards evolve from collective contributions, becomes a natural choice. Not surprisingly, most of the open source programs use open standards wherever available. A suitably detailed perspective of open source was provided in *The Cathedral and the Bazaar* by Eric S. Raymond. Since then, there have been sporadic attempts to reconcile the principles of open source with established practices in software development.

Open source initiative claims that this rapid evolutionary process produces better software than the traditional closed model, in which only very few programmers can see the source and everybody else must blindly use an opaque block of bits. The OSI is actively involved in open source community-building, education, and public advocacy to promote awareness and the importance of non-proprietary software. Open source initiative board members frequently travel the world to attend open source conferences and events, meet with open source developers and users, and to discuss with executives from the public and private sectors about how open source technologies, licenses, and models of development can provide economic and strategic advantages.

Almost every large corporation is involved in the movement in one way or another, with IBM, Sun, Intel, HP, etc. leading the way. The movement has got into the legislative bodies of many countries, creating pressure at the level of policies, guidelines, etc. to support and adopt FOSS wherever possible. Many countries have taken explicit steps to nurture a FOSS ecosystem, by training programs, certifications, resource centres, and so on. The FOSS activities have a number of dimensions, stretching well beyond the availability of software with source code. One builds on the knowledge created by others, and shares the enriched knowledge with others to let it grow further. In a broad sense, software and content can be seen as embodiments of knowledge, differing perhaps, in the way the knowledge is captured and represented is the basis on which open software is being practised.

Open source software (OSS) is developed in a way different from proprietary software. Development is done inside communities of developers that work on code for their personal satisfaction or need. However, lately, a trend has formed inside large companies, that pay their employees to contribute to OSS projects, using this as a platform that enables them to affect the introduction of new software features so that the final OSS product is better aligned with the company's interests and needs. Open source software communities have significant similarities with professional software engineering teams utilized by software houses. However, these two organizational structures also portray critical differences that show that they stand quite far apart.

Open source software has a definition that focuses on specific characteristics that software has to serve in order to be labelled as open source. The open source initiative is a non-profit corporation dedicated to managing and promoting the OSS definition for the good of the community. It acts as the official organization behind OSS. The current OSS landscape presents a very interesting picture. Although the idea behind OSS dates back to the 1960s and the UNIX era in the 1980s, the official term of OSS was coined in 1998 and, at the same time, the OSI was created. Since then, the OSS movement has evolved at a very fast pace. Prime examples of successful OSS projects include operating systems (Linux, FreeBSD, OpenBSD, NetBSD), web browsers (Firefox, Konqueror), graphical environments (KDE, Gnome), productivity applications (OpenOffice), programming languages and infrastructure (Apache, MySQL), and development tools (GNU toolchain, Eclipse). These widely accepted OSS endeavors show that, today, a wide range of OSS applications are available and they present a viable and robust alternative to proprietary software solutions.

3.6.1 Open Source License

The emergence of open source software and the rapid expansion of the Internet have brought new software licensing practices to mass markets. The new entrants have challenged incumbents in the expanding software markets with the help of innovative copyright licensing strategies and courageous anti-patent policies.

Software licenses are contractual documents, which define how copyright and patent rights are used. A licensor is typically a software developer (software company) who licenses more or less of these rights to licensees. A licensee can be either another developer or end-user. The term 'open source' is defined as a set of software licenses, which follow certain criteria further defined in Open Source Definition. GNU/Linux operating system, Apache web server and MySQL database are perhaps the best-known examples of open source software. On their part, the biggest companies in the industry from IBM to Apple have adapted to the changing environment with different open source licensing and operation strategies.

3.7 Linux Operating System

Linux traces its ancestry back to a mainframe operating system known as Multics (multiplexed information and computing service). Linus Torvalds, the founder of Linux, was inspired in part by the GNU project when he adopted an open community approach for the development of the Linux kernel. He announced in 1991 that he had created a very experimental operating system core called a kernel, based on a clone of UNIX called Minux. This new operating system kernel later was known as Linux. From the very beginning, contributions to kernel code from various developers spread over the world ensured improvements in Linux kernel.

Linux is an operating system, a software program that controls the computer. The core component to the operating system is called the kernel in UNIX and UNIX-like operating systems. The heart of the Linux operating system is the Linux kernel, which is responsible for allocating the computer's resources and scheduling user jobs so that each one gets its fair share of system resources, including access to the CPU; peripheral devices, such as disk, DVD, and CD-ROM storage; printers; and tape drives. In a Linux system, several concurrent processes attend to different tasks. Each process competes for system resources, be it computing power, memory, network connectivity, or some other resource.

Linux operating system kernel was refined for maximum performance on the Intel 386 microprocessor, which made this new Linux kernel platform specific. Linux kernels come in two forms: stable or development. Stable kernels are production-level releases suitable for widespread deployment. New stable kernel versions are released typically only to provide bug fixes or new drivers. Development kernels, on the other hand, undergo rapid change. Linux kernels distinguish between stable and development kernels with a simple naming scheme. Three numbers, each separated by a dot, represent Linux kernels. The first value is the major release, the second is the minor release, and the third is the revision.

Linux is a fully protected multitasking operating system allowing each user to run more than one job at a time. A process is a program in execution. It is an active program and related resources.

Linux has a unique implementation of threads. To the Linux kernel, there is no concept of a thread. Linux does not recognize a distinction between threads and processes. Linux implements all threads as standard processes. Using a mechanism similar to the lightweight processes of Solaris, user-level threads are mapped into kernel-level processes. This approach to threads contrasts greatly with operating systems such as Microsoft Windows or Sun Solaris, which have explicit kernel support for threads.

Depending on the hardware and task type, a Linux system can support multiple users, each concurrently running a different set of programs. Linux includes a family of several hundred utility programs, often referred to as commands. For example, the sort utility is an important programming tool and is part of the standard Linux system. A pipe sends the output of one program to another program as input. A filter is a special form of a pipe that processes a stream of input data to yield a stream of output data.

The Linux filesystem provides a structure whereby files are arranged under directories, which are like folders or boxes. Each directory has a name and can hold other files and directories.

GNU/Linux comes in many different flavours, apart from the fact that each individual distribution has a new release almost every six months, if not less. All Linux installers use two files to boot a computer: a kernel and an initial root filesystem, also known as the RAM disk or initrd image. This initrd image contains a set of executables and drivers that are needed to mount the real root filesystem. When the real root filesystem mounts, the initrd is unmounted and its memory is freed.

Today, Linux is one of the fastest growing operating systems in history. From a few dedicated fanatics in 1991-92 to millions of general users at present, it is certainly a remarkable journey. The big businesses have discovered Linux, and have poured millions of dollars into the development effort, denouncing the anti-business myth of the open-source movement. A full discussion on the various aspects of Linux operating system is beyond the scope of this chapter, and the reader is referred to the references at the end of the chapter for further information.

C-DAC Launches Advanced Version of BOSS

C-DAC has launched its Bharat Operating Systems Solutions (BOSS) Linux software version 3.0, developed by NRCFOSS (National Resource Centre for Free/Open Source Software). BOSS v3.0 is coupled with GNOME and KDE, and comes with wide Indian language support and packages that are relevant for use in the government domain. The software is also endowed with Bluetooth for short-range communications, along with features like a RSS Feed reader and PDF viewer to edit documents.

3.7.1 Debian Linux

It all started with the Debian project. The Debian Project is an association of individuals who have made common cause to create a free operating system. Debian is a free operating system. The Debian community is a community of volunteers. Debian uses the Linux kernel, but most of the basic OS tools come from the GNU project. Hence, it takes the name GNU/Linux. Debian will run on almost all personal computers, including most older models. The Debian system owes much of its power to numerous free software projects and movements, most notably GNU and Linux. Debian uses the Linux kernel, so anything that is possible with Linux itself is possible with Debian GNU as well. The Debian GNU/Linux operating system is a fully-featured operating system for servers, workstations, and home desktop machines alike.

The Debian project has a strong commitment to the free software community and makes all its work available for the benefit of others, just as it uses the produce of others for its own good. In August 2000, Debian released version 2.2, code named potato, which featured almost 4,000 packages and additionally supported the PowerPC and ARM architectures. Debian 3.0 woody was scheduled for release in two years time. The release following sarge in 2005 was code named etch. Several of the developers want to move towards a time-based release cycle for etch and its successors.

Installing the Debian GNU/Linux operating system on a computer is no different than installing any other operating system by following straightforward guidelines. Debian takes a different approach to installation and provides the basics needed to pull up a minimal system, queries for the essential configuration data of the base system, and then leaves the user to the grace of the package management system. Debian provides various different types of installation images. The official CD image includes everything needed for a standard installation of Debian, and furthermore provides some of the most popular packages in what is called a full installation. The *netinst* image available in the CD is optimized for installations with network access. It provides everything needed to run the installer and setup a standard Debian system. The *netboot* image allows a machine to boot and pull the installation over the network, using PXE and Boot Protocol. The *hd-media* image allows for booting from a USB stick. The installation process is more or less the same in all the installation options.

The installer is based on a modular architecture. Modules are simple Debian packages, called deb files. The packages use *debconf* to interface with the user, and simple hooks to register with the installer, which then allows access to their functionality from within the installer menu.

A full discussion on the various aspects of Debian OS is beyond the scope of this chapter, and the reader is referred to the references at the end of the chapter for further information.

3.7.2 OpenSolaris

OpenSolaris is a rapidly evolving operating system with roots in Solaris 10, suitable for deployment on laptops, desktop workstations, storage appliances, and data centre servers from the smallest single purpose systems to the largest enterprise class systems. The growing OpenSolaris community now has hundreds of thousands of participants and users in government agencies, commercial businesses, and universities, with more than 100 user groups around the world contributing to the use and advancement of OpenSolaris.

> **The New XenServer Release is Free for All**
>
> Citrix Systems has unveiled the new version of XenServer that will be offered free of charge to any user for unlimited production deployment. With this new release, XenServer adds powerful new features like centralised multinode management, multi-server resource sharing and full live motion. Powerful centralised management enables full multinode management for an unlimited number of servers and virtual machines; includes easy physical-to-virtual and virtual-to-virtual conversion tools, centralised configuration management and a resilient distributed management architecture.

3.8 Real-time Operating System (RTOS)

General-purpose microprocessors are increasingly being used for control applications due to their widespread availability and software support for non-control functions like networking and operator interfaces. Real-time systems are classified as hard, firm, or soft systems. In hard real-time systems, failure to meet response-time constraints leads to system failure. An embedded system is a specialized real-time computer system that is part of a larger system. In the past, it was designed for specialized applications, but reconfigurable and programmable embedded systems are becoming popular. Some examples of embedded systems are: the microprocessor system used to control the fuel/air mixture in the carburetor of automobiles, software embedded in airplanes, missiles, industrial machines, microwave ovens, dryers, vending machines, medical equipment, and cameras. Clearly the choice of an RTOS in the design process is important for support of priorities, interrupts, timers, inter-task communication, synchronization, multiprocessing, and memory management.

Two classes of RTOS exist for these systems. The traditional RTOS serves as the sole operating system, and provides all OS services. Examples include ETS, LynxOS, QNX, Windows CE, and VxWorks.

LynxOS is a POSIX compatible, multithreaded OS designed for complex real-time applications that require fast, deterministic response. It is scalable from large switching systems down to small-embedded products. The micro-kernel can schedule, dispatch interrupts, and synchronize tasks. Other services offered by the kernel lightweight service modules are TCP/IP streams, I/O and file systems, sockets, etc. In response to an interrupt, the kernel dispatches a kernel thread, which can be prioritized and scheduled similar to other threads. The OS depends upon hardware memory management units for memory protection, but does offer optional demand paging. It uses scheduling policies such as prioritized FIFO, dynamic deadline monotonic scheduling, time-slicing, etc.

VxWorks is a widely adopted RTOS in the embedded industry with a visual development environment. It is scalable with over 1800 APIs and is available on popular CPU platforms. It offers network support, file system, and I/O management. The micro-kernel supports 256 priority levels, multitasking, deterministic context switching and preemptive and round robin scheduling, semaphores and mutual exclusion with inheritance. Transmission control protocol (TCP), UDP, sockets, and standard Berkeley network services can all be scaled in or out of the networking stack as necessary.

Real-time operating system extensions add real-time scheduling capabilities to non-real-time OSs, and provide minimal services needed for the time-critical portions of an application. Examples include RTAI and RTL for Linux, and HyperKernel, OnTime and RTX for Windows NT.

> On 15 November 2007, IBM announced Blue Cloud, a set of technologies that enable development of a cloud computing infrastructure. The initial offering includes an IBM BladeCenter hardware platform running Xen and Linux on X64 and Power blades. It also includes Apache Hadoop, an open-source implementation of Google's
>
> *Contd*

> MapReduce parallel computing environment. IBM, Google, and the University of Washington offer clouds based on this technology. Blue Cloud is IBM's commercialized version of the technology that underpins this joint-initiative. Since IBM does not provide Blue Cloud as a set of external infrastructure services, Gartner does not consider Blue Cloud itself to be a 'cloud computing offering' but in the future, we do expect IBM to deliver such an offering based on Blue Cloud. The infrastructure services provided on Blue Cloud start with a utility infrastructure model. Utility infrastructure users can obtain a virtual Linux server or cluster of servers, on which they can 'install' whatever Linux-based software they desire. Blue Cloud's provisioning and management environment is one if its key features. Via a Web-based interface built using Tivoli, WebSphere and DB2, users can request access to infrastructure resources from the cloud.
>
> *Source:* Gartner, 'IBM Moves Toward a Cloud Computing Infrastructure,' 20 November 2007

3.9 Programming Languages

A programming language is a set of words, codes, and symbols that allow a programmer to give instructions to the computer. The language a computer can understand (called 'machine code') is composed of strings of zeros and ones. This smallest element of a computer's language is called 'a bit'—0 or 1. Four bits are a nibble. Two nibbles (8 bits) equal a byte. The 'words' of a computer language are the size of a single instruction encoded in a sequence of bits (for example, many computers speak a language with words that are '32-bits' long). Many programming languages exist, each with their own rules, or syntax, for writing these instructions.

Programming languages can be classified as low-level and high-level languages. Low-level programming languages include machine language and assembly language. Machine language, which is referred to as a first-generation programming language, can be used to communicate directly with the computer.

3.9.1 Low-level Programming Language

Assembly language is directly influenced by the instruction set and architecture of the processor [Sivarama et al. 2005]. Assembly language programming is referred to as low-level programming because each assembly language instruction performs a much lower-level task compared to an instruction in a high-level language. As a consequence, to perform the same task, assembly language code tends to be much larger than the equivalent high-level language code. Assembly language programs are created out of three different classes of statements. Statements in the first class tell the processor what to do. These statements are called executable instructions, or instructions for short. The second class of statements provides information to the assembler on various aspects of the assembly process. These instructions are called assembler directives. The last class of statements, called macros, are used as a shorthand notation for a group of statements.

3.9.2 High-level Programming Language

High-level programming languages, are often referred to as third generation programming languages (3GL). High-level programming languages allow the specification of a problem solution in terms closer to those used by human beings. These languages were designed to make programming far easier, less error-prone, and to remove the programmer from having to know the details of the internal structure of a particular computer. These high-level languages were much closer to human language. High-level programming languages have English-like instructions and are easier to use than machine language. Some of the high-level languages are as outlined below.

FORTRAN FORTRAN (formula translating), released in 1957 after a three-year developmental period, is well-suited for math, science, and engineering programs because of its ability to perform numeric computations. The language was developed in New York by IBM's John Backus. A new standard has been designed and widely implemented in recent years. It is unofficially called Fortran 90, and adds many powerful extensions to FORTRAN 77. Fortran has been around for a long time. It was one of the first widely used high-level languages, as well as the first programming language to be standardized. Although Fortran has been enhanced many times, the enhancements almost always have been upward compatible; old programs continue to work with new compilers.

COBOL Computer business oriented language (COBOL), was released in April of 1959 and has been updated several times since then. Shortly after the introduction of FORTRAN, users from different fields, including academia and manufacturing, convened at the University of Pennsylvania to discuss the need for a standardized business language that could be used on a wide variety of computers. The eventual result was COBOL, a programming language well-suited for creating business applications. It started out with a language called FLOWMATIC in 1955 and this language influenced the birth of COBOL-60 in 1959. Over the years, improvements were made to this language and

COBOL 61, 65, 68, 70 were developed, being recognized as a standard in 1961. Now, the new COBOL 97 has included new features, such as object oriented programming, to keep up with the current languages.

C The C Programming language was developed from 1969-1973 at Bell labs, at the same time when the UNIX operating system was being developed there. C was a direct descendant of the language B, which was developed by Ken Thompson as a system/s programming language for the fledgling UNIX operating system. B, in turn, descended from the language Basic Combined Programming Language (BCPL) which was designed in the 1960s by Martin Richards while at MIT. C is a programming language which was born at 'AT&T's Bell Laboratories'. It was written by Dennis Ritchie. This language was created for a specific purpose—to design the UNIX operating system. The inclusion of types, its handling, as well as the improvement of arrays and pointers, along with the demonstrated capacity of portability, contributed to the expansion of the C language. Because C is such a powerful, dominant, and supple language, its use quickly spread beyond Bell Labs.

C++ In the early 1980s, also at Bell Laboratories, another programming language was created which was based upon the C language. This new language was developed by Bjarne Stroustrup and was called C++. Stroustrup states that the purpose of C++ is to make writing good programs easier and more pleasant for the individual programmer. When he designed C++, he added OOP (object-oriented programming) features to C without significantly changing the C component. C++ was designed for the UNIX system environment. Programmers could improve the quality of code they produced and reusable code was easier to write with C++.

In the 1980s, object-oriented programming evolved out of the need to better develop complex programs in a systematic, organized approach.

Java Java was designed by Sun Microsystems in the early 1990s to solve the problem of connecting many household machines together. When the World Wide Web became popular in 1994, Sun realized that Java was the perfect programming language for the Web. The Java programming language is a general-purpose object-oriented concurrent language. Its syntax is similar to C and C++. The Java virtual machine (JVM) is the cornerstone of the Java and Java 2 platforms. It is the component of the technology responsible for its hardware- and operating system- independence, the small size of its compiled code, and its ability to protect users from malicious programs. The Java virtual machine is an abstract computing machine. Like a real computing machine, it has an instruction set and manipulates the various memory areas at run time.

Perl Perl is abbreviation of 'practical extraction and report language'. The first version of Perl was developed by Larry Wall around 1987 and like many other languages the development was partially financed by military. Perl was designed to assist the programmer with common tasks that are probably too heavy or too portability-sensitive for the shell, and yet too weird or short-lived or complicated to code in C or some other UNIX glue language. Perl is distributed under the GNU Public License.

PHP PHP (Personal home page) tools is an 'HTML-embedded scripting language' primarily used for dynamic Web applications. The first part of this definition means that PHP code can be interspersed with HTML, making it simple to generate dynamic pieces of Web pages on the fly. As a scripting language, PHP code requires the presence of the PHP processor. PHP takes most of its syntax from C, Java, and Perl. It is an open source technology and runs on most operating systems and with most Web servers. PHP was written in the C programming language by Rasmus Lerdorf in 1994.

Python Python is a programming language that is freely available and that makes solving a computer problem almost as easy as writing out one's thoughts about the solution. Python 2.0 was released on 16 October 2000, with many major new features.

There are more than 2,000 different programming languages in existence. Depending on the kind of program being written, the computer it will run on, the experience of the programmer, and the way in which the program will be used, the suitability of one programming language over another will vary. Once a program is written in a high-level language, a program called a compiler or an interpreter is used to convert it to a computer's specific machine language, much like an assembler converts assembly code into machine language.

SUMMARY

This chapter highlighted the various elements of the computing infrastructure with emphasis on software. Intel's 32-bit instruction set architecture is referred to as IA-32 architecture. Programmers should have some basic knowledge about the processor and the system architecture in order to effectively program in the assembly language. The imperative model of computation is used by a programming language to provide a structured way of writing programs. In 1972, the Unix operating system was written in C. C was more structured than the assembly language most operating systems were written in at the time. It was portable and could be compiled to efficient machine code. C++ was designed by Bjarne Stroustrup. For several years, there had been ideas around about how to write code in object-oriented form. Simula, created by Ole-Johan Dahl and Kristen Nygaard around 1967, was an early example of a language that supported object-oriented design and Modula-2, created by Niklaus Wirth around 1978, was also taking advantage of these ideas. Smalltalk, an interpreted language, was object-oriented and was also developed in the mid 1970's and released in 1980. C++ was designed as a superset of C which was an immensely popular language in the seventies and eighties and still is today.

Scripting languages are languages that are interpreted and allow the programmer to quickly build a program and test it as it is written. Ruby is an object-oriented scripting language. Ruby is a relatively new addition as a programming language. Yukihiro Matsumoto created Ruby in 1993 and released it to the public in 1995.

The emergence of FOSS continues to have great impact in the way software is developed, supported, maintained, and distributed. It is collaboratively built software that is shared by developers and users and can be freely downloaded with or without the source code for use, modification, and further distribution. It is making a tremendous impact on the way people work in private and public administrations and is redefining the software industry. The increased adoption of FOSS by well-known organizations has led to more enterprise recognition and hence consideration.

Key Terms

API Application programming interface (API) is a set of software functions used by one application as a means for providing access to another application or operating system's capabilities.

Applet An applet is a small application that executed inside a larger application, such as a web browser. On the Web, an applet is typically sent along with a Web page to a user. Applets are more limited than standard applications because they do not have access to some computer resources (such as a printer), nor network resources (such as other computers).

ASCII American Standard Code for Information Interchange (ASCII) is a seven bit code used to represent alphanumeric characters. In an ASCII (ass-key) file, each character is represented with a 7-bit binary number and 128 possible characters are defined. ASCII was developed by the American National Standards Institute (ANSI).

Boot Boot (bootstrapping) is the process of initializing the computer and loading the operating system. This usually occurs when the computer is powered up or reset.

Client A client is a software application that contacts and obtains data from a server software application. A client makes a request and the server fulfils the request. An example of a client would be an e-mail program connecting to a mail server or an Internet browser client connecting to a web server.

Component object model Component object model (COM) is a Microsoft proprietary technology that enables Windows software components (DLLs) or executable applications (EXEs) to communicate with each other on the same user account. When COM is used in a network with a directory and additional support, COM becomes the distributed COM (DCOM). OPC communication depends on COM and DCOM.

COTS It refers to a commercial off-the-shelf (COTS) product. COTS products are purchased and integrated with little configuration. Typically COTS products reduce integration cost and time.

DCOM Distributed component object model (DCOM) is a Microsoft proprietary technology that enables Windows executable applications (EXEs) to communicate with each other between two user accounts. OPC communication depends on COM and DCOM. DCOMCNFG enables users to configure DCOM.

DDE Dynamic data exchange (DDE) is a form of inter-process communication (IPC) in the Microsoft Windows operating environment. When two or more applications that support DDE are running simultaneously, they can exchange information, data and commands. DDE has been enhanced with object linking and embedding (OLE) technology.

Driver Software that enables a computer to communicate with a specific client device, such as a printer.

Firmware Software application or sequences of instructions, that is hard-coded into non-volatile memory devices. Firmware is essentially software that is embedded in memory and does not change.

J2EE Java 2 Platform, Enterprise Edition, takes advantage of many features of the Java 2 Platform, Standard Edition, such as portability, JDBC for database access, CORBA support, and a security model. J2EE adds support for Enterprise JavaBeans, JavaServer Pages, Java Servlets, and XML

JVM The Java Virtual Machine (JVM) is the cornerstone of the Java programming language. It is the component of the Java technology responsible for cross-platform delivery. The JVM is an abstract computing machine and (just like a real computing machine) has a defined instruction set. The JVM knows nothing of the Java language, only of a particular file format, the class file format. A class file contains JVM instructions (or bytecodes) and a symbol table, as well as other required information.

Kernel During normal operations of a computer system, some portions of the OS remain in main the memory to provide critical operations services, such as interrupt handling, dispatching, or (critical) resource management. These portions of the OS are collectively called the operating system kernel.

Operating system Set of software instructions that controls the interaction of the computer's hardware components so that they work together in an orderly and efficient way; the user interface (U.I.); and the interaction between the computer and software applications and other hardware devices.

Software license Rights granted by the publisher of a commercial software product that establish the terms of use for that product. Software licenses may stipulate how many users or devices can use the software.

Review Questions

3.1 What are the types of operating systems?
3.2 Differentiate between an operating system and real-time operating system.
3.3 Compare the various versions of Windows operating systems.
3.4 What are the variants of Linux operating system?
3.5 Discuss the higher level languages.
3.6 Compare the real-time operating system with conventional operating system.

Projects

3.1 Visit a couple of IT firms and understand the various operating systems that are deployed. Prepare a report of the same.
3.2 You are aware ABC Unlimited is moving to a new data centre. As you will come to know ABC is planning to implement ERP and CRM solutions to take care of their supply chain. It is also interested to have real-time view of the IT infrastructure.
3.3 Suggest suitable selection of operating systems which provide a good availability, and return on investment (ROI). Justify your selection.

References

Adams, Jeanne C. et al. 2003, *The Fortran 2003, Handbook: The Complete Syntax, Features and Procedures*, Springer.

Dandamudi, Sivarama P. 2005, *Introduction to Assembly Language Programming For Pentium and RISC Processors*, Springer.

Glass, Graham, King Ables 2006, *Linux For Programmers And Users,* Prentice Hall.

Gomaa, H. 1993, *Software Design Methods for Concurrent and Real-time Systems,* First Edition, Addison-Wesley.

Laplante, P.A. 1997, *Real-Time Systems Design and Analysis: An Engineer's Handbook,* Second Edition, IEEE Press.

Lytras, Miltiadis 2007, *Open Source for Knowledge and Learning Management: Strategies Beyond Tools,* Idea Group.

Silberschatz, A., P.B. Galvin, and G. Gagne 2001, *Operating Systems Concepts,* Sixth Edition, John Wiley.

Walls C. 1996, *RTOS for Microcontroller Applications,* Electronic Engineering, vol. 68, no. 831, pp. 57-61.

CHAPTER 4

Networking Infrastructure

Computing is rapidly approaching an inflection point where science fiction writers' predictions of helpful, ubiquitous and seamless technology will become a reality.
–Richard Rashid
Senior Vice-President, Microsoft Research

> **Learning Objectives**
>
> After reading this chapter, you should be able to understand:
> - the need for network infrastructure
> - the basic elements of network infrastructure
> - the differences between the various network infrastructure

4.1 Introduction

The popularity of the Internet has caused the traffic on the Internet to grow drastically every year for the last several years. Increased traffic envisages careful capacity planning and crafted information and communication technology (ICT) architecture. Architecture can be viewed at various levels, including hardware, network, system, application, and enterprise level. Planning a network is a little more complex that sitting down with a piece of paper and drawing lines between the cheapest boxes, yet this is basically the way a lot of networks are planned. The most common errors made in planning the average corporate network are in relation to capacity planning, or provisioning in telecommunications terms. Unclear capacity planning results in the proliferation of incompatible systems leading to legacy issues. The ICT infrastructure of a company is equally fraught with legacy issues. Mergers and acquisitions often result in legacy systems. The challenges arising from unplanned growth in legacy architecture of a firm results as impediments to business process clarity and capacity to change business processes to reflect new business models. Chief information officers (CIOs) and ICT managers are very sensitive to the risks of making sudden changes in systems due to mission critical dependency of businesses. Maintaining and upgrading legacy systems is one of the most difficult challenges CIOs face today. Constant technological change often weakens the business value of legacy systems, which have been developed over the years through huge investments. Chief information officers struggle with the problem of modernizing these systems while keeping their functionality intact. In fact, the strategic use of technology within the enterprise is no longer just an issue for the chief information officer or chief technology officer (CTO). Business spending on technology is such a large part of corporate capital and ongoing expenditure that all business executives, including the chief executive officer (CEO) and board members, need to have a clear understanding of how it can be most effectively leveraged and exploited within their organization.

As technology continues to evolve at an accelerating rate, non-trivial hardware and software will remain diverse and heterogeneous. There is a need for a more interoperable and flexible ICT architecture that can more efficiently and effectively meet these future requirements. All business applications rely on a stable, secure, functional, scalable, and well-performing network infrastructure to balance the networking ecosystem where an ecosystem is a collection of interdependent networking elements. Often one network is built per application. It is important to establish a general physical IT infrastructure in which the capacity of the network can be successively increased by upgrading the communications equipment.

The network infrastructure is an important component in a network design. It is important simply because, at the end of the day, it is those wires that carry the information. A well-thought-out network infrastructure not only provides reliable and fast delivery of that information, but it is also able to adapt to changes, and grow as your business expands. Building a network infrastructure is a complex task, requiring work such as information gathering, planning, designing, and modelling. Though it deals mainly with bits and bytes, it is more of an art than a science, because there are no fast rules to building one. It is, therefore, important to understand the fundamentals of how data is transmitted in an IP network, so that the difference in how the various technologies work can be better understood.

In the emerging environment of high-performance IP networks, it is expected that local and campus area backbones, enterprise networks, and Internet service providers (ISPs) will use multi gigabit and tera-bit networking technologies where IP routers and switches will be used not only to interconnect backbone segments but also to act as points of attachments to high performance wide area links.

4.2 Protocols—The Communication Enabler

In information technology, a protocol is the special set of rules that end points in a connection use when they communicate. The protocols provide a standardized way for computers to format and transmit data to one another. A set of formal rules governs the exchange of information amongst the networked peers, and the communication activity fails if the protocol is not correctly followed. The exchange of data may take place between two or more parties. When there are n parties to the exchange, we shall talk of N-peer communication, and speak of the protocol as an N-peer protocol. Protocols ensure that computers communicating over data communications lines speak the same language. A variety of protocols work in unison to achieve a high rate of communication amongst peers. Each group of protocols that work together is known as a protocol stack.

The procedures used to transfer messages over a network are known as network protocols, specifications of how a computer will format and transfer data. If a computer contains implementations of a set of protocols, theoretically it can communicate with any other computer that has implementations of the same protocols. It is possible for one computer to use more than one protocol stack at the same time.

4.3 Open System Interconnection

To understand and appreciate the networking technology, a brief overview and understanding of fundamental concepts is presented in the successive sections.

A computer system communicates with another system by sending a stream of bytes. A byte is a sequence of 8 bits. A checksum is the arithmetic sum of a sequence of numbers used to detect errors that may have altered some of the numbers in the sequence. The communication is actually between a process running on one system with one running on the other system. The two processes communicate information in a pre-agreed form known as protocol.

For our purposes, a network is a composition of communication devices and links that connect at least two nodes that consist of hardware and software. These connected communications devices and links perform interactions between nodes using exact prescriptions, including protocols.

The open system interconnection (OSI) architecture and family of standards, which specify the services and protocols for interchange of information between systems, have been defined to provide an operating environment for the implementation of distributed information systems from a multi-vendor marketplace. Interchange of information is in digital form and can convey data, voice, and image communications. The interconnected telecommunication resources can be dedicated transmission paths or switched services on a demand basis. Switched paths interconnecting communicating users can be on a fully reserved basis or on a demand basis using various switching technologies. Computer networking is easier to understand as a stack of layers, each layer providing the functionality needed by the layer above it. The OSI model of computer networks has seven layers. Each layer provides functionality that the next higher layer depends on. During the 1970s, the International Organization for Standards (ISO) established a framework for standardizing communications systems. This framework was called the open system interconnection reference model and defines an architecture in which communication functions are divided into seven distinct layers, with specific functions becoming the responsibility of a particular layer. Open system interconnection gives users of data networks the freedom and flexibility to choose equipment, software, and systems from any vendor. It aims to sweep away proprietary systems which oblige a user to build a system with kit from a single vendor. It is a concept which relies upon the emergence of common standards to which components and systems must conform.

Open system interconnection standards define only those functions that are necessary to facilitate communications between systems. They do not describe the specific implementation, design, or technology that is used. These are left to the innovation of the system developer.

4.3.1 N-layer Service

The OSI reference model may be thought of as a series of conceptual layers. Layer N provides service N to layer N+1. In order for layer N to fulfil its service to layer N+1, it in turn requests a service from layer N-1 (Figure 4.1) The interfaces or boundaries between adjacent layers or services are known as N-layer Service Access Points (SAPs).

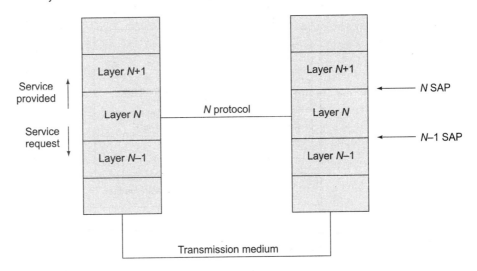

FIGURE 4.1 N-Layer Service

4.3.2 OSI Layers

The physical layer represents the lowest layer in the OSI reference model (Figure 4.2). The physical layer is responsible for specifying the electrical and physical connection between communication devices that connect to different types of media.

The second layer in the OSI reference model is the data link layer. This layer is responsible for defining the manner by which a device gains access to the medium specified in the physical layer.

The third layer, the network layer, is responsible for arranging a logical connection between a source and destination on the network to include the selection and management of a route for the flow of information between source and destination based on the available paths within a network. This layer provides the IP protocol.

The fourth layer, the transport layer, is responsible for governing the transfer of information after a route has been established through the network

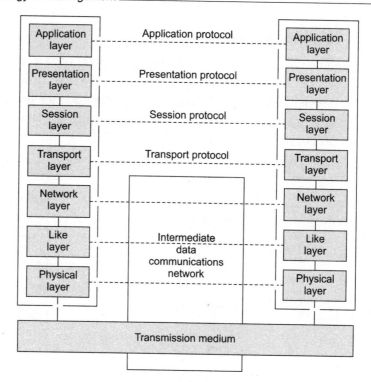

FIGURE 4.2 OSI

by the network layer protocol. There are two general types of transport layer protocols: connection-oriented and connectionless. A connection-oriented protocol first requires the establishment of a connection prior to data transfer occurring. A second type of transport layer protocol operates as a connectionless and best-effort delivery mode.

The fifth layer, the session layer, is responsible for providing a set of rules that govern the establishment and termination of data streams flowing between nodes in a network.

The sixth layer, the presentation layer, is concerned with the conversion of transmitted data into a display format appropriate for a receiving device.

The seventh layer, the application layer, functions as a window through which the application gains access to all of the services. Examples of functions performed at the application layer include electronic mail, file transfers, resource sharing, and database access.

4.4 TCP/IP Suite

Transmission control protocol (TCP) and Internet protocol (IP) were developed by a US Department of Defense (DOD) research project to connect a number different networks designed by different vendors into a network of networks. The Internet and the World Wide Web are based on TCP/IP (Figure 4.3). The term 'TCP/IP' refers to not only TCP and IP, but also includes other protocols, applications, and even the network medium. The TCP/IP protocols have been designed to operate over nearly any underlying local or wide area network technology. Although certain accommodations may need to be made, IP messages can be transported over all of the technologies. The Internet protocol (RFC 791), provides services that are roughly equivalent to the OSI network layer.

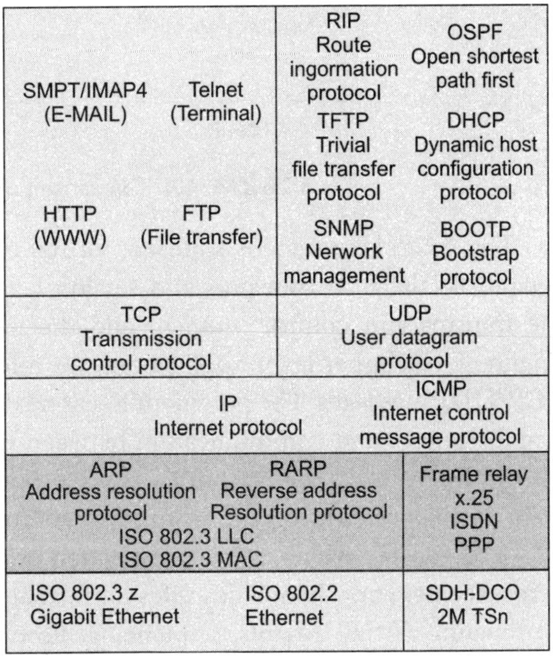

SMPT/IMAP4 (E-MAIL)	Telnet (Terminal)	RIP Route ingormation protocol	OSPF Open shortest path first
		TFTP Trivial file transfer protocol	DHCP Dynamic host configuration protocol
HTTP (WWW)	FTP (File transfer)	SNMP Nerwork management	BOOTP Bootstrap protocol
TCP Transmission control protocol		UDP User datagram protocol	
IP Internet protocol		ICMP Internet control message protocol	
ARP Address resolution protocol	RARP Reverse address Resolution protocol		Frame relay x.25 ISDN PPP
ISO 802.3 LLC ISO 802.3 MAC			
ISO 802.3 z Gigabit Ethernet	ISO 802.2 Ethernet		SDH-DCO 2M TSn

FIGURE 4.3 TCP/IP Suite

The Internet protocol provides IP addresses which are hierarchical for routing purposes and are subdivided into two subfields. The Network Identifier (NET-ID) subfield identifies the TCP/IP subnetwork connected to the Internet. The NET-ID is used for high-level routing between networks, much the same way as the country code, city code, or area code is used in the telephone network. The host identifier (HOST-ID) subfield indicates the specific host within a subnetwork. To accommodate different size networks, IP defines several address classes. Classes A, B, and C are used for host addressing and the only difference between the classes is the length of the NET-ID subfield.

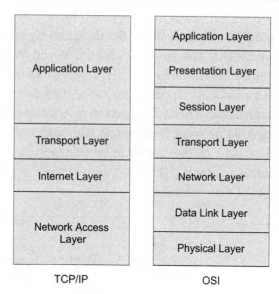

FIGURE 4.4 Comparison of OSI and TCP/IP Suite

The TCP/IP protocol suite comprises two protocols that correspond roughly to the OSI transport and session layers; these protocols are called the transmission control protocol and the user datagram protocol (UDP) (Figure 4.4). Higher-layer applications are referred to by a port identifier in TCP/UDP messages. The port identifier and IP address together form a socket, and the end-to-end communication between two hosts is uniquely identified on the Internet by the four-tuple (source port, source address, destination port, destination address). Transmission control protocol, described in RFC 793, provides a virtual circuit (connection-oriented) communication service across the network. It includes rules for formatting messages, establishing and terminating virtual circuits, sequencing, flow control, and error correction. Most of the applications in the TCP/IP suite operate over the reliable transport service provided by TCP.

For additional information and insight, readers are urged to read two excellent histories of the Internet: *Casting The Net: From ARPANET to INTERNET and Beyond...* by Peter Salus (Addison-Wesley, 1995) and *Where Wizards Stay Up Late: The Origins of the Internet* by Katie Hafner and Matthew Lyon (Simon and Schuster 1997)

4.4.1 Ethernet

Ethernet was invented in 1972 at Xerox PARC by Metcalfe and his colleagues. They named it so as to emphasize the capability of the physical medium to carry bits to all hosts, analogous to the 'luminiferous ether' of old physics. Although Ethernet is no longer a new technology, it has the potential to solve some of legacy data access methods' biggest deployment and performance headaches. Access to the shared medium is governed by the MAC mechanism based on the carrier sense multiple access with collision detection (CSMA/CD) system. This ensures that access to the network channel is fair, and that no single host can lock out other hosts.

Ethernet interface is implemented in an electronic hardware called network interface card (NIC). All networked devices come with this electronic hardware interface. In a shared media network, one station sends a message to another station in an envelope called packet. The packet consists of message data surrounded by a header and trailer that carry special information used by the network software. The packet is delivered to NIC. The NIC then transmits the packet onto the local area network (LAN). The packet is transmitted as a stream of data bits represented by changes in electrical signals. As it travels along the shared cable, all of the stations attached to it see the packet. As it goes by the NIC in each station, the NIC checks the destination address in the packet header to determine if the packet is addressed to it. When the packet passes the station it is addressed to, the NIC at that station copies the packet and then takes the data out of the envelope and gives it to the computer.

Fast Ethernet (100Base-T) is an extension to the IEEE 802.3 Ethernet standard to support service at 100M bps. It is virtually identical to 10Base-T in that it uses the same media access control layer, frame format, and carrier sense multiple access with collision detection (CSMA/CD) protocol. This means that network managers can use 100Base-T to improve bandwidth and still make maximum use of investments in equipment, management tools, applications, and network support personnel. Gigabit Ethernet is a logical backbone technology to connect existing networks built with both Fast Ethernet and 10Base-T. Gigabit Ethernet extends the ISO/IEC 8802-3 Ethernet family of networking technologies beyond 100M bps to 1000M bps. Gigabit Ethernet is an expansion of the popular 10 Mbps (10BASE-T) Ethernet and 100 Mbps (100BASE-T) Fast Ethernet standards for network connectivity. It builds on top of the Ethernet protocol, and increases speed tenfold over Fast Ethernet to 1000 Mbps, or one gigabit per second (Gbps). This protocol was introduced by the 10 GEA (Gigabit Ethernet Alliance) in June 1998, mainly for use in high-speed local area network backbones and server connectivity.

The main purpose of Gigabit Ethernet is to build on the existing installed base of Ethernet and fast Ethernet. The idea is to facilitate a high-speed network without forcing customers to throw away existing networking equipment. The initial market for gigabit Ethernet will be primarily to upgrade network infrastructures by providing high bandwidth links for backbones and server connections. A second market consists of vertical markets with specialized applications that require high bandwidth to the end user station.

While most CIOs are not bothered about wide area network (WAN) bottlenecks bottlenecks, they are not averse to the idea of deploying WAN optimization technologies in the future.

4.5 LAN and WAN

On the information technology side, there has been a focus on improved business management of IT in order to extract the most value from existing resources, and a general realignment of business and IT priorities from those of previous years. Today's businesses are focused on defending and safeguarding their existing market positions in addition to targeting market growth. Cost-constrained businesses are generally focused on achieving more from the same amount of resources in terms of people, knowledge, and systems, and in optimizing their existing operations and business processes. The rising adoption of enterprise applications such as enterprise resource planning (ERP), supply chain management (SCM), customer relationship management (CRM) and databases along with the growth of customers, geographical locations, business units, sales personnel, and business executives has made a nationwide enterprise WAN, or a connectivity backbone a necessity for enterprises.

Local area networks are defined in terms of the protocol and the topology used for accessing the network. Prior to the development of LAN technology, individual computers were isolated from each other and limited in their range of applications. By linking these individual computers over LANs, their usefulness and productivity have been increased enormously. But a LAN by its very nature, is a local network confined to a fairly small area such as a building or even a single floor of a building. However, today's LANs and LAN internetworks are powerful, flexible, and easy to use technologies and incorporate many sophisticated mechanisms that must work together flawlessly.

Local area networks are deployed based on deployment topologies. Some of the topologies are as shown in Figure 4.5.

A gateway provides a LAN with access to a different type of network, an internetwork, a mainframe computer, or a particular type of operating

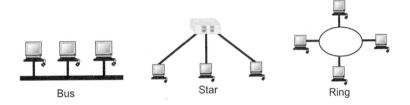

FIGURE 4.5 Deployment Topologies

environment. A gateway serves to connect networks with very different architectures. Gateways also provide protocol conversions, since the different environments connected by a gateway will generally use different protocol families. Gateways are often bottlenecks in network communications.

University campus networks or networks in large companies are examples of LAN networks.

The most widely used LAN technology today is Ethernet. It strikes a good balance between speed, price, ease of installation, and supportability. Approximately 80 per cent of all LAN connections installed use Ethernet. The most common way to connect machines into a LAN is to use Ethernet.

Topology is the configuration formed by the connections between devices on a LAN or between two or more LANs. Topology can include such aspects as the transmission media, network adapters, and physical design of the network. A network topology is the method in which nodes of a network are connected by links. A given node has one or more links to others, and the links can appear in a variety of different shapes. The simplest connection is a one-way link between two devices. A second return link can be added for two-way communication. Modern communication cables usually include more than one wire in order to facilitate this, although very simple bus-based networks have two-way communication on a single wire.

Topologies specify which of these devices are used to connect systems on the network. The most common topologies are bus, star, ring, and mesh (Figure 4.6). There are advantages and disadvantages for each type of topology, and careful consideration should be used when choosing which type of network to install. The characteristics of our topology determine how the network functions and affects aspects such as installation and troubleshooting of the network.

As data centre proliferate and companies shift to a centralized set-up, the networking ecosystem consists not just of servers that process and dispatch data

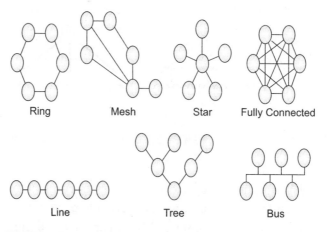

FIGURE 4.6 Topology

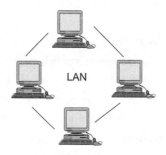

FIGURE 4.7 Local Area Network

over the network, but also of intermediate routers, file servers, application servers, and satellite communications. Network bottlenecks are a common occurrence. Bottlenecks occur for a variety of reasons but a common one is improper design of the network and bloating applications. Remote locations linked to the data centre by WANs are under pressure to support the sheer volume of packets that need to be transported. When a packet is transported over the network, latency comes into play contributing to bottlenecks. Latency worsens when there are multiple technologies communicating within a network. Applications are generally designed to work in a LAN (Figure 4.7) environment. A topology of a network shows the locations of nodes and links of a network.

The requirements and choices of WAN technologies are different from LAN technologies. The main reason is that WAN technologies are usually a subscribed service offered by carriers, and they are very costly. A WAN

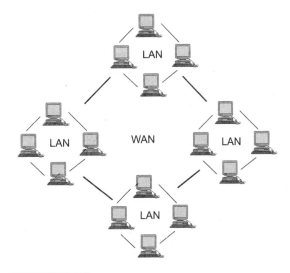

FIGURE 4.8 Wide Area Network

(Figure 4.8) also differs from LAN technologies in the area of speed. It covers larger distances and hence latency also becomes a critical factor when dealing with it.

One of the many debates around technology in the business community has been the question of the extent to which technology has actually improved productivity, both within individual businesses and on a macroeconomic basis. An important point to bear in mind is that new technology affects the economy only when it has been broadly adopted and utilized.

4.6 Communication Medium

Business executives and CIOs are placing more emphasis on the business management of information technology, i.e., placing more controls on how IT departments are managed and which initiatives get funded. With large percentages of organizations' capital and recurring expenditures going into computing infrastructure and applications, IT departments are increasingly required to justify each and every investment and maximize their returns. To run IT departments like businesses, CIOs need to know what they have in place already in terms of people and technology resources, and they need to minimize their cost of ownership. One of the recurring expense identified in any Internet-enabled organization is towards bandwidth and connectivity. Clear understanding of connectivity options will enable the CTO to plan his/her connectivity.

The average number of nodes on a network segment has decreased dramatically, while the number of applications and the size of the data transferred has increased dramatically. Applications are becoming more complex, and the amount of network bandwidth required by the typical user is increasing. Bandwidth is a measure of the capacity of the medium to transmit data. The higher the bandwidth, the faster the data can be injected while maintaining acceptable error rates at the point of reception.

4.6.1 ISP

The Internet is a large network formed out of more than 30,000 autonomous systems (AS), in which each AS is a collection of IP networks sharing a common routing strategy. These networks are operated by thousands of Internet Service Providers (ISPs). The term Internet service provider is used as an umbrella term for information providers, server service providers, Internet network service providers. On one hand, the ISPs compete with each other for customers and traffic, on the other, they have to cooperate and exchange traffic, otherwise the worldwide connectivity would be lost. The ISP market is characterized by a huge diversity of offered services and business connections, differing significantly from the traditional telecommunications market. Progression of Internet technologies, business innovations, and regulatory and policy factors are adding to the complexity. The diversity of ISP market services and interactions is also reflected by the companies involved, ranging from niche market ISPs to global players.

The ISP market is characterized by serious competition and is currently in a phase of consolidation. Never before had ISPs been as competitive as they have to be today. It is highly important for ISPs to operate their network efficiently. In addition, they have to strive for successful business practices. Traditional successful business practices in a competitive environment are cost leadership, market segmentation, and differentiation.

End-users, for example, access the access provider's network via a modem or digital subscriber line (DSL) connection to a point-of-presence (POP) or a virtual point-of-presence (VPOP) of the access provider. A POP can be described as a node in the ISP's network topology. The routers, switches, servers, and other equipment of an ISP are located at its POPs. Typically, these POPs are geographically distributed to keep the distances to customers and interconnection partners short. The size of an Internet network service provider (INSP) is often measured by the number of POPs it is operating.

The network medium, which physically carries data packets from one point to another, happens to be the most critical component in WANs. Reliability of the network hinges around reliability of the medium, as it is the weakest link in the overall network architecture. Average WAN links in at most organizations are in the range of 2 Mbps.

The Merriam-Webster's Dictionary defines efficiency as an effective operation as measured by a comparison of production with cost (as in energy, time, and money). In the context of network services, the "production" of an ISP's network can be described by the amount of traffic transported by the ISP's. Therefore, we define the efficiency of a network as

$$\text{Network efficiency} = \frac{\text{Transported traffic}}{\text{Costs}} \quad (4.1)$$

Depending on the level of abstraction, the traffic can be measured by

- the volume of traffic carried through the network,
- the number of flows or sessions transported through the network, or
- the number of customers served.

Internet being a heterogeneous network of networks; the ISPs that control these networks differ from very small regional niche providers to huge multinational backbone providers. Where along the optimal performance boundary an individual ISP should operate, is basically its own decision.

There are a variety of options providing the requisite bandwidth for various business requirements across the industry. Some of the technologies are listed below.

4.6.2 Digital Subscriber Line (DSL) Network

Since 1980, public switched telephone network (PSTN) users have begun to use their subscriber lines to access data network, and more particularly the Internet. For that purpose, a voiceband modem must be installed at the customer premises. A residential user accesses the Internet by means of an ISP. In the most current configuration, such access is carried out by means of an analog voiceband modem, also called a dial-up modem, to connect a PC to a subscriber line. The DSL technology is a way of transporting data over a normal phone line at a higher speed. The xDSL technology is capable of providing a downstream bandwidth of 30 Mbps and an upstream bandwidth of around 600 kbps. But in commercial deployment, it is usually 1.5 Mbps downstream and maybe 256 kbps upstream. Subscribers of xDSL technology are connected to a device called multiplexer (MUX) in a point-to-point manner. The MUX aggregates a number of subscribers (usually 48, some may go as high as 100) and has an uplink to a networking device, typically a switch.

In this context, access to the Internet means access to the network of a regional Internet carrier by way of a connection through a PSTN. Once the user has typed on his or her keyboard the uniform resource locator (URL) of the web server to which he or she wants to be connected, a phone connection is automatically set up by the user's voiceband modem to connect his or her PC to a network access server (NAS).

Network access servers, which are controlled by ISPs, carry out three types of functionalities. The first one is to control the identity of residential users before allowing them to be connected to the distant web server. The second one is to record eventually the duration of the connection of end users to a web server. In most cases, this second functionality is not activated, as most of the ISPs offer non-metered access. The third functionality of a NAS is to allocate a temporary IP address to the end-users.

4.6.3 Leased Lines

Traditionally, the leased line has been the preferred mode of wireline connectivity between company locations in different cities. It is basically a permanent circuit leased from the carrier and connects in a point-to-point manner. The leased line technology has been around for quite some time and many network managers are familiar with it. Leased lines provide the last mile access from the user premises to the ISP. They provide permanent connection as compared to the temporary connectivity through dial-up access. Leased lines are typically 64 or 128 Kbps links going up to E1 and beyond. A leased line is an ideal medium for bandwidth hungry and low-latency applications. However WAN connectivity through leased lines had several disadvantages. The lines were costly and one had to apply for a leased line and wait for a connection based on ISP's backbone availability. Leased lines have bandwidth slabs and cannot be shared on a network basis. Thus, the capacity is under-utilized. Leased line remains the preferred medium for connectivity among enterprise users, followed by ISDN, dial-up, DSL, satellite, and cable.

Leased line provides permanent, reliable, high-speed connectivity as compared to the temporary connectivity of dial-up access. The quality of the connection is far superior to what is normally available through dial-up, because of the digital signaling, less noise, fewer exchanges, etc.

4.6.4 Integrated Services Digital Network

Integrated services digital network (ISDN) is a subscribed service offered by phone companies. It makes use of digital technology to transport various information, including data, voice and video, by using phone lines. There are two types of ISDN interfaces, the basic rate interface (BRI) and the primary rate interface (PRI). The BRI provides 2 x 64 kbps for data transmission (called the B channels) and 1 x 16 kbps for control transmission (called the D channel). Integrated services digital network provides a 'dial-on-demand' service that means a circuit is only connected when there is a requirement for it. The charging scheme of a fixed rate plus charges based on connections makes ISDN ideal for situations where a permanent connection is not necessary.

> One of the national banks in the country needed to share data between its head office, nationwide offices, branches, and ATMs. The bank decided to link its nationwide locations on a WAN, and share data across systems. The bank had a widespread reach. Considering the organization's size, the entire exercise was no mean task.
>
> The bank came up with a layered design. The design is such that every RO is connected to the HO through a 2 Mbps primary leased line link, and multiple 64 Kbps leased lines. This adds redundancy in the connectivity between the HO and the ROs, and acts as a fail-over path to the primary link.
>
> At the lower layer, the nationwide branches connected to the nearest RO with 64 Kbps leased lines. This reduced the per-branch cost of deploying a leased line link to the HO. Along with the CTO, the top management was involved in the design and deployment of the project, right from the first phase. This helped to resolve all critical issues at the earliest. The bank has enabled the network to offer ATM services to a vast majority of its customers.

4.6.5 Terrestrial Microwave

Terrestrial microwave links are widely used to provide communication links when it is impractical or too expensive to install physical transmission media, for example, across a river or perhaps a swamp or desert. As the collimated microwave beam travels through the Earth's atmosphere, it can be disturbed by such factors as manufactured structures and adverse weather conditions. With a satellite link, on the other hand, the beam travels mainly through free space and is therefore less prone to such effects. Nevertheless, line-of-sight microwave communication through the Earth's atmosphere can be used reliably for the transmission of relatively high bit rates over distances in excess of 50 km.

4.6.6 Very Small Aperture Terminal

In a large network, sharing of bandwidth helps in optimal use of resources. Very small aperture terminal (VSAT) technology provides speedy, cost-effective, and reliable solutions to the organization. A VSAT is a small fixed earth station which provides a communication link required to setup a satellite-based communication network. It represents a cost-effective solution for those who want an independent communication network to connect a large number of geographically dispersed sites, especially sites where any other connectivity options are not possible to implement. As many business verticals are beginning to reach closer to the consumer in the remotest corners of the country, they find VSAT the most suitable medium considering its ability to provide ubiquitous coverage, high reliability, ease of network management, and cost-effectiveness.

Availability of satellite bandwidth is critical in the growth of VSAT market; demand for high capacity satellites preferably in Ka band is on the rise. Very small aperture terminal is used in remote places, hilly terrain areas, and for places that are not feasible to be connected through wireline.

4.7 Racks

Rack is a standardized (EIA 310-D, IEC 60297 and DIN 41494 SC48D) frame or enclosure for mounting multiple electronic modules. It is measured in rack Us. A rack unit or U is an electronic industries alliance (EIA) or more commonly EIA standard measuring unit for rack mount type equipment. It is equal to 1.75" in height. Most of the racks are sold in the 42U form.

Equipment designed to be placed in a rack is typically described as rack-mount equipment. The various rack mountable electronic modules are standardized in terms of rack Us. Rack-mountable equipment are mounted simply by bolting its front panel to the rack. Heavier equipment is designed to use a second pair of mounting strips as applicable.

4.8 Networking Elements

Keeping track of IT assets is becoming increasingly difficult as the number and variety of computing devices and applications proliferate. Employees now utilize a variety of devices including personal computers, laptops, personal digital assistants, cell phones, and pagers. The computer and communication network is composed of multiple elements. A brief overview of the various prominent elements are discussed in the following sections.

4.8.1 Network Interface Card

A computer is connected to the network with an adapter called a network interface card (NIC) (Figure 4.9). Typically, the NIC is installed inside the computer. It plugs directly into one of the computer's internal expansion slots. The card usually also contains the protocol control firmware and Ethernet Controller needed to support the Medium Access Control (MAC) data link protocol used by Ethernet.

The MACs on the network are used to direct traffic between the computers. The back plate of the network interface card features a port that looks similar to a phone jack, but is slightly larger. This port accommodates an Ethernet cable, which resembles a thicker version of a standard telephone line.

It is both an OSI layer 1 (physical layer) and layer 2 (data link layer) device, as it provides physical access to a networking medium and provides a low-level addressing system through the use of MAC addresses. Every Ethernet network card has a unique 48-bit serial number called a MAC address, which is stored in read only memory carried on the card. Every computer on an Ethernet network must have a card with a unique MAC address.

FIGURE 4.9 Network Interface Card

4.8.2 Overview of Routers

A router is a hardware device in computer networking that forwards data packets to their destinations, based on their addresses. It routes data from a LAN to another network connection (Figure 4.10). Generally, routers consist of several network interfaces to the attached networks, processing module(s), buffering module(s), and an internal interconnection unit called switching fabric. Typically, packets are received at an inbound network interface (line card), processed by the processing module and, possibly, stored in the buffering module. Then, they are forwarded through the internal interconnection unit to the outbound interface that transmits them to the next hop on their journey to the final destination. The interface cards consist of adapters that perform inbound and outbound packet forwarding. The router backplane is responsible for transferring packets between the cards. Routing protocols are used by routers to gain information about the network.

Line cards are a category of cards that contain the logical functions for processing data. A line card contains functions that can be associated with a port or interface card, a processor card, and optionally a traffic management engine, a segmentation and reassembly (SAR) engine, and a switch fabric interface.

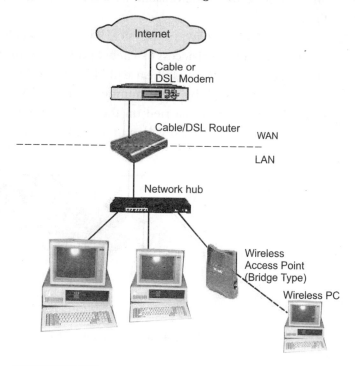

FIGURE 4.10 Router Connected in a Network

In general, routers can be classified into one of two design categories: single-board or modular. Single-board design, typically, are called 'pizza boxes,' whereas the modular designs consist of midplane or backplane architectures. Their generalized deployment scenarios differ significantly, and their prices also differ. Single-board designs typically trade the ultimate throughput, reliability, and serviceability for cost. While the focus of single-board designs clearly is on providing advanced functions at a low cost, the modular designs highest priority is high availability at high throughput.

Routers with modular architectures share a variety of design elements. They both are deployed in environments in which high availability and high throughput are of crucial importance. Enterprise routers supporting moderate to large numbers of ports operating at gigabit data rates and higher need to process tens to hundreds of millions of packets per second in real time. Chassis-based architectures are predominant in applications where it is not yet possible to integrate all functions into one board, or where system uptime requirements make redundant and modular systems a necessity. Chassis-based routers fall into two categories: backplane and midplane architectures. Midplane designs and backplane designs are typically used for routers that are deployed in corporate data centers or central offices. The midplane design is preferred when there is a common processor card design for a wide variety of line or interface cards. Backplane design type routers are typically installed in temperature-controlled environments. Since the power consumption and therefore the heat dissipation of the modules is significantly higher than in other architectures, a reasonably designed chassis will incorporate a few temperature sensors in the fan trays, the power supplies, and near other known heat sources.

Routers and Routing

The Internet can be described as a collection of networks interconnected by routers using a set of communication standards known as TCP/IP suite. The IP provides the glue that connects multiple networks to form a single internetwork, where each pair of hosts connected to the internetwork can exchange IP datagrams. The systems that connect networks are called IP routers, or simply routers (Figure 4.11). Routers form the basic infrastructure for all IP networks. A router generally connects with a set of input links through which a packet can come in and a set of output links through which a packet can be sent out. The Internet comprises a mesh of routers interconnected by links, in which routers forward packets to their destinations, and physical links transport packets from one router to another. Because of the scalable and distributed nature of the Internet, there are more and more users connected to it and more and more intensive applications over it.

Routing is the process of moving packets through an internetwork. Each packet contains a destination IP address; the packet has to follow a path through the Internet to its destination. Once a router receives a packet at

FIGURE 4.11 A View of Router
(*Source*: Cisco.com)

an input link, it must determine the appropriate output link by looking at the destination address of the packet. The packet is transferred router by router so that it eventually ends up at its destination. Therefore, the primary functionality of the router is to transfer packets from a set of input links to a set of output links.

Routing Software

Router can also be viewed as a computer whose software and hardware are usually tailored to the tasks of routing and forwarding information. The software contained within the router is a specialized operating system. The software is also called as firmware.

Routing can be accomplished by manually entering the information necessary for packets to reach any part of the inter-network into each router. This is called static routing. In static routing, the routing table entries are created by default when an interface is configured. With static routing, a router may issue an alarm when it recognizes that a link has gone down, but will not automatically reconfigure the routing table to reroute the traffic around the disabled link.

Dynamic routing, on the other hand, uses routing protocols to create and maintain the routing tables automatically. Dynamic routing responds more quickly to network changes (and network failures) than static routing. Dynamic routing, used in internetworking across WANs, automatically reconfigures the routing table and recalculates the least expensive path.

Know Your Firm–Cisco

Computer scientists, Len Bosack and Sandy Lerner, from Stanford University, found Cisco Systems. At Cisco, customers come first creating long-lasting customer partnerships and working with them to identify their needs and provide solutions that support their success. Beginning in 1997 and the reality that voice and video would all be one, moving to the networks of networks in 2000 and the network becoming the platform for all related technologies and the core of customer solutions, and the most recent market transition of collaboration and Web 2.0 technologies.

Contd

- 1986: Cisco forever changes the networking communications industry and the Internet by launching its first routing innovation, the AGS multi-protocol router.
- 1989: With only three products and 111 employees for FY89, Cisco reports revenues of 27 million dollars.
- 1990: Cisco goes public on February 16, 1990, listed as "CSCO" on the NASDAQ. Cisco celebrates the public offering with a market capitalization of 224 million dollars.
- 1997: Cisco launches Networking Academy program for high school and college students to learn how to design, install, and maintain computer networks, and the Cisco Powered Network program for service providers offering end-to-end, high-quality network services based on Cisco equipment.
- 2000: Cisco earns a key patent in IP technology, No. 6,097,718, for a method and system for maintaining and updating routing information in a packet switching network for a set of quasi-dynamic routes, in which intermittent routing updates are permitted, so that routes are no longer "always static" or "always dynamic," but may change over time between static and dynamic, and are treated accordingly.
- 2006: Cisco selects India as site for the Cisco Globalization Center.
- 2008: Cisco takes consumers beyond connectivity to a 'Visual Networking' with Internet Protocol Next-Generation Home Gateway.

Source: www.cisco.com

Following are a few of the competing open source routing software packages.

Quagga Quagga is a routing software suite, providing implementations of OSPFv2, OSPFv3, RIP v1 and v2, RIPv3 and BGPv4 for Unix platforms, particularly FreeBSD, Linux, Solaris, and NetBSD. Quagga is a fork of GNU Zebra which was developed by Kunihiro Ishiguro. The Quagga tree aims to build a more involved community around Quagga than the current centralised model of GNU Zebra.

XORP XORP is the eXtensible Open Router Platform. The goal of the XORP project is to develop an open source software router platform that is stable and fully featured enough for production use, and flexible and extensible enough to enable network research. Currently, XORP implements routing protocols for IPv4 and IPv6 and a unified means to configure them. XORP is a free software. It is covered by a BSD-style license and is publicly available for research, development, and use.

Zebra GNU Zebra is a free software that manages TCP/IP based routing protocols. Zebra is released as part of the GNU Project, and is distributed under the GNU General Public License. Zebra supports the BGP-4 protocol as described in RFC1771 (a border gateway protocol 4) as well as RIPv1, RIPv2, and OSPFv2. Unlike traditional, monolithic architectures and even the so-called 'new modular architectures' that remove the burden of processing routing functions from the CPU and utilize special ASIC chips instead, Zebra software offers true modularity.

OpenBGPD OpenBGPD is a free implementation of the border gateway protocol, Version 4. It allows ordinary machines to be used as routers exchanging routes with other systems speaking the BGP protocol. Started out of dissatisfaction with other implementations, OpenBGPD nowadays is a fairly complete BGP implementation, powering many sites. Users often praise its ease of use and high performance, as well as its reliability. OpenBGPD's companion, OpenOSPFD, adds support for the OSPF protocol suite.

Routing Hardware

A router is a hardware component used to interconnect networks. It connects the local area network to the wide area network, has interfaces on multiple networks. The work a router does is called routing. Each router keeps information about its neighbours. Routers forward traffic that enters on an input interface and leaves on an output interface, subject to filtering and other local rules.

A routing interface is a connection to an external network. Each interface is identified by a unique interface index. Interfaces are manageable objects. Interfaces that are active have an adapter that provides connectivity to the network they represent.

A router is usually located at any gateway.

Routing Table

A routing table is an electronic document that stores the routes to the various nodes in a computer network. It is usually stored in a router or networked computer in the form of a database or file. When data needs to be sent from one node to another on the network, the routing table is referred to in order to find the best possible route for the transfer of information.

Routing tables contain a list of IP addresses. Each IP address identifies a remote router that the local router is configured to recognize.

The routing table can be seen on Windows and Unix/Linux computers. The netstat -r command displays the contents of the routing table configured on the local computer.

Operation of Router

Routers used in home networking usually include an Ethernet hub for wired networking. In general, an IP network will consist of a set of core routers that are interconnected. The user sends data packets, which have embedded in them a destination address within the header. Typically, packets are received at an inbound network interface, processed by the processing module and, possibly, stored in the buffering module. The router examines the header, does a lookup in a routing or address table to find the next hop, updates the header, and then sends the packet towards its destination. The packet will generally pass through several routers before reaching its destination.

4.8.3 Switches

Networks are commonly used to interconnect computers or other devices. Today, performance of the network matters more than ever before and the criterion for buying and operating a network has changed dramatically. A detailed discussion on the topic of switching is out of the scope of this book. A switch is composed of network elements and is governed by OSI principles. It consists of NIC cards. A media access control address which is present in the NIC, uniquely represents any network hardware entity in an Ethernet network. An IP address uniquely represents a host in the Internet. The port on which the switch receives a packet is called the ingress port, and the port on which the switch sends the packet out is called the egress port.

High-performance businesses are driving richer and ubiquitous applications, which, in turn, are driving the need for faster, reliable, and more secure networks. Each network generally includes two or more computers, often referred to as nodes or stations, which are coupled together through selected media and various other network devices for relaying, transmitting, repeating, translating, filtering, etc. A network system is a communication system that links two or more computers and peripheral devices, and allows users to access resources on other computers and exchange messages with other users. Traditional internetworks consist of a collection of workstations, hosts, and LANs connected by WAN links. They are interconnected by switching technologies and services that connect the end user equipment to the network. The heart of a modern communication system is the digital switch.

A network switch is a small hardware device that joins multiple computers together within one LAN. Local area network that use switches to join segments are called switched LANs or, in the case of Ethernet networks, switched Ethernet LANs. Switches are hardware/software devices capable of creating temporary connections between two or more devices. A switch provides a transparent signal path between any pair of attached devices. The path is set up between the participating nodes on demand.

With the increase in the total number of endpoints connected to the LAN, increasing diversity of traffic flows on the LAN through network convergence, and more awareness regarding security caused due to an underlying technology shift in the Ethernet switch market has resulted in LANs becoming more intelligent and with increased bandwidth.

Case of General Bankers

General Bankers, one of the largest banks with a network of about 540 branches and offices and over 1,000 ATMs offers banking products and financial services to corporate and retail customers through a variety of delivery channels. The legacy systems available within the bank were stand-alone systems, networked only for basic e-mail and none of the core applications were linked to the network. The bank realized that to improve its operations and increase efficiency, it needed to centralize its core banking applications.

The traditional systems available within the bank were centric to the branch. The banking transactions were thus limited to the respective branch offices as customer data was not available in other branches. This made banking a limited service and very branch-specific. General bankers realized the importance of offering nationwide banking but this would be possible only by having a centralized data repository. From a business perspective, the main reason to go in for a network was centralization of data, provide all channels of communication and at the same time provide anytime, anywhere banking.

The network comprised of a mix of servers running different applications at various branches of the bank. This also resulted in duplication of backend services and procedures, as the systems were not centralized for the core banking applications. The centralization procedure started around late 2000. Utmost care was taken to design a network with a strong backbone. General bank realized that there was a need for the right solution whereby they could offer services across the country. Centralizing the operations was not the solution, but centralization of data was the key to the problem. Apart from leased lines the bank decided to go for backups on ISDN and VSATs. The bank decided to go for hub and spoke architecture with a mix of VSATs, leased lines, ISDN, and radio links. Today, the network supports general banks group offices, regional offices, branches, and over 1,000 ATMs. High-end Nortel routers and switches have been deployed for switching and connectivity. The CTO of the bank said that the real challenge was present while designing and deploying the network, as the team had to view business processes at a very micro level.

Types of Switching

Communication technologies use different types of switching technologies. This section provides basic information on the fundamental switching types, namely circuit switching, packet switching, and cell switching.

Circuit-switched telephone networks have provided voice communications for more than a century. Over much of the period, their development has been relatively slow, with major changes taking place only rarely. Circuit switching creates a direct physical connection between two devices. Circuit switching networks pre-allocate transmission bandwidth for an entire call or session. The best-known example of a circuit switching network is the public telephone switching network.

Packet switching is a technology that splits data in network communications into manageable small pieces, called packets. It divides a larger message into smaller pieces called payload and then encapsulates them with routing and protocol information to generate a data packet. Each packet contains relevant information about its destination and the data content. The routing information is used to move the packet across the network. Each packet is then transmitted individually and can even follow different routes to its destination. Once all the packets forming a message arrive at the destination, they are recompiled into the original message. The PSTN has been described as the largest machine in the world. It relies on faithful transmission of packets from one telephone to another, whether they are in the same town or on opposite sides of the world. It relies on packet signaling systems that allow any telephone user to send instructions into the network. These can set up links to a wanted destination or make reliable connections via a switching system.

Packet switching is the alternative to circuit switching protocols used historically for telephone (voice) networks. Packet-switched networks improve circuit switching by allowing dynamic sharing of the available LAN or WAN bandwidth.

Switching concepts revolve around actual switching blocks or functional groups of switches. The functional groups are classified as core, distribution, and access layers.

The access layer is the network entry point for end devices such as workstations and printers. It is known as the edge of the network. The access layer switches connect, in turn, to the distribution layer. The distribution layer interconnects the edges to the central core switch. The core layer of switches generally resides at the centre of the network. The function of the core layer is to aggregate all of the distribution layer traffic. It is responsible for fast and reliable transportation of data across a network.

Overview of Switches

A switch forwards data packets only to the appropriate port for the intended recipient, based on information in each packet header. As switches have moved from (what we would now consider) primitive bridges to more modern implementations, there have been continual changes and improvements in the level of integration. A network switch is a data switching device. The switch typically includes a set of input ports for receiving packets arriving on the buses, a set of output ports for forwarding packets outward on the buses, and a switch fabric such as a crosspoint switch for routing packets from each input switch port to the output switch ports that are to forward them. The switch provides the switching function for transferring information, such as data frames, among entities of the network. The switching function provided by the switch typically comprises receiving data at a source port from a network entity, transferring the data over the backplane to a destination port and, thereafter, transmitting that data over a medium to another entity of the network.

The network switch functions as the traffic management system within the network, directing data packets to the correct destination.

An unmanaged switch is the cheapest option and is typically used in a small office or business. These network switches perform the basic functions of managing the data flow between a shared printer and multiple computers.

A managed switch has a user interface or software offering that allows users to modify the settings of the switch. A managed switch allows the control of individual ports of the switch. A detailed view of the various aspects of switches are discussed later in the book.

Switch fabrics can be judged based on a few parameters, aside from the obvious number of ports and the nominal data rate on each of the ports. Some of the parameters are:

- Net bit rate or link rate utilization
- Throughput system availability
- System uptime
- Reliability
- Logical connection setup time
- Logical connection teardown time

- Delay and latency
- Round-trip delay
- Cell delay variation
- Scalability
- Field upgradability
- Resource utilization on the network processor
- Cost structure
- Feasibility

Some of the important parameters which affect the overall performance are outlined. Throughput of a switch fabric is crucial to the performance of a switch. As a result, the throughput of a switch fabric, under the given circumstances, determines the throughput of the switch. The throughput depends on the ability of the switch fabric to dynamically switch cells, and on the rate at which it can load (saturate) the links. A high degree of system availability through redundancy and fail-safe operation can only be achieved by architecture and proper design. An incapable architecture cannot be retrofitted to provide fail-safe operations and redundancy. Therefore, the architecture must be set up such that it supports a fail-safe mode and independent operation of the switch fabric planes. The system uptime is mostly determined by the mean time between failure (MTBF) of any subsystem and the ability of the system to cope with the outage of one subsystem. In a redundant system, uptime is not affected by the outage of one part of a redundant subsystem. The reliability of a switch fabric is determined by its internal architecture and by its input/output (I/O) links. The latency t_D of a cell switched through a switch fabric is determined by a variety of parameters. This latency is defined as the period of time it takes for any given cell from the incoming port to be switched through the switch fabric and arrive at the outgoing port. Scalability is of foremost importance during the expected life cycle of a router or a switch. There are two major components to scalability. One is that the switch fabric is scalable, and the other is that the node must be able to make use of scaled-up switch fabric components. Scalability, additionally, means that the number of switch fabric ports can be increased without impacting the line cards that are already installed. This is especially an issue in installations that will grow linearly over time.

Based on these parameters, an informed and educated decision can be made as to which switch fabric is right for the project.

Types of switches

Crosspoint or crossbar switches are switches that are based on analog switches in a crossbar pattern that can connect any one input to any one output at a time. Crossbar networks allow any processor in the system to connect to any other processor or memory unit so that many processors can communicate simultaneously without contention. A new connection can be established at any time as long as the requested input and output ports are free. Crossbar networks are used in the design of high-performance small-scale multiprocessors, in the design of routers for direct networks, and as basic components in the design of large-scale indirect networks. A crossbar can be defined as a switching network with N inputs and M outputs, which allows up to $\min N, M$ one-to-one interconnections without contention.

Self-routing switches have a built-in scheduler that allows for autonomous connection setup and tear down. Shared memory switches and input-buffered output queued switches as well as (combined) virtual output queued switches fall in this category. Shared memory switches are based on arrays of dual-ported memory cells that are accessible through a centralized scheduler. Non-buffered and non-queued switch fabrics are crosspoint switches that have a built-in scheduler.

Typically, a switch can either be unmanaged or managed. A managed switch can either be a stackable switch or a chassis switch.

Managed Switches

A managed switch is simply a network switch that allows the network administrator to monitor, configure, and manage certain network features such as what computers are allowed to access the LAN via the switch. A web-managed switch is a relatively new concept in the market. Managed switches come with web interface and are easy to configure. A managed switch is administered through a command line interface (CLI), while a web-managed switch, as the name suggests, is managed through a web browser. Access to the management features for the switch is password-protected so that only the network administrators can gain entry.

Case of an Educational Institute

One of the leading educational institutions in India decided to deliver educational content to its students in a more flexible and organized manner, and decided to use IT as the enabler.

Contd

> The organization recognized that information networks have emerged as strategic assets, and are a critical element for delivering education and services. The organization felt the need to provide state-of-the-art IP networking which could enhance the quality of the educational experience. The objectives were to deliver live lectures and screen events with the help of IP multicasting, deliver high-quality video-on-demand to students' hostels without affecting the network performance and bandwidth, standardise repetitive courses through video-on-demand, and provide opportunities to students and staff to pursue fields of individual interest in addition to standard course curriculum.
>
> With help from its alumni, the institute set up a campus-wide converged IP network with over 4,000 access points covering all the hostel rooms. The network offers multi-layered switched QoS, video-on-demand, multicast video, and the ability to host IP telephony in future.
>
> There are about 150-odd servers which converge to the core of the network based on Catalyst switches. This is connected to multiple distribution switches. Each hostel has a distribution switch, and the entire switching capacity is around 32 Gbps. The uniqueness of this exercise is the scale and the range of applications.
>
> Source: 'Delivering education at wire speed, *Network* magazine, July 2005

Ports

A port present in the switch is an interface to a LAN or WAN link. In most devices, there are two types of ports to consider. They are discussed below.

Attachment ports are used to connect to devices at the hierarchical level at which the device is being used; for example, the attachment ports on a desktop switch are those ports connecting to desktop devices.

Uplink ports are used either to connect to the next tier in the hierarchy or to connect to a high-performance device (for example, a server) at the current level. While not always the case, uplink ports usually operate at a higher data rate than attachment ports.

Depending on the level in the hierarchy, there are differing needs for both attachment and uplink ports.

Layer 2 Switches

The switch operates at layer 2 of the OSI model and therefore uses the MAC or Ethernet address for making decisions for forwarding data packets. The layer 2 switch is an improved network technology that addresses the issue of providing direct data connections, minimizing data collisions, and maximizing the use of a LAN's bandwidth; in other words, that improves the efficiency of the data transfer in the network. The switch has multiple ports similar to the hub and can switch in a data connection from any port to any other port. The switch monitors data traffic on its ports and collects MAC address information in the same way the bridge does to build a table of MAC addresses for the devices connected to its ports.

Layer 3 Switches

The switch operates at layer 3 of the OSI model. A Layer 3 switch is a high-performance network device that forwards traffic based on OSI stack layer 3 information. The key difference between Layer 3 switches and routers lies in the hardware technology used to build the unit. Figure 4.12 depicts Nortel's L3 Switch.

Stackable Switch

A stackable switch provides a means whereby a user can grow a system without having to discard hardware at each upgrade. Stackable configurations allow multiple physically separate switches to be clustered such that they function as a single device. A network planner can purchase a switch supporting the number of ports appropriate for a given environment. If the number of devices or users grows beyond the capabilities of the original switch, the switch can be expanded through stacking. A typical stackable switch may provide up to 48 ports per switch and be stackable upwards. Properly designed, a stackable switch configuration appears as if it were a single switch, regardless of the number of physical devices in the stack. As a result, the entire stack can be managed as if it were a single device. All of the switches in the stack can be more easily and consistently configured.

FIGURE 4.12 ERS 5530
(*Courtesy*: Nortel)

Nortel Ethernet Routing Switch 5530-24TFD is a next-generation stackable 10/100/1000/10000 Mbps Ethernet Layer 3 routing switch designed to provide high-density Gigabit desktop connectivity and Gigabit and 10 Gigabit fiber connectivity for aggregation for mid-size and large enterprise customers' wiring closets. It combines higher flexibility of deployment using Gigabit copper or fibre connections coupled with exceptional performance utilizing dual 10 Gigabit uplinks.

Chassis Switch

A chassis switch provides the ultimate flexibility. The switch is designed to accept a variety of plug-in modules in its slots. The exact configuration is a function of the plug-in modules deployed. The user purchases a base chassis, which usually includes a power supply, backplane, and possibly a management processor and other necessary features, and selects modules based on the application environment. Modules can be easily changed, often without powering down the chassis or disrupting the network. Chassis switches provide the highest performance and maximum flexibility; of course, they usually command the highest price. Figure 4.13 depicts the parts of Nortel ERS8600 Chassis Switch.

The Ethernet Routing Switch 8600 combines Terabit performance with Nortel's unique reliability technologies to create the basis of a Resilient Terabit cluster solution and to provide enterprises with a truly unified communications network. The Ethernet Routing Switch 8600 takes the complexity out of network design by simplifying the network architecture and increasing value per port per slot. It is a chassis-based platform that offers intelligent switching and network security. It scales up to 384 Gigabit Ethernet ports and 9610 Gigabit Ethernet ports.

KVM Switches

KVM is an acronym for keyboard, video, and mouse. It is a hardware device which enables the single use of a single keyboard video monitor and a mouse to control more than one computer at a time. KVM switches are often found in data centers where multiple servers are placed in a single rack and a single KVM switch attached to all the servers helps to control the servers. The switches come in different sizes and can be used in a variety of environments. Most of the KVM switches occupy 1U rack space. A KVM switch that supports N servers requires $3N + 3$ cables - 3 between each server and the KVM switch and 3 from the KVM switch to the shared keyboard, mouse, and display.

The first-generation KVM switches are mechanical in nature and act like on or off switches. The second-generation intelligent KVMs perform the connection electronically inside the KVM such that all systems connected to the KVM recognize they are always connected to end user devices. Electronic KVMs do this by spoofing or simulating the devices so that the attached systems never know that a switch occurs. Figure 4.14 depicts a KVM switch

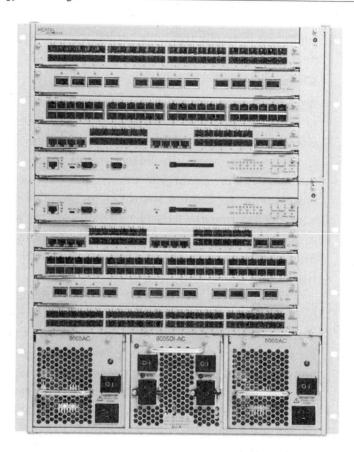

FIGURE 4.13 ERS 8600
(*Courtesy*: Nortel)

from ATEN. The KN2132 is an IP-based KVM control unit that allows both local and remote operators to monitor and access multiple computers from a single console. KVM allows up to three administrators, one logged in at a local console, the other two logged in remotely from any IP connected web browser and up to 32 users to securely monitor, manage, troubleshoot, or run applications from up to 32 connected devices respectively.

KVM comes with 'port scan' feature. It is often included with most, if not all, models of KVM switches. A simple hot-key command will permit users to view a selected list or all ports attached to the KVM switch and show the video of each for a selected amount of time.

A KVM splitter allows the administrator to control one computer from more than one control location. For instance, a server could be located in a

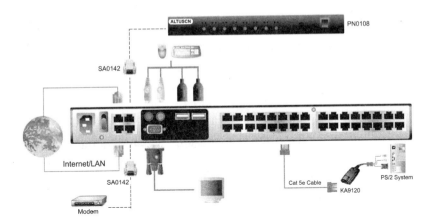

FIGURE 4.14 KVM 2132
(*Courtesy*: ATEN)

clean environment with control locations in the network administrator's office, the factory floor, the plant manager's office, and the boardroom.

In addition to standard KVM switches, there are a number of KVM extenders as well as KVM-to-Ethernet concentrators that have appeared in the market.

Aten's innovative KVM switches featuring a dual rail 19-inch (different form factors are available) LCD monitor and keyboard with built-in touchpad allows users to connect up to 8 or 16 servers of multiple platforms and to control them directly from the KVM switch console.

> **Let Us Ping!**
>
> Ping is a universal command that is available on every operating system to test the reachability of a network. When you type the command ping on your terminal which is connected in the network along with an IP address as its argument, the machine will try to send some bits of raw data towards the machine owning that IP address. If some machine exists with that IP address, it will send back certain bits. Thus, the machine receives the bits and it confirms that a path is available from the current machine to the other through a network.

4.8.4 Availability

At its simplest level, availability is a measure of the time a network element is functioning normally. The standard simple equation to calculate availability is

$$\text{Availability A} = \frac{\text{MTBF}}{\text{MTTR} + \text{MTBF}} \tag{4.2}$$

Where, A is the degree of availability expressed as a percentage, MTBF is the mean time between failures, and MTTR is the maximum time to repair or resolve a particular problem.

The technologies that systems need to achieve high availability are not automatically included by system and operating system vendors. High availability cannot be achieved by merely installing failover software and walking away. High availability is planned and achieved as a design.

4.8.5 Business Benefits

The recent major trends in switching includes migration from 10/100 to Giga switches and also a lot of up-gradations from unmanaged to managed switches. The access layer contains devices that allow workgroups and users to use services provided by the distribution and core layers. From an access layer point of view, access speeds would triple.

4.8.6 Application Accelerator

Network managers are constantly under pressure to provide better bandwidth and better response time on various hosted applications. Applications can be slow for reasons other than lack of bandwidth. Application acceleration addresses non-bandwidth congestion problems caused by TCP and application-layer protocols, significantly reduces the size of the data being sent along with the number of packets it takes to complete a transaction, and takes other actions to speed up the entire process.

A new technology called the 'application acceleration' has emerged, which accelerates the Internet applications over WANs using the same Internet infrastructure, circumventing to some extent the problems caused due to lack of bandwidth. Application accelerators, as the name suggests, are appliances that accelerate applications by re-engineering the way data, video, and voice is sent/transmitted over networks. Application acceleration addresses non-bandwidth congestion problems caused by TCP and application-layer protocols, thereby, significantly reducing the size of the data being sent along with the number of packets it takes to complete a transaction, and performs other actions to speed up the entire process.

The application acceleration market comprises two segments: application delivery controllers (ADCs) and wide area network optimization controllers (WOCs). The market has consolidated along two lines: application- and server-focused ADCs and network-focused WOCs. Figure 4.15 depicts an application acclerator 610 from Nortel. The Nortel Application Accelerator 510 and 610 were designed to improve web-based application response time by as much as 40 times, resulting in increased employee satisfaction and productivity.

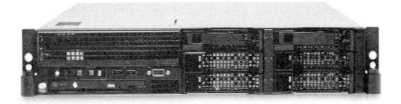

FIGURE 4.15 Application Accelerator 610 (*Courtesy*: Nortel)

Application delivery controllers reside in the data centre, generally in front of front-line web servers, are deployed asymmetrically (only at the data centre end), and are deployed to improve the performance of web/IP-based applications. They accelerate end-user performance of web-based and related applications by providing a suite of services at the network and application layers.

Wide area network optimization controllers are designed for networks, are typically deployed symmetrically (in data centre and remote locations) and improve the performance of applications that are accessed across a WAN. They address application performance problems caused by bandwidth constraints and by latency and protocol limitations.

4.8.7 Modems

Modem, short for modulator-demodulator, is an electronic device that converts a computer's digital signals into specific frequencies to travel over telephone lines (Figure 4.16). Since 1980, PSTN users have begun to use their subscriber lines to access data network, and more particularly the Internet. For that purpose, a voiceband modem was required to be installed at the CPE.

A faster service than voice-grade modems that is beginning to be offered by telephone companies is the digital subscriber line (DSL) [Brian et al. 2003]. Digital subscriber line services are growing rapidly as local access carriers continue to install DSL access modules (DSLAMs). Most service providers break down their DSL services into three main categories: residential, SOHO (small office/home office), and enterprise. ADSL (asymmetric DSL, where upstream and downstream speeds differ) appears to be the offering of choice for residential customers, while SDSL (symmetric DSL) is usually marketed to businesses because it has T1 or more speeds both ways. Depending on the geographic location, installation time for DSL circuits can range anywhere from one week to over ten weeks. Quality of service (QoS) is an issue. Most service providers, because of the multiple risk factors, do not guarantee QoS with DSL service. A standard modem of today contains two functional parts: an analog section for generating the signals and operating the phone, and a digital section for set-up and control. This functionality is actually incorporated into a single chip.

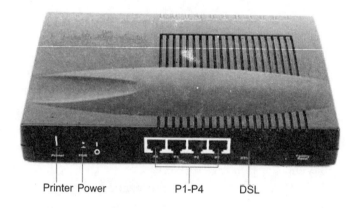

FIGURE 4.16 A Modem

Asymmetric digital subscriber line (ADSL) modems, a more recent development, are not limited to the telephone's voiceband audio frequencies. Asymmetric digital subscriber line offers high-speed downstream access to the customer site, and a lower speed upstream access from the customer. An ADSL modem has significant, immediate advantages over the dial-up modem. Though it uses a standard telephone line like a dial-up modem, it does not tie up the line, making it possible to use the telephone while cruising the Internet. Asymmetric digital subscriber line service is 'always on' connection, unlike a dial-up modem that cannot be left connected indefinitely. Most users find asymmetric speeds to be acceptable, since upstream traffic frequently consists of keystrokes or the transmission of short e-mail messages, whereas downstream traffic may include web pages, or large amounts of data. While most DSL and cable providers offer a modem with Internet service, the subscriber can also opt to provide his or her own modem. Often the Internet provider will make a list of compatible modems available to access the service.

Cable modems are a service offered by cable television companies. Often, the cable television or telephone company operates as both a transmission carrier and a network provider. As with ADSL, the downstream speed of cable modem is much faster than the upstream speed. The upstream speeds are similar to ADSL, but the downstream speeds can be several times faster. However, multiple customers on the same cable line share the capacity.

4.8.8 Access Points

Wireless networking is accomplished by sending a signal from one computer to another over radio waves. The most common form of wireless computing today uses the IEEE 802.11b standard. This popular standard, also called Wi-Fi or wireless fidelity, is now supported directly by newer laptops and PDAs, and most computer accessory manufacturers.

An access point is the essential component for setting up a typical wireless network. In a wireless local area network (WLAN), an access point is a station that transmits and receives data. It is a network gadget attached to an Ethernet network, and provides a wired network gateway to wireless clients. Wireless access points (APs or WAPs) are specially configured nodes on wireless local area networks. It connects users to other users within the network and also serves as the point of interconnection between the WLAN and a fixed wire network.

4.9 Network Resilience

A high-availability network is the foundation of network resilience. This requires substantial investment in both capital and operational expenditures, but pays off with significantly higher uptime, greater customer satisfaction, increased revenues and reduced exposure to regulatory penalties. Deploying a resilient network provides the flexibility to adapt an infrastructure to future services and applications with minimal disruption. A resilient network has end-to-end intelligence that segments (through virtual LANs, virtual storage area networks, or WANs), prioritises (through quality of service), and protects (through encryption) traffic without requiring major upgrades, only configuration of existing features.

A resilient network increases applications resilience. Network intelligence complements server and storage technologies to maintain application availability. Offloading processor-intensive tasks, such as encryption, compression, and load balancing, into the network increases application resilience and scalability by freeing server and storage processors to perform their core duties.

Summary

The chapter highlighted the various elements of the networking infrastructure.

Today, the Internet offers the fastest and easiest information access ever, with services such as online banking, shopping, stock trading, news posting, e-mail, and more. Businesses are capitalizing on the Internet with e-commerce opportunities, while the home networking market consumes those services through high-speed cable modem and DSL connections shared over Ethernet networks. With everyone exchanging data almost instantly, it is networking that keeps everyone connected.

Just a few years ago, building a corporate network was a daunting task undertaken only by seasoned professionals. Small businesses that could not afford to spend lavishly on technical assistance either had to struggle along without a network or try to put the pieces together themselves, often on a trial-and-error basis, until they got it right. Recognizing the growing importance of this market, many interconnect vendors have started to design products that are easy for ordinary people to install and use. In some cases, the equipment is ready to use out of the box or is self-configuring after installation.

One must manage infrastructure systems separately. If any of these systems fails, one will certainly have to respond more quickly than for most application systems. These failures often affect many users and even other stakeholders. One may want to set priorities on some systems even within the infrastructure, because some outages are more visible than others. The organization can enforce the priority of individual components by creating operations to manage them.

Networking components form the backbone of the network connectivity. Network connectivity options available for an organization was discussed. A non-technical overview on routers and switches were presented for the benefit of the reader. Other networking gadgets like the application accelerator and access points were discussed. Although this chapter has focused on the LAN connectivity options for business environment, the same connectivity options are available for home, where it is increasingly common for several desktop computers to reside.

Key Terms

Ethernet It is a network protocol invented by Xerox Corporation and developed jointly by Xerox, Intel, and Digital Equipment Corporation. Ethernet networks use CSMA/CD and run over a variety of cable types at 10 Mbps (megabits per second).

Fast Ethernet It is a new Ethernet standard that supports 100 Mbps using category 5 twisted pair or fiber optic cable.

Integrated services digital network (ISDN) It combines voice and digital network services in a single medium.

Modems These are devices that convert digital signals to the analog signals and analog signals (telephone wires) to the digital (computer) devices that convert digital and analog signals.

Network interface card It is a board that provides data communication capabilities to and from computers.

Port A connection point for a cable is called port.

Routers It is a device that routes information between interconnected networks. It can select the best path to route a message, as well as translate information from one router to another. It is similar to a superintelligent bridge.

Switches These are network hardware that route packets or cells (either ATM or voice) based on the address of the virtual circuit.

REVIEW QUESTIONS

4.1 What is the need for planning the ICT infrastructure?

4.2 Can sudden changes be introduced in the network by the CTOs? Justify your answer.

4.3 Briefly outline the OSI model.

4.4 Differentiate between fast Ethernet and gigabit Ethernet technology.

4.5 Discuss routing.

4.6 What are the various types of switching?

4.7 What is a switch? Differentiate between a stackable switch and a chassis switch.

4.8 What is the business advantage a KVM switch will provide to a business?

4.9 What is an application accelerator? Why do we need one?

4.10 What are access points?

Projects

4.1 A bank has planned to set up core and access switching to serve its customers. Identify a list of chassis grade switches available in the market and compare the internal architecture. Prepare a design and provide your recommendations.

4.2 Visit the nearest ISP and identify the various leased line and VSAT options available with the ISP. Prepare a comparison statement on the various options for your top management.

4.3 The office of a small independent realty company has four real estate agents, an executive secretary, and two assistant secretaries. They agreed to keep all common documents in a central location to which everyone has access. They also need a place to store their individual listings so that only the agent who has the listing should have access to. The executive secretary has considerable network administration knowledge, but everyone else does not. It is expected that in the near future the office will employ 10 more agents and three assistant secretaries. Would you recommend implementation of peer-to-peer or server-based architectural networking?

REFERENCES

Beasley, Jeffrey S. 1995, *Networking*, Pearson Education.

Di Marsico, Brian, Thomas Phelps IV, and William A. Yarberry Jr 2003, *Telecommunications Cost Management*, Auerbach.

Duck, Michael and Read Richard, 2003, *Data Communications and Computer Networks for Computer Scientists and Engineers*, Pearson Education.

Duato, Jose, Yalamanchili Sudhakar, and Lionel Ni 2003, *Interconnection Networks An Engineering Approach*, Morgan Kaufmann.

Evans, Nicholas D. 2003, *Business Innovation and Disruptive Technology*, Pearson Education.

Gutiérrez, Jairo 2007, *Business Data Communications and Networking: A Research Perspective*, Idea Group Publishing.

Halsall, Fred 2005, *Computer Networking and the Internet*, Pearson Education.

Held, Gilbert 2000, *Network Design Principles and Applications*, Auerbach.

Held, Gilbert 2001, *TCP/IP Professional Reference Guide*, Auerbach.

Ilyas, Mohammad and Mahgoub Imad 2005, *Mobile Computing Handbook*, Auerbach.

Marcus, Evan and Hal Stern 2003, *Blue Prints for High Availability*, Wiley.

Tricker, Ray 2007, *Wiring Regulations in Brief*, Elsevier.

Weidong Wu 2008, *Packet Forwarding Technologies*, Auerbach.

Wetteroth, Debbra 2003, *OSI Reference Model for Telecommunications*, McGraw-Hill.

CHAPTER

5 Cabling Infrastructure

My only hope is that we never lose sight of one thing that it all started with a mouse.
–Walt Disney

Learning Objectives

After reading this chapter, you should be able to understand:

- structured cabling system
- the types of cables
- the development of optical fibre as a communication method
- the basic principles of optical-fibre systems
- the application of fibre-optical technology in LAN/WAN systems
- the need for testing and certification
- the use of testing a cable distribution system for signal leakage

5.1 Introduction

Information technology (IT) has come a long way over the past decade. Over the years, IT has improved the performance by streamlining application portfolios, reducing infrastructure costs, improving governance, consolidating vendors, and outsourcing many activities. Information technology capabilities have fostered new sales channels, defined new customer segments, and even helped create new business models. IT systems are built on IT infrastructure. An IT architecture is the framework that guides and moderates the evolution of a complex system with interrelated elements. It provides order and rules so that IT hardware, IT software, electronic data, and data communications can work together.

In continuation of earlier discussions on networking infrastructure, there is a need to interconnect the networking infrastructure. Interconnection of the network infrastructure is achieved through cabling systems. It is important to understand the building blocks of cabling system for adoption in an organization. Cabling systems address the networking requirements based on data and voice cabling approaches. Data and voice cabling has come a long way to reach its current state. The cabling philosophy of a company is now a central communications issue and represents a substantial investment, not merely supporting today's equipment but to service a wide range of equipment for an extended period of time. Words and phrases such as attenuation, crosstalk, twisted pair, modular connectors, and multi-mode optical-fibre cable may be completely foreign to the reader. Just as the world of different computing and industrial segments have evolved its own industry buzzwords, so does the cabling business. But it is not really that mysterious. As such, cabling is no longer an invisible overhead within a computer package purchase but rather a major capital expense, which must show effective return on investment and exhibit true extended operational lifetime. Cabling represents the vast majority of the total investment applied to these frequently complex transmission paths interconnecting the various computing and networking elements.

Historically, the value of the communicated data was much less crucial than it is now or will be in the future. If a domestic or office telephone line fails, then voice data is interrupted and alternative arrangements need to be made. However, if a main telecommunication link fails, the cost can be significant both in terms of the data lost at the moment of failure and, more importantly, the cost of extended downtime. It is, hence, essential that we understand the cabling infrastructure powering organizations across the globe.

Wired (cable) media are made up of a central conductor (usually copper) surrounded by a jacket material. They are suitable for local area networks (LANs) because they offer high speed, good security, and low cost. Before we discuss the cable types, we will review some basic electricity concepts and properties.

We must install the proper network cable because different types of cables have different electrical properties. These properties also determine the maximum allowable distance and the transmission rate in Mbps. We must be familiar with the following electrical properties.

Resistance When electrons move through a cable, they must overcome resistance. Pure resistance affects the transmission of direct current (DC), and it is measured in ohms. When more resistance is met, more electric power is lost during transmission. The resistance causes the electric energy to be converted to heat. Cables with small diameters have more resistance than cables with large diameters.

Impedance The loss of energy from an alternating current (AC) is impedance. Like resistance, it is measured in ohms.

Noise Noise is a serious problem in cables and in most cases is hard to isolate. As the signals pass through a communications channel, the atomic particles and molecules in the transmission medium vibrate and emit random electromagnetic signals as noise. The strength of the transmitted signal is normally large relative to the noise signal. However, as the signal travels through the channel and is attenuated, its level can approach that of the noise. Noise can be caused by radio frequency interference (RFI) or electromagnetic interference (EMI). Many things can cause noise in a cable. Common causes are problems experienced by adverse weather conditions, fluorescent lights, transformers, and anything else that creates an electric field. Noise can be minimized if we plan our cable installation properly. We should route new cables away from lights and other EMI sources, use shielded cable when necessary, and ground all equipment.

Attenuation Attenuation is the fading of the electrical signal over a distance. As the signal travels along a communication channel, its amplitude decreases as the physical medium resists the flow of the electrical or electromagnetic energy. This effect is known as attenuation. All the above properties affect the rate of attenuation in a cable. After a certain distance, devices at the other end of a cable are unable to distinguish between the real signal and induced noise.

Cross Talk Cross talk occurs when a signal from one cable is leaked to another by an electrical field. An electrical field is created whenever an electrical signal is sent through a wire. If two wires are close enough and do not have enough

EMI protection, the signal may leak from one wire and cause noise on the other.

5.2 Ethernet—The Common Denominator

In today's business world, reliable and efficient access to information has become an important asset in the quest to achieve a competitive advantage. Originally, Ethernet was the brainchild of one person, Metcalfe. In the early 1970s, while working at Xerox PARC on the office of the future project, the concept of Ethernet was conceived by Metacalfe. One of the concepts behind Ethernet, that of allocating the use of a shared channel, can be traced to the pioneering efforts of the research team at the University of Hawaii during the early 1970s. Using a ground-based radio broadcasting system to connect different locations through the use of a shared channel, the team developed the concept of listening to the channel before transmission called ALOHA. ALOHA formed the basis for the development of numerous channel contention systems including Ethernet.

Table 5.1 Bandwidth Requirements for Different IP Services

Application	Bandwidth	Qos
Video (SDTV)	3.5 Mbps	Low loss, low jitter, constant bit rate
Video (HDTV)	15 Mbps	Same as above
Telecommuting	10 Mbps	Best effort, bursty
Video gaming	10 Mbps	Low loss, low jitter, bursty
Voice	64 kbps	Low loss, low latency, constant bit rate
Peer-to-peer downloading	100 kbps–100 Mbps	Best effort

Ethernet is actually a collection of networking protocol standards. This is the most widely used LAN technology in the world today. Ethernet has since become the most popular and most widely deployed network technology in the world. The Ethernet standard has grown to encompass new technologies as computer networking has matured, but the mechanics of operation for every Ethernet network today stem from Metcalfe's original design.

Ethernet is taken as the base for current communication technologies especially due the the concept of Internet protocol (IP) convergence. The Institute of Electrical and Electronics Engineers (IEEE) adopted Ethernet as a LAN standard and published its initial specifications as 10BASE5. Ethernet was one way to regulate physical access to network transmission media. The types of Ethernet are defined in terms of their maximum transmission speeds. The bandwidth requirements for common IP services are listed in Table 5.1. Fast Ethernet transfers data at a maximum of 100 Mbps. Gigabit Ethernet transfers data at a maximum of 1 gigabit per second.

5.2.1 10 Gigabit Ethernet

Currently, the industry is emerging with a standard called 10Gig Ethernet standard. This standard is an extension of the basic IEEE 802.3 standard protocols to a wire speed of 10 Gbps. Ethernet is 10 times faster than gigabit Ethernet. As an extension, 10G is still fully Ethernet-compatible and retains the key Ethernet architecture, including the media access control protocol, the Ethernet frame format, and the minimum and maximum frame size. Local area network backbones are expected to adopt this technology and 10GbE will encompass all IP networks across the enterprise. Refer to the development trend of Ethernet technologies in Figure 5.1. It is emerging as the preferred interconnect for most enterprises and is expected to be deployed for applications such as video streaming to the desktop, aggregations between server farm and service provider data centres, data centre communication from LAN switches to storage networks, and grid computing.

Applications are expected to include backbones, campus size networks, and metropolitan and wide area networks. This latter application is aided by the fact that the 10 Gbps data rate is comparable with a basic SONET fibre-optic transmission standard rate. In fact, 10 Gbps Ethernet will be a competitor to ATM high-speed packet switching technology.

Any discussion of Ethernet networks requires knowledge of both existing and pending cabling and cabling standards adopted in Ethernet data communications industry. The following sections will provide an overview on cabling and cabling standards.

5.3 Banking on Structured Cabling

Network cabling infrastructure is the most important part of a network, and an essential building block without which, a system may fail to function properly or at optimum efficiency. Early cabling systems were unstructured, proprietary, and often worked only with a specific vendor's equipment. They were designed and installed for mainframes and were a combination of thicknet cable and terminal cables. The concept of an extended cabling

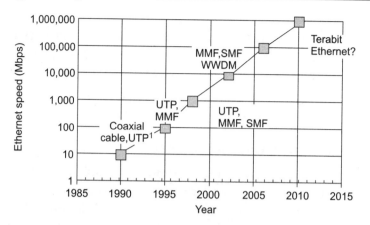

FIGURE 5.1 Development Trend of Ethernet Technologies

infrastructure is, therefore, no longer a series of strands of wire linking one component with another. It is rather a carefully designed network of cables installed to provide high-speed communication paths which have been designed to be reliable with minimal mean-time-to-repair. With networks becoming more complex and their requirements changing constantly, having a good foundation for these networks is crucial.

5.3.1 Structured Cabling as a Solution

Cabling, the backbone of any data communications system, is vital when it comes to determining performance and reliability of any network. Cabling is generally regarded as the lowest cost component in networking, having a longer upgrade cycle compared to other components such as switches, PCs, and other hardware. The growing size of networks and the introduction of higher speed access methods create an overwhelming need for reliable and manageable cabling systems. The channel is the entire cabling system comprising of all the cable, connecting hardware such as outlets and patch panels, and all the cords between the mated plug/socket in the network interface card (NIC), used in the PC at one end and the communication room equipment, typically a switch, at the other end.

Structured cabling is the first step towards achieving optimum performance from the network. A structured cabling system (SCS) can be described as a system that comprises a set of transmission products, applied with engineering design rules that allow the user to apply voice, data, and signals in a manner that maximizes data rates. An SCS consists of an open architecture, standardized media and layout, standard connection interfaces, adherence to national and international standards, and total system design and installation. An SCS installation is neither vendor-specific nor topology-dependent. It offers uniformity, flexibility, scalability, easy and low-cost changes, and investment protection of an asset with a very long life, usually, 10 years or more. Structured cabling divides the entire infrastructure into manageable blocks and then attempts to integrate these blocks to produce high-performance networks. Often regarded as the lowest cost component in networking, structured cables have a longer upgrade cycle compared to other components as switches, PCs, and other hardware. Structured cabling is perhaps that element of the infrastructure which does not need any upgradation for several years. Standardized cabling architecture allows a single delivery method to be designed for supporting the various horizontal cables in the work space. Because of the long life of this network component, there is always a demand for cabling solutions.

5.3.2 Advantages of Structured Cable

The advantages of structured cabling are as follows.

Consistency A structured cabling system means the same cabling systems for data, voice, and video.

Support for multi-vendor equipment A standard-based cable system will support applications and hardware even with mix and match vendors.

Simplify moves/adds/changes Structured cabling systems can support any changes within the systems.

Simplify troubleshooting With structured cabling systems, problems are less likely to bring down the entire network, is easier to isolate, and easier to fix.

Support for future applications Structured cabling system supports future applications, such as multimedia, videoconferencing, etc. with little or no upgrade pain.

A structured cabling system provides a universal platform upon which an overall information system's strategy is built. With a flexible cabling infrastructure, a structured cabling system can support multiple voice, data, video, and multimedia systems regardless of their manufacturer. A well-designed cabling plant may include several independent cabling solutions of different media types, installed at each workstation to support multiple system performance requirements.

5.3.3 Design of Structured Cabling

Selection of structured cabling components as per their designed performance levels is a complex task. Before an enterprise chooses to go for cabling, it needs to understand the purpose for cabling. It is important to have a foresight of business growth as SCS will be a long-term commitment. All too often, organizations fail to identify the importance of structured cabling systems and do not allow for future expansion as more and more bandwidth is required to cope with ever more demanding business applications. The system integrator or cable installer considers key issues such as size of enterprise, expansion growth rate expected, physical spread, and transmission speeds and capacities needed.

A structured cabling system consists of outlets, which provide the user with an RJ45 extension. The outlets are either one or two RJ45 connectors mounted in a standard single gang faceplate, or as single snap in modules which can be fitted into floor boxes, single gang faceplates or dual gang faceplates.

Structured cabling provides a flexible cabling plan and can support computers and telephone systems from any vendor. Structured cabling is hierarchy based on backbone cables that carry signals between telecommunication closets and floors of a building, and on horizontal cables that deliver services from telecom closets to work areas. For example, ADC Krone's TrueNet PLM drastically speeds and simplifies daily network provisioning, maintenance, and security providing network managers with full visibility and control of the network and its assets. While Molex claims that its premise networks real-time patching system comprises intelligent patch panels, patch cords, integration strips, cable and physical layer management hardware and software.

5.3.4 Need for Cabling Standards

Installation of structured cabling product is equally important like the product selection. Following the right installation practices and adherence to standards is very important when it comes to new applications that require higher data rate transfer. Following standards pays in terms of life of the cabling, durability, and return on investment (ROI) as one cannot possibly lay and remove cables as it were a LAN switch. The whole point of cabling standards is to be able to plug together cabling components from different manufacturers and still meet the overall channel requirement. A consortium of telecommunications vendors and consultants worked in conjunction with the American National Standards Institute (ANSI), Electronic Industries Alliance (EIA), and the Telecommunications Industry Association (TIA) to create a standard originally known as the Commercial Building Telecommunications Cabling Standard or ANSI/TIA/EIA-568-1991 [Voicendata, 2004]. The American standard, TIA/EIA-568-B, defines a compliant channel as a channel which is constructed of individually compliant cabling components. This standard has been revised and updated several times. Qualified installers know how to handle the wire during installation and to be aware of such factors as pull strength, minimum bend radius, proper termination techniques, separation of communication cables from electrical wiring, and the importance of maintaining tight twists.

Compliance with international standards, ability to support future technologies with ease, and smooth management of IT infrastructure are the key reasons why companies are spending on cabling.

5.3.5 Roles of Management

As the need for bandwidth and to run more applications on the network increases, the cabling infrastructure assumes importance, throwing challenges at chief information officers (CIO) as they go in for a particular solution. Before evaluating a structured cabling solution for deployment, the CIO or the chief technical officer (CTO) is expected to know his/her network with a clear understanding of expected performance requirements. Besides, he/she is expected to have a fair idea of the requirements that are going to emerge in the future. Having understood the load that his/her network may have to face, he/she should decide on a solution that best meets his/her requirements. At the same time, this futuristic solution should not cost a lot in terms of purchase, installation, and maintenance. The total cost of ownership (TCO), which is sum of the cost of acquisition and recurring costs, should be low at all times. Within a converged network, unless the expectations of cost, speed, performance, and reliability are met, an optimum ROI cannot be realized. While the design and engineering of the product contributes to cost and speed, it is the customer experience that defines performance and reliability. It is, therefore, essential that solutions be designed to deliver the desired quality of service (QoS) consistently.

Enterprises should take into account reliability, manageability, and flexibility of the cabling network. The management of cabling infrastructure can be taken care by opting for a good cable management solution. Cabling is a complex deployment and not too many people know about the proper installation of the equipment. The complex nature of deployment of structured cabling compels the vendors to provide extensive training on their technologies and solutions.

5.3.6 Cable Solutions in the Market

Cables are used to meet all sorts of power and signaling requirements. The demands made on a cable depend on the location in which the cable is used and the function for which the cable is intended. These demands, in turn, determine the features a cable should have. Cabling is not just about buying good quality cable and connecting with the hardware. In order to ensure the use of full bandwidth, a structured cabling system needs to be properly designed, installed, and administered.

With an escalation in the cabling systems and standards, a large number of players are foraying into the cabling solutions market. ADC Krone, Systimax, R&M, D-Link, and Molex are some of the key players in the cabling arena.

5.4 Technology Drivers

The pressure to reduce costs amongst budget constraints and competitive pressure is making organizations search for innovative ways of designing, building, and managing networks. The emergence of converged networks has been a key driver of growth, bringing along a new set of capabilities and deployment and management issues.

The technology driver for large-scale deployment in optical fibre is quite vivid. Enterprise wide applications (EWA) are the biggest drivers of bandwidth and hence cabling market. Data centre cabling has emerged as another niche area for structured cabling solutions vendors. Laser diodes, optical amplifiers, dense wavelength division multiplexing (DWDM) tunable lasers, new types of fibres, and optical switches are the key drivers for delivering huge bandwidth over a long distance with high degree of reliability at radically lower price point. Another key driver is wireless with wireless local area network (WLAN) and other deployments rising exponentially. The upgrade market, organizations that deployed their existing networks a few years ago in the networking boom is also hot, with rapid migration to multiprotocol label switching (MPLS) for wide area networks (WANs), and wireless and gigabit Ethernet for LANs.

5.5 Classification of Cables

All copper cable types have the following components in common.

- One or more conductors to provide a medium for the signal.
- Insulation of some sort around the conductors to help keep the signal in and interference out.

- An outer sheath, or jacket, to encase the cable elements. The sheath keeps the cable components together, and may also help protect the cable components from water, pressure, or other types of damage.

Many types of communications media are available for data networking today. The TIA/EIA 568 recognizes three different kinds of cables: unshielded twisted pair (UTP), shielded or screened twisted pair (STP or ScTP), and fibre optic (FO). Globally, structured cabling trends get guided by the bandwidth crave in all the three media. Each of the cable types is subdivided into more specialized categories and it has its own design and specifications, standards, advantages, and disadvantages. Cable types differ in price, transmission speed, and recommended transmission distance.

5.5.1 Twisted-pair Cable

Twisted-pair cable is lightweight, easy to install, inexpensive, and supports many different types of networks. Twisted-pair cable is widely used, inexpensive, and easy to install. A twisted pair is characterized by two parameters: the diameter of its copper wire and its twisted period. In a twisted-pair cable, two conductor wires are wrapped around each other. Each pair is twisted with a different number of twists per inch to help eliminate interference from adjacent pairs and other electrical devices. Twisted pairs are made from two identical insulated conductors, which are twisted together along their length at a specified number of twists per metre, typically 40 twists per metre (12 twists per foot). The wires are twisted to reduce the effect of electromagnetic and electrostatic induction. Twisted-pair lines are suitable, with appropriate line driver and receiver circuits that exploit the potential advantages gained by using such a geometry, for bit rates in order of 1 Mbps over short distances (less than 100 m) and lower bit rates over longer distances. Twisted-pair cables come in two main varieties, namely unshielded twisted pair and shielded twisted pair (STP).

> **Case of Chemical Company's Network Infrastructure**
>
> A large Asian chemical company had a small fleet of airplanes for use by its executives. The company established a local airport office away from the headquarters (HQ), which would be responsible for all aspects of flight operations. The network in this small office consisted of a file server, 20 PCs, and 4 printers in an Ethernet network connected by Category 5 UTP.
>
> *Contd*

> Soon after installation, the network began having problems with network failures. According to the IT director, the netware operating system crashed on average 4-5 times per day. Since the small office was at a remote site, it caused tremendous frustration on the part of the IT department and also to the users of the network.
>
> While cabling problems were suspected, isolating the problem proved difficult. The physical layer appeared to be truly Category 5 compliant. All the network components were properly specified and when the cabling was tested using a hand-held tester, it passed. The rationale behind suspecting cables was due to the fact that it might be cabling related was that the average run in this 'L' shaped building was 250-300 feet—near the 568 length limit of 100 metres. This IT director, like most end-users, had been told that UTP was a balanced system and was, therefore, immune to outside interference.
>
> To test the suspicion that the problem might be cabling-related they tried Screened Category 5 cable connected between the server and one workstation. The result was success and cautious optimism. Before re-cabling the office, they considered fibre-to-the-desk because of its complete EMI immunity. However, according to the IT director, while the cost of fibre itself was not prohibitive, the network interface cards were terribly expensive and he could not justify spending the money when another solution was available at much lower cost.
>
> The decision was made to re-cable the office using Screened Category 5 cabling and connectors. While the new cabling runs were just as long as the old, this action solved their problem completely. The new Ethernet network connected by Screened Category 5 cable was truly immune to the airport's stray electromagnetic radiation.

5.5.2 Unshielded Twisted-pair

Unshielded twisted pair (UTP) is one of the media types (Figure 5.2). It can be classified into categories depending on the performance. To distinguish between varieties of UTP, the United States EIA/TIA has formulated several categories. The electrical specifications for these cables are detailed in EIA/TIA- 568A, TSB-36, TSB-40, and their successor SP2840.

FIGURE 5.2 UTP Cable

Category 1 (Cat1) Voice-grade UTP telephone cable. This describes the cable that has been used for years in North America for telephone communications. Officially, such a cable is not considered suitable for data-grade transmissions.

Category 2 (Cat2) Voice-grade UTP, capable of supporting transmission rates of up to 4 Mbps.

Category 3 (Cat3) Data-grade UTP, used extensively for supporting data transmission rates of up to 10 Mbps. An Ethernet 10BASE-T network cabled with twisted pair requires at least this category of cable.

Category 4 (Cat4) Data-grade UTP, capable of supporting transmission rates of up to 16 Mbps. An IBM Token Ring network transmitting at 16 Mbps requires this type of cable.

Category 5 (Cat5) Cat5 or Category 5 is the de-facto standard now as it is easy to install and has a lower installation cost. This is a data-grade UTP capable for supporting transmission rates of up to 155 Mbps.

Enhanced Category 5 (Cat5e) Cat5 was stable and dominant for quite a while. As bandwidth demands grew with the introduction of Fast Ethernet and ATM, changes were made to the Cat5 standard to support the higher levels of throughput was required. Enhanced Cat5 called Cat5e emerged as a solution. This standard specifies transmission performance that exceeds Cat5. This standard is used for gigabit Ethernet.

Category 6 (Cat6) Cat6 includes all of the Cat5e parameters but extends the test frequency to 200 MHz, greatly exceeding current Cat5 requirements. Cat6 is a higher performing system than its predecessor at the physical level where it supports frequencies of upto 250 MHz vs. 100 MHz for Cat5e and cabling performance is higher than that of its predecessor Cat5e. Cat6 is also the best available standard at present for Gigabit Ethernet. It has double the bandwidth of Cat5e, and improved signal to noise ratio. Cat6 provides a mechanism to certify to a level much greater than Cat5, providing assurance that the cabling infrastructure will support future networks that may require greater than 100 MHz bandwidth.

Augumented Cat6 Augmented Cat6 (Cat6a) is the latest buzzword doing the rounds in cabling. This standard attempts to beef up Cat6 to make it suitable for 10 gigabit Ethernet (10 GbE) utilization. For example, Seimon [CXO Today 2005] has announced that the full 10G 6A F/UTP channel consists of a smaller diameter F/UTP cable, screened tool-less MAX modules and TERA-MAX patch panels as well as screened 10G 6A MC patch cords and the screened version of Siemon's innovative high-density BladePatch patch cord. Gigabit Ethernet is an expansion of the popular 10 Mbps (10BASE-T) Ethernet and 100 Mbps (100BASE-T) Fast Ethernet standards for network connectivity. It builds on top of the Ethernet protocol, and increases speed tenfold over fast Ethernet to 1000 Mbps, or one gigabit per second (Gbps). 'No single 10Gb/s copper cabling option is right for every customer,' said Robert Carlson, Siemon's VP of global marketing [CXO Today, 2005]. 'As people become more savvy about potential solutions, the strength and viability of each system becomes clear and decisions are based on actual needs as opposed to habit or comfort.'

Category 7 (Cat7) Cat7 is designed to operate at frequencies up to 600 MHz and is beyond the reach of UTP. A filler often runs through the centre of the cable to keep the pairs apart. This is typically a soft plastic strip which is X-shaped to separate the cable space into four compartments each holding a cable pair.

5.5.3 Shielded Twisted Pair (STP)

Shielded twisted pair (STP) (Figure 5.3) also known as the IBM Type 1 is the other cable type. The 568-A standard for STP is known as STP-A. This standard extends the system's rating through 300 MHz. This provides improved protection against external EMI and also protects against crosstalk between each pair and allows the cable to operate at high frequencies. When installed properly, STP-A structured cabling system can run a 16 Mbps token ring signal and a 550 MHz broadband video signal at the same time. Shields achieve their beneficial effects by first providing a reflective barrier which prevents radio frequency interference from coupling with the differential mode signals on twisted pairs (signal ingress), and then by providing a reflective barrier which prevents energy from broadcasting out of the cable.

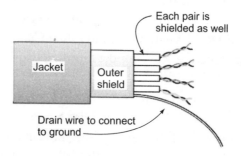

FIGURE 5.3 STP cable

Advantages of Twisted-pair Cable

Twisted-pair cable has the following advantages over other types of cables for networks.

- It is easy to connect devices to twisted-pair cable.
- If an already installed cable system, such as telephone cable, has extra, unused wires, you may be able to use a pair of wires from that system – but see the warnings on this above.
- STP does a reasonably good job of blocking interference.
- UTP is quite inexpensive.
- UTP is very easy to install.

While no data transmission method is forever, an argument can be made that shielded cables are more future proof than UTP. This speaks to the noise immunity inherent in shielded cabling and the emissions immunity provided by the shield. Twisted pair cable has gained a prominent position in the networking market.

5.6 Fibre-optic Cable (FOC)

Fibre optics is uniting continents, and therefore the world. Fibre optic transmission has become one of the most exciting and rapidly changing fields in telecommunications engineering. Fibre optics was in the right place at the right time to take advantage of the boom in competitive long-distance carriers in the 1980s, so fibre became the backbone of the new digital global telecommunication network. In reality, we are crossing the oceans with a few clicks or numbers to communicate with almost whomever we want, whenever we want. Submarine optical fibre cable, now being installed every day on the ocean floor, is providing the bandwidth needed to unite the nations of the world, helping to bring down the walls that once divided people. There are now dozens of submarine fibre optic systems embedded on ocean floors. Why did our oceans become full of optical fibre cables? The answer is that people, organizations, and governments crave bandwidth, the ability to communicate through a variety of tools (telephone, Internet, multimedia, videoconference, fax, e-mail, and so forth), and nothing can deliver bandwidth like fibre optics. Global Crossing Ltd, a telecommunications company is betting heavily that an expanded customer base and additional traffic due to the Internet will ensure its success. So far it has been a winning bet, and it may remain so for many years to come.

The huge investment in fibre-optic cables and wavelength division multiplexing (WDM) systems left the global telecommunications network with much more transmission capacity than it needed, bringing down prices for both long-distance telephone calls and Internet traffic. It is one of the reasons for the massive increase in international telecommunications and arguably the perception of the apparent shrinking planet. Optical component prices have come down dramatically as volumes have increased and manufacturing technology has improved. Costs are low enough, and hence carriers around the world are installing fibres all the way to homes to provide premium broadband services, using singlemode fibre and coarse WDM.

In the 1970s, the Japanese correctly spotlighted optical communications as an industry of the future. They focused thousands of best engineers of their nation on fibre optics and laser development. Verizon's fibre-to-the-home cables run parallel to its standard copper telephone wires in many communities. But fibre is playing an increasing role as data rates continue to soar.

5.6.1 Optical Fibre Media

The fibre optics medium has been in use for a long time and it has certain abilities that are not there in the copper medium. Fibre-optic communication uses light signals guided through a fibre core. An optical fibre is simply a very thin piece of glass which acts as a pipe, through which light can pass. The light that is passed down the glass fibre can be turned on and off to represent digital information or it can be gradually changed in amplitude, frequency, or phase to represent analog information. The light is guided by principles of refractive index and internal reflections. Data transmission using a fibre-optic cable is many times faster than with electrical methods, and speeds of over 10 Gbps are possible. Little of the light signal is absorbed in the glass core, so fibre-optic cables can be used for longer distances before the signal must be amplified or repeated. Some fibre-optic segments can be many kilometres long before a repeater is needed. Data transmission using a fibre-optic cable is many times faster than with electrical methods, and speeds of over 10 Gbps are possible.

The major components of a fibre-optic cable are the core, cladding, buffer, strength members, and jacket (refer to Figure 5.4). The core has a refractive index, n_1, higher than the refractive index, n_2, of the cladding. It is used as a transmission medium for guided optical waves.

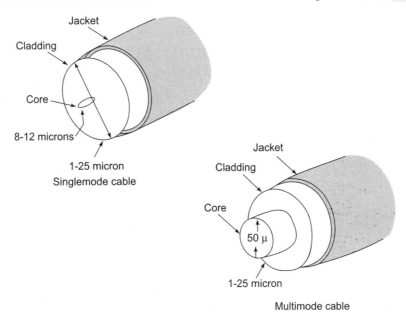

FIGURE 5.4 Fibre-Optic Cable

Fibre Core

The core of a fibre-optic telecommunications cable consists of a glass fibre through which the light signal travels. The most common core sizes are 50 and 62.5 microns, which are used in multimode cables. In singlemode systems, 8.5 micron fibres are used (Figure 5.5).

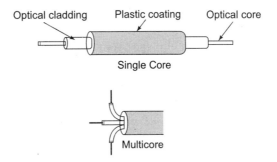

FIGURE 5.5 Fibre-Optic Cable Compared

Fibre Cladding

The cladding is a protective layer with a lower index of refraction than the core. A lower index means any light that hits the core walls will be redirected back to continue on its path. The cladding diameter is typically 125 microns.

Sheath

The cable is usually covered with a heavy plastic sheath. It is the outer casing surrounding the fibre that provides primary mechanical protection. It also has external layers, which are strands of very tough material (such as steel, fibreglass, or Kevlar) that provide tensile strength for the cable. The sheath provides primary protection against abrasion, cut resistance, crushing resistance, and additional resistance to excessive bending. Light-duty cables can utilize polyvinyl chloride (PVC), or polyurethane sheath materials (Figure 5.6). The more durable polyethylene (PE) materials are used on outdoor and heavier duty cables. Special cable sheath materials are sometimes used for cables installed in air-handling spaces or 'plenum' areas. These are defined as the spaces above suspended ceilings, below elevated floors, and in heating and ventilation ducts.

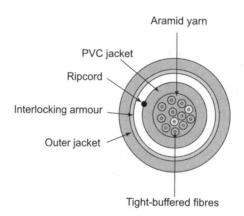

FIGURE 5.6 Cross Section of Fibre-optic Cable

The buffer of a fibre-optic cable is made of one or more layers of plastic surrounding the cladding (Figure 5.7). The buffer helps strengthen the cable, thereby decreasing the likelihood of micro cracks, which can eventually break the fibre. The buffer also protects the core and cladding from potential invasion by water or other materials in the operating environment. The buffer typically doubles the diameter of the fibre. A tight buffer fits snugly around the fibre. A tight buffer can protect the fibres from stress due to pressure and impact, but not from changes in temperature. A loose buffer is a rigid tube of plastic with one or more fibres (consisting of core and cladding) running through it.

Fibre-optic cables have the highest data-carrying capacity of any wired medium. A typical fibre has a capacity of 50 Tbps (terabits per second or

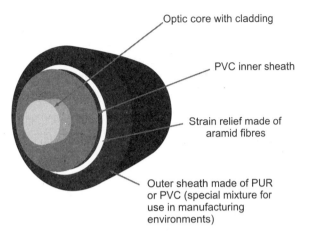

FIGURE 5.7 Parts of Fibre-optic Cable

50×10^{12} bits per second). In fact, this data rate for years has been much higher than the speed at which standard electronics could load the fibre. This mismatch between fibre and nodal electronics speed is called the electronic bottleneck.

Pulse shapes are more accurately preserved in singlemode fibre, lending to a higher potential data rate. However, the cost of multimode and singlemode fibre is comparable. The real difference in pricing is in the optoelectronics needed at each end of the fibre.

Know Your Firm–AT&T Corporation

AT&T is a communications company offering long-distance telephone service, among other services. In 1996, the company divided itself into three separate operations: AT&T Corp. (telecommunications), NCR Corporation (computers), and Lucent Technologies Inc. (network products). The two latter divisions were then sold. With increased long-distance competition during the 1990s, AT&T was forced to focus on retaining its market share in that core business area. Refining their strategy in 1997, the AT&T Board of Directors announced its intent to sell two of the company's profitable but non-strategic businesses: AT&T Universal Card Services and the Customer Care unit (formerly known as American Transtech) of AT&T Solutions.

- 1925: Bell Labs is formed
- 1969: Allows other companies such as MCI to connect to their phone network
- 1990: Buys the electronic mail service division of Western Union
- 1991: Acquires Teradata and NCR Corp

Contd

> - 1995: Purchases McCaw Cellular
> - 1996: Company is divided into three separate operations: AT&T Corp., NCR Corporation, and Lucent Technologies
> - 1997: Forms a partnership with Bell Atlantic Corp. and Nynex Corp. to provide customers with lower rates
>
> *Source:* www.att.com

5.6.2 Strengths of Fibre-optic Cables

Some of the common benefits of fibre-optic cables are listed below.

- A fibre-optic cable offers near infinite bandwidth and perfect immunity to noise. It is ideal for high data-rate systems such as gigabit Ethernet, fibre distributed data interface (FDDI), multimedia, asynchronous transfer mode (ATM), synchronous optical networking (SONET), fibre channel, or any other network that requires the transfer of large, bandwidth-consuming data files, particularly over long distances.
- A common application for FOC is as a network backbone, where huge amounts of data are transmitted. Fibre offers more bandwidth than copper cables. In fact, a single pair of fibres can handle the same amount of voice traffic as 1,400 pairs of copper.
- Fibre provides extremely reliable data transmission. It is completely immune to many environmental factors that affect copper cable.
- Fibre is made of glass, which is an insulator, so no electric current can flow through. It is immune to EMI/RFI, crosstalk, impedance problems, and more. It is possible to run fibre cable next to industrial equipment without issues.
- Fibre is also less susceptible to temperature fluctuations than copper and can be submerged in water.

Benefits of Fibre

The vision of carrying the triple-play of voice, data, and video services over a single convergent infrastructure has given birth to the idea of the next-generation network. One key driving force is the ever-increasing level of data traffic and bandwidth-intensive applications.

Fibre-optic cables offer the following advantages over other types of transmission media.

- Light signals are impervious to interference from EMI or electrical crosstalk.
- Light signals do not interfere with other signals. As a result, fibre optic connections can be used in extremely adverse environments, such as in elevator shafts or assembly plants, where powerful motors produce lots of electrical noise.
- Optical fibres have a much wider, flat bandwidth than coaxial cables and equalization of the signals is not required.
- Fibre has a much lower attenuation, so signals can be transmitted much further than with coaxial or twisted-pair cable before amplification is necessary.
- Optical fibre cables do not conduct electricity and so eliminates problems of ground loops, lightning damage, and electrical shock when cabling in high-voltage areas.
- Fibre-optic cables are generally much thinner and lighter than copper cable.
- Fibre-optic cables have greater data security than copper cables.
- Licensing is not required, although a right-of-way for laying the cable is needed.

Types of Fibre

One reason optical fibre makes such a good transmission medium is because the different indices of refraction for the cladding and core help to contain the light signal within the core, producing a waveguide for the light. Two general types of fibre have emerged to meet user requirement: multimode and singlemode. Singlemode fibres can propagate only the fundamental transmission mode. Multimode fibres can propagate hundreds of transmission modes.

Singlemode Fibre

A singlemode fibre is basically a step index fibre with a very small core diameter. It has almost unlimited bandwidth due to the small core supporting only one light mode. But, this requires very high precision alignment in both joints and connectors and the need to use expensive laser technology to drive the fibre. Singlemode is used over distances longer than a few miles. They offer the highest performance and are the most reliable option for high-speed applications such as carrier networks and corporate backbones.

Multimode Fibre

In a multimode stepped index fibre, the cladding and core material each has a different but uniform refractive index. All the light emitted by the diode at an angle less than the critical angle is reflected at the cladding interface and propagates along the core by means of multiple (internal) reflections. Depending on the angle at which it is emitted by the diode, the light will take a variable amount of time to propagate along the cable. Therefore, the received signal has a wider pulse width than the input signal with a corresponding decrease in the maximum permissible bit rate. This effect is known as dispersion and means this type of cable is used primarily for modest bit rates with relatively inexpensive light emitting diodes (LEDs) compared to laser diodes.

Alternatively, multiple high bit rate transmission channels can be derived from the same fibre by using different portions of the optical bandwidth for each channel. This mode of operation is known as WDM and, when using this, bit rates in excess of tens of Gbps can be achieved.

A multimode cable comes with two different core sizes: 50 micron or 62.5 micron. Multimode fibres have a large core which allows less critical alignment and can be used with low-cost LED technology.

Step-index Fibre

Fibres with an abrupt change in refraction index are called step-index fibres. In step-index fibres, changes in refractive index are made in single steps. This is the simplest, least-expensive type of fibre-optic cable. It is also the easiest to install. The core is usually 50 or 62.5 microns in diameter; the cladding is normally 125 microns.

Graded-index Fibre

A cable with a gradual change in refraction index is called graded-index cable, or graded-index multimode. This FOC type has a relatively wide core, like single-step multimode cable. The change occurs gradually and involves several layers, each with a slightly lower index of refraction. A gradation of refraction indexes controls the light signal better than the step-index method (Figure 5.8).

Tata Consultancy Services decided to standardize the network cabling system for its Bangalore Office Center of Excellence. TCS preferred to go with a standard network cabling system and passive components from ADC Krone. It also went by best practices and colour-coded and labelled the cabling infrastructure. The most important benefit from the installation was the establishment of a robust structured network cabling system. The colour-coded cables ensured efficient manageability of the network infrastructure while complimenting the architecture of the facility.

Source: dqindia07

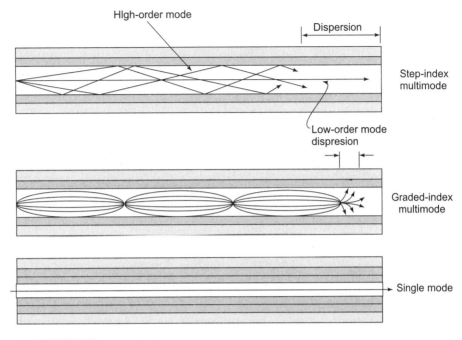

FIGURE 5.8 Types of Fibre

5.6.3 Packing Density

There are four broad application areas into which FOC can be classified: aerial cable, underground cable, sub-aqueous cable, and indoor cable. The two most common definitions used in the cable industry for packing density are

- the ratio of total fibre cross-sectional area to cross-sectional area within the core tube and
- the number of fibres per cable cross sectional area.

5.7 Cables in Buildings

When a building is being designed, the designer must include pathways that are actually the containers for the cables or their access routes. It is important that telecommunication routes or pathways are carefully thought out and included in the initial design and costing, as well as providing scope for later modifications. Cabling types for residential installations can include UTP for data and voice, coax for home theaters, security cameras, and cable modems, and specific gauge wire for speaker systems and alarm systems. Commercial data cabling networks include cabling types like UTP/STP and fibre for data and voice, and must support a very high level of network traffic. It should guarantee support for future applications on the horizon. For commercial data communication and telecom applications, certification of cabling to performance standards such as TIA-568-B is absolutely necessary to guarantee the cabling system's ability to support the required bandwidth. Certification refers to the process of making measurements and then comparing the results to rigid pre-defined standards requirements, so that a pass/fail determination can be made.

The cabling for the building starts at the demarcation point (DP). This is the point where the responsibility passes from the access provider to the owner of the building.

The Electronics Industry Association/Telecommunications Industries Association 'Commercial Building Telecommunications Standard,' commonly referred to as EIA/TIA-568, specifies a variety of building cabling parameters, ranging from backbone cabling used to connect a building's telecommunication closets to an equipment room, to horizontal cabling used to cable individual users to the equipment closet. The standard defines the performance characteristics of both backbone and horizontal cables, as well as different types of connectors used with different types of cable.

5.7.1 Horizontal Cabling

Horizontal cabling includes all the cabling between the telecom outlet (TO) in the work area and the cable connections in the telecommunication room (TR). Horizontal cables run from the communication closets to the wall outlets.

To connect to the LAN, we have to plug the patch cable into a socket called a telecommunications outlet (TO). This outlet may be a wall socket, a surface-mounted socket, or a floor-mounted socket. They should be positioned so that the patch cable or cord does not need to be any more than 5 m (16 ft) in length and since it is highly unlikely that we will need a data feed to any device that is not electrically powered, it makes good sense for a power socket to be within 1 m (3 ft) of the outlet and at about the same height. Good cable management starts with the design of the cabling infrastructure. When installing horizontal cable, cable trays or J hooks in the ceiling are recommended to run the cable. They will prevent the cable from resting on ceiling tiles, power conduits, or air-conditioning ducts, all of which are not allowed according to ANSI/TIA/EIA-568-B.

5.7.2 Backbone Cabling

Backbone cabling is the connection between the telecom room and where the cable first comes into the building and between interconnected buildings. It is important to distinguish between backbone cables and horizontal cables. Backbone cables connect network equipment such as servers, switches, and routers and connect equipment rooms and communication closets.

5.7.3 Pathways

Pathways are space allocated for routing the cables. They are generally taken as a part of network design. Enough care is taken to plan the cable routing design so that hairpin bends are avoided. The design is translated to pathways. There are a couple of commercial raceways available to accommodate the cables and accessories. Pathways include all the structures that keep the cabling safe by supporting and protecting it and must also allow easy and clear access to the cables.

5.7.4 Glides

Cable management system is the solution to long-term efficiency in a building distributor, floor distributor, or any IT environment where patching is required. The glide vertical managers attach to each side of the 42U rack, ensuring that the front of the rack is left available for patch panels and other patching equipment.

FIGURE 5.9 Glide

The glide system (Figure 5.9) works by managing excess lengths of cable through a series of vertical spools, called slack managers, located within the glide verticals. Each glide vertical also has a rib cage, which is its own set of outer cable managers. The patented rib cages are designed to accommodate patch cords at the front of the rack and terminated solid cables at the rear of the rack. While decreasing the initial installation time, this provides greater access for technicians when maintenance is required.

5.7.5 Connectors

Connectors are used to make flexible interconnection between cables and optical devices. Connectors have significantly greater losses than splices since it is more difficult to repeatedly align the fibres with the required degree of precision.

Twisted-pair cables can have many different types of connectors. The two most common are Registered Jack (RJ) 11 and RJ45. RJ11 connectors (Figure 5.10) are commonly used in US telephones with four or six contact points. RJ45 connectors (Figure 5.11) are similar but wider, with eight contact points, and are generally used in heavy-duty computing environments.

FIGURE 5.10 RJ11 Connector

RJ45 is a standard type of connector for network cables. It looks much like a normal telephone cable connector, but larger. RJ45 connectors are most commonly seen with Ethernet cables and networks. They are typically used to terminate twisted-pair cable, RJ45 connectors are commonly used in computer networking, where the plug on each end is an RJ45 modular plug wired according to a TIA/EIA standard. The cable pairs are assigned to specific pin numbers. The pin numbers are numbered from left to right if you are looking into the modular jack outlet or down on the top of the modular plug.

The Bayonet Neill-Concelman (BNC) connector is a small, round cylinder with two small prongs on the outside that allows a connector to be attached to it. A small hole for a copper wire is provided to connect the cable to the connector. A T connector is used to connect the network adapter to the two pieces of coaxial cable.

FIGURE 5.11 RJ45 Connector

Most connector designs produce a butt joint with the fibre ends as close together as possible. The fibre is mounted in a ferrule with a central hole which is sized closely to match the diameter of the fibre cladding. The ferrule is typically made of metal or ceramic and its purpose is to centre and align the fibre as well as provide mechanical protection to the end of the fibre. The fibre is normally glued into the ferrule, then the end is cut and polished to be flushed with the face of the ferrule. The two most common connectors are the SC and ST.

SC connector This is built with a cylindrical ceramic ferrule which mates with a coupling receptacle. The connector has a square cross section for high-packing density on equipment, and has a push–pull latching mechanism. The ISO and TIA have adopted a polarized duplex version as standard and this is now being used as a low-cost FDDI connector.

ST connector This is an older standard used for data communications. This is also built with a cylindrical ceramic ferrule which mates with a coupling receptacle. The connector has a round cross section and is secured by twisting to engage it in the spring-loaded bayonet coupling. Since it relies on the internal spring to hold the ferrules together, optical contact can be lost if a force of greater than about 1 kg is applied to the connector.

Most fibre-optic connectors are designed for indoor use. The optical performance can be badly degraded by the presence of dirt or dust on the fibre ends. A single dust particle could be 10 microns in diameter, which would scatter or absorb the light and could totally disrupt a single-mode system. Connectors and patch panels are normally supplied with protective caps. These protective caps should always be fitted whenever the connectors are not mated. They not only protect patch panels from dust and dirt, but also provide protection to the vulnerable, polished end of the fibre.

5.7.6 Wiring Patterns

T568A and T568B are the two colour codes used for wiring eight-position RJ45 modular plugs. Both are allowed under the ANSI/TIA/EIA wiring standards. The ANSI/TIA/EIA-568-B standard recommends one of the two wiring patterns for modular jacks and plugs: T568-A (Figure 5.12) and T568-B (Figure 5.13). The only difference between these wiring patterns is that pin assignments for pairs 2 and 3 are reversed. T568A wiring pattern is recognized as the preferred wiring pattern for this standard because it provides backward compatibility to both one-pair and two-pair USOC wiring schemes.

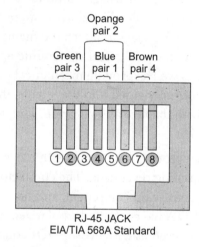

FIGURE 5.12 T568-A

5.7.7 Wall Plates

Wall plates include one or more jacks. A jack is the connector outlet in the wall plate that allows a workstation to make a physical and electrical connection to the network cabling system. The terms 'jack' and 'outlet' are often used interchangeably. Wall plates come in many different styles, types, and brands. A wall-plate system consists of a wall plate and its associated jacks. Wall plates should be placed horizontally so that they are as close as possible to work-area equipment. The vertical positioning of the wall plate depends on the national electrical code conventions. The most common configuration of a fixed-design wall plate is the single six-position (RJ-11) or eight-position (RJ-45) jack.

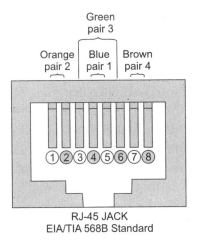

FIGURE 5.13 T568-B

Other types of fixed-design wall plates can include any combination of socket connectors, based on market demand and the whims of the manufacturer. Modular wall plates have individual components that can be installed in varying configurations depending on your cabling needs. The number of jacks a plate can have is based on the size of the plate. Fixed-design wall plates mainly come in one size. Modular plates come in a couple of different sizes.

5.8 Testing and Certification

Local area network validation is the final task associated with a network installation or upgrade project. It consists of testing, measuring, and assessing those network installation aspects that are critical for acceptable LAN operation. Ethernet cabling is generally of high quality, and so network faults are more likely to lie elsewhere. The reality is that networks do not work all the time. There are downtime and failure issues. The CIO of a company may think that the infrastructure is good and plugs in servers, routers, and other infrastructure, only to see that the cable performance does not support such features. However, there are times when the cabling itself is at fault. There may be high error rate, return loss, and crosstalk in the cables that cause issues. This calls for cable testing. Whether in installing new cable, or troubleshooting existing cable, cable testing plays an important role in the process. Common tests for data communication cabling include length, attenuation, NEXT, DC loop resistance, and return loss. The system integrator is aware of the cable parameters mentioned above. Validation provides assurance that the LAN will perform as designed, resulting in fewer trouble tickets and more satisfied users. By providing confirmation of reliable network operation, validation helps to reduce the anxiety that comes from uncertainty. Network validation provides the system integrator with assurance that there will be fewer midnight phone calls.

With the right tools, testing is quick and accurate. Continuity testing tells you if signals are making it all the way down the wire, this is the most common test. Performance testing is more suited for testing a new installation. Ethernet performance testers measure key end-to-end link characteristics.

Cable testing provides a level of assurance that the installed cabling links provide the desired transmission capability to support the data communication desired by the users. Cable test instruments are designed with a variety of focused feature sets for particular fieldwork tasks. They vary in price, performance, and application. Depending on the task the field test instrument performs, it can be classified into one of the three hierarchical groups: certification, qualification, or verification. While some features overlap between test tools, each group answers a unique testing need and provides a different level of operational assurance. Certification test tools determine whether a link is compliant with a category (TIA) or class (ISO); for example, category 6 or class E in cat6 testing. Certification is the final step required by many structured cabling manufacturers to grant their warranties for a new cabling installation.

Cable verification testers can quickly measure basic characteristics including length and faults. Certification testers can perform more in-depth cable measurements. Testers are available for both UTP and fibre-optic cables. Network testers can quickly verify Ethernet link configurations, connectivity, and service availability. Configuration details include speed and duplex, IP address and subnet mask, virtual LAN or VLAN membership and PoE provisioning. A continuity tester is a nice low-budget test tool to validate the installation at a vary basic level. Having network cabling installed by someone untrained in the requirements results in cable installations in excess of the maximum allowed length in metres.

Insertion loss, more commonly known as attenuation, is usually associated with cable length. The amount of signal lost grows proportionally to the length of the cable. Thus, the first thing to check when trying to solve this problem is the overall length of the cable.

Return loss is a measure of all reflections that are caused by the impedance mismatches at all locations along the link. It indicates how well the cabling's characteristic impedance matches its rated impedance over a

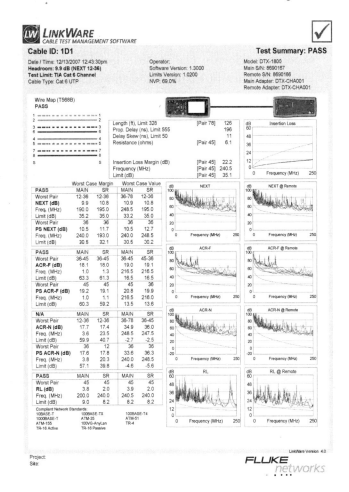

FIGURE 5.14 Fluke Test Report Sheet,
(*Courtesy*: Gemini Communication Ltd)

range of frequencies. The characteristic impedance of links tends to vary from higher values at low frequencies to lower values at the higher frequencies.

There are three general types of noise:

- Impulse noise that is more commonly referred to as voltage or current spikes induced on the cabling
- Random (white) noise distributed over the frequency spectrum
- Alien crosstalk (crosstalk from one cable to another adjacent cable)

Of the three, impulse noise is most likely to cause network disruptions.

Most cable analysers have impulse noise test capabilities. Impulse and random noise sources include nearby electric cables and devices, usually with high current loads. These may include large electric motors, elevators, photocopiers, coffee makers, fans, heaters, welders, compressors, and so on. Another less obvious source is radiated emissions from transmitters, including TV, radio, microwave, cell-phone towers, hand-held radios, building security systems, avionics, and anything else that includes a transmitter more powerful than a cell phone. Some cable analysers will average this sort of noise out of the test results. Figure 5.14 shows the impact of radiated emissions on cabling.

Summary

This chapter introduced the basics of cabling, types of cabling, structured cabling, and horizontal cabling. The network infrastructure is an important component in IP network design. A well-thought out network infrastructure not only provides reliable and fast delivery of that information, but it is also able to adapt to changes, and grow as the business expands. Four types of transmission media are available for data communication, namely copper wire, coaxial cable, optical fibre, and air. Optical fibre is the most efficient medium, as it has the lowest signal attenuation and is insensitive to electrical noise. Light-emitting diodes are the simplest and lowest-cost light source used for fibre optic communications. They have the broadest spectrum of the sources used, which limits the distance due to fibre dispersion, and have relatively slow transient behaviour.

The communication networks in use today started with different purposes. Commercial LAN/WAN systems have been expanded to provide computer voice and video to their users. Optical-fibre systems, developed mostly in telephone company laboratories, became an important contribution to the communications industry. Someone once said 'to measure is to know'. The development of optical fibre systems as a method of signal transmission and reception provided a much needed solution for cable systems faced with the problem of extending their plant beyond the coaxial cable system capabilities.

Key Terms

Cable It is the transmission medium of copper wire or optical fibre wrapped in a protective cover.

10Base2 It is the Ethernet specification for thin coaxial cable, which transmits signals at 10 Mbps (megabits per second) with a distance limit of 185 metres per segment.

10Base5 It is the Ethernet specification for thick coaxial cable, which transmits signals at 10 Mbps (megabits per second) with a distance limit of 500 metres per segment.

10BaseF It is the Ethernet specification for fibre optic -cable, which transmits signals at 10 Mbps (megabits per second) with a distance limit of 1000 metres per segment.

10BaseT It is the Ethernet specification for unshielded twisted pair cable (category 3, 4, or 5), which transmits signals at 10 Mbps (megabits per second) with a distance limit of 100 metres per segment.

Fibre-optic cable It is a cable consisting of a centre glass core surrounded by layers of plastic, that transmits data using light rather than electricity. It has the ability to carry more information over much longer distances.

RJ-45 It is a standard connector used for unshielded twisted-pair cable.

Twisted pair Network cabling that consists of four pairs of wires that are manufactured with the wires twisted to certain specifications is called twisted pair. Available in shielded and unshielded versions.

REVIEW QUESTIONS

5.1 What is the data rate for gigabit Ethernet?
5.2 When was the first major standard describing a structured cabling system released?
5.3 Identify the subsystems of a structured cabling system.
5.4 What is the difference in CAT 5 and CAT 5e?
5.5 What improvements will CAT6 and CAT7 cable provide?
5.6 What is a benefit of using shielded twisted-pair cabling?
5.7 Which cable, UTP or STP, is preferred by the industry?
5.8 What are the components of fibre optic system?
5.9 Describe the various types of FOC.
5.10 What is the need for testing installed cable?
5.11 Differentiate between return loss and insertion loss.
5.12 How is the ROI calculated for cabling systems?

Projects

5.1 An organization dealing with uranium and uranium isotopes wants to expand its office. Information technology requires a couple of servers and about 50 machines and five printers to be installed in its premises. Some of the machines need to be interfaced with radiation readers available in the company. The company prefers to go with structured cabling. As a CIO you are requested to design the structured cabling system and arrive at the ROI for the organization.

5.2 The administrator in charge of all computer users in a small office has decided on the implementation of a peer-to-peer network architecture and desires to allow access to some files on his/her computer station. The administrator has created user accounts for everyone else. Using Microsoft Windows for Networking, the administrator assigned the appropriate permissions for these accounts. With this arrangement, will each user have access to all files?

5.3 Your company is expanding and needs to install a new network cable system into the same building. However, immediately below the area where the new networks will be located is a manufacturing company that uses many machinery equipment. It is estimated that about 500 metres of new wiring will be required to interconnect the existing network with the new network. What type of cable would you recommend?

REFERENCES

Bailey, David, Edwin Wright 2003, *Practical Fibre Optics*, Elsevier.

CXO Today 2005, http://www.cxotoday.com/cxo/ jsp/article.jsp?article id=69016 cat id=910, dt, Nov 7, downloaded on 3 February 2009, 07:00 a.m.

Decusatis, Casimer 2007, *Handbook of Fibre Optic Communication*, Third Edition, Academic Press.

DQIndia, 2007 http://dqindia.ciol.com/content/structuredcabling/2007/107052203.asp, dt May 2007, downloaded on 3 Febuary 2009 07:30 a.m.

Evans, Nicholas D. 2003, *Business Innovation and Disruptive Technology*, Pearson Education.

Hill, Goff 2007, *The Cable and Telecommunications Professionals' Reference*, Elsevier.

Kaminow, Ivan P., Tingye Li, Alan E. Willner 2008, *Optical Fibre Telecommunications*, Elsevier.

Kaminow, Ivan P., Thomas L. Koch 1997, *Optical Fibre Telecommunication IIIA* Academic Press.

Kaminow, Ivan P., Thomas L. Koch 1997, *Optical Fibre Telecommunication IIIB* Academic Press.

Lam, Cedric 2007, *Passive Optical Networks, Principles and Practice*, Elseivier.

Reynders, Deon, Steve Mackay, and Edwin Wright 2005, *Practical Industrial Data Communications Best Practice Techniques*, Elsevier.

Salah, Aidarous, Thomas Plevyak 2003, *Managing IP Networks*, Wiley Inter Science.

Voicendata 2004, http://voicendata.ciol.com/content/structuredcabling/104120403.asp downloaded on 3 February 2009, 07:00 a.m.

CHAPTER 6

Wireless Infrastructure

When life seems hard, the courageous do not lie down and accept defeat.
–Queen Elizabeth II

Learning Objectives

After reading this chapter, you should be able to understand:

- the need for wireless connectivity
- the different types of wireless options
- the evolution of IEEE 802.11 standards

6.1 Introduction

This chapter provides a brief overview of wireless systems and standards. The evolution of wireless systems from voice-centric circuit-switched systems to data-centric packet-switched systems is discussed. Radio-frequency (RF) wireless communication systems have been around for many years with applications ranging from garage door openers to satellite communication. The technology has been advancing at an unprecedented rate and its impact is evident in our daily lives. In many parts of the world, wireless communication is the fastest growing area of the communication industry, providing a valuable supplement and alternative to existing wired networks. Wireless systems encompass a wide range of information transmission mechanisms, including cordless phones, paging systems, cellphones, satellite communication systems, maritime mobile systems, industrial and medical monitoring systems, infrared (IR) remote controls, and so on. These systems have unique operating frequency bands. The choice of the frequencies is often determined by the range of operation, the medium through which the signal traverses, and the amount of data to be transmitted.

Unlike wired/fixed networks that require a physical connection to a network termination point, the user devices associated with wireless networks—mobile phones, personal digital assistants (PDAs), notebook computers, pocket personal computers (PCs), laptops, etc.—communicate with the network by radio transmissions. In some instances, wireless networks are used to replace existing wired local area networks (LANs). IEEE 802.16 standard-based systems, commonly referred to as WiMAX, are being proposed for carrying data traffic. Popular wireless LAN and wireless personal area network (PAN) standards are also discussed.

Wireless systems have been around for over a century. Guglielmo Marconi successfully transmitted Morse code from Cornwall, England to St-John's, Canada in 1901. The wireless technology has come a long way since then. Radio waves are electromagnetic waves propagated through space. Because of their varying characteristics, radio waves of different wavelengths are employed for different purposes and are usually identified by their frequency. The shortest waves have the highest frequency while the longest waves have the lowest frequency. In honour of German radio pioneer Heinrich Hertz, his name has been given to the cycle per second (hertz, Hz); 1 kilohertz (kHz) is 1000 Hz, 1 megahertz (MHz) is 1 million Hz, and 1 gigahertz (GHz) is 1 billion Hz. Radio waves range between frequencies of 10 kHz to 1 GHz.

6.1.1 Radio Waves

Radio frequencies or radio waves constitute the portion of the electromagnetic spectrum extending from 30 kHz to 300 GHz.

The entire RF spectrum is classified into different bands and ranges, based on propagation properties. Radio waves (Table 6.1) include the following ranges.

Table 6.1 Radio Waves and Frequency

Frequency Band	Frequency Range	Propagation Characteristics
Very-low frequency (VLF)	< 30 kHz	Low attenuation day and night; high atmospheric noise level
Low frequency (LF)	30–300 kHz	Slightly less reliable than VLF, absorption in daytime
Medium frequency (MF)	0.3–3 Mhz	Attenuation low at night, high in day; atmospheric noise
High frequency (HF)	3.0–30 Mhz	Omni-directional energy radiation; quality varies with time of day, season, and frequency
Very-high frequency (VHF)	30–300 Mhz	Direct and ground waves; cosmic noise; antenna design and height is critical
Ultra-high frequency (UHF)	0.3–3 Ghz	LOS; repeaters are used to cover greater distances; cosmic noise
Super-high frequency (SHF)	3.0–30 Ghz	LOS; atmospheric attenuation due to rain (> 10 Ghz), oxygen and water vapour
Extremely-high frequency (EHF)	30–300 Ghz	LOS; millimetre wave; atmospheric attenuation due to rain, oxygen and water vapour

- Short-wave: An electromagnetic wave with a wavelength of approximately 200 metres or less, especially a radio wave in the 20 to 200 metre range.
- Very-high frequency (VHF): This is the band of radio signal frequencies, ranging from about 30 MHz to 300 MHz, with corresponding wavelengths ranging from about 1 to 10 metres. This frequency is used for FM and amateur radio broadcasting and for television transmission.
- Ultra-high frequency (UHF): This is a short-wave radio frequency ranging from about 300 MHz to 3000 MHz, with corresponding wavelengths ranging from about 10 to 100 cm. This frequency is used mainly for communication with guided missiles, in airplane navigation, radar, and in the transmission of television signals.

Details of the entire range can be had from the above table. Radio waves propagate in space in various forms. The characteristics of the propagating waves are of interest in many wireless communication systems designs. Propagating radio waves can be classified as direct (or free space), ground (or surface), tropospheric, and ionospheric. Because the wireless channel is not a reliable propagation medium, techniques to achieve reliable and efficient communication are necessary and have been established.

Currently, cellular mobile communication is undoubtedly the most popular RF wireless communication system. In cellular systems, instead of using a single large coverage area with one high-power transceiver (used in traditional mobile systems), the coverage area is divided into small, localized coverage areas called cells. A cell is the basic geographic unit of a cellular system, commonly represented as a hexagon. Each cell has a base station (BS) or cell site, which in comparison uses much less power. The BS can communicate with mobiles as long as they are within range. To prevent interference, adjacent cells are assigned different portions of the available frequencies

6.1.2 ISM Band

This is the globally available unlicensed industrial, scientific, and medical (ISM) band of 2.4 GHz. Bluetooth shares this band with 802.11b wireless LAN or wireless fidelity (Wi-Fi). Frequency hopping spread-spectrum transmission scheme is used to minimize the interference risks. The band is divided into 79 channels, with an inter-space of 1 MHz. The frequency is shared by using the time division duplex (TDD) technology. Because of the strict regulations existing in the ISM band and due to the principle of power consumption, there are limitations in the power level.

The important features of ISM band are the frequencies of operation—transmit power limitation of 1 W and low power with any modulation. The rules restricted the transmit power to 1 W and enforced the modems radiating more than 1mW to employ spread spectrum technology.

6.1.3 Antenna

In electronics, an antenna is a device used to propagate radio or electromagnetic waves or to capture radio waves. Antennas are necessary to transmit and receive radio, television, microwave, telephone, and radar signals. Most antennas for radio and television consist of metal wires or rods connected to the transmitter or receiver. Many communication systems require radiation characteristics that are not achievable by a single antenna element.

6.1.4 Spread Spectrum

Spread spectrum is a communication technology. Spread spectrum development began during World War II, with the earliest studies dating from the 1920s. Details on spread spectrum remained classified until the 1980s. Spread spectrum was originally used by the military to provide reliable data transmissions that are resistant to jamming. It is still used today in military communication systems. The term 'spread spectrum' refers to the broad spectral shape of the transmitted signal. A spread spectrum radio infrastructure consists of transmission towers or radio transceivers mounted on buildings, poles, or street lights that are equipped with directional antennas for transporting signals via narrow beams directly to destination sites. Spread spectrum technology distributes a transmitted signal over a much wider frequency than the minimum bandwidth necessary for signal transport. Spread spectrum networks employ direct-sequence spread spectrum (DSSS) and frequency-hopping spread spectrum (FHSS) technologies for enabling wireless networking operations. Frequency-hopping spread spectrum technology safeguards information transport by employing pseudo-random algorithms that enable signals to jump or hop from frequency-to-frequency. Direct-sequence spread spectrum implementations support broadband applications and services at rates reaching 11 Mbps. With DSSS transmission, a transceiver or sending station transmits a data signal in conjunction with a higher data rate bit sequence to counteract signal interference.

Spread spectrum technology supports a diverse array of applications, including television broadcasts, air traffic control, Citizens' Band (CB) radio programs, radar operations, and freight transport. In addition, spread spectrum networks enable digital cellular communications services, web browsing, emergency medical assistance, critical government functions, SMS, and e-mail relay.

6.2 Wireless Networks and Connectivity

Data transmissions across the network can occur in two forms, either analog or digital. Analog signals exist in an infinite number of values. Just like an analog watch, in which the hands glide across the seconds and minutes, there are many time values displayed. Digital signals exist in a finite number of values. Digital watches display the time down to the second or fraction of a second, but these values are always displayed as definite amounts.

Analog transmissions are displayed using a continuous curve in which the signal changes from one value to another; as a value increases from 0 to 1, it becomes every value along the way. Digital transmissions are displayed using pulses to indicate that the change from one value to another is instant; the value changes instantly from 1 to 0.

When the signal is moved over the media air, we claim wireless data movement. This is accomplished by a combination of different technologies.

Wireless networks are evolving on two major fronts. First, radio access systems are evolving to third and fourth generation systems that can support significantly higher system capacity and per-user data rates with enhanced quality-of-service (QoS) support capabilities. Second, wireless Internet protocol (IP) networking technologies are profoundly changing the overall wireless network architectures and protocols. Wireless Internet protocol (IP) networks are more suitable for supporting the rapidly growing mobile data and multimedia applications. Internet protocol technologies bring the globally successful Internet service creation and offering paradigm to wireless networks, bringing the vast array of Internet services to mobile users, and providing a successful platform for fostering future mobile services. IP-based wireless networks offer a range of advantages over traditional circuit switched wireless networks. For example, IP-based networks are more suitable for supporting the rapidly growing mobile data and multimedia applications. IP-based wireless networks bring the globally successful Internet service creation and offering paradigm into wireless networks. This not only makes Internet services available to mobile users but also provides a proven successful platform for fostering future mobile services. Furthermore, IP-based protocols are independent of the underlying radio technologies, and therefore are better-suited for supporting services seamlessly over different radio technologies and for achieving global roaming.

6.2.1 Overview of Standards

Standards are developed because of a need to connect systems together. Standards evolve due to the search for compatibility between the systems. There are many standards available, and many more are in development. Choosing the best ones for your system can be a challenge, because there are so many alternatives. It is sometimes difficult to select standards that are compatible with systems and technologies you already have in place. The standards continue to evolve while your systems might be deployed and running for years. A standard is usually developed by a group of like-minded people who may operate commercially in competition with one another.

Standards create business relationships and drive commercial activity. That may mean agreeing with one of your competitors to deliver consistent output from each of your competing products. This is driven by the hope that the overall marketplace will grow larger, and your own competitive slice of the cake will be bigger even if your market share remains the same.

Suppliers use standards to make their systems more flexible and capable for integration. Purchasers like standards because they help them to avoid lock-in. We all need standards to ensure systems and software are interoperable. Different methods and standards of wireless communication have developed across the world, based on various commercially-driven requirements. These technologies can roughly be classified into four individual categories, based on their specific application and transmission range.

The family of 802.11 standards from the Institute of Electrical and Electronics Engineers (IEEE) includes several varations on high-speed wireless communication protocols. In 1997, the IEEE created the first wireless local area network (WLAN) standard. They called it 802.11 after the name of the group formed to oversee its development. 802.11 is the generic name of a family of standards for wireless networking. 802.11 standards define rules for communication on WLANs. Popular 802.11 standards include 802.11a, 802.11b, and 802.11g.

The IEEE expanded on the original 802.11 standard in July 1999, creating the 802.11b specification. 802.11b supports bandwidth upto 11 Mbps, comparable to traditional Ethernet. 802.11b uses the same unregulated radio signaling frequency (2.4 GHz) as the original 802.11 standard.

While 802.11b was in development, the IEEE created a second extension to the original 802.11 standard called 802.11a. In fact, 802.11a was created at the same time. Due to its higher cost, 802.11a is usually found on business networks, whereas 802.11b better serves the home market. 802.11a supports bandwidth up to 54 Mbps and signals in a regulated frequency spectrum around 5 GHz. This higher frequency compared to 802.11b shortens the range of 802.11a networks. The higher frequency also means 802.11a signals have more difficulty penetrating walls and other obstructions.

In 2002 and 2003, WLAN products supporting a newer standard called 802.11g emerged on the market. 802.11g attempts to combine the best of both 802.11a and 802.11b. 802.11g supports bandwidth up to 54 Mbps, and it uses the 2.4 Ghz frequency for greater range

The newest IEEE standard in the Wi-Fi category is 802.11n. It was designed to improve on 802.11g in the amount of bandwidth supported by utilizing multiple wireless signals and antennas (called MIMO technology) instead of one. 802.11n will work by utilizing multiple wireless antennas in tandem to transmit and receive data. The associated term MIMO (multiple input, multiple output) refers to the ability of 802.11n and similar technologies to coordinate multiple simultaneous radio signals. It increases both the range and throughput of a wireless network.

When this standard is finalized, 802.11n connections should support data rates of over 100 Mbps. 802.11n also offers a somewhat better range over earlier Wi-Fi standards due to its increased signal intensity.

6.2.2 Types of Wireless Deployments

Demand for remote data access, web services, pervasive communications links, and ubiquitous computing capabilities contributes to the implementation of a broad spectrum of wireless network solutions. The effectiveness of wireless network solutions in accommodating current and expected user requirements depends on factors such as budget allocations, geographical coverage, information flow, service availability, and wireless network operations and performance.

Wireless networks are usually optimized to fit different coverage areas and communication needs. Based on radio coverage ranges, wireless networks can be categorized into wireless PANs, WLANs, low-tier wireless systems, public wide-area (high-tier) cellular radio systems, and mobile satellite systems. Personal area networks use short-range low-power radios to allow a person or device to communicate with other people or devices nearby. They allow a person to communicate wirelessly with devices inside a vehicle or a room. People with PDA or laptop (notebook) computers may walk into a meeting room and form an ad hoc network among themselves dynamically.

Wireless networks have two different architectures in which they may be set up: infrastructure and ad hoc. Choosing between these two architectures depends on whether the wireless network needs to share data or peripherals with a wired network or not. The basis of infrastructure mode centres around an access point, which serves as the main point of communications in a wireless network. Access points transmit data to PCs equipped with wireless network adapters, which can roam within a certain radial range of the access point. We can use access points to extend the effective range of the wireless network. Placing access points in overlapping ranges, will enable our wireless-equipped PCs to reach remotely located users on the network. Some access points also incorporate other features which can enhance network efficiency.

Wireless network configurations employ infrared, laser, spread spectrum, microwave, and satellite technologies for supporting voice, video, and data transmission in local area and wide area environments.

6.3 Wireless Local Area Network

Wireless local area networks (WLANs) are rapidly replacing wires within homes and offices. The most common WLANs are based on the IEEE

802.11 standard. They are commonly referred to as Wi-Fi networks. Wireless LAN is a flexible data communication system implemented as an extension to, or as an alternative for, a wired LAN within a building or campus. It enables individual users to access a corporate network without hindering their mobility. A wireless network is a data communication system which uses electromagnetic media, such as radio or infrared waves, instead of cables. Wireless LAN was the first technology that was examined for wide-band local access. It was implemented based on a variety of innovative technologies and raised a lot of hopes in the technology market. Currently, WLAN technologies are comprised of infrared, UHF radio, spread spectrum, and microwave radio. These technologies can range from frequencies in the MHz (USA), GHz (Europe), to infrared frequencies. Until recently, WLANs operated in the 900 MHz ISM band. The other two ISM bands are 2.4 GHz band and 5.7 GHz band. Today, the major standards for WLANs are IEEE 802.11 and HiperLAN.

Wireless LANs can be connected to a wired LAN as an extension to the system or it can be operated independently. To help ensure Wi-Fi products perform correctly and are interoperable with each other, the Wi-Fi Alliance was created in 1999. The Wi-Fi Alliance is a non-profit organization that certifies whether products conform to the industry specification and interoperate with each other. Wi-Fi is a registered trademark of the Wi-Fi Alliance and the indication that the product is Wi-Fi certified indicates products have been tested and should be interoperable with other products regardless of who manufactured the product.

Wireless LAN technology is a form of wireless Ethernet networking standardized by the IEEE 802.11 Working Group (WG). The goal of the 802.11 WG was to create a set of standards for WLAN operation in the unlicensed portion of the ISM frequency spectrum. Using electromagnetic waves, WLANs transmit and receive data over the air, minimizing the need for wired connections. Thus, WLANs combine data connectivity with user mobility, and through simplified configuration, enable movable LANs. Wireless LAN provides a shared radio media for users to communicate with each other and to access an IP network. Its performance is dependent on the radio-propagation environment in which the WLAN operates.

Some of the other factors that can be attributed to their widespread adaptation are their low cost and the ease of installation. They can be easily deployed by individuals within buildings without a license. Over the years, WLANs have gained strong popularity in a number of vertical markets, including health care, retail, manufacturing, warehousing, and academic arenas. These industries have profited from the productivity gains of using hand-held terminals and notebook computers to transmit real-time information to centralized hosts for processing. While IEEE 802.11 technology is adequate for residential and small office/home office (SOHO) customers, the same is not always true for enterprise customers. For example, wireless-enabled print server helps enable sharing of a broad range of network-capable printers and multi-function peripherals across 802.11b wireless networks.

In fact, some chief information officers (CIOs) and information technology managers are reluctant to deploy wireless LANs. Among their concerns are security, reliability, availability, performance under heavy load, deployment, mobility, and network management. While security is often mentioned as managers' greatest worry about wireless, some of their other concerns, such as reliability, availability, performance and deployment, can be addressed through radio resource management techniques.

Advantages of WLAN

The advantages of WLAN are as follows.

- Flexibility: The nodes present within the radio coverage region can communicate. Radio waves can penetrate walls. Users can access shared information without looking for a place to plug in. Network managers can set up or increase the size of network with more machines without installing or moving wires.
- Planning: Wireless ad hoc networks allow communication with out planning. Wired networks require prior planning.
- Design: Only wireless networks allow for design of small independent devices.
- Robustness: Wireless networks can survive disasters. When the wireless devices survive, people can still communicate.

Disadvantages of WLAN

The disadvantages of WLAN are as follows.

- Quality of Services: WLANs offer lower quality as compared to the wired networks.
- Proprietary solutions: Many companies have come up with proprietary solutions due to slow standardization procedures.

- Restrictions: All wireless products should comply with the national regulations.
- Cost: The cost of wireless interfaces are more compared to wired interfaces.

The Wi-Fi technology is inexpensive due to factors like the use of free, unlicensed radio spectrum at 2.4 GHz and 5.8 GHz bands which obviates heavy investment in private, dedicated radio spectrum, and the widespread use of Wi-Fi equipped PDAs, laptops, and even phones provides significant business opportunities and justification for deploying ad hoc Wi-Fi networks. Since all Wi-Fi devices must comply with the IEEE standard, Wi-Fi products from multiple vendors can be mixed and matched for seamless operation. This has driven the cost of individual Wi-Fi devices low, which in turn, made deployment of Wi-Fi-based networks covering medium to large areas an attractive and, even necessary, business investment. The reasons for the popularity of wireless networks over the wired ones are highlighted below.

- Mobility: Wireless LANs can provide users with real-time information within their organization without the restrictions inherent with physical cable connections.
- Installation speed and simplicity: The installation of WLANs does not involve the tedious work of pulling cables through walls and ceilings.
- Installation flexibility: Wireless LANs allow access from places unreachable by network cables.
- Cost of ownership: Overall installation expenses and life-cycle costs of WLANs are significantly lower than wired LAN. The discrepancy is even higher in dynamic environments requiring frequent moves and changes.
- Scalability: Wireless LANs can be configured relatively easily since no physical arranging of network cables is required.

Wireless cards are an essential component of Wi-Fi and perform same functions as Ethernet cards. These use Personal Computer Memory Card International Association (PCMCIA) to connect and thus can be used in desktop computer. Peripheral component interconnect (PCI) adopters are available to accommodate these cards to a free PCI slot in a PC.

6.4 Wi-Mesh

Wireless mesh networks (WMNs) (IEEE 802.11s) are composed of routers and mesh clients. Router nodes have minimum mobility forming a distribution backbone. Clients may be mesh (i.e., act as router) or conventional clients. Mesh clients may be fixed or mobile and can interconnect between them or with routers. Mesh networks improve the restrictions in range, bandwidth, reliability, and performance.

A wireless mesh inherently is more resilient and fault-tolerant than a centralized infrastructure network. Provided there are a sufficient number of nodes, the network is able to sustain temporary congestion, individual node failure, and localized interference. The built-in ability to find neighbour nodes, set-up connections, find optimal traffic paths—all these standard features make the products based on 802.11s mesh less failure-prone. Mesh devices can be set up to create a high-bandwidth network among themselves without the need for a central access point. This ability for peer-to-peer connectivity opens a host of new applications in the enterprise and home markets.

6.5 WiMAX

There have been growing demands for broadband wireless applications, such as audio and video streaming, near real-time downloading services, broadband Internet access services, and voice over Internet protocol (VoIP) services. On the other hand, mobile devices, such as smart phones, PDAs, and laptops are all quite ready to embrace such services. While all these developments are almost ready to support broadband services, the wireless access bandwidth has become a bottleneck. Worldwide interoperability for microwave access (WiMAX) technology is one of the potential solutions to respond to this demand. The most successful wireless metroplotian area network (MAN) technology is based on the IEEE 802.16 standard, which is often referred to as WiMAX networks. Worldwide interoperability for microwave access is an exciting new technology that delivers wireless high-speed access, enabling fixed and mobile broadband services over large coverage areas. It is an alternative to third-generation cellular systems to provide broadband connections over long distances, and it is considered to be a fourth-generation technology.

WiMAX has an IP-based wireless access architecture, which contains three parts: user terminal devices, access service network (ASN), and core service network (CSN). A user terminal device can be a fixed or portable/mobile terminal device, which supports the fixed/nomadic/mobile usage scenarios. Each device can establish a connection link to a WiMAX Base Station (BS), and perform authentication and registration through an access gateway in the CSN.

WiMAX operates in one of two ways. The first is through line of site from one tower to another. A steady stream of data is beamed from one tower directly to another, up to 30 miles under ideal conditions. The effective distance can be affected depending on weather conditions and other obstacles that may be in the way. The second way WiMAX operates is through non-line-of-sight, similar to the way Wi-Fi works. When WiMAX is not relying on line-of-sight, its effective distance is cut to about a 5 mile radius. The 802.16-based implementations, named WiMAX, are used for applications that include mobile broadband, last mile fixed broadband connections, hotspots and cellular backhaul, and high-speed connectivity for businesses. WiMAX provides high-throughput broadband connections over long distances. WiMAX provides metropolitan area network connectivity at speeds of up to 75 Mbps covering quite a few kilometres.

IEEE 802.16 and 802.16e are standards for broadband wireless access technology. They are mainly aimed at providing broadband wireless access (BWA) and thus it may be considered as an attractive alternative solution to wired broadband technologies like digital subscriber line (xDSL) and cable modem access. Its main advantage is fast deployment which results in cost savings. Such installation can be beneficial in very crowded geographical areas like cities and in rural areas where there is no wired infrastructure. IEEE 802.16 was formally approved by the IEEE in 2001. The low-cost, all-IP network architecture and backwards compatibility with existing 2G and 3G cellular network deployments makes WiMAX easier and more cost-effective to deploy and operate than current mobile wireless data solutions.

Video streaming applications, including IPTV, VoD, and contemporary web- and P2P-based video services, require significantly more network resources than VoIP. Video is considered a premium service. Delivering high data rates and constraining jitter consistently, over both short and long periods, are difficult challenges to meet, in particular when wireless communication is involved. In order to deliver high-quality video over a network, support for high data rates, bounded latencies and low delay jitter are typically required. Of course, the exact bounds on all of these parameters depend on the specifics of the video streaming application. Different video applications have slightly different demands with respect to the network capabilities. For example, live video streaming is the most demanding: the maximum acceptable end-to-end delay and jitter need to be low, so that the interactivity and real-time qualities of a live video feed are preserved and smooth playback of the video content is possible.

WiMAX comes in two flavours: fixed WiMAX and mobile WiMAX. Fixed WiMAX, based on IEEE 802.16-2004 standard, is ideally suited for delivering wireless last mile access for fixed broadband services, similar to DSL or cable modem service. Mobile WiMAX, based on IEEE 802.16-2005 standard, supports both fixed and mobile applications with improved performance and capacity while adding full mobility.

6.5.1 Fixed WiMAX

Fixed WiMAX, based on the IEEE 802.16-2004 Air Interface Standard, has proven to be a cost-effective fixed wireless alternative to cable and DSL services.

6.5.2 Mobile WiMAX

Mobile WiMAX is the next revolution in wireless technology that will enable pervasive, high-speed connectivity to meet the ever-increasing demand for broadband Internet on the go. Mobile WiMAX or 802.16e standard was ratified by the IEEE in late 2005 as a potential to emerge as a real viable competitor to existing 3G technologies. Hundreds of companies have contributed to the development of the technology and many companies have announced product plans for this technology. This addresses another important requirement for the success of the technology—the low cost of subscription services for mobile internet. The broad industry participation will ensure economies of scale that will help drive down the costs of subscription and enable the deployment of mobile Internet services globally, including emerging countries.

The Mobile WiMAX air interface utilizes orthogonal frequency Division multiple access (OFDMA) for improved multipath performance in non-line-of-sight environments and high flexibility in allocating resources to users with different data rate requirements. A variety of antenna techniques are supported by mobile WiMAX to increase throughput. Mobile WiMAX technology incorporates the most advanced security features currently used in IEEE 802 wireless access systems. These include extensible authentication protocol (EAP) based authentication, advanced encryption standard (AES)–based authenticated encryption, cipher-based message authentication code (CMAC) and hashed message authentication code (HMAC)-based control message protection schemes.

Access to the mobile WiMAX network is possible through a wide selection of devices and form factors which support varying degrees of mobility such as indoor units, also called CPE or customer premise equipment, outdoor units which are mounted outside a building and use high-gain antennas, PC cards that fit into slots in a laptop, embedded WiMAX network modules integrated into a laptop and hand-helds and PDAs with WiMAX interfaces integrated on the system board. The scalable architecture, high-data throughput and low-cost deployment make mobile WiMAX a leading solution for wireless broadband services. Mobile WiMAX vendors include baseband silicon makers, chipset vendors, component manufacturers, along with base station and customer premises equipment makers, and carriers. Intel, for example, plans to incorporate a mobile WiMAX chipset into its Centrino wireless package, which today supports only Wi-Fi connections.

6.5.3 Wireless Personal Area Network

The concept of personal area network (PAN) is relatively new. A PAN, basically, is a network that supports the inter-operation of devices in personal space. In this sense, it is a network solution that enhances our personal environment, either work or private, by networking a variety of personal and wearable devices within the space surrounding a person and providing the communication capabilities within that space and with the outside world. A wireless PAN (WPAN) is the natural evolution of this concept. Since a WPAN has by definition a limited range, compatible devices that are encountered along its path can either link to it or leave it when they go out of its range.

Of all current wireless technologies, Bluetooth is the most promising and employed for many real-life applications. The Bluetooth standard is an open standard published by an industry-based association called the Bluetooth Special Interest Group (SIG). Applications using Bluetooth have become important in hot spots such as at hotels, shopping malls, railway stations, airports, and so forth. Bluetooth is a well-established communications standard for short distance wireless connections. A wide range of peripherals, such as printers, personal computers, keyboards, mouse, fax machines, and any other digital device, can be part of a Bluetooth network. The basic form of a Bluetooth network is the piconet. The system control for Bluetooth requires one device to operate as the coordinating device (a master) and all the other devices are slaves. This is a network with a star topology with up to eight nodes participating in a master/slave arrangement. More specifically, the master is at the centre and transmits to the slaves; a slave can only transmit to the master, provided it has been given prior permission by the master. Bluetooth devices may have different power classification levels.

6.5.4 Business Benefits

Some of the advantages of Bluetooth as a technology include:

- low cost,
- considerable degree of interference-free operation,
- speed,
- appropriate range,
- low power,
- connectivity,
- provision for both synchronous and asynchronous links, and
- wide availability in mobile phones, PDAs, and other devices.

6.5.5 Wireless System

A typical wireless communication system consists of a source of information, a hardware subsystem called the transmitter, the channel or means by which the signal travels, another hardware subsystem called the receiver, and a destination of the information (the sink). The source supplies the information to the transmitter in the form of audio, video, data, or a combination of the three. The Tx and Rx combination is used to convert the signal into a form suitable for transmission and then to convert the signal back to its original form. This is achieved through the process of modulation (or encoding) at the Tx side and demodulation (or decoding) at the Rx side. The channel is the medium by which the signal propagates, such as free space, unshielded twisted pair (UTP), coaxial cable, or fiber-optic cable (FOC). In wireless communication, the channel is free space.

6.6 Bluetooth

The computing, communications, and consumer electronics industries have introduced many benefits to today's consumers. Of course, they have also introduced many problems, not the least of which is the necessity of connecting all these devices to each other, usually with a phalanx of cables and wires that are both annoyingly messy and confusing.

Engineers saw a need for the wireless transmission technology that would be cheap, robust, and flexible and consume little power. There was a need for technology that from one side could replace cables, but also offer new possibilities and cross the boundaries set by the existing solutions. The development of Bluetooth technology industry standard started by 1998 when the major companies—Ericsson, IBM, Intel, Nokia, and Toshiba—formed the Bluetooth SIG to develop and promote a global solution for short-range wireless communication operating in the unlicensed 2.4 G.Hz ISM. The Bluetooth specification was developed by the Bluetooth SIG, an industry consortium. Bluetooth is a specification designed to replace cables and enable wireless communication between small mobile devices. Bluetooth wireless technology was created to enable many different usage models from networking to cable replacement.

To promote the widespread use of this new technology, the SIG offered all the intellectual property included in the Bluetooth specification royalty free to members willing to adopt it. The aim of enabling seamless voice and data communication via short-range links and allowing users to connect a large range of devices easily and quickly, without the need for cables was the goal of Bluetooth.

Bluetooth technology is essentially a protocol for wireless connectivity of a diverse set of devices ranging from PDAs, mobile phones, notebook computers, to cooking ovens, refrigerators, thermostats, etc. in home, enterprise, mobile, and other like environments. Implementation of these usage models and of issues specific to Bluetooth, such as service discovery and security, involve mapping the Bluetooth protocol stack into operating systems frameworks in order to ensure seamless integration. The Bluetooth specification defines a low-power short-range radio and protocol stack supporting a personal communication bubble. The radio has an operational range of 10 metres at 0 dbm and operates in the unlicensed 2.4 GHz ISM frequency band.

The Bluetooth wireless specification includes radio frequency, link layer and application layer definitions for product developers for data, voice, and content-centric applications. The specification documentation contains the information necessary to ensure that diverse devices supporting the Bluetooth wireless technology can communicate with each other worldwide. The Bluetooth specification, currently at version 1.1, comprised of two parts: one being the core specification and the other being the profile specification. The core specification defined the radio characteristics and the communication protocols for exchanging data between devices over Bluetooth radio links. The profile specification defined the usage of Blue tooth protocols to realize a number of selected applications. The Bluetooth specification was primarily written as an implementation manual. Using Bluetooth wireless technology, devices will have the ability to form networks and exchange information.

6.6.1 Bluetooth Connections

Thus, Bluetooth is a standard for short-range, low-cost, low-power wireless access technology. Today, Bluetooth interface offers a full wireless networking solution for ad hoc networks. Bluetooth is a short-range wireless communication standard defined as cable replacement for a PAN. Bluetooth enables portable devices to connect and communicate in a wireless way via short-range ad hoc networks. The main advantages of Bluetooth are its robustness, low complexity, low power emission, low cost and universality. The range of a typical Bluetooth device is 10m. Stability is ensured by implementing a frequency hopping spread spectrum scheme which makes Bluetooth robust against interference from other piconets. Two or more devices sharing the same channel form a piconet. There is one master device and up to seven active slave devices in a piconet. Piconet formation takes place by establishing a connection between a master device and slave device through 'inquiry' and 'page' stages. The 'inquiry' stage identifies the available devices, while the 'page' process connects the master with the slaves discovered. Once a link is formed and a connection is made, a service discovery protocol (SDP) is established, allowing Bluetooth devices to discover what other services Bluetooth devices can offer.

Despite all the hype over the past few years, the very first Bluetooth-enabled devices have hit the market very recently; it will take several years for Bluetooth-enabled products to become both common and affordable, and even more years for Bluetooth products to replace traditional wired products. Bluetooth can make the real world a much more mobile, much more flexible, much more user-friendly place.

6.6.2 Benefits of Bluetooth

Bluetooth is more than just a cable-replacement technology. It is also a technology that enables any electronic device to communicate with any other electronic device, automatically. This means that, over short distances (30 feet or so), a cellphone or PDA can connect to, synchronize with, and even control the other electronic devices in a home or office such as personal computer, printer, television set, home alarm system, or home/office telephone system. All of this communication can take place in an ad hoc fashion, without the human intervention. If Bluetooth truly becomes the enabling technology for wireless connections and communications, we may expect many new and innovative applications to emerge—applications that could have the same impact on our future lives as the first computers and mobile phones had on our recent past.

Summary

The wireless revolution is just beginning. The confluence of communication and computing has created many new applications for wireless systems. Two main development directions in untethered communications can be identified, wide-area communications, with the omni-present cellular systems as the most representative example, and short-range communications, involving an array of networking technologies for providing wireless connectivity over short distances, for instance WLANs, WPANs, wireless body area networks (WBANs), wireless sensor networks (WSNs), Bluetooth, etc. Recent years have witnessed an enormous growth in interest in the metropolitan wireless networks. The throughput, capacity, and range of wireless systems are constantly being improved. The IEEE 802.11 WLAN standard has been gaining popularity worldwide in recent years.

The IEEE 802.11b, commonly referred to as wireless fidelity (Wi-Fi), is by far the most successful commercially. There are other standards in the IEEE 802.11 family being developed to overcome the limitations of IEEE 802.11b. WiMAX, based on the IEEE 802.16 standard, defines wireless networks combining key characteristics of wide-area cellular networks as well as short-range networks, namely mobility and high data throughput. WiMAX products and certification follow the IEEE 802.16 air interface specifications. Bluetooth consists of the most promising and well-established wireless technology for a vast number of applications.

Key Terms

Access point It is a device that 'connects' wireless communication devices to create a wireless network. A wireless access point acts as the network's arbitrator, negotiating when each nearby client device can transmit.

Access point device It is a device that bridges wireless networking components and a wired network.

Antenna it is a device used for receiving or transmitting signals.

Ad hoc It is a class of wireless networking architectures in which there is no fixed infrastructure or wireless access points.

Ad hoc network It is a peer-to-peer 802.11 network formed automatically when several computers come together without an access point.

Cellular It is a wireless communication technique used in mobile phones.

IEEE The IEEE (pronounced Eye-triple-E) is a non-profit, professional organization that develops international standards. The IEEE Standards Association has a portfolio of some 900 active standards and more than 400 standards in development.

MAC address The MAC (media access control) address is the hardware address of a device connected to a network. The MAC address is a computer's unique hardware number for the network interface card (NIC). Note that the MAC address belongs to the NIC rather than to the computer itself. The MAC address contains 48 bits.

Spectrum It is the range of electromagnetic radio frequencies used in signal transmission.

Review Questions

6.1 Discuss the radio spectrum and list out the uses of each spectrum.
6.2 Briefly outline the advantages of spread spectrum.
6.3 What is ISM band and what is the need for ISM band?
6.4 Briefly outline the WiMAX technology and the corresponding standard.
6.5 What is a WLAN and what are the advantages of WLAN?
6.6 Differentiate between fixed WiMAX and mobile WiMAX.
6.7 What is the use of Bluetooth.

Projects

6.1 Visit the nearest IT firm and understand the use of access points within the organization. Write a brief report on the usage of access points within the organization you have visited.
6.2 With the help of Internet, identify three different access point vendors and compare the product offerings on technical and commercial grounds.
6.3 As you are aware, ABC Unlimited is a supermarket chain. Refer the case in Chapter 1. As a part of modernization and to ensure adequate stock levels, the CIO of ABC plans to deploy access points within each supermarket. He wants to attempt the deployment of SMART DISPLAY RACK. He has been asked to submit a report on the use of Wi-Fi to gather real-time data on the movement of goods placed in the display rack in order to take care of appropriate refilling of the racks.

You are expected to help the CIO prepare a suitable report on the SMART Display Rack planned for the supermarket which can provide real-time information on the movement of goods to the central store. Submit a report with suitable justifications.

References

Katz, Marcos D. and Frank H.P. Fitzek 2009, *WiMAX Evolution Emerging Technologies and Applications*, John Wiley, Great Britain.

Syed, Ahson and Mohamad Ilyas 2008, *Wimax Standards and Security*, CRC Press, USA.

Yan Zhang, Hsiao-Hwa Chen 2008, *MOBILE WiMAX, Toward Broadband Wireless Metropolitan Area Networks*, Aurabach, USA.

Yang Xiao 2008, *WiMAX/MobileFi Advanced Research and Technology*, Auerbach, USA.

CHAPTER

7 Storage Infrastructure

Crude classification and false generalizations are the curse of organized life.
–George Bernard Shaw

> **Learning Objectives**
>
> After reading this chapter, you should be able to understand:
>
> - the need for storage infrastructure
> - the differences between the types of storage infrastructure
> - the differences between NAS and SAN
> - the concept of RAID
> - the differences between the types of RAID
> - the different types of storage protocols
> - the differences between the various types of storage protocols

7.1 A Quick History of Data Storage

A chapter on the subject of storage infrastructure is especially important during these fast-moving times. The technology for storage networking is basically bringing with it new structures and procedures that will remain topical in the foreseeable future, regardless of incremental differences and changes in products.

Information has become a commodity in today's world, and protecting that information from being lost is mission critical. The Internet has helped push this information age forward. For many companies, data storage has become one of the fastest-growing parts of the information technology (IT) budget, thanks to enterprise-wide transactional systems, massive data warehouses, and explosive growth in e-mail traffic. The shift from a seemingly controllable to uncontrollable by-product occurred as we began to network our computers. In the pre-Internet era, a significant amount of information exchanged in the business environment passed through human gatekeepers. Orders, inventory, customer support, and corporate correspondence, to name a few, typically flowed through an organization only as fast as a person could send or receive it. Today, the human gatekeepers no longer place artificial bottlenecks on the flow of information, and computers within and across organizations now generate communication and data at astronomical rates. Behind the scenes of every corporation's mission-critical IT operations, lies the foundation layer of data storage. Typically, in the form of large disk subsystems and tape libraries, the media upon which the bits and bytes are stored empower the information flow. Couple this core with an effective software layer, the appropriate network interconnect, and precise administration, and you have the makings of a complete data storage system to drive organizational success.

New applications, more complex business analytics, and the need to meet regulatory requirements are major contributors to the demand for additional storage capacity. In many enterprises, the storage environment has grown so rapidly that it has outrun the IT department's ability to manage it effectively. If information is power, storage is the reservoir of that power. Enterprises can do better in their business with a storage solution that takes care of not only their present needs but also future ones. The incredibly obvious observation is that data storage disks contain data, which in most companies is one of their two most valuable assets, the other being their people. Disks themselves have very little intrinsic value (and that value is dropping all the time, as disks get cheaper and cheaper); their primary value comes from the data that is stored on them.

Large enterprises are forced to manage storage more efficiently if they are to exploit opportunities created by new forms of information. In many cases, companies find it quicker and easier to back up information by making multiple copies. But across a large organization, this approach increases storage volumes dramatically as companies keep too many copies of business data. Very few companies archive and purge outdated information aggressively, either because they have an aversion to risk or they believe it is not worth the expense, further adding to the storage burden. Poor storage decisions stem from a variety of underlying issues, most notably time pressure, information gaps, and breakdowns in the relationship between business users and IT. Many organizations underestimate costs by failing to include in the analysis the hidden expenses such as network hardware, short-term back up, and long-term archiving. It is easy to miscalculate because storage costs are relatively complex and depend on pricing options, configuration decisions, and labour productivity across multiple organizations, including the vendor, IT, and business.

Information technology must implement practices and policies that clearly define the storage options available to business users and that help them set priorities based on storage economics and business needs. So how can IT departments improve the way they manage data and deliver solutions to meet the rising demands of the business without allowing costs to spiral out of control?

The first step towards a storage network is to quantify data and its growth rate. Once the amount of data on a network has been assessed, the chief information officer (CIO) has to identify how much information needs to be shared and by whom. The information is then broken down into various categories according to its importance. This gives the CIO an idea of the online data at any point in time.

The storage management capability versus the total data to be managed, from an overall industry standpoint, has continued to widen since the birth of recent data storage technologies over the last decade. The growth in unstructured data, such as large media files, has been a driving force in adopting modern ways to manage, de-duplicate, safely store, and move the large media files that are now commonplace in the enterprise environment. An example of tackling this problem is the use of clustered storage systems for virtualized storage with storage area networks to provide high-speed connections between hosts and disks.

To properly provision (allocate) storage space in an enterprise data solution, a clear picture must be formed in order to properly assess the needs and limits of data resources to effectively meet the needs of an organization. Previously, the costly approach of fat provisioning, where data storage space is allocated beyond current needs in anticipation of future storage requirements, was often employed. The inability to combine storage resources, often due to interoperability and compatibility issues in heterogeneous equipment approaches, complicated the abilities of data administrators to formulate this clear picture. Fat provisioning also proved costly and poorly utilized available storage resources. Through the combination of storage virtualization with their storage resources, many organizations are implementing thin provisioning throughout their storage system to take better advantage and gain better utilization of these resources. Through the storage area network (SAN) storage virtualization relationship, resources are better managed and scaled in accordance to business need better than without the combination.

Data storage administration, with advances in utilizing storage networks and storage virtualization, has emerged as a specialized role in the enterprise storage environment. Proof of this fact can be seen in the rapid increase of scale in all things stored, the very complicated methodologies and changes in elements available in data administration today. This also includes the need for understanding of the broad spectrum of possibilities and available products and uses for storage solutions that can meet the needs to protect and manage hardware and data assets of an enterprise. Although there are many reasons that adoption of virtualization and SAN is now a cost-effective and realistic solution to meeting the storage challenges faced in many enterprise environments, there are still problems and barriers to adoption that hinder widespread acceptance of the two technologies. For example, Mumbai-based WNS Global Services posted healthy revenues in 2008's fourth quarter and financial year despite the global meltdown that has crippled the business process outsourcing (BPO) industry. Part of the reason for this performance is the company's Rs 2,700-crore virtualization initiative.

7.2 Storage I/O Basics

Data storage uses a variety of technology to store data. Some devices, such as hard drives, floppy drives, and tape, use magnetism to encode information on the media. Others, such as CD-ROM/RW and DVD-ROM/RW, use lasers to burn information in the substrate of a disk. Still others, such as solid state storage, use a charge in a solid state device to store data. Whatever the media, all data is stored as bits (1s and 0s) on some device that can be read back later. To be found, stored data must have a structure that can be understood by an operating system (OS). If this structure does not exist, data cannot be written to the storage media. Reading back data would be a futile exercise. At the lowest level, the drive and protocol determine how data is organized and transferred to and from a device. Unfortunately, the high-level structure is imposed by the OS. This makes file systems unique to the operating environment. It is why a file written to a hard drive by Microsoft Windows cannot be read by any other OS, such as a Macintosh, without a translation program that understands both file systems. I/O stands for input/output.

With the exception of solid state storage, data is stored by encoding a bit onto a material, called the *media*. The electromechanical devices that are used to read and write data are called the *heads*. The read and write heads are mounted on an arm called the *head assembly* or *actuator assembly*.

The amount of time it takes for a controller to find and read data on a media is called *access time*. Access time is an important performance measure because it directly measures the time it takes to retrieve data from the media. With disk drives, access time is a measure of the latency and transfer time of the disk.

Storage devices that are mechanical in nature take time to find the data on the media and place the read or write heads over it. The time it takes to place the heads over the spot where the data is stored or will be stored is called *latency*.

Transfer time is the amount of time it takes for the disk, tape, or other storage media to transfer data off the media and onto the data bus or network. It is a function of a number of variables, including how fast the heads can detect the magnetic field on the disk or tape (in the case of magnetic media), the performance of the drive mechanisms, and the speed of the electronic components.

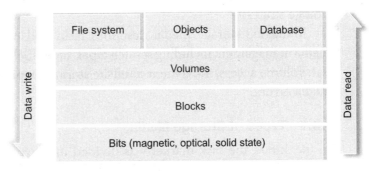

FIGURE 7.1 I/O Stack

When a protocol accesses data on a disk as blocks, it is referred to as block input/output (I/O) (refer to Figure 7.1). Certain types of applications prefer to access data directly as blocks. Accessing data through a file system is called file I/O. It is easier for an application to use the existing mechanisms in a file system for most I/O than to deal with data at a block level.

The small computer systems interface (SCSI) enables host computer systems to perform block data (I/O) operations to a variety of peripheral devices. Target devices may include disk and tape devices, optical storage devices, as well as printers and scanners. The traditional SCSI connection between a host system and peripheral devices is based on parallel cabling.

The SCSI architecture is based on a client/server model. The client is typically a host system such as a file server that issues requests to read or write data. The server is a resource, such as a disk array, that responds to client requests. In storage parlance, the client is an initiator and it plays the active role in issuing commands. The server is a target and has a passive role in fulfilling client requests. A target has one or more logical units that process initiator commands. Logical units are assigned identifying numbers, or logical unit numbers (LUNs). A LUN is a logical grouping of physical disks into a redundant array of inexpensive disks (RAID) construct. When the grouping is done inside an intelligent disk array, and passed to connected hosts as a unit, that unit is called a LUN.

7.3 Storage Media

Crashing disk and hardware prices meant more and more businesses could afford sophisticated storage infrastructure. An interesting side effect of cheaper disks was that it encouraged large and medium enterprises to go in for disk-based data replication solutions, abandoning tapes altogether. Steadily-decreasing margins on hardware forced major vendors to focus more on storage strategies.

Managing storage media describes tasks and concepts associated with media management including scratch tapes, tape labelling, check-in and check-out, volume access, movement to offsite storage, expiration, reclamation, and media errors.

7.3.1 Types of Storage Media

In the 1980s, the disk market was divided into two major portions. On the one hand, large diameter disks for mainframes and minicomputers, and on the other, small-diameter disks for personal computers (PCs). Because of technological advances and the economic arguments from very high demand for the small-format disks, the smaller media soon came to equal and surpass the large-format disks in capacity, performance, and reliability. With the first generations of small-format disks, servers had to use groups of disks to obtain sufficient capacity. This affected reliability to a great extent, since the failure rate increased linearly with the number of disks used.

Hard disks and tapes are currently the most important media for the storage of data. When storage networks are introduced, the existing small storage devices are replaced by a few large storage systems. Success in the media creation and distribution industry is dependent on two primary factors: speed and access to data. The most significant obstacle to developing and delivering new content is the time and effort required to manage and move large media files. Most storage has been developed for the transactional environment faced by a majority of IT applications. This ignores the special needs that media applications have for large files, large groups of files, and consistent bandwidth over long periods of time. These requirements are seen in applications that archive content for reuse, render video, or capture live events. Content developers lose time while waiting for large media files to be retrieved, created, or to move via the local area network (LAN) from one workstation to another.

Data is stored in files. There are a variety of file formats in the market. A file format is often described as either proprietary or non-proprietary.

Proprietary formats Proprietary file formats are controlled and supported by just one software developer. Proprietary formats, such as Microsoft Word files and WordPerfect files, carry the extension of the software in which they were created.

Non-proprietary formats These formats are supported by more than one developer and can be accessed with different software systems. For example, extensible markup language (XML) is a popular non-proprietary format.

There are three basic records storage options. These are discussed as follows.

Online storage Records are available for immediate access and retrieval. Online storage devices include mainframe storage and network-attached storage. Online storage provides the fastest access and regular integrity checks.

Nearline storage Records are stored on media such as network-attached storage, optical disks in jukeboxes, or tapes in automated libraries. Nearline storage provides faster data access than offline storage at a lower cost than online storage.

Offline storage Records are stored on removable media such as magnetic tape or optical disk. Because human intervention is necessary, this option provides the slowest access.

The storage capacity of digital media is measured in bytes, the basic unit of measurement:

- 1,024 bytes = 1 kilobyte (KB)
- 1,024 KBs = 1 megabyte (MB)
- 1,024 MBs = 1 gigabyte (GB)
- 1,024 GBs = 1 terabyte (TB)
- 1,024 TBs = 1 petabyte (PB)
- 1,024 PBs = 1 exabyte (EB)

Digital media are divided into the following three types.

Magnetic Electronic information is stored on computer drives, disks, or tapes by magnetizing particles imbedded in the material. Magnetic media include the following.

- Magnetic disks, such as computer hard drives that store programs and files, are randomly accessed. Fixed disks reside permanently in a drive while removable disks are encased in plug-in cartridges, allowing for storage and transfer of data.
- Magnetic tape is a sequential storage medium used for data collection, backup, and archiving. Common magnetic tape formats include digital audio tape (DAT), digital linear tape (DLT), and linear tape open (LTO).

Optical Digital data is encoded by creating microscopic holes in the surface of the medium. Optical media options include the following.

- Compact discs (CD) can be read only (CD-ROM), write once read many (CDR), and rewritable (CD RW). Since the introduction of the audio CD in 1982 and the CD-ROM in 1985, the CD has become a universally accepted carrier for music, data, and multimedia entertainment. CD-Audio was by many measures the most successful new consumer entertainment format of its time. CD-ROM has enhanced and enabled many aspects of the desktop computer revolution, and is now included as a standard item on most desktop and laptop PCs. Compact disks can hold roughly 700 MB of data.

- Digital versatile disc (DVD) is also called digital video discs. Since its introduction in the autumn of 1996, DVD has been growing at an astronomical rate. By far the most successful launch ever of a consumer electronics format, the adoption of DVD has outpaced that of CD, video home system (VHS), and LaserDisc (LD). With the finalization of the DVD-Audio specification, DVD is now a family encompassing six closely-related formats: DVD-ROM, DVD-Video, DVDAudio, DVD-Recordable (DVD-R), DVD-Random Access Memory (DVD-RAM), and DVD-RW. While DVD-Video was the first of the DVD formats to grab the attention of consumers, DVD-ROM has found ready acceptance in the personal computing market, and the more recently introduced DVD-RAM format is also catching on. DVD-R, meanwhile, has established itself as an attractive medium for professional applications such as presentations and kiosks. However, the data they store do not have to be in video form. DVDs can hold between 4.7 GB and 17.0 GB of data. Common types of DVDs include the following.
  - DVD-RAM is a rewritable disc that provides 4.7 GB per side storage capacity.
  - DVD-R has the same storage capacity as DVD-RAM, but can only be written to one time.
  - DVD+R is a writable disc with 4.7 GB of storage capacity on either side.
  - DVD-RW offer 4.7 GB per side, but can be overwritten 1,000 times. The DVD-RW technology is mainly used for video.
  - DVD+RW is an alternative rewritable format that has a capacity of 4.7 GB per side and is used for both data and video content.

Solid state With no moving parts, a solid-state device uses electronics instead of mechanics. These devices are much faster and more reliable than magnetic and optical media. Solid state devices include the following.

- A computer's basic input/output system (BIOS) chip
- Personal Computer Memory Card International Association (PCMCIA) Type I and Type II memory cards, which are used as solid state disks in laptops
- Flash and dynamic random access memory (DRAM) based solid state drives

Within information life cycle management (ILM) applications, data archival becomes a critical tier within the storage mix. It is a long-term repository for information that is important for the organization to maintain but does not justify the investment required for online hard disk drive (HDD) storage. Archival data must not only remain accessible for extended periods of time (50 years and beyond), but must also be protected from accidental or deliberate alteration.

7.3.2 Holographic Storage

Holography has long held promise as a data storage technology with the potential for vast capacity and high data rates. Recent advances in materials, multiplexing architectures and components are finally making this vision a reality. Holographic storage not only provides a solution for the archive tier of ILM but also for the nearline storage tier. With an archive life of at least 50 years, true write once/read many (WORM) capability and the availability of low-cost storage media, holographic storage is an ideal archive solution. Holographic technology also addresses the nearline tier by providing fast and reliable access to data, but at a price point that is significantly lower than that of HDDs. Data stored anywhere on holographic media is accessible in milliseconds compared to the 10s of seconds or even minutes if stored on tape. Holography breaks through the density limits of conventional storage by going beyond recording only on the surface to recording through the full depth of the medium. Unlike other technologies that record one data bit at a time, holography allows a million bits of data to be written and read in parallel with a single flash of light. This enables transfer rates significantly higher than current optical-storage devices. Combining high storage densities and fast transfer rates, with durable reliable low-cost media, holography is poised to become a compelling choice for next-generation storage and content distribution needs.

Storage Gets United and Flexible with Tyrone Opslag FS2

Netweb Technologies, a server, storage, and HPC solution provider, has launched its new product the Tyrone Opslag FS2 series of storage solutions. Unifying the traditional storage area network (SAS) and network attached storage (NAS) into a single box, FS2 offers higher flexibility and allows users to have as much of both. The product provides scalability from a few terabytes to 576 TB, and could perform up to 2GB/s, claims a note in Linux For You, February, 2009.

7.4 Storage Protocols

A disk is not worth much if there is no way to access it. In order to access a disk, there must be one or more I/O paths to it that the OS can use. Over the course of time, many types of paths have been developed. Storage has been detached from the servers and combined to form a separate storage network. This has resulted in a fundamentally different approach to dealing with storage. The new procedures required will continue to be developed into the near future. Data storage, therefore, has a value of its own. It is no longer a matter of attaching another disk drive to a server. Today, stored data and the information it contains are the crown jewels of a company. The computers (servers) needed for processing data can be purchased by the dozen or in larger quantities—individually as server blades or packed into cabinets—at any time, integrated into a LAN or a wide area network (WAN) or exchanged for defective units. However, if stored data is lost, restoring it is very expensive and time-consuming, assuming that all or some of it can even be recovered. As a rule, data must be available 'around the clock'. Data networks must, therefore, be designed with redundancy and high availability. The requirements of storage networks are fundamentally different from those of the familiar local networks. Storage networks have, therefore, almost exclusively been using fibre channel (FC) technology, which was specially developed as a connection technology for company-critical applications. Storage networking is not a short-term trend and efforts are, therefore, currently underway to use other existing network technologies as well as new ones that are coming on the market.

Protocols are a set of rules for using an interconnect or network, so that information conveyed on the interconnect can be correctly interpreted by all parties to the communication. Protocols include such aspects of communication as data representation, data item ordering, message formats, message and response sequencing rules, block data transmission conventions, timing requirements, and so forth. Internet protocol (IP) storage refers to a group of technologies that allows block-level storage data to be transmitted over an IP-based network. There are two key concepts in this definition: the use of IP and block-level storage. The concept of IP storage has emerged since networked storage requirements have grown and since IP has become firmly established as the predominant general purpose networking protocol.

Currently two major approaches to file transfer in a storage network exist. These approaches are called 'block I/O' and 'file I/O'.

For the 'block IO' approach, the SCSI protocol is the protocol of choice in the storage management arena. Different transport mechanisms, such as parallel SCSI, FC, or Internet SCSI (iSCSI), can be used to transfer SCSI protocol data.

Fibre channel is the technique most frequently used for implementing storage networks. Interestingly, fibre channel was originally developed as a backbone technology for the connection of LANs. The original development objective for fibre channel was to supersede fast Ethernet (100 Mbit/s) and fibre distributed data interface (FDDI). Meanwhile, gigabit Ethernet and 10-gigabit Ethernet have become prevalent or will become prevalent in this market segment.

7.4.1 Internet SCSI

The Internet SCSI (iSCSI) protocol defines a means to enable block storage applications over transmission control protocol/Internet protocol (TCP/IP) networks. The iSCSI protocol is a relatively new protocol, and, in essence, sends SCSI commands over the Internet via TCP/IP. Among emerging storage protocols, iSCSI is a very promising mechanism for SCSI command transfer using IP network. Based on TCP/IP protocol, iSCSI uses TCP flow control, congestion control, segmentation mechanisms, and it is built upon the IP addressing and discovery mechanisms. Using existing gigabit Ethernet infrastructure, iSCSI can provide gigabit speeds comparable to FC. Internet SCSI is designed to be a host-to-storage end-to-end solution. It uses TCP/IP for reliable data transmission over potentially unreliable networks. Internet SCSI technologies will include iSCSI-enabled hosts that will communicate through IP switches to iSCSI-enabled storage arrays. The iSCSI protocol has two halves to it—the initiator resides on a server or desktop computer, and sends commands to the iSCSI target, which resides on the storage appliance. The target performs the work requested by the initiator, and sends a reply. All communications take place via TCP/IP. Device discovery and device ownership in iSCSI is done by means of a separate component called iSNS. The SCSI command is encapsulated into an iSCSI protocol data unit (PDU). As defined by the Internet Engineering Task Force (IETF), the iSCSI protocol will use TCP as its underlying transport layer to provide a reliable transport with guaranteed in-order delivery. Once the TCP/IP headers are added, the encapsulated SCSI command is treated the same as any other IP packet. It can be routed to its end destination (based on its IP address) over standard

IP infrastructure. Once the destination device receives the packet, it strips off each layer until it eventually returns the SCSI command to the SCSI layer.

7.4.2 Fibre Channel

Fibre channel is a newer technology for connecting servers to disks. At this writing, FC supports speeds up to 2GB/second. The 2GB/second data transfer rate is expected to increase to, and probably exceed 10GB/second over time. Since FC uses a 24-bit addressing scheme, approximately 16 million devices can be connected to a single FC network or fabric. The formal FC specification says that devices can be up to 10 kilometres apart, but vendors are exceeding the specification, and spreads of 80 kilometres and more are being achieved when long-wave optics are used to enhance the optical signal sent on the fibre.

Storage area networks were initially developed based on FC technologies. The key benefit of FC networks is that they provide a high performance connection to storage that behaves like a dedicated link. The trade-off, however, is that they are not as widely implemented as IP networks. Organizations have to consider the costs of deploying a new infrastructure and acquiring a new set of skills to effectively manage and operate FC networks. To date, FC reach limitations and lack of interoperability between FC vendors have also been issues as customers strive to evolve from a centralized storage architecture to a more geographically distributed one that can provide additional benefits of efficiency and secure redundancy. Fibre channel is a well-known and widely-deployed medium for storage applications. However, it requires custom, proprietary drivers, and often demands custom and proprietary communications media.

7.4.3 Types of Fibre Channel

Fibre channel is a set of standards for a serial I/O bus capable of transferring data between two ports at up to 100 MBytes/second, with standards proposals to go to higher speeds. Fibre channel supports point-to-point, arbitrated loop, and switched topologies. Fibre channel was completely developed through industry cooperation, unlike SCSI, which was developed by a vendor and submitted for standardization after the fact. Work is being undertaken to provide fibre channel over IP, using two distinct paths: Internet fibre channel protocol (iFCP), and fibre channel internet protocol (FCIP), both of which provide existing FCP FC-4 commands over TCP/IP.

As its name implies, the aim of the FCIP protocol is to transport FC frames over an IP infrastructure. FCIP is a simple tunneling protocol for encapsulating entire FC frames within TCP/IP. FCIP provides the mechanisms to allow islands of FC SANs to be interconnected over IP-based networks to form a single, unified FC SAN fabric. The extended FC SAN fabric continues to use standard FC addressing.

Internet fibre channel protocol encapsulates FC frames to be sent over the IP infrastructure. Because of this, the IETF chose to specify a common FC encapsulation format. The main difference between the two protocols lies in their addressing schemes. The FCIP protocol establishes point-to-point tunnels that can be used to connect two FC SANs together, with Ethernet, to create a single, larger SAN. In contrast, iFCP is a gateway-to-gateway protocol that combines FC and IP addressing to allow the FC frames to be routed to the appropriate destination address. Unlike the addressing scheme of the FCIP protocol, the current iFCP addressing scheme allows each interconnected SAN to retain its own independent namespace.

Request for comment (RFC) 3723 discusses the use of the IPsec protocol suite for protecting block storage protocols over IP networks (including iSCSI, iFCP, and FCIP), as well as storage discovery protocols (iSNS and SLPv2).

7.5 Storage Models

The storage industry has been trying hard to move away from the scattered environment and creating a single pool of data, which can be accessed by all. Instead of spreading horizontally, effort has been made to have multiple layers of data storage based on the behaviour and importance of data. Storage solutions for small to medium sized businesses generally consist of two options: direct attached storage or network attached storage.

7.5.1 Direct Attached Storage

Direct attached storage (DAS) refers to a digital storage system directly attached to a server or workstation, without a storage network in between. The disks may be internal to the server or they may be in an array that is connected directly to the server. Either way, the storage can be accessed only through that server. The main protocols used in DAS are advanced technology attachment (ATA), serial advanced technology attachment (SATA), small computer system interface (SCSI), serial attached small computer system interface (SAS), and FC.

Although the implementation of networked storage is growing at a faster rate than that of DAS, it is still a viable option by virtue of being simple to

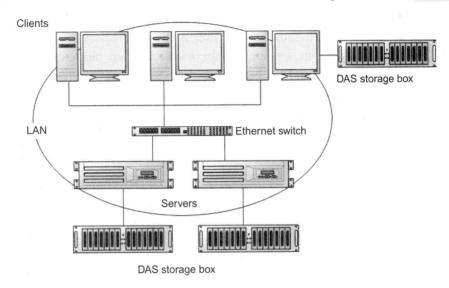

FIGURE 7.2 Direct Attached Storage

deploy and having a lower initial cost when compared to networked storage. From an economical perspective, the initial investment in DAS is cheaper. This is a great benefit for IT managers faced with shrinking budgets, who can quickly add storage capacity without the planning, expense, and greater complexity involved with networked storage. Direct attached storage can also serve as an interim solution for those planning to migrate to networked storage in the future.

Backups must either be performed on each individual server with a dedicated tape drive or across the LAN to a shared tape device consuming a significant amount of bandwidth (refer to Figure 7.2). Backups become more challenging, and because storage is not shared anywhere, storage utilization is typically very low in some servers and overflowing in others. Disk storage is physically added to a server based on the predicted needs of the application. If that application is underutilized, then capital cost gets unnecessarily tied up. If the application runs out of storage, it must be taken down and rebuilt after more disk storage has been added. An obstacle that many IT managers are facing is the inability to manage their DAS connections. Since DAS systems are unable to be networked together easily or efficiently, they are not capable of supporting a company that uses multiple enterprise-wide applications, such as enterprise resource planning (ERP), customer relationship management (CRM), and supply chain management (SCM). Network bottlenecks and slowdowns in data availability may occur as server bandwidth is consumed by applications, especially if there is a lot of data being shared from workstation to workstation. Disadvantages of DAS include inability to share data or unused resources with other servers.

> The Nexans Group leads the world market in the cable industry and offers customers an extensive range of cables and cabling systems. Nexans is a major player in the infrastructure, industrial, and construction, markets, developing solutions for a series of market segments, ranging from energy, transport, and telecoms networks, to shipbuilding, oil and gas, the nuclear industry, wind power, the automotive industry, railways, electronics, air travel, goods processing, and automation.

7.5.2 Network Attached Storage

The development of networked storage has its roots in early computing models and has been enabled by technology developments of the past 10 years. Network attached storage (NAS) is defined by Storage Networking Industry Association (SNIA) as storage elements that connect to a network and provide file access services to computer systems. Network attached storage is a class of systems that provide file services to host computers. A host system that uses network-attached storage uses a file system device driver to access data using file access protocols such as network file system (NFS) or common Internet file system (CIFS). Common Internet File System is a network file system access protocol originally designed and implemented by Microsoft Corporation under the name Server Message Block protocol, and primarily used by Windows clients to communicate file access requests to Windows servers.

Network attached storage, a special purpose device, is comprised of both hard disks and management software, which is 100 per cent dedicated to serving files over a network. Network attached storage systems interpret these commands and perform the internal file and device I/O operations necessary to execute them. A NAS device can be attached anywhere in the network independent of the server (refer to Figure 7.3).

A NAS appliance is a simplified form of file server; it is optimized for file sharing in an organization. Authorized clients can see folders and files on the NAS device, just as they can on their local hard drive. Network attached storage appliances are so-called because they have all of the required software preloaded and they are easy to install and simple to use. Installation consists of rack mounting, connecting power and Ethernet, and configuring via a simple browser-based tool. Installation is typically achieved in less

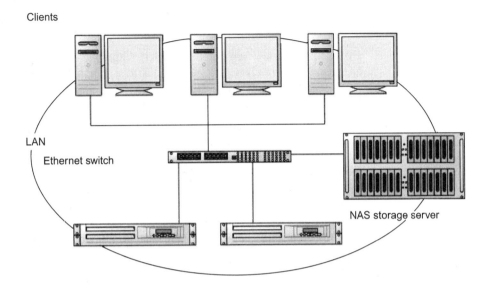

FIGURE 7.3 Networked Attached Storage

than half an hour. Administrators manage the NAS device from anywhere they have network access. Network attached storage is an ideal choice for organizations looking for a simple and cost-effective way to achieve fast data access for multiple clients at the file level. In recent years, NAS has developed more sophisticated functionality, leading to its growing adoption in enterprise departments and workgroups. Network attached storage systems provide terabytes of storage in high-density form factors, making efficient use of data centre space. As the volume of digital information continues to grow, organizations with high-scalability requirements will find it much more cost-effective to expand upon NAS than DAS. Multiple NAS systems can also be centrally managed, conserving time and resources. Historians generally credit grammarian and scholar Jernej Kopitar with formalizing the Slovene language in the early 1800s. Kopitar's priceless collections of medieval Slavic writings—called codices—reside in Slovenia's National and University Library (NUL). Until recently, only qualified scholars could access these writings on a limited basis. However, due to an NAS/SAN solution, scanned images of these documents will soon be available to the public over the Web and from local library computers.

> **EMC's Virtual Matrix Architecture**
>
> EMC Corporation unveiled a breakthrough new approach to high-end data storage with an innovative new architecture purpose-built to support virtual data centres. EMC also announced the first storage system based on this architecture, which will serve as a cornerstone of virtual computing infrastructures that are transforming the technology landscape.
>
> The new EMC Virtual Matrix Architecture integrates industry-standard components with unique EMC Symmetrix capabilities to deliver massive scalability-enabling systems that scale to hundreds of thousands of terabytes of storage and tens of millions of IOPS (input/output per second) supporting hundreds of thousands of VMware and other virtual machines in a single federated storage infrastructure. It is the first storage architecture that combines the performance and efficiency of a scale-up architecture and the cost-effective flexibility of a scale-out architecture. It was designed and built from the ground up to break the physical boundaries of data centre storage, incorporates automation to simplify storage management, enables resources to be scaled on demand, and uses less energy per terabyte of data stored than traditional high-end storage systems.
>
> *Source*: http://www.emc.com/about/news/press/2009/20090414-01.htm

7.5.3 Storage Area Network (SAN)

A storage area network (SAN) is a networked high-speed infrastructure (subnetwork) that establishes direct access by servers to an interconnected group of heterogeneous storage devices such as optical disks, RAID arrays, and tape backups, which are effective for storing large amounts of information and backing up data online in e-commerce, online transaction processing, electronic vaulting, data warehousing, data mining, multimedia Internet/intranet browsing, and enterprise database managing applications. A SAN consists of a communication infrastructure, which provides physical connections, and a management layer, which organizes the connections, storage elements, and computer systems so that data transfer is secure and robust. Another definition of a SAN is that it is a storage system consisting of storage elements, storage devices, computer systems and/or appliances, plus all control software, communicating over a network. A SAN supports centralized storage management. Storage area network make it possible to move data between various storage devices, share data between multiple servers, and back-up and restore data rapidly and efficiently. In addition, a properly configured SAN facilitates both disaster recovery and high-availability.

In its simplest form, a SAN consists of one or more servers: (a) attached to a storage array, (b) using one or more SAN switches. All hosts connect to the storage devices on the SAN through the SAN fabric. The common communication and data transfer mechanism for a given deployment is commonly known as the storage fabric. The network portion of the SAN consists of fabric components like SAN switches. Storage area network switches connect various elements of the SAN. These switches can connect to servers, storage devices, and other switches, and thus provide the connection points for the SAN fabric. The type of SAN switch, its design features, and its port capacity all contribute to its overall capacity, performance, and fault-tolerance. The number of switches, types of switches, and manner in which the switches are interconnected define the fabric topology. In particular, they might connect hosts to storage arrays. The switches also allow administrators to set up path redundancy in the event of a path failure from host server to switch or from storage array to switch. The storage components of a SAN are the storage arrays. Storage arrays include storage processors (SPs). The SPs are the front-end of the storage array. Storage processors communicate with the disk array (which includes all the disks in the storage array) and provide the RAID/LUN functionality. Storage processors provide front-side host attachments to the storage devices from the servers, either directly or through a switch.

A SAN allows more than one application server to share storage. In contrast to DAS or NAS, which is optimized for data sharing at the file level, the strength of SANs lies in its ability to move large blocks of data. This is especially important for bandwidth-intensive applications such as database, imaging, and transaction processing. Data is stored at a block level and can, therefore, be accessed by an application, and not directly by clients.

7.5.4 Fibre Channel Topologies

Fundamentally, fibre channel defines three configurations:

- Point-to-point
- Fibre channel arbitrated loop (FC-AL)
- Switched fibre channel fabrics (FC-SW)

Backups can be performed centrally and can be more easily managed to avoid interrupting the applications. The primary advantages of a SAN are its scalability and flexibility. Storage can be added without disrupting the applications, and different types of storage can be added to the pool.

The storage is increasingly moving from DAS to NAS. And within NAS, storage area networks are getting a large chunk of the market share. Storage area network methodology has its roots in two low-cost technologies: SCSI-based storage and the NAS concept. They both successfully implement storage-network links, but are limited to a low volume of data flows and rates. A SAN system consists of software and hardware components that establish logical and physical paths between stored data and applications that request them. Storage area networks are the best way to ensure predictable performance and 24x7 data availability and reliability. The importance of this is obvious for companies that conduct business on the Web and require high volume transaction processing.

A SAN system consists of three architectural components: interfaces, interconnects or network infrastructure components, and fabrics.

The SAN becomes a key element of the enterprise environment where data availability, serviceability, and reliability are critical for a company's business. The SAN architectures have changed over time, adapting to new application demands and expanding capacities. The original FC-based SANs were simple loop configurations based on the FC-AL standard. Requirements of scalability and new functionality had transformed SANs into fabric-based switching systems.

A SAN makes physical storage capacity a single, scalable resource and allows the flexible allocation of virtualized storage volumes. Even with all the benefits of SANs, several factors have slowed their adoption, including cost, management complexity, and a lack of standardization. The backbone of a SAN is management software. A large investment is required to design, develop, and deploy a SAN, which has limited its market to the enterprise space. The rise in data volume has transformed storage from being a tactical decision to a strategic one. Companies now see storage network as an asset and the investment made in it as more of CAPEX.

Zoning provides access control in the SAN topology. When a SAN is configured using zoning, the devices outside a zone are not visible to the devices inside the zone. In addition, SAN traffic within each zone is isolated from the other zones. Typically, zones are created for each group of servers that access a shared group of storage devices and logical units (LUNs). Logical unit masking is commonly used for permission management. It is also referred to as selective storage presentation, access control, and partitioning, depending on the vendor.

> **IT Health Coverage for Insurance Brokerage Firm**
>
> Zenith installed and configured the AiO600 in a storage area network and migrated storage capacity from thirteen servers to the new system. A single, centralized, high-performance data storage system is essential to business efficiency at Zenith. For example, Zenith uses the PaperClip document-management system to collect and store images from many types of documents (including applications, policies, medical records, and correspondence), from many sources (mail, e-mail, fax, and scanned images), into individual client case files. One file might have hundreds of pages of images. With the AiO600, Zenith underwriters can look at all the information on a particular case in one place—so they do not have to search for documents in different files—and can easily exchange information among agents and carriers.
>
> The all-in-one Storage System gave Zenith the centralized, enterprise-wide storage solution the company needed, without requiring a rebuild of the existing infrastructure.

7.5.5 Data Backup

A lot of tools are available to assist in data backup. Some of the prominent tools include Amanda and the Advanced Maryland Automated Network Disk Archiver, which is the most well-known open-source backup software. Amanda was initially developed at the University of Maryland in 1991 with the goal of protecting files on a large number of client workstations with a single backup server. The Amanda project was registered on SourceForge.net in 1999. Jean-Louis Martineau of the University of Montreal has been the gatekeeper and leader of Amanda development in recent years. Over the years more than 250 developers have contributed to the Amanda codebase, and thousands of users provided testing and feedback, resulting in a stable and robust package. Amanda is included with every major Linux distribution. As of April 2006, more than 20,000 sites worldwide use Amanda.

7.6 RAID

In contrast to a conventional file server, a disk subsystem can be visualized as a hard disk server. Servers are connected to the connection port of the disk subsystem using standard I/O techniques such as SCSI. In most disk subsystems, there is a controller between the connection ports and the hard disks. The controller can significantly increase the data-availability and data access performance with the aid of a so-called RAID procedure. A disk subsystem with a RAID controller offers greater functional scope than a just a bunch of disk (JBOD) disk subsystem. RAID was originally developed at a time when hard disks were still very expensive and less reliable than they are today.

7.6.1 RAID Defined

RAID is a category of disk drives that employ two or more drives in combination for fault, tolerance and performance. It is an acronym to describe a redundant array of inexpensive disks. A RAID system is a collection of hard drives joined together using a RAID-level definition. RAID is now used as an umbrella term for computer data storage schemes that can divide and replicate data among multiple hard disk drives.

Redundant array of inexpensive disks was developed to increase the performance and reliability of data storage by spreading data across multiple drives. It is actually a technology which is implemented with the intellectual capability to improve the performance of the system. It is a specific system that utilizes the normal hardware but the technological implementations make it most complex. RAID is a technique that was developed to provide speed, reliability, and increased storage capacity using multiple disks, rather than single-disk solutions. The RAID standard describes several ways to combine and manage a set of independent disks so that the resultant combination provides a level of disk-redundancy. RAID basically takes multiple hard drives and allows them to be used as one large hard drive with benefits depending on the scheme or level of RAID being used. Simply put, RAID technology either divides or duplicates the task of one hard disk between many (or as few as two) disks to either improve performance or create data redundancy in case of a drive failure.

The controller of the disk subsystem must ultimately store all data on physical hard disks. Standard hard disks that range in size from 36 GB to 1 TB are used for this purpose. Since the maximum number of hard disks that can be used is often limited, the size of the hard disk used gives an indication of the maximum capacity of the overall disk subsystem. When selecting the size of the internal physical hard disks, it is necessary to weigh the requirements of maximum performance against those of the maximum capacity of the overall system. With regard to performance, it is often beneficial to use smaller hard disks at the expense of the maximum capacity. Given the same capacity, if more hard disks are available in a disk subsystem, the data is distributed over several hard disks and thus the overall load is spread over more arms and read/write heads and usually over more I/O channels.

Disk array is a set of disks from one or more commonly accessible disk subsystems, combined with a body of control software. The array may use SCSI or FC internally to address its disks (refer to Figure 7.4), and may use SCSI or FC to connect to its host(s). The same array may have SCSI disks internally, while connecting to its host(s) via FC. The control software presents the disks' storage capacity to hosts as one or more virtual disks. Control software is often called firmware or microcode when it runs in a disk controller. Control software that runs in a host computer is usually called a volume manager.

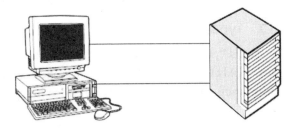

FIGURE 7.4 Disk Array

RAID may be implemented by hardware or software-based method, differentiated by the presence or absence of a RAID controller. Hardware RAID services take the physical disks and organize them into logical units (LUNs). Software RAID and volume management, which are usually so closely connected as to be inseparable, allow the host that uses the storage to take the disks, in whatever form it sees them, and build RAID configurations on them. The disks may be seen as simple disks or LUNs. The host system does not generally differentiate between the two, so software RAID and volume management products cannot differentiate either. Basically, a number of independent hard disks are connected to form a single and often larger virtual volume. The term RAID applies to an architecture that safeguards data—if a disk fails, data is reconstructed. Data is 'striped' across several disks. An extra disk is used to store parity information, which is used to reconstruct data. This architecture ensures that users can always access the data they need at any time.

7.6.2 Need for RAID

Intelligent RAID solutions utilize an I/O subsystem that is separate from the host CPU to provide greater performance and reliability. With an intelligent

RAID implementation, a hard drive crash will not result in lost data and the users may never know the failure even occurred. Intelligent RAID solutions utilize a specialized I/O processor residing on a RAID controller card or on the server motherboard. This processor offloads I/O processing tasks from the computer's main processor, resulting in increased server scalability and operating system independent data protection. If the operating system becomes unstable, the I/O processor maintains the capability to finish all disk writes, resulting in a deterministic state upon system reboot. In these cases, the risk of data loss or corruption is greatly reduced and little or no user intervention is required, even if the disk holding the operating system fails.

7.6.3 Types of RAID

The RAID Advisory Board has defined six official levels of RAID. Five of them (RAID-1 through RAID-5) provide various levels of increased disk-redundancy. One RAID (RAID-0) does not provide any redundancy.

7.6.4 RAID 0

This is the simplest level of RAID. Striping is the splitting of data between multiple drives. In a striping model, each chunk of data to be written to disk is broken up into smaller segments, with each highly available data management segment written to a separate disk (refer to Figure 7.5). Because each disk can complete the smaller writes at the same time, write performance is improved over writes to a single disk. Performance can be enhanced further by striping the writes between separate controllers too. Striped RAID arrays generally aim to merge maximum capacity into one single volume. Requiring at least two drives, RAID 0 stripes data onto each disk. The available capacities of each disk are added together so that one logical volume mounts on the computer.

RAID 0 is ideal for users who need maximum speed and capacity. Video editors working with very large files may use RAID 0 when editing multiple streams of video for optimal playback performance.

Each disk in a RAID 0 system should have the same capacity. Storage capacity in a RAID level 0 configuration is calculated by multiplying the number of drives by the disk capacity, or

$C = n \times d$

where

C = available capacity
n = number of disks
d = disk capacity

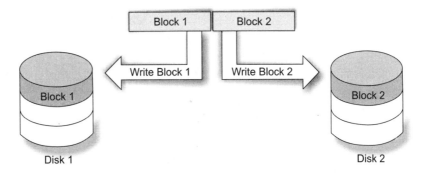

FIGURE 7.5 RAID 0

For example, in a RAID 0 array with four drives each with a capacity of 1,000 GB, the total capacity of the array would be

$4,000 \text{ GB} : C = (4 \times 1000)$

If one physical disk in the array fails, the data of all disks becomes inaccessible because parts of the data have been written to all disks.

7.6.5 RAID 1

RAID 1 requires at least two drives and works with pairs of drives. One logical volume mounts on the computer and the combined available capacity of both drives is limited to the capacity of the lowest-capacity disk (refer to Figure 7.6). If one physical disk fails, the data is available immediately on the second disk. No data is lost if one disk fails. This level is usually implemented as mirroring. Mirroring involves having two copies of the same data on separate hard drives or drive arrays. So, basically, the data is effectively mirrored on another drive. The system basically writes data simultaneously to both hard drives.

This is one of the two data redundancy techniques used in RAID to protect from data loss. The benefit is that when one hard drive or array fails, the system can still continue to operate since there are two copies of data. RAID 1 provides maximum data safety in the event of a single disk failure, but because data is written twice, performance is reduced slightly when writing. RAID 1 is a good choice when safety is more important than speed. The downtime is minimal and data recovery is relatively simple.

Each disk in a RAID 1 system should have the same capacity. Storage capacity in a RAID level 1 configuration is calculated by multiplying the number of drives by the disk capacity and dividing by 2, or

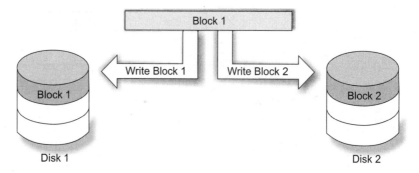

FIGURE 7.6 RAID 1

$$C = n \times \frac{d}{2}$$

where
C = available capacity
n = number of disks
d = disk capacity

For example, in a RAID 1 array with four drives each with a capacity of 1000 GB, the total capacity of the array would be

$$2,000 \text{ GB} : C = \frac{4 \times 1000}{2}$$

7.6.6 RAID 2

This level uses bit-level striping with Hamming code error correcting code (ECC). The technique used here is somewhat similar to striping with parity but not really. The data is split at the bit level and spread over a number of data and ECC disks.

7.6.7 RAID 3

RAID 3 uses byte-level striping with a dedicated parity disk. In other words, data is striped across the array at the byte level, with one dedicated parity drive holding the redundancy information. A RAID 3 array can tolerate a single disk failure without data loss. If one physical disk fails, the data from the failed disk can be rebuilt onto a replacement disk. If a second disk fails before data can be rebuilt to a replacement drive, all data in the array will be lost. RAID 3 provides good data safety for environments where long, sequential files are being read, such as video files. Disk failure does not result in a service interruption because data is read from parity blocks.

Each disk in a RAID 3 system should have the same capacity. Storage capacity in a RAID level 3 configuration is calculated by subtracting the number of drives by 1 and multiplying by the disk capacity, or

$$C = (n - 1) \times d$$

where
$C =$ available capacity
$n =$ number of disks,
$d =$ disk capacity

For example, in a RAID 3 array with four drives each with a capacity of 1,000 GB, the total capacity of the array would be

3000 GB: $C = (4 - 1) \times 1000$

7.6.8 RAID 4

This level is very similar to RAID 3. The only difference is that it uses block-level striping instead of byte-level striping.

7.6.9 RAID 5

RAID 5 combines the striping of RAID 0 with data redundancy in an array with a minimum of three disks. RAID 5 uses block-level striping and distributed parity (refer to Figure 7.7). This level tries to remove the bottleneck of the dedicated parity drive. With the use of a distributed parity algorithm, this level writes the data and parity data across all the drives. The difference between RAID 3 and RAID 5 is that a RAID 3 configuration will offer better performance at the expense of slightly less overall capacity. Data is striped across all disks and a parity block (P) for each data block is written on the same stripe. If one physical disk fails, the data from the failed disk can be rebuilt onto a replacement disk. No data is lost in the case of a single disk failure, but if a second disk fails before data can be rebuilt to a replacement drive, all data in the array will be lost. RAID 5 combines data safety with efficient use of disk space. Disk failure does not result in a service interruption because data is read from parity blocks. RAID 5 is useful for archiving and for people who need performance and constant access to their data, like video editors.

Each disk in a RAID 5 system should have the same capacity. Storage capacity in a RAID level 5 configuration is calculated by subtracting the number of drives by one and multiplying by the disk capacity, or

$$C = (n - 1) \times d$$

where
$C =$ available capacity

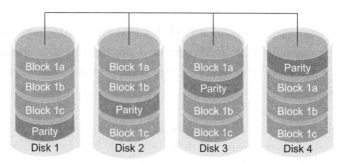

FIGURE 7.7 RAID 5

n = number of disks
d = disk capacity

For example, in a RAID 5 array with four drives each with a capacity of 1000 GB, the total capacity of the array would be

3,000 GB: $C = (4 - 1) \times 1000$

AAI Choose SANAT for its Data Consolidation

To assist Airport Authority of India, to consolidate its data generated by security systems - Video Surveillance. Solution Migrate from a DAS environment to an unified solution using the SANAT storage system for a low TCO and long-term cost savings.

In Chennai, the security systems of Airports Authority of India has more than 15 IP cameras deployed across the airport. These cameras send the recorded image to a server provisioned with direct attached storage. Performing storage management, backup and restore, and disaster recovery functions become a nightmare for the IT administrators. Looking to consolidate its storage resources and to implement a comprehensive, online data management system for compliance, Airports Authority of India chose a SANAT (Figure 7.8) unified solution.

'We were evaluating other solutions when SANAT technologies approached to us. We learnt of their solution offerings and understood that we had to have a unified storage solution in place using IPSAN and NAS on the same Storage. SANAT MaxStor blended beautifully with our requirements and also taking into account the following of our needs scalability, upgradeability, seamless integration with our existing infra, feature rich solution, An aggressive price line that fitted to budget (Said by Mr Vincent).

The SANAT topology offers an application server that is independent of storage and services. Instead of loading applications on each server, as was the case under the previous DAS environment IT administrators can now install application directory onto the IP SAN. This way, even if an application server does go down there would be no service outage that is visible to the users on the network.

Contd

Administrators could simply, with a click of a mouse, re-provision the application that is still available on the UNIFIED system to another host on the network and the users would continue to enjoy access to the application, seamlessly. Utilizing the UNIFIED architecture, administrators can run full backups quickly and directly onto the storage by creating 'Snappict' or 'point-in-time' copies of the data. In addition to simplifying backup, this approach enables instantaneous restore from disk via mounting the snappict volume. The IP SAN also enables the continually growing backup log files to be kept online, thereby facilitating quick identification of archived data for restoration. Furthermore, the snappict home directory can now be set up so that users can complete restoration on their own instantly, significantly reducing help desk requests and thereby allowing IT technicians to spend their time on other projects. Additionally, with backups being run directly on the back-end storage network rather than the user-facing front-end network, as was the case in the previous DAS environment, there is no network congestion or slow down of IT infrastructure.

Source: www.sanat.data.net

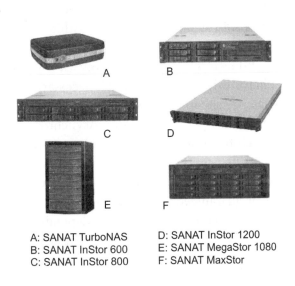

A: SANAT TurboNAS
B: SANAT InStor 600
C: SANAT InStor 800
D: SANAT InStor 1200
E: SANAT MegaStor 1080
F: SANAT MaxStor

FIGURE 7.8 SANAT

7.6.10 RAID Controller

The RAID controller is an electronic device that provides the interface between the host computer and the array of disks. From the viewpoint of the host computer, the RAID controller makes the array of disks look like one virtual disk that is very fast, very large, and very reliable. An FC RAID controller that is external to the host computer provides disk operation that is seamless and transparent to the host computer; i.e., the host computer does not require changes to software to realize the benefits a RAID system offers.

7.7 Backup Solutions

Despite the fact that plenty of storage products are now available, the challenge remains for companies to understand how to align the software to the various departments, divisions, geographical locations, employees, and business processes to achieve improved efficiency and profitability. Businesses now rely almost entirely on applications and databases, causing data and storage needs to increase at astounding rates. It is, therefore, imperative for a company to optimize and simplify the complexity of managing its data resources. Data-replication technologies fall into the categories of synchronous and asynchronous, and it is important to understand the difference between the two.

Synchronous replication: This technique is closely related to traditional RAID-1 mirror implementations within a storage array. The big difference is that the source and target of the mirror operation can be separated by a good distance. There are problems with synchronous replication. First of all, to minimize the write penalty, it is necessary to use expensive, high-bandwidth, low-latency site-to-site links.

Asynchronous replication: Asynchronous replication takes a different approach by allowing the replication process to be separated from the local write so that the application server does not have to suffer a performance penalty. The downside of the asynchronous approach is that the remote copy is never guaranteed to be a perfect copy of the source, it lags behind at any given point in time. This lag varies depending on the bandwidth of the network and the resources available at the remote end to commit the writes to disk.

7.7.1 Tape-based Solutions

Since the dawn of the digital computer age, long-term data storage and backup have been the province of a single primary medium: magnetic tape. Magnetic tape has been the backup medium of choice for a long time. Tapes are the most widely used solutions when it comes to data backup. Tape has compelling advantages. It is inexpensive to operate and buy, and even cheaper to store, whether it exists on reels, inside cartridges, or as part of an automated tape library system. Tape also has the benefit of separating the portable and inexpensive storage medium from the larger, more costly recording machinery. The introduction of tape made it possible to back up everything, keep copies off-site, and restore older or deleted files as needed. Tape arrays are the entry-level solutions, whereas tape libraries form the higher end in tape-based backup. Tape arrays and tape libraries use the same kind of data backup technology and the only difference is in their capacities. Tape arrays are usually used for storage in the case of special applications or for subsystems such as a section in an enterprise. Tape libraries are usually used for backing up large volumes of data such as data in a storage network. They are the most popular solutions for larger application.

Tapes offer the following advantages.

- Tapes offer the lowest costs per gigabyte of storage.
- Current tape technology is extremely reliable.
- Tapes can take overwrites amounting up to thousands of load/unload cycles.
- Tapes are easy to use, store, and manage.
- The tape backup automation system is easy to use.
- Hardware from multiple vendors is easily integrated in an existing system.

However, the trade-off is performance. As the amount of data that organizations have and need to back up has grown, the amount of time it takes to back up all that data to tape has become increasingly inconvenient.

No change has been more dramatic or rapid than the shift from traditional tape-based backup technology to disk-to-disk (D2D) backup. Disk-based backup has enabled shorter backup windows and more rapid data recovery, which has opened the way for more sophisticated backup and recovery software technologies that were not possible with tape backup systems. Beyond the pure speed advantages, disk backup is also the right technology at the right time to address the convergence of two business trends: the necessity for 24×7 data access in a global wired economy and the increasing use and importance of remote offices, meaning that more remote data is at risk than ever. Disk-based backup has largely removed the issue of the backup window.

Driven by business objectives in terms of application and data availability, and the disaster recovery and business continuity planning, as well as compliance with government regulations and legal discovery, data backup processes are now focused on two new metrics: recovery time objective (RTO) and recovery point objective (RPO). The RTO establishes a maximum duration for how long the restore process will take, while the RPO sets a goal for the maximum age of the data that will be used for a restore operation. The challenge is to drive the RTO as close to zero as possible and to have minimal RPO, while being able to afford it.

7.7.2 Virtual tape library

A tape library is a storage device which contains one or more tape drives, a number of slots to hold tape cartridges, a barcode reader to identify tape cartridges, and an automated method for loading tapes (a robot). A virtual tape library (VTL) is a data storage virtualization technology, used typically for backup and recovery purposes.

7.8 Continuous Data Protection

Traditional backup and recovery technologies are tape-based solutions and primarily address data centre disaster scenarios with recovery time measured in days. However, new business continuity (BC) and compliance requirements beyond traditional disaster recovery (DR) are driving IT organizations (ITOs) to evaluate more timely solutions that can reduce the potential outages and data loss to hours, minutes or seconds, depending on the application, the recovery situation, and its business criticality. In addition, CXOs realize that there is an enormous risk of losing critical data, and the inability to recover in a timely manner transcends the loss of a data centre. Therefore, one size no longer applies, and solution sets must address a multitude of requirements and be readily adaptable to changing business requirements.

According to the Storage Networking Industry Association (SNIA), snapshots are copies of data as it appeared at a specific point in time. The many implementations of snapshot functionality fall into one of two general categories: full or differential copy.

A full copy snapshot replicates the data set in its entirety. Often implemented via a process called mirror splitting, a full copy snapshot of a 3 terabyte database consumes an additional 3 terabytes of disk space.

A differential snapshot manages only deltas that have occurred since the snapshot was last taken. It uses less disk space but requires more processing overhead. There are a number of differential snapshot technical implementations: copy on write (CoW), redirect on write (RoW), and write anywhere (WA).

The SNIA continuous data protection (CDP) special interest group's working definition for CDP is a methodology that continuously captures or tracks data modifications and stores changes independent of the primary data, enabling recovery points from any point in the past. Continuous data protection systems may be block, file, or application-based and can provide fine granularities of restorable objects to infinitely variable recovery points. Continuous data protection, also called continuous backup or real-time backup, refers to backup of computer data by automatically saving a copy of every change made to that data, essentially capturing every version of the data that the user saves. Continuous data protection is different from traditional backup. In effect, CDP creates an electronic journal of complete storage snapshots. A major advantage of CDP is the fact that it preserves a record of every transaction that takes place in the enterprise.

Continuous data protection technologies have moved from the cutting edge of data protection deployments to the mainstream, driven, in part, by products such as Microsoft System Centre Data Protection Manager (DPM). But other storage technologies also are making a big impact on the way businesses handle data, and one of the most important is iSCSI. Charged with protecting their company's information, IT organizations have established aggressive service level agreements (SLAs) that impact the manner in which they implement data protection by setting RPOs and RTOs.

SUMMARY

The world depends on storage. Large disk subsystems have a storage capacity of up to a petabyte, that is often shared by several servers. The administration of a few large disk subsystems that are used by several servers is more flexible and cheaper than the administration of many individual disks or many small disk stacks. Storage systems and servers are connected together by means of block-oriented I/O techniques such as SCSI, fibre channel, iSCSI and FCoE. Storage area networks (SANs) are being implemented in enterprises of all sizes. The separation of the storage of data from the application or file server has numerous benefits. Fibre channel, a set of standards defined over a period of years to support high speeds and ubiquitous connectivity, offers the enterprise a variety of different topologies. However, in some cases, the SAN must be extended over a wide area data network. When this happens, the impact to the data network can be severe if proper planning and tools are not put in place. RAID is an excellent and proven technology for protecting data against the possibility of hard disk failure; in addition, data striping offers improved system performance. When a hardware RAID solution is used, the range of RAID levels enables the user to tradeoff reliability, performance, and cost. This capability is lost with software RAID, where the user is typically limited to only a single level of RAID.

Storage virtualization has helped SAN technologies realize the goals of simplification in storage management that were initially unable to fully be realized due to the administration conflicts existing in first generation hardware. The relationship between networked storage and storage virtualization is not only a possibility, but now an industry accepted reality that is pushing acceptance as a fully realized solution into the IT industry.

Organizations are finding problems in the creation of storage administrators who are educated, trained, and specialized to manage and implement storage resources. Not only is it important to accept the need to specialized storage administrators, but also for a firm to have their own in-house storage expertise in order to implement a storage solution that best fits the organization's need. With a global shortage in storage administrators and individuals with the variety of disciplines needed for their roles, in-house storage expertise is quickly becoming more difficult to obtain. Thus, organizations are forced to turn to vendors and service providers for storage solutions, which can diminish purchasing power and overall control of a company's own IT department.

Key Terms

Cache A cache is a temporary storage place. Computers typically use a cache for frequently accessed information. Retrieval of the information is faster from the cache than from the original source. There are many types of cache, including RAM cache, secondary cache, and cache memory.

Network attached storage (NAS) Storage (usually disks but possibly tapes or memory) attached to and distributed (shared) over a network. Usually the environment of a NAS is many computers sharing a lot of data; the typical bottleneck in performance is the file manager or object store. A common example is NFS over a LAN.

Network data management protocol (NDMP) An open protocol for network-based backup of data.

RAID A set of multiple disks that exchange data in such a way as to permit the failure of at least one without losing any data. There are multiple levels of RAID.

RAID 0 Striping, where data is simply striped across multiple disks to make one larger logical file system; no parity information is used; one dead disk takes out the whole array.

RAID 1 Mirroring, where one disk is duplicated onto another disk; if either disk fails, the other disk has all the data (no data is lost).

RAID 2 Parity (proprietary), special case. Patented by Thinking Machines, Inc. Hamming code disk array which consists of data drives and multiple parity drives.

RAID 3 Striping with parity, where one disk in an n-way stripe is reserved for parity information (also known as byte striping).

RAID 4 Striping with parity, where one disk in an n-way stripe is reserved for parity information, using a variable striping unit size (also known as segment striping).

RAID 5 Striping with parity, where information (including both data and parity) is striped across all disks in an n-way stripe.

RAID S Striping with parity (proprietary), where information (including both data and parity) is striped across all disks in an n-way stripe; proprietary to EMC Corp. (no longer in use).

RAID 6 Segment striping with multiple parity, an improvement on RAID 5 by having multiple parity drives, allowing for multiple disk failures with no data loss.

RAID 7 Striping with parity (proprietary), a method patented by Storage Computer Corp. where data is striped across multiple drives with one or more parity drives (an unofficial RAID level).

Small computer systems interface (SCSI) Pronounced 'scuzzy,' SCSI is a parallel interface standard used by Apple Macintosh computers, PCs, and many Unix systems for attaching peripheral devices to computers. SCSI interfaces provide faster data transmission rates (up to 80 megabytes per second) than standard serial and parallel ports. In addition, you can attach many devices to a single SCSI port, so that SCSI is really an I/O bus rather than simply an interface.

Storage area network (SAN) A network where a small number of computers share a large amount of data, usually within a single server room, where performance is critical. Examples of SAN protocols include fibre channel and the small computer systems Interface (SCSI).

REVIEW QUESTIONS

7.1 What does planning mean in data storage for corporations?
7.2 What are the roles of file systems in the context of storage?
7.3 Briefly outline the various storage media available for a user.
7.4 What is the need for a DAS in an organization?
7.5 Differentiate between DAS and NAS.
7.6 What is the role of SAN in an organization?
7.7 Why according to you should an organization plan for SAN?
7.8 Compare and contrast the features of SAN and NAS.
7.9 List out the storage protocols.
7.10 Differentiate between the various storage protocols.

7.11 What do we understand by the term RAID?
7.12 What is the need for RAID in storage?
7.13 What are the types of RAID solutions available in the market?
7.14 Differentiate between the various types of RAID solutions available in the market.
7.15 What is the need for backup in an organization?
7.16 Differentiate between RTO and RPO.
7.17 Differentiate between snapshot and CDP.

Projects

7.1 Get an appointment with a chief technical officer (CTO) of an organization and document the total storage space available within the company using an Excel sheet. Physically verify the actual storage occupied versus the free space available. Identify the ratio of used versus unused space. Estimate the rate of data growth for that organization.

7.2 Get an appointment with a CIO of an organization and identify the list of backup devices being used by the organization. Based on the discussion with the CIO, calculate the RTO and RPO for the organization.

7.3 New York-based ABC Unlimited was founded in 1965 by an Indian named Pooja as a venture capital attempt with an initial investment of about 1 million dollars as a family business. As you are aware, ABC has set up an excellent data centre facility to take care of its data consolidation. (Refer to Chapter 1 for details of the case)

However, CIO Nayak is now stuck with the new mandate put up by his CEO David. Nayak was thinking aloud on the new mandate of understanding storage consolidation including the implications of SAN for his organization. He had actually planned for expansion possibilities in the data centre, and hence he was not worried on the issues of space or for that matter other infrastructure requirements.

Like any other organization, ABC had invested in storage over the years and has a couple of storage systems spread across the organization. ABC had a couple of entry-level storage boxes for each of the 120 locations. He was contemplating the possibilities of SAN deployment for storage consolidation and business recovery.

Discuss the technical and management plans which enabled ABC to carry out the storage migration with no detectable disruptions and with zero data loss. Also help the organization to build a resilient business continuity plan with an appropriate RPO and RTO.

Note: AS/400 is a mid range server.

References

Barker, Richard 2001, *Storage Area Networking Essentials: A Complete Guide to Understanding and Implementing SANs*, Wiley, Canada.

Clark, Tom 2003, *Designing Storage Area Networks A Practical Reference for Implementing Fibre Channel and IP SANs*, Second Edition, Addison-Wesley Professional, USA.

Clark, Tom 2005, *Storage Virtualization: Technologies for Simplifying Data Storage and Management*, Addison-Wesley Professional, USA.

Farley, Marc 2004, *Storage Networking Fundamentals An Introduction to Storage Devices, Subsystems, Applications, Management, and File Systems*, Cisco Press.

Hufferd, John L., 2002, *iSCSI: The Universal Storage Connection*, Addison Wesley, USA.

Kaplan, James M., Rishi Roy, and Rajesh Srinivasaraghavan 2008, *Meeting the Demand for Data Storage*, Mckinsey & Company.

Marshall, David, Stephen S. Beaver, and Jason W McCarty 2009, *VMware ESX Essentials in the Virtual Data Center*, CRC Press, USA.

Naik, Dilip C. 2003, *Inside Windows Storage: Server Storage Technologies for Windows Server 2003, Windows 2000 and Beyond*, Addison-Wesley Professional.

NIIT 2002, *Special Edition Using Storage Area Networks*, Que, USA.

Orenstein, Gary 2003, *IP Storage Networking: Straight to the Core*, Addison-Wesley Professional USA.

Simitci, Huseylin 2003, *Storage Network Performance Analysis*, Wiley, Indiana, USA.

Spalding, Robert 2003, *Storage Networks: The Complete Reference* McGraw-Hill.

Toigo, Jon William 2003, *The Holy Grail of Network Storage Management*, Prentice Hall, USA.

Vengurlekar, Nitin 2007, Murali Vallath, and Rich Long 2007, *Oracle Automatic Storage Management Under-the-Hood and Practical Deployment Guide*, McGraw-Hill Osborne Media, USA.

PART II

IT Production Tools

- **Chapter 8** Security Infrastructure
- **Chapter 9** Office Tools
- **Chapter 10** Data Management Tools
- **Chapter 11** Web Tools

PART II

CHAPTER 8

Security Infrastructure

All warfare is based on deception . . . hold out baits to entice the enemy. Feign disorder, and crush him.

–Sun Tzu
The Art of War

Learning Objectives

After reading this chapter, you should be able to understand:

- the basics of security
- the differences between the various security gadgets
- the need of a firewall
- the differences between a firewall and IDS
- the use of RFID technologies
- the need of endpoint security

8.1 Introduction

The Internet revolution has dramatically changed the way individuals, firms, and the government communicate and conduct business. For example, the telecommunications, banking and finance, energy, and transportation industries, as well as the military and other essential government services, depend on the Internet and communication technology (ICT) infrastructure to conduct most of their day-to-day operations. The ICT security and security awareness domains are dynamic and are constantly influenced by new developments and technologies. Evolving ICT poses ongoing challenges to both researchers and practitioners. However, this evolving widespread interconnectivity has increased the vulnerability of computer systems, and more importantly, of the critical infrastructures they support to information security breaches.

Security is generally defined as the freedom from danger or as the condition of safety. Computer security, specifically, is the protection of data in a system against unauthorized disclosure, modification, or destruction and protection of the computer system itself against unauthorized use, modification, or denial of service.

Security is achieved through security gadgets. A spectrum of security gadgets has plagued the market. A worldwide study by *CIO* and PricewaterhouseCoopers reveals a digital landscape ablaze, with thousands of security leaders fighting the flames to come up with solutions. For the purpose of our discussion, let us collectively refer to them as security gadgets. These security gadgets are broadly classified under the groups of gateway security, network security, and endpoint security. These security gadgets operate on ICT infrastructure to provide various levels of security and access control to the end-user. These types of access open up the network to a myriad of potential web threats.

Security for open information systems, allows access to information assets by outsiders. Security deployments extends, enhances, and complements the security measures provided by point solutions, regardless of whether the perimeter is known or whether data is in transit. Implementing such security measures to meet the requirements of computer security and corporate applications presents a distinct set of challenges. Installing appropriate security measures is generally a daunting process, even when networks are closed.

The appealing manner in which security solutions are marketed, along with claims of easy installation and management, can lead organizations to make the decision to implement a security solution without taking time to thoroughly examine the solution. With the large number of firewall solutions available today, firewall selection and implementation can be a time-consuming and overwhelming process.

At a general level, security threat concerns include, security threat through Internet (distributed denial of service, virus/worm, spam) internal threats such as disclosure of proprietary information and sabotage; user identity impersonation, unauthorized modification of data or other sensitive information, improper software configuration and physical protection of the network elements. Identity theft and online fraud are on the rise. Between December 2007 and February 2008, researchers measured a 70 percent increase in fraud acts called phishing, in which e-criminals use convincing-looking emails to lead consumers to fraudulent, but just as convincing, web pages. When Internet users fall for phishing scams, they can unwittingly hand over an array of sensitive personal information, including user names, passwords, credit card numbers, and social security numbers.

For example, software configuration refers to the process of setting software quality parameters to meet specific user requirements. Proper configuration is particularly critical for information technology (IT) security software as evidenced by frequent warnings by security experts about risks from using default out-of-the-box settings. In this environment, control and availability gaps continue to grow as infrastructure complexities increase.

The top six security concerns reported in the literature are unauthorized systems access, auditability or compliance concerns, customer data breaches, sabotage (internal and external), theft of intellectual property, and cost of administration.

Transmitting data across the Internet link, through secured and unsecured channel is always a challenge. With increased convenience of online banking, the threat of unscrupulous elements getting into the networks and perpetrating frauds looms large. While Internet banking is still evolving, banks are surely getting networked to enhance operational efficiency and serve customers faster and better. This essentially means that they are building networks to run business-critical applications and interconnect branches throwing up several security challenges. Some of the widely prevalent security threats in the financial sector are denial-of-service, virus, insider net abuse, unauthorized access, theft of proprietary information, financial fraud, sabotage, and system penetration. In this context, the security governance, security policy and its implementation, and the well-tested and reliable business continuity measures assume importance.

The enterprise perimeter has expanded with mobile devices such as laptops, PDAs, Universal Serial Bus (USB) memory sticks constantly travelling outside the corporate firewall. In the past, users could only access the network through a few ingress or egress points usually where the Internet connected to the enterprise network. Enterprises stacked security at the Internet perimeter using firewalls and intrusion detection systems. However, different means of gaining entry to the network exist. With the perimeter having been extended and distributed, enhanced security needs to be applied at each of these new ingress and egress points to avoid damaging threats, thus complicating security architectures.

Computing system security relies mostly on users, operators, and administrators, and even the best designed system, if badly operated, would be unsecure. Most authentication and protection mechanisms can be diverted by malicious or careless users, then allowing possible intruders to perform security breaches.

Internal networks are complex with homegrown applications, client-to-client applications, loose adherence to protocols, and no central security coordinator. Unlike perimeter networks, where all traffic is blocked unless explicitly allowed, internal networks need to allow all traffic unless it is explicitly blocked. The inability of administrators to keep pace with hundreds of new vulnerabilities in applications, operating systems (OS) and even network infrastructure has led to networks with more potential holes than barriers. Coupled with the fact that these vulnerabilities and tools to exploit them are blatantly advertised on the Internet, enterprises are facing the daunting task of building walls to protect their network while keeping it open for their business. Network administrators are, therefore, finding traditional perimeter security solutions inadequate in preventing the spread of worms and viruses inside their networks. An effective internal and external threat prevention and containment strategy is to deploy multiple lines of defense.

Increasing compliance demands and insider risk mitigation, coupled with accelerated opening of the network, are shifting the focus to internal user-based access control in the enterprise. This necessitates not only true application recognition, but more granular application-specific controls to balance adherence to business/mission policy, compliance and security needs. Over the past years one has seen many examples of breach notifications that affect hundreds to crores of victims. Studying the business impact of post-breach processes, it was observed that the way an organization reacts to a breach can make the difference between a minor financial impact and a complete corporate meltdown.

Today, security management is a vast domain on its own comprising threat management, identity and access management, and security incident management. Security management is the process of increasing the confidentiality, integrity, and availability of the services. Confidentiality is the protection of information in the system so that unauthorized persons cannot access it. Many believe this type of protection is of utmost importance to military and government organizations that need to keep plans and capabilities secret from potential enemies. The IT environment is now so dynamic that chief information officers (CIOs) can never say that security is a 'solved' problem. The only certainty lying in front of CIOs is that security threats will continue to grow and they would need to find newer and better ways to ensure immunity to their organization. Addressing the audience at the CIO's year ahead program held in Malaysia, Teo Choo Siong [Teo 2009], IBM has recommended that while investments on IT security continue to rise to address the ever-evolving nature of threat, every new technology solution that is considered must deliver significant return on investment (ROI) and the ROI on security spending needs to take into account any possible productivity impact. It is time organizations consider switching from the best-of-breed to the best-of-need in security products and they are getting commoditized, says Neil MacDonald, vice-president with research firm Gartner. The broad drivers for growth of security vertical were infrastructure build-out, regulatory compliance, and a growing awareness of the emerging security threats.

8.2 Vulnerability—Gaps in Computing System

With regards to the use of IT, companies are at a greater risk today than in any other time in their history. This comes from the speed and variety of threats that are propagated from all over the world, 24 hours a day. Attacks and threats to networks have not only increased in the last couple of years but the nature of these attacks has also changed. Threats in cyberspace deserve increased attention not only because of the dangers they pose to commercial and government entities but also to the individual user. From simple virus infections or malicious codes snooping around the network, the threats have become blended, combining the characteristics of viruses, worms, trojans, and malicious codes. Blended threats exploit the vulnerabilities to initiate, transmit, and spread an attack by using multiple methods and techniques and cause widespread damage. As the complexity of threats and the affected areas increased, demand for security solutions also went up. Compounding the security challenge is the proliferation of wireless devices, the fastest-growing class of devices that need to be managed and secured, increasing the security burden substantially. These smart devices run Internet protocol (IP) services and offer access to corporate networks and the Web.

Vulnerabilities often exist at the network layer in the form of firmware loopholes, badly configured sample network management protocol (SNMP) access control and non-existent access lists on critical devices. Vulnerability analyzers find holes in websites, gaps in virus or network management coverage, and shortcomings in user passwords. Quantifying and valuing risk is much harder, because diagnostic tool results are devoid of organizational context and business domain knowledge.

> **Client-side Exploits**
>
> Client-side exploits are some of the most commonly seen exploits and this is mainly due to the fact that traditional perimeter security (firewalls, router access lists) offer little or no protection against these kinds of exploits. This is due to the fact that client-side exploits target vulnerabilities on the client applications.

8.2.1 Plugging Vulnerability—Management Game

One vital process is reducing the exposure a company presents to adversaries. It is called attack surface reduction. The term attack surface refers to a program's susceptibility to various avenues of attack or to systems as a whole. Companies often use a combination of network design exercises, access management, and configuration management to reduce attack surfaces. A system's attack surface is reduced by exposing only required services to the network, disabling or removing unnecessary software, or limiting the number of users authorized to log on to a system.

A vulnerability is defined as a bug, flaw, behaviour, output, outcome, or event within an application, system, device, or service that could lead to an implicit or explicit failure of confidentiality, integrity, or availability. Vulnerabilities may result from weak passwords, software bugs, a computer virus or other malware, a script code injection, or standard query language (SQL) injection. Today's vulnerability landscape is mined with custom application exposures, infrastructure deficiencies such as improperly secured wireless networks, and desktop and end-user-centric attack methods. Vulnerability assessment (VA) provides baseline and discovery functions in support of vulnerability management. An accurate vulnerability assessment requires a deep understanding of failure modes and effects on each of the network components, and the knowledge of how these components are interrelated at each point in time to various applications in a networked system.

When a vulnerability discovery is performed via network scanning, passive network monitoring or patch auditing procedure, the discovered vulnerabilities can each be classified if they were newly discovered, or if they were previously known about. Vulnerability disclosure drives the majority of the content in vulnerability monitoring solutions. Vulnerability assessment products scan an endpoint and attempt to determine vulnerable conditions based on a database of known vulnerabilities. Vulnerability assessment products can also determine many other aspects of the endpoint, including open ports, running services and protocols, applications, and operating system. This information provides the security groups with the data they need to measure security postures. A vulnerability rated as a low risk in the morning could turn into a worst nightmare that very night. To meet the ever-increasing speed with which exploits are written and propagated, traditional network-based vulnerability scanners have morphed into more full-scale vulnerability management products. Vulnerability scanning technology has matured in recent years, and the tools' accuracy, speed, and safety have improved dramatically. Business information is vital for vulnerability prioritization, since it ties the technical threat and vulnerability data into the business function. Every organization is different and thus has different critical assets and applications. Vulnerability management is a process that can be implemented to make IT environments more secure and to improve an organization's regulatory compliance posture.

8.2.2 Common Vulnerabilities and Exposures

Common vulnerabilities and exposures (CVE) is a dictionary of publicly known information security vulnerabilities and exposures. CVE Identifiers are unique, common identifiers for publicly known information security vulnerabilities. CVE Identifiers are used by information security product/service vendors and researchers as a standard method for identifying vulnerabilities and for cross-linking with other repositories that also use CVE Identifiers.

The ability to score information system vulnerabilities is extremely important to the professional computing world. Common vulnerability scoring system (CVSS), an open standard for scoring vulnerabilities is designed to rank information system vulnerabilities and provide the end-user with a composite score representing the overall severity and risk the vulnerability presents. A vulnerability metric is a constituent component or characteristic of a vulnerability that can be quantitatively or qualitatively measured. These atomic values are clustered together in three separate areas: a base group, a temporal group, and an environmental group. When using CVSS, security professionals, executives, and end-users will have a common language with which to discuss security vulnerability severity.

8.2.3 Adaptive Management for changing Threats

Vulnerability assessment comprises three basic approaches. These are as follows.

- Active network scanning, also referred to as network VA, will remotely scan devices over the network without requiring agents; deeper inspection of endpoints can be performed through credentialed access.
- Passive observation of network traffic does not actively scan endpoints, but it captures traffic between endpoints to determine their state based on those traffic patterns. Although passive observation can provide information about endpoints that cannot be actively scanned (for example, systems with personal firewalls), this technique alone does not provide sufficient data to support remediation activity.
- Persistent agents reside on the endpoints, collecting state information in real time. They can determine aspects of the endpoint that cannot be determined remotely, such as applications or services that are installed but not running. Agent-based approaches must be augmented with discovery and baseline functions that can be applied to unmanaged endpoints.

Transforming chaos into clear and manageable security policy is essential, which is why future network security systems need to focus on convergence and consolidation. For robust information security for an enterprise, a proactive architectural and system approach is critical. The idea is to accurately identify and stop attacks as early and as far from the destination host as possible, while simultaneously simplifying the security architectures required to do this.

According to Gartner, enterprises that implement a vulnerability management process will experience 90 per cent fewer successful attacks. Vulnerability management was born out of a need to intelligently prioritize and fix discovered vulnerabilities, whether by patching or other means. Managing vulnerabilities requires a well-thought-out process that aligns to business needs and provides a solid framework for the IT department. The goal of vulnerability management is to have a system that helps reduce the time and money invested in dealing with vulnerabilities and reduces the risk of vulnerability exposure. Effective vulnerability management programs require the right balance of technology, business intelligence, and process. According to Gartner analysts, the vulnerability management process includes policy definition, environment baselining, prioritization, shielding, mitigation, as well as maintenance and monitoring. Successful network vulnerability management balances the demands of security against the demands of individual business units. An effective vulnerability management program helps in managing a company's overall security posture and risk tolerance. An effective vulnerability management program helps to demonstrate compliance with established controls, as well as alert management to compliance problems.

For example, Ryan Miller, director of global information assurance for Federal Mogul, says that from a manufacturing perspective all the equipment on the shop floors that used to be dumb are becoming more intelligent. Everything has an operating system and is basically becoming an intelligent multifunction device. Those kinds of devices are rapidly becoming a concern where they were not in the past.

The impact of a major consumer data breach may be more muted on the corporate relationships than on consumer relationships. It can take a year or more to move business from one company to another, given contract restrictions, partner sourcing, and the ability to find similar products for a competitive price. According to a panel of vulnerability research experts, enterprises should test vendor software for vulnerabilities before deploying, much like they should be testing their home-grown applications.

Perimeter hardening is the process of removing vulnerabilities from the customer's IT environment that may be exploited by hackers for unauthorized access. To ensure that customers' servers and workstations have been sufficiently hardened and unauthorized access is denied, the perimeter hardening service will review existing security configurations to determine the level of security required and then develop the appropriate configurations on an ongoing basis.

Attack Vectors

There are many ways to exploit a vulnerable system. Attack vectors define the ways in which anyone can gain access to the system or server in order to deliver the exploit. Exploit writers choose their attack vectors based on the number of systems that they wish to target.

8.3 Penetration Testing

Today, penetration testing is an often confused term. At its simplest definition, a penetration test is the process of actively evaluating the information security measures of an organization. The term penetration testing is the security-oriented probing of a computer system or network to seek out vulnerabilities that attackers could exploit. It is a method of evaluating the security of a computer system through an active analysis of the system for any potential vulnerabilities that may result from poor or improper system configuration. These tests are conducted in several ways. At a macro level, the tests are classified into black-box and white-box testing.

Black-box testing simulates an attack from someone new to the system. A black-box testing assumes no prior knowledge of the infrastructure. A white-box testing simulates the possible internal paths and leak of information. The services offered by firms range from a simple scan of IP address and open ports to a full audit of source code.

Penetration testing is usually invaluable to an organization's information security program. A great deal of technical effort is spent on the testing the real value of this analysis is in the report and debriefing which is done at the end.

8.4 Patch Management

With new vulnerabilities being discovered every day and hackers launching flash attacks, patch management started gaining importance. Patch management began to involve scanning of the network for vulnerabilities, understanding the threats, downloading the relevant patches, and installing them over the network; often done remotely from a security operating centre.

In response to new vulnerability, organizations have created an arsenal of technical weapons to combat computer security breaches. This arsenal includes firewalls, encryption techniques, access control mechanisms, and intrusion detection systems.

8.5 Integrated Network Security

Nortel's definition of integrated network security is based on a key tenet known as 'security in the DNA'. Nortel Networks strategy for enterprise security called 'unified security architecture' assumes that all components of an IT infrastructure are targets. It is not a one-size-fits-all prescription, but rather a framework of functionality that offers multiple implementation choices suitable for closed, extended, and open enterprises in different industries and for diverse application requirements within all enterprise types, says Nortel.

Today's cyber threat landscape is far less defined than it was a few years ago. The most serious network attacks are focussed on thefts, fraud, and organized crime. Bruce Johnson, the adaptive security appliance (ASA) product manager at Cisco Systems Inc., commented that the increasing number and complexity of network threats has lead to a corresponding increase in the number and complexity of network security devices required to provide adequate defenses.

8.6 Types of Attack Vectors

An attack is an intentional threat and is an action performed by an entity with the intention to violate security. Examples of attacks are destruction, modification, fabrication, interruption, or interception of data. An attack is a violation of data integrity and often results in disclosure of information, a violation of the confidentiality of the information, or in modification of the data. The exact number of attacks cannot be readily determined because only a small portion are actually detected and reported. A threat is a potential violation of the security of a system – an event that may have some negative impact while vulnerabilities are actual security weaknesses or flaws that make a system susceptible to an attack.

Any attack on the security of a system can be a direct and an indirect attack. A direct attack aims directly at the desired part of the data or resources. An indirect attack looks for information received from the desired data/resource without directly attacking that resource.

Passive attacks are made by monitoring a system performing its tasks and collecting information. In general, it is very hard to detect passive attacks since they do not interact or disturb normal system functions. Monitoring network traffic, central processing unit (CPU) and disk usage, etc. are examples of passive attacks. An active attack changes the system's behaviour in some way. Examples of an active attack can be to insert new data, to modify, duplicate or delete existing data in a database, to deliberately abuse system software causing it to fail and to steal magnetic tapes, etc.

8.6.1 Computer Worms

A computer worm is a self-replicating computer program. It uses a network to send copies of itself to other nodes (computer terminals on the network) and it may do so without any user intervention. A computer work carries a payload. A payload is code designed to do more than spreading the worm. The payload might delete files on a host system (for example, the ExploreZip worm), encrypt files in a cryptoviral extortion attack, or send documents via e-mail. A very common payload for worms is to install a backdoor in the infected computer to allow the creation of a 'zombie' under control of the worm. Worms spread by exploiting vulnerabilities in operating systems.

8.6.2 Denial of Service and Distributed Denial of Service

One of the basic forms of a denial of service (DoS) attack involves flooding a target system with so much data, traffic, or commands that it can no longer perform its core functions. When multiple machines are gathered together to launch such an attack, it is known as a distributed denial of service attack, or DDoS.

8.6.3 SPAM

On any given day, e-mail users across the world receive hundreds of spam and junk mails that eat into precious bandwidth and make for difficult management. Spammers send out millions of e-mails taking advantage of people using computers for online shopping and sending the wishes. The spammers' main aim is trying to fool people into buying things or trick them into providing the ID. The financial costs associated with spam are large and growing by every passing day. Spam leads to loss of employee productivity due to time spent managing their inboxes and junk e-mail folders, requiring employees to delete spam and block senders. The plenitude of spam can block business communications systems as the e-mail flow at the workplace can be clogged for hours, if not days. Experts estimate that spam cost approximately 17 billion annually in the USA. It includes lost of productivity and the expense of fighting it. The worldwide cost was estimated at 50 billion.

Large volume of spam enters the company's networks, thereby choking mail servers and occupying expensive space in e-mail quarantines and storage archives. The magnanimity of the problem becomes quite evident as a 2006 study by the Radicati Group estimated that spam constituted 70 per cent of the total worldwide messaging traffic, and this figure is expected to increase to 79 per cent by 2010. Spam has rapidly changed from a mere nuisance to a major security threat and financial drain for organizations worldwide, as they attempt to stem the flood of unsolicited bulk e-mail while ensuring that legitimate correspondence is delivered correctly.

The impact of spam is multi-dimensional. It cannot be treated in isolation and it needs to be enmeshed with the security policy. Based on the threat perception and the levels of protection, enterprises need to go in for solutions that best work for them.

8.6.4 BOTS

One of the biggest threats to the Internet is the presence of large pools of compromised computers, also known as botnets, or zombie (drone) armies, sitting in homes, schools, businesses, and governments around the world. Bot is a generic term and is used to describe an automatom or automated process in both the real world and the computer world. A bot is a type of malware which allows an attacker to gain complete control over the affected computer. Computers that are infected with a bot are generally referred to as 'zombies'. A botnet refers to a pool of compromised computers that are under the command of a single hacker, or a small group of hackers, known as a botmaster. The beauty of bot, according to the attacker, is the ability to control that specific bot and hundreds of others via simple Internet relay chart (IRC) commands. Internet bots (derived from the word web robots) are software applications that run automated tasks over the Internet. They are used for both commercial as well as malicious purposes. Bots are controlled by a single person called bot master. Botnets are the largest growing threat in cyber security today. They have been successful in proliferating themselves throughout the Internet and internal networks by exploiting common vulnerabilities, commented Carl Banzhof, the chief technology officer (CTO) at Citadel Security Software.

Analysing traffic may be enough to determine if bots are present. Unusually high rates of outgoing traffic could signal the presence of bots. Traffic flowing through Port 6667 (used for IRC) in corporations is usually a strong indication of the presence of bots, as bots often receive instructions on how to act from a 'master bot' communicating through IRC. Other ports to watch include Port 25 (e-mail or spam relay) and Port 1080 (often used for proxy servers such as Socks, which manages connections between clients and servers).

8.6.5 Insider Threat

An insider can be a current or former employee, a contractor, service provider, software vendor, and anyone who has the same or similar access rights into a network, system, or application. The term 'insider' is often narrowly defined as part- and full-time employees. While employees certainly comprise the bulk of insiders, organizations often fail to consider the security implications caused by consultants, contractors, and others who also get trusted access to data and network resources. Insiders, by virtue of legal access to their organizations' information, systems, and networks, pose a significant risk to employers.

A malicious insider (MI) is a person motivated to adversely impact an organization's mission through a range of actions that compromise information CIA (confidentiality, integrity, and availability) Triad. Malicious insiders increasingly will be able to leverage the vulnerabilities on remote networks, personal computers (PCs) in home offices, notebooks, and intelligent hand-held devices as they attempt to gain access to proprietary and regulated information.

There are two notable types of malicious employees who may have the potential to cause significant harm or damage: one is the IT expert with a hacker mentality and intent to cause harm and the other is the disgruntled employee who causes harm typically out of the desire to 'get even' with his or her employer. 'Exploring the Mind of the Spy' is an excellent paper written for the Naval Criminal Investigative Service by Dr Mike Gelles. Dr Gelles points out three criteria that usually have to be met for a previously trustworthy and loyal employee to commit a serious crime.

- The presence of a personality or character weakness that manifests itself in anti-social tendencies that can lead to malicious behaviour.
- The presence of a personal, financial, or career crisis that exposes the individual to suffering and extreme stress. The behaviour related to this stress is often observable in the workplace.
- The absence of appropriate assistance in a crisis. Others may fail to recognize the person's problems, or they may recognize them and refuse to become involved.

Inside employees enjoy a wealth of technology-enabling remote connectivity such as laptops, thumb drives, PDAs, smartphones, iPods, digital cameras with powerful processing and enormous storage capabilities. There are, of course, legitimate business needs for all of these new gadgets. These cutting edge technologies have increased productivity, but by extending the 'mobile edge' of the organization, which has in turn created new security risks. The extended office has introduced much larger deployed software. The deployed software is bundled with software bugs, also called vulnerability, and requires patches. Advances have been made in patch management and software distribution that enable organizations to maintain secure PCs and data transmission, but most organizations have failed to adequately manage this new requirement. These additional services or so-called new overheads provide enough opportunities for insiders.

Insiders are in a unique position with the privileges entrusted on them and the knowledge about their computational environment. Insiders can cause damage either by: remaining within their default set of privileges, or exceeding them by seeking new information and capability through a repertoire which contains not only common attacks but also unconventional ones such as social engineering. This translates directly to a special advantage and a certain amount of capability to the insider. An insider has the choice of abusing the system within this default capability. Since insiders have access privileges to use the computational infrastructure that can be used against the parent organization. An organization should, therefore, secure and audit the access rights periodically. It also necessitates the need for appropriate security controls. Security controls and validation controls are required on the process flow. The problem of insider threat is precisely the problem of evaluating the damage which can potentially occur in these two cases. The insider threat is a problem faced by all industries and sectors today.

In recent years, employees, both current and former, have been suspected, indicted, and convicted of all types of crimes—selling of usernames and passwords, planting destructive logic bombs, sabotaging critical servers and applications, stealing credit reports, and other personally identifiable financial information used to conduct identity theft and fraud, and more. Several more recent trends have become evident regarding changes in the motivation, processes, and techniques for compromising remote systems.

The potential damage an insider can commit has also been increased within the last decade by two related trends in information systems consolidation and, for all intents and purposes, the elimination of the need-to-know principle. The most damaging insider threat to deal with relates to the individual on the inside who captures and then exfiltrates information in some manner while avoiding detection alarm conditions. Unfortunately, these malicious individuals attempt to mask their activities by operating within normal or abnormal but acceptable behaviour. The consequences of insider incidents can include lost staff hours, negative publicity, and financial damage so extensive that a business may be forced to lay-off employees or close its doors.

The emergence of insider threats is difficult to identify and difficult to document because insiders have intimate knowledge of internal control and security systems, allowing them to cover their tracks and disguise their attacks as innocent mistakes. Insider threat problems seem to be roughly partitioned into cases that are motivated by personal gain and cases that are motivated by a grudge against the organization. Cases due to grudge are usually crimes of emotion. Cases due to personal gain are often much better thought out and can occur over a longer time span.

Employee theft of tangible assets can be easily detected and prevented using common security measures such as door locks, fences, guards, and closed-circuit television (CCTV) cameras. External threats, such as viruses and malware, can be adequately addressed by installing and properly configuring firewalls and anti-virus software. Unlawful or dangerous insider behaviour, however, is difficult to detect, deter, and address.

There are two schools of thought in evaluating and analysing insider crime fraud (ICF) activities, including an evaluation of profiling the behavioural aspects of the insider, based on some type of empirical study, and the evaluation of what motivates people to do certain things given a set of variables. Ostensibly, in the first method of evaluating the insider threat, people and their behavioural characteristics and subsequent nefarious actions are profiled. The second precept of the evaluation of the insider threat is largely predicated upon profiling data versus people, based on the previously described behavioural traits and circumstances.

There are two schools of thought in understanding ICF activity. The first involves understanding the behaviour of data; the second involves behavioural analysis of those involved in committing the ICF.

8.6.6 Social Engineering

Social engineering is the human side of breaking into a corporate network. It is defined generally as the process by which a hacker deceives others into disclosing valuable data that will benefit the hacker in some way. Social engineering stems from the application of psychological techniques for interacting with and manipulating the victim to obtain the desired information.

Social engineering is an attack method used by many attackers taking advantage of trust and complacency at work. Such attacks are mostly financially driven, with the attacker looking for confidential information. Some of the common tactics used for such attacks are forging identities, exploiting the inability of people to realize the value of the data held by them or the know-how to protect data. Social engineering attacks take place on two levels: the physical and the psychological. The most prevalent type of social engineering attack is conducted by phone. A hacker will call up and imitate someone in a position of authority or relevance and gradually pull information out of the user. Help desks are particularly prone to this type of attack.

Office snooping and dirty desks lead to the leakage of information. If not controlled, both are vulnerabilities that can be detected. When people leave work at night some lock up their desk and offices, some do not. Because some people do not lock up before they leave, anyone could copy and steal information, put the original back and the victim would never know that he/she had been attacked.

Dumpster diving, also known as trashing, is another popular method of social engineering. A huge amount of information can be collected through company dumpsters. The Internet is a fertile ground for social engineers looking to harvest passwords. The primary weakness is that many users often repeat the use of one simple password on every account. Another way hackers may obtain information online is by pretending to be the network administrator, sending e-mail through the network and asking for a user's password.

Social engineering robustness is the ability to reveal and report malicious requests over time that intentionally manipulate the receiver to perform actions in order to jeopardize the confidentiality, integrity, or availability of information. It is a characteristic that is difficult to measure directly.

8.6.7 Countermeasures

A security prevention mechanism is one that enforces security during the operation of a system by preventing a security violation from occurring. Usually, prevention involves implementation of mechanisms that users cannot override and that are trusted to be implemented in a correct, unalterable way, so that the attacker cannot defeat the mechanism by changing it. A detection mechanism is used to detect both attempts to violate security and successful security violations.

8.7 Firewall

The security industry abounds with a wide variety of manufacturers and service providers filling up particular niches segments of the market. The current security defense paradigm is to deploy more and more of the existing security technologies throughout every segment of the network. This includes firewalls and access control lists (ACLs) to block access and perform application inspection, intrusion protection system (IPS) technology to provide very granular traffic inspection and identify known threats, encryption software to counter eavesdropping, anomaly detection to detect worms or DoS attacks, and antivirus software to battle viruses.

Business connections are becoming more complex, driving firewalls to have increasingly rich management and configuration solutions, as well as to provide deeper and broader inspection and blocking capabilities. It is becoming increasingly difficult for IT managers to safeguard the corporate network. Employees, partners, and contractors increasingly have access to the network through remote access, local area network (LANs), or via wireless connections.

Organizations and individuals are becoming increasingly dependent on the information stored and transmitted over advanced communications and computer networks which passes through the enterprise gateway. One of the prominent gateway security element is firewall. A firewall is a network's first line of defense. The term firewall has been around for quite some time and originally was used to define a barrier constructed to prevent the spread of fire from one part of a building or structure to another.

Network security starts from authenticating any user, most likely using a username and a password. Once authenticated, a firewall enforces access policies such as what services are allowed to be accessed by the network users. A firewall is configured to keep unauthorized or outside users from gaining access to internal or private networks and services. All messages entering or leaving the intranet pass through the firewall, which examines each message and blocks those that do not meet the specified security criteria. They work by analysing each network access request. Transport requests are compared to a list or database of approved source IP addresses and other parameters. When implemented correctly, firewalls can control access both to and from a network. In addition to controlling and blocking potentially dangerous traffic, firewalls usually log traffic, based on configurations. Such logs contain the source IP address of the blocked packet, the destination address, the date and time the packet arrived at the firewall, the port for which the packet was destined, and the disposition of the packet (transmitted or blocked).

Firewalls can appear in many areas of a network topology but they are almost always found guarding the perimeter of the network. Firewalls are often the first line of defense and the primary implementers of a positive security model policy of denying all except that which is expressly allowed. They are the enforcement points for creating demilitarized zones (DMZ) for external connections.

The enterprise firewall market is one of the largest and most mature security markets. It is driven primarily by the requirement to provide network policy enforcement and intrusion prevention at trusted boundary points. It is populated with mature vendors, and product availability is fairly homogeneous among horizontal and vertical markets.

Innovation has been limited, and opportunities for reducing firewall unit costs have increased because of fewer points of differentiation between competing products and virtualization.

8.7.1 Firewall Defined

When a network is connected to a public network, it is potentially exposed to a number of threats including, hackers, spyware, and Trojan horse programs. The increasing ubiquity of 'always on' broadband Internet connections means users need to be increasingly vigilant of security issues, as network traffic coming into the computer can cause damage to files and programs even when the user is away from the computer and the computer is idle. In a system that is not protected with any security measures, malicious codes such as viruses can infect systems and cause damage that may be difficult to repair. The loss of financial records, e-mails, and customer files can be devastating to a business or to an individual.

A firewall (Figure 8.1) is a system designed to prevent unauthorized access to or from a private network. It is a system composed of hardware and software utilized to enforce access control between two network entities. A firewall typically consists of several layers of protection designed to intercept and prevent penetration by intruders.

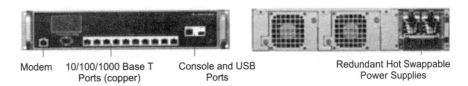

Modem | 10/100/1000 Base T Ports (copper) | Console and USB Ports | Redundant Hot Swappable Power Supplies

FIGURE 8.1 Firewall

Firewalls can be implemented in both hardware and software, or a combination of both. They implement basic network security. Firewalls enforce security restrictions and prevent inappropriate access to internal networks. They block packets destined for specific software ports, filter traffic based on IP addresses, or even block packets destined for specific applications.

Firewalls are frequently used to prevent unauthorized Internet users from accessing private networks connected to the Internet, especially intranets. A firewall appliance is a dedicated hardware and software system whose sole purpose is to function as the implementer of the defined access control policy. It is a device used to implement a security policy between networks. A firewall has multiple network interfaces, and is typically used to create a secure boundary between untrusted external networks and trusted internal networks. The security policy defines what type of access is allowed between the connected networks. Firewalls are a necessary security control for policy enforcement at any network trust boundary, but changing business and threat conditions are putting pressure on growth in the firewall market.

Case Study—Firewall Implications

Founded in 1998, the XYZ Corporation Centre is a not-for-profit research institute with a global vision to improve health science research. Research at the centre will improve the research conditions and will provide an IT platform.

The centre started looking at alternatives to its discontinued and functionally-limited network security appliance (firewall) after it began having problems preventing SQL injection attacks, which exploit security vulnerabilities at an application's database layer. The malicious SQL injection attacks defaced the Centre's public web pages, critically undermining the site's integrity. The Centre's activities require collaboration between researchers attached to the Centre, universities allied with the Centre and scientists from around the world. Restoring the pages was very time-intensive and consumed the Centre's limited IT resources. Simultaneously, the centre was seeing an increase in virus and spyware attacks.

Additionally, as a global research centre, the centre receives e-mail from people around the world, but was being overwhelmed by spam attacks. The centre needed an alternative to its existing solution to ensure reliable e-mail security.

The Centre conducted extensive research before selecting a replacement solution for their network security appliance. The new solution was engineered to combat the evolving threats to the enterprise network by providing administrators with a high performance, scalable, multifunction threat prevention appliance. The selected solution combined the traffic processing capabilities with deep packet inspection engine. The solution empowered administrators with customizable firewall tools for precise control and inspection over network traffic. The results were instant and effective.

8.7.2 Connection Table

Connection table is a dynamic table in which the firewalls maintain the information regarding traffic flowing in and out of your network. It has information about the characteristics of all the packets (source address, destination address, source port, destination port, translated address, translated port, etc.).

8.7.3 Stateful Firewall

A new trend in home networking firewalls is called stateful packet inspection, an advanced form of firewall that examines each and every packet of data as it travels through the firewall. A stateful firewall keeps track of the state of communications sessions. Stateful packet inspection firewalls examine protocol packet header fields while proxy firewalls filter services at the application level. They monitor both the incoming and outgoing packets in each transmission control protocol (TCP) connection. These firewalls learn and remember connection states and evaluate new traffic transactions against prior connection histories.

8.7.4 Packet Filtering Firewall

Packet filtering firewalls work at levels 3 and 4 of the TCP/IP protocol stack, filtering TCP and user datagram protocol (UDP) packets based on any combination of source IP address, destination IP address, source port, or destination port. Packet filtering firewalls, especially those running on routers or on stand-alone appliances, also provide network address translation (NAT).

However, packet filtering firewall do not examine the payload of a packet, and do not keep track of what happens to a packet once it gets through the firewall.

8.7.5 Application Proxy Firewall

Proxy firewalls are able to create virtual connections and can hide the Internal client IP address making it more difficult to discern the topology of the protected intranet. Circuit-level gateways monitor TCP or UDP sessions. Once a session has been established, it leaves the port open to allow all other packets belonging to that session to pass. The port is closed when the session is terminated.

8.7.6 Software Firewall

Software-based firewalls are software packages containing firewall software that you install on top of an existing operating system and hardware platform. Software-based firewalls come in both small office/home office (SOHO) models and enterprise models.

8.7.7 Firewall Deployments

Firewalls have remained the most popular security tool deployed across multiple segments. As usual most of the enterprises are using these for perimeter security. An enterprise may want to give its database of customers a little more protection than just a few firewalls. This is achieved through the deployment of a firewall in any of the standard architectures. For large organizations and data centres, the concept of layered or redundant security still holds good.

The security specialists interpret and integrate the design of the firewall into the IT environment, in accordance with the organization's security policy. Operating procedures are developed, and a firewall implementation report is generated to ensure that the firewall is managed securely.

Load-balancing firewalls run specialized software/programs that check for load in all firewalls and distribute the traffic to a least loaded firewall. All firewalls work in sync. In case a firewall in the cluster fails, the traffic that was flowing through that firewall is distributed to other firewalls transparently.

8.7.8 Demilitarized Zone

In military terms, a demilitarized zone (DMZ) is an area, usually the frontier or boundary between two or more groups, where military activity is not permitted. In computer network terms, a DMZ is a network or part of a network, separated from other systems by a firewall.

8.7.9 UTM

In an era of convergence, various security tools are being packed into one box. The integrated boxes, by their very nature of having everything at one place, make it easier to manage. With the bundling of security products, a single package called universal threat management (UTM), was floated by companies such as Fortinet, Watchguard, and Sonicwall. These vendors brought in multiple-function boxes for the price sensitive companies, who did not want to spend on multiple equipment and the management of these boxes.

Unified threat management (UTM) appliances, combine firewall, antivirus, and intrusion detection and prevention capabilities into one offering. The basic UTM has rapidly changed from its basic definition of Firewall, Gateway AV IPS, and virtual private network (VPN) to many more features such as content filtering, multi-ISP load sharing, anti-spam, traffic management, Virtual LAN (VLAN) support and secure socket layer virtual private network (SSL VPN). Emerging trends comprise enabling administrators to prevent users from visiting certain websites, secure social networking applications, and roll out of (SSL/VPN) connections to remote sites were some of the drivers enabling this change. The trend of router consolidation into UTM devices also brought about prominent shift in customer buying patterns eliminating the need for purchasing and managing separate routing devices. This led to reduced total cost of ownership (TCOs). Fortinet, CheckPoint, Cisco, Juniper, SonicWall, and Cyberoam emerged as the prominent vendors on the UTM scene. Universal threat management appliances quickly became a network security favourite for SMB, mid-market (SME), and enterprise branch office environments. It also received widespread adoption across all the verticals.

Know Your Firm—Fortinet

Fortinet is a leading provider of network security appliances and the leader of the unified threat management (UTM) market worldwide. Fortinet's award-winning portfolio of security gateways, subscription services, and complementary products delivers the highest level of network, content, and application security for enterprises of all sizes, managed service providers, and telecommunications carriers, while reducing total cost of ownership and providing a flexible, scalable path for expansion.

Fortinet was founded in 2000 by Ken Xie, the visionary founder and former president and CEO of NetScreen (later sold to Juniper for more than 3.5 billion dollars) and is led by a strong and seasoned management team with deep experience in networking and security.

Source: www.fortinet.com

The single-box approach would simplify product selection, product integration, and ongoing support. As most enterprises find it difficult to retain their security staff, single-box solutions are the best way out as most of them can be easily installed and managed by even non-technical people. They can be easily managed remotely also. Another important benefit that 'all-in-one' box solution could entail is that it could help them overcome the problem of supporting too many different operating systems and heterogeneous platforms.

Limited skill sets to manage growing networks, and is an equally important driver for new deployments such as UTM.

According to Nobert Kiss, VP-APAC Sales, WatchGuard Technologies, Gartner predicted that the extended threat management (XTM) would be the next generation UTM due to the customers need of extended security features. XTM will deliver to customers a solution that would address threats more quickly.

8.7.10 Challenges

While firewalls of today are doing a good job of protecting the networks, firewalls for voice over Internet protocol (VoIP) will need application-level gateways in for protocols such as session initiation protocol (SIP) or H.323. These special requirements crop up due to issues like these protocols using more than one port in a session and the extremely small size of the VoIP packets. A VoIP packet is one of the smallest packets in IP and presents some very unique challenges to the network security equipment.

> For over 60 years, a national bank has offered safe and sound financial services to its members. Today, with assets over 500 million, the bank provides a full range of services to its customers.
>
> Previously, the bank had deployed firewalls from vendor A, but experienced difficulty obtaining adequate support from that vendor. As a result, the bank's IT staff found that additions, moves, or changes to security policy were prohibitively difficult and time-consuming.
>
> The bank maintains two primary gateway firewalls, one for securing internal traffic generated by the staff, and a second for securing external traffic generated by external members and a couple of internal firewall appliances for controlling various departments. Initially, the bank had not configured these network security appliances for redundant failover. Though the bank wanted to provide a real-time failover of the gateway firewall, they were unable to proceed. They opted for a UTM solution in the form of deep packet inspection engine with parallel processing architecture. The opted solution supported high-availability architecture. The bank designed a high-availability architecture for the deployed firewalls with a concurrent heartbeat.

8.8 Intrusion detection and Prevention Systems

Attacks on current corporate networks have highlighted the importance of proactive network protection. The ability of enterprises to proactively safeguard against security threats has in-fact become a considerable business challenge. Currently, enterprises seriously consider integrating intrusion prevention systems. Leading security vendors are working to replace intrusion detection system (IDS) with intrusion detection and prevention systems (IPS). As a proactive tool, IPS would not only help detect an attack but also halt one in progress. Today's IPS products support multiple detection methods (signature, protocol anomaly, behavioural anomaly)as well as address a range of performance needs. Security companies are also positioning vulnerability assessment tools as successor to IDS, because they scan a company's networks and machines and suggest patches and fixes.

Intrusion detection system deployment has been more successful as a technology for securing information in corporate sectors. The sole purpose and advantage of using IDS is its ability to track the inbound traffic and alert the network users against hacker attacks.

8.8.1 IDS Defined

Intrusion detection is the process of monitoring the events occurring in an IT system and analysing them for signs of intrusions. These intrusions are defined as attempts to compromise confidentiality, integrity, or availability, or to bypass the security mechanisms of an IT system. The primary goal for deploying IDS is to detect, identify, and monitor unauthorized use, misuse, and abuse of IT systems by both internal network users and external attackers.

8.8.2 Types of IDS

The traditional host-based IDS (HIDS) is a type of IDS that watches for processes inside the host and monitors log files and data for suspicious activity. The network IDS is a commonly used type of IDS that works better than host-based IDS solution in terms of critical packet inspection capabilities. It consists of one or more sensors and a console to aggregate and analyse data from the sensors. Hybrid IDS is a combination of host-based IDS and network IDS technologies. Hybrid intrusion detection is system-based and provides attack recognition on the network packets flowing to or from a single host.

8.9 Network Access Control

Dealing with the damage caused by dangerous and vulnerable PCs and servers after they connect to corporate networks has made network access control (NAC) a critical security process for enterprises. Network access control, also called network admission control, is a method of bolstering the security of a proprietary network by restricting the availability of network resources to endpoint devices that comply with a defined security policy.

Gartner defined three approaches to implementing network access control functionality. These are as follows.

- Infrastructure-based: Primarily driven by Cisco and Microsoft, this approach involves upgrading your network or client/server operating system infrastructure to provide integrated (and generally proprietary) NAC functionality. An emerging infrastructure NAC approach utilizes dynamic host configuration protocol (DHCP) as a mechanism for enforcing NAC policies.
- Endpoint software-based: Network access control can also be accomplished by installing, both permanently and as part of a temporary download to unmanaged devices, endpoint software that implements NAC baselining and access-control functionality.
- Network security appliance-based: The above approaches can be expensive and complex, or require waiting for an incumbent vendor to provide NAC support. Network-based NAC appliances can be used to implement NAC functionality (or subsets of full NAC). Often, these appliances are the simplest solutions to derive some of the key benefits of NAC if the first two approaches are not feasible or are too expensive.

8.10 Endpoint Security

In the whole security chain, endpoints play a critical role in safeguarding enterprise data. The need for a well-managed infrastructure, specifically around endpoint security, is a key component of a security strategy. While personal firewalls provide a solid frontline defence, not all endpoints that connect to the internal network are protected. Very often customers, partners, and consultants access the internal network without endpoint integrity verification. Infected endpoints can proliferate threats instantly across the corporate network.

As the threat landscape has evolved beyond viruses and worms, customers now require a more comprehensive endpoint solution that combines antivirus, anti-spyware, firewall, intrusion prevention, and device and application control in a way that is more easily manageable. Although antivirus, antispyware, and other signature-based protection measures were sufficient to protect organizations in the past, small businesses now need proactive endpoint security measures that can protect against zero-day attacks and even unknown threats. They also need to take a structured approach to endpoint security, implementing a comprehensive solution that not only protects from threats on all levels but also provides interoperability, seamless implementation, and centralized management.

8.11 Security Surveillance

Security is no longer about protecting your data against virus and malicious software attacks, nor is it about protecting it against hacking attempts or against preventing disgruntled employees from stealing information. These you need to do anyways. Today, security also means protection against infrastructure misuse and information leakage. How do you know that the person you have recently recruited is the person who he/she claims to be? Maybe he/she a terrorist. In today's world, it does not really sound that absurd. How do you know that your network is not being misused by terrorists for communication?

Today's security systems generate vast amounts of data from a variety of sources in diverse formats. Powerful, IP-based platforms empower enterprises, their security operations, and the bottom line performance of the organization.

Leveraging multiple components in a security infrastructure for more proactive event response is commonly referred to as multi-factor alarm architecture (MFAA). It combines security system and sensor data with mission-critical video and powerful analytics to pinpoint significant events, images, and data. Then the intelligence is sent to the right people, wherever they are located: to consoles and Web browsers, physical security systems, video walls, mobile phones, and even personal digital assistants (PDAs). Multi-factor alarm architecture ensures the long-term investment protection of the systems that comprise a security network by harnessing their aggregate power and the resulting actionable intelligence to enhance the overall enforcement of security procedures. One of the major reasons why surveillance solutions are steadily gaining acceptance in India is due to the awareness factor. As more and more businesses shift their operations to India or outsource their work from this hub, there is a need to provide world-class features and facilities.

8.11.1 Surveillance Defined

Digital video transmission is fast becoming the standard requirement for security and surveillance systems. Both wired and wireless links are of interest for security and surveillance architects.

8.11.2 Surveillance Cameras

Video Surveillance System is an important security requirement to be provided at waiting halls, railway yards, workshops, reservation counter, parking area, main entrance/exit, platforms, foot over bridges, etc. of railway station and others railway establishments to capture images of commuters and public and to carry out analysis. It mainly consists of indoor and outdoor fixed cameras, indoor and outdoor pan /tilt /zoom dome cameras, indoor and outdoor IP Cameras, single/multi- channel video encoders, video management hardware and software, recording servers, switches, from monitor, etc. for surveillance of different locations of an organization and other establishments from centralized location.

Surveillance cameras can be connected over the Internet to allow monitoring from anywhere in the world. With the popularity of wireless cameras, IP surveillance can be extended to areas where wiring is not possible or not cost-effective. To implement an IP- based surveillance three things are required. One needs a network that can be wireless or wired, cameras that will define quality of feed, and monitoring and recording system that will store and analyse recorded footage.

Internet protocol cameras are closed-circuit television (CCTV) cameras that utilize IP to transmit image data and control signals over a fast Ethernet link. As such, IP cameras are also commonly referred to as network cameras. These cameras are primarily used for surveillance in the same manner as analog closed-circuit television. A network camera can be described as a camera and computer combined in one unit. It captures and transmits live images directly over an IP network, enabling authorized users to locally or remotely view, store, and manage video over standard IP-based network infrastructure.

A IP camera has its own IP address. It is connected to the network and has a built-in web server, file transfer protocol (FTP) server, FTP client, e-mail client, alarm management, programmability, and much more. A network camera does not need to be connected to a PC; it operates independently and can be placed wherever there is an IP network connection. A web camera, on the other hand, requires connection to a PC via a USB or IEEE1394 port and a PC to operate.

IP cameras are cameras that are designed to hook up to a network, and after they are hooked up to a network, they can be monitored from any computer with internet access. The market provides a variety of IP cameras with different features and capabilities, and that includes wireless IP cameras and Ethernet IP cameras. The difference between these two cameras is how they hook up to a network. Wireless models can hook up to a network through a wireless connection, while Ethernet cameras need to hook up to a network through an Ethernet cable.

The need for surveillance has been due to reasons such as the need to meet compliance standards or the fact that enterprises can easily deploy IP cameras and use their existing cabling for carrying video signals. Enterprises are readily deploying web-based IP surveillance solutions as they provide several advantages over traditional CCTV solutions. Administrators can remotely view and manage IP surveillance systems, the system itself is fast, ease of use, etc. Further, IP video surveillance is also being used as proof in lawsuits and at times is required to meet compliance.

> **Real-time Messaging Protocol**
>
> Adobe has announced plans to publish the real-time messaging protocol (RTMP) specification, which is designed for high-performance transmission of audio, video, and data between Adobe Flash Platform technologies. Providing developers and companies open and free access to RTMP is the latest advancement of the Open Screen Project. This is an industry-wide initiative to enable the delivery of rich multi-screen experiences built on a consistent runtime environment for web browsing and stand-alone applications across PCs, mobile devices, and consumer electronics.

8.11.3 IP—CCTV

Closed-circuit television (CCTV) is a television system that sends a signal to one or more monitors rather than broadcasting over a public network, hence closed-circuit. A standard CCTV system will normally include a CCTV camera (for capturing video), transmitters and receivers (to transfer the video from the source to where it is recorded), a recording system (for video playback), and a monitor (for video monitoring). These systems are primarily used for security purposes inside and outside buildings. Closed-circuit television is generally used in areas where there is an increased need for security, such as banks, airports, and town centres. It was initially developed as a means to increase security for banks but over time it has developed into a cost-effective means of general surveillance and home security. However, they can also be used for specialist applications such as mobile police use and interrogation.

A web camera is connected to a PC, normally through universal serial bus (USB), and uses the PC to make video available to the network and other viewers. It will not work properly without a PC connected between the camera and the network. An IP camera on the other hand is a stand-alone device, and combines the functionality of a high-end web camera, the PC and network interface into one network-ready product.

8.11.4 Video Walls

The growth of video wall technology has come a long way since the early years of its conception to today's incarnations. Originally developed for the military as a method for space command to track orbiting debris and shuttle missions, it was quickly implemented throughout the military's other branches and has since made its way into the public mainstream markets.

A video wall consists of several video screens, such as TVs, cathode ray tube (CRT) monitors, or liquid crystal displays (LCDs), that are placed on top of each other or side by side. Each of the available screens is configured to only show the section of a video frame that corresponds to its position within the array of screens.

8.11.5 Camera Deployments

Surveillance, be it CCTV or IP, has become an integral part of an enterprise security strategy. In terms of battling terror through surveillance, there are two major factors. One is the placement of cameras and second is the storage of surveillance footage. Again with the placement of cameras there are various factors in play, such as the needs of an enterprise, their type, resolution, and the number of cameras to be deployed. The most commonly monitored locations are data centers and entrance of an enterprise. Ideally, all peripheral walls of a company should be monitored along with mission critical areas.

8.12 Biometric Systems

Biometrics is the automated method of recognizing a person based on a physiological or behavioural characteristic. Biometric technologies are becoming the foundation of an extensive array of highly secure identification and personal verification solutions. Biometrics refers to the automated recognition of individuals based on their behaviour and biological characteristics.

With security paramount on their checklist, enterprises are increasingly eyeing new technology areas that can help tackle the growing complexity of the threat landscape. There has especially been a rise in demand for technologies in areas of secure identification and personal verification. This has resulted in biometrics gaining prominence among the newer security technologies.

The technology measures and analyses human physical characteristics as well as behavioral characteristics for security purposes. Who you are refers to identification by recognition of unique physical characteristics. This is the natural way people identify one another with nearly total certainty. When accomplished (or attempted) by technological means, it is called biometrics. Biometric scanning techniques have been developed for a number of human features that lend themselves to quantitative scrutiny and analysis. Some of the features include fingerprint (hand shape of fingers and thickness of hand), iris (pattern of colours), face (relative position of eyes, nose, and mouth), retina (pattern of blood vessels), handwriting (dynamics of the pen as it moves). Biometric devices are generally very reliable, if recognition is achieved – that is, if the device thinks it recognizes you, then it almost certainly is you.

Globally, biometrics is finding its way into application areas such as access control, attendance system, national ID, ePassports, consumer ID, and surveillance, with access control and attendance applications dominating the lot. Presently, commercial applications with consumer applications, such as access control for residential security, are beginning to emerge. Within the worldwide commercial space as well, the corporate market is fairly nascent as compared to applications in sectors, such as defence, police, security agencies, and the government.

Know Your Firm–Gemini Communication Ltd

Gemini Communication Ltd offers a wide spectrum of networking, services, and security solutions. The company's product portfolio includes solutions for networking, security, storage, supervision, and IT services. Gemini's solutions are centered on IT Infrastructure and management and carry the message of Innovation and Leadership.

The services portfolio are broadly organized into Infrastructure Implementation Services (IIS), Infrastructure Managed Services, (IMS), Infrastructure Outsourcing Services (IOS), and Infrastructure Consultancy services (ICS) besides its LAN, WAN, and Telecom solutions. Gemini is serving various verticals such as education, banking/financial services/insurance, government, health care, IT/ITeS, call centres/BPOs/KPOs/SDCs, and various service providers like cable operators, mobile operators, and wireless carriers.

- 1995: Gemini Communication Ltd incorporated, and goes through Initial Public Offer
- 1996: Gemini Communication Ltd gets registered in Chennai and Delhi Stock Exchanges

Contd

> - 1998: Widens the products and service offering
> - 1999: One of the very few companies to offer the last mile connectivity for ISPs
> - 2000: Diversifies into security portfolios
> - 2002: Conceptualized RFID market and establishes RFID business
> - 2006: Gemini acquires Point Red Technologies, stationed out of United States
> - 2007: Gemini bags a good number of government orders in the areas of Networking and WiMAX deployments
>
> Source: www.gcl.in

8.12.1 Biometric Access

Biometric-based authentication applications include workstation, network, and domain access, single sign-on, application logon, data protection, remote access to resources, transaction security, and web security. Utilizing biometrics for personal authentication is becoming convenient and considerably more accurate than current methods. This is because biometrics links the event to a particular individual, is convenient, accurate, can provide an audit trail and is becoming socially acceptable and inexpensive.

8.13 RFID

Radio frequency Identification or RFID has revolutionized the way many industries work in other countries. It remains a promising technology not just for retail but even for sectors such as oil and gas, animal husbandry, vehicle tracking, e-Passport, etc. Wal-Mart and other large retailers have been a major driving force behind radio frequency identification (RFID) adoption, causing it to be viewed merely as a more effective alternative to bar codes where it saves billions through reduced labour costs, out-of-stock expenses, theft, warehouse management costs and inventory levels.

Gemini Traze RFID, a subsidiary of Gemini Communications Ltd was launched five years ago to capture the emerging needs on RFID. Germini Traze RFID Pvt. Ltd, a subsidiary of Gemini Communications Ltd, opened India's first tag manufacturing plant at the Seriperumbudur electronic park.

8.13.1 Associations and Forums

The RFID Association of India (RFIDAI) has been formed as a not-for-profit society under the Companies Act to promote the adoption of RFID technology, standards, and applications across industry, government, and academia.

8.14 Contact-less Card-based Systems

Contact-less cards are special type of cards which are used for security purposes where the card need not be in contact with a card reader. An example of this system is the RFID system.

8.14.1 RFID System

Radio frequency identification is rooted in discoveries made by Faraday during the mid-nineteenth century and discoveries made between 1900 and 1940 in radio and radar technologies. Faraday discovered the concept of mutual induction, which forms the basis for powering passive tags operating in the near field. The birth of RFID technology is credited to the 1948 research paper by H. Stockman on 'Communication by Means of Reflected Power'. Use of Automatic Identification Technology (AIT) first began in 1966. Since then, the development of both standards and the proper utilization of the technology have become the focus of today's logistics professionals. With the advent of electronic product code (EPC), AIT matured into Radio Frequency Identification Technology (RFID). The RF part in RFID in particular points to the fact that this capability is enabled by wireless communication between the computing system and the item identified.

A global standard for use of RFID in inventory control and supply chain contexts is emerging in the form of 'EPC global'. Electronic Product Code is a globally unique reference number found on an RFID tag. It is divided into numbers which identify the manufacturer, product type, and unique item. 'Gemini' opened up on RFID initiatives and was closely following the standards since its early entry in India.

EPCglobal was formed to commercialize and set up a user-driven standards process based on intellectual property developed by the Auto-ID Center at the Massachusetts Institute of Technology (MIT). In October 2003, the Auto-ID Center ceased to operate and the Intellectual Property developed by the six laboratories was licensed to the newly formed joint venture, EPCglobal Inc. The EPCglobal Architecture Framework specification (EPCglobal, 2005a) is a collection of interrelated standards for hardware, software, and data interfaces, together with core services that have a common goal of enhancing the supply chain through the use of EPCs. EPCglobal works with retailers, manufacturers, and hardware, software, and integration solution providers to create and share intellectual property that will benefit the entire subscriber base.

Radio frequency identification is the name for a set of automation technologies that allow relatively large amounts of data to be associated with objects by attaching a tag to them. The history of radio frequency engineering can be traced to 1864 when James Clerk Maxwell predicted the existence of electromagnetic waves, of which microwaves are a part, through Maxwell's equations. By 1888, Heinrich Hertz had demonstrated the existence of electromagnetic waves by building an apparatus that produced and detected microwaves in the UHF region, the radio frequency selected by the Auto-ID Center at MIT for its passive RFID initiative a century and a half later. Radio frequencies, like other physical signals in nature, are analog, as is also the case for voltage, current, pressure, temperature, and velocity. A precursor to passive RFID was the electronic article surveillance (EAS) systems deployed in retail stores in the 1970s that used dedicated short-range communication (DSRC) RF technology for anti-theft detection.

Auto-ID RFID technology builds on automatic identification and data capture (AIDC) barcode standards for identifying products that, together with the standardization of shipping container dimensions, have so dramatically lowered the cost of transportation in recent decades. Companies that seized the opportunity to optimize their supply chains with this technology have become some of the largest companies in the world, including such retailers as Wal-Mart, Metro, Target, and Carrefour. Radio frequency identification systems are now being installed to expedite non-line-of-sight data capture using radio frequency (RF) to read the EPC on RFID tags. The emerging RFID technologies are based on a number of wireless interoperability standards, which specify varied requirements as to power level, modulation technique, occupied bandwidth, frequency of operation, and so forth.

Radio frequency identification technology has broad applicability across a variety of industries. It also provides for sightless or non-line-of-sight identification of items. It is a combination of radio-frequency-based technology and microchip technology. It is a wireless system that works in conjunction with an organization's IT infrastructure to improve business processes such as inventory management and efficiency in supply chain management. As the EPC/RFID technology becomes ubiquitous and the implementations mature, more and more demands will be made on the integration landscape leading to an increased complexity of business processes. Understanding the need for better supply chain visibility and advanced asset management requirements, 'Gemini' driven by the urge to enable small and medium enterprises (SMEs) to exploit RFID in their business operations and true to its brand, has set up an RFID tag manufacturing unit in Chennai.

8.14.2 RFID Components

An RFID system consists of three components: the tag, the reader, and the application or middle ware, that makes use of the data the reader reads on the tag. Radio frequency identification facilitates the tracking of objects, primarily for inventory tracking, via a three part technology comprised of a reader, a transceiver with a decoder and a transponder (RF tag).

An RFID chip may also contain information other than an EPC, such as biometric data. The antenna attached to the chip is responsible for transmitting information from the chip to the reader, using radio waves. Gemini has established a research and development unit to better understand tag antennas. The chip and antenna combination is referred to as a transponder or, more commonly, as a tag.

With RFID (Figure 8.2), automatic identification or contact-less identification is described as follows. The actors or objects are marked with a unique identifier code with the use of an RFID tag. The tag is in someway attached to or embedded in the target. In turn, a computing device that needs to identify the target employs an RFID reader to search for tags. When it receives an indication that a tag is present in its vicinity, it instructs the reader to request the code. The retrieved data are then recorded or otherwise processed in whatever way is suitable for the particular application. For example, ITC's Lifestyle Retailing Business Division has established a nationwide retailing presence through its Wills Lifestyle chain of exclusive specialty stores. Wills Lifestyle was named Superbrand 2006 by the Superbrands Council of India and has been twice declare 'The Most Admired Exclusive Brand Retail Chain of the Year'. ITC Lifestyle Retailing identified RFID as the solution that could best increase the company's responsiveness. The aim of moving to RFID was to

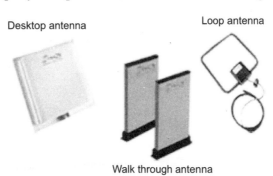

FIGURE 8.2 RFID Antenna

speed up existing processes, reduce time-to-market, handle material efficiently, and bring more accuracy of books versus physical stock. Programmable logic controller (PLC) based RFID tunnels for the warehouse and smart, customized point-of-sale software (POS) for the stores did the job.

Case Study–Automated Car Parking Solutions at IT Infrastructure Developer

IT INFRASTRUCTURE DEVELOPER focuses on providing state-of-the-art infrastructure and professionally-managed services to attract global and local investors to Bangalore. The IT park was envisaged as a professionally managed, state-of-the-art facility for organizations in IT, IT-Enabled Services (including BPOs), Research and Development, Bioinformatics, Electronics, Telecommunication, and Financial Services, and so on.

The park provides infrastructure, amenities, and services to rival the best in the world, giving organizations a conducive environment to set up and expand their businesses.

Objective of deploying RFID solution was to collect the parking fee from the staffs and visitors who are using the parking area of the IT park. For visitors, the cards are dispensed as they enter the parking gate using an RFID card dispenser. The visitors flash the RFID card to the reader installed near the gate immediately triggering the boom barrier installed at the entry point. The visitor goes inside the parking slot and the date and time of the entry is recorded in the reader. At the exit gate, the cards are dropped on drop boxes at the exit gate. The software calculates the time the visitor has parked his vehicle, and generates a bill using a slip printer. All the staff in the IT park who uses the parking are given a RFID card on monthly rechargeable basis. The staff would flash the card on the reader as they enter the parking area. The RFID card reader immediately triggers a relay to open the boom barrier which is connected to the reader. The date and time is recorded as the staff gets in to the parking lot and goes out.

Benefits

- Reduction in time for the entry and exit
- Manpower (staff) reduction in the parking lot
- Able to know the availability of the space in the parking area

Source: www.traze.in

8.14.3 RFID inlays

The RFID network begins with a tag, which carries an EPC. Tags are also called transponders. The tag consists of an antenna and an integrated circuit (IC) embedded silicon chip encapsulated in glass or plastic.

The tag permits storage of information through external write devices. Tags can be read only (RO), write once read many (WORM), or the tag permits storage of information through external write devices. Tags can be Read Only (RO), write once read many (WORM) or read write (RW). A read-only tag, records information during the manufacturing process and cannot be typically modified or erased. The data stored normally is used as a reference to look up more details about a particular item in a host system database. A read/write tag, permits recording and erasure of data on demand at the point of application. Since a rewriteable tag can be updated numerous times, its reusability can help to reduce the number of tags that need to be purchased and add greater flexibility and intelligence to the application. Additionally, data can be added as the item moves through the supply chain, providing better traceability and updated information.

Information stored in the memory of the RFID tag is transmitted by the antenna circuit embedded in the RFID inlay via radio frequencies, to an RFID reader.

8.14.4 Passive Tags

Passive tags rely on the radio signal sent by the reader for power. Passive tags derive the energy to power up the micro-circuit from the interrogating RF field, and then use the same RF field to send back information, including the unique identity of the item. The information is sent back by reflecting the RF energy back to the interrogator. Passive tags are consequently much lighter than active tags, less expensive, and offer a virtually unlimited operational lifetime. Most RFID applications today utilize passive tags because they are so much cheaper to manufacture. However, the lack of power poses significant restrictions on the tag's ability to perform computations and communicate with the reader. It must be within the range of the reader to function.

8.14.5 Active Tags

An active tag has a small battery attached to it, and can transmit information under its own power to a reading device. Active tags are often readable over much greater distances. They usually contain a cell that exhibits a high power-to-weight ratio and are usually capable of operating over a temperature range of -50°C to +70°C. The use of a battery means that a sealed active transponder has a finite lifetime. The performance characteristics of the RFID tag will then be determined by factors such as the type of IC used, the read/write capability, the radio frequency, power settings, environment, etc. Examples of active tags are transponders attached to aircraft to identify their national origin, and Lo-Jack devices attached to cars that incorporate cellular technology along with a Global Positioning System (GPS), communicating the location of a car if stolen.

Despite the advantages of active tags, the current interest in RFID is solely due to passive tags, which do not depend on batteries and thus do not require recharging or replacement.

8.14.6 RFID Readers

A system requires, in addition to tags, a means of reading or interrogating the tags and some means of communicating the data to a host computer or information management system. RFID readers (Figure 8.3) or receivers are composed of a radio frequency module, a control unit, and an antenna to interrogate electronic tags via RF communication. Readers vary in size, weight, and power, and may be mobile or stationary. Many also include an interface that communicates with an application. Reading tags refers to the communication between the tag and reader via radio waves operating at a certain frequency. Readers can be hand-held or mounted in strategic locations so as to ensure they are able to read the tags as the tags pass through an interrogation zone. The interrogation zone is the area within which a reader can read the realtag. The size of the interrogation zone varies depending on the type of tag and the power of the reader.

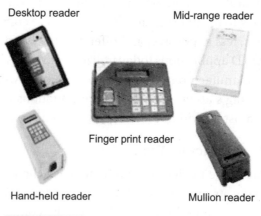

FIGURE 8.3 RFID Readers

The RFID reader sends out electromagnetic waves and the tag antenna is enabled to receive these waves. When the tag antenna enters the RF field, the

the reader. The chip then modulates the waves and the tag sends them back to the reader. The reader converts the signals received from the tag into digital data and sends it to a computer. It has a range of readers to its portfolio which includes Gemini mullion reader (DTS-100), multi-functional reader (DTS-PK), fingerprint reader (RTS-FP), Desktop reader (DTS-DR), hand-held reader (DS-HHR), long-range reader (FLR-100), mid-range reader (FMR-101A).

8.14.7 RFID Deployments

Radio frequency identification has the potential to improve the productivity of a business. RFID offers greater flexibility, higher data storage capacities, increased data collection throughput, and greater immediacy and accuracy of data collection. It's ability to increase data collection throughput and accuracy enable companies to identify materials, products and trends in supply chain with greater accuracy in real-time, compared to data collection technologies utilized to date. Once RFID technology is fully integrated, minimal human effort is required in this process thus reducing errors and costs. By providing accurate, realtag time data and information, RFID solutions enable companies to capture live data, converting it to meaningful information and automating all associated transactions and processes. Radio frequency identification technology itself can be used for a variety of applications, from contact less identification cards that can be scanned no farther than inches away from a reader, to highway systems utilizing active RFID tags that can initiate communication with a scanner 100 feet away. It holds great promise for all parties in the extended supply chain.

Case Study–PEOPLE TRACKING using RFID at Technology Institute

Introduction

A technology institute at Hyderabad is a premier R&D Institute in India. With over 450 highly professional and dedicated scientists and technical officers/technicians, excellent laboratory and instrument facilities for research in chemical sciences and technology and allied sciences, TECHNOLOGY INSTITUTE is known nationally as well as internationally for its contributions both in basic and applied research The research scholars and employees at TECHNOLOGY INSTITUTE had to go through a strict verification process as they enter the TECHNOLOGY INSTITUTE campus. Guards who were positioned in the four gates manually had to check the identities of everyone entering the premises. This creates a bottle neck in people movement where

Contd

the scholars and employees had to wait at the gates for verification. The management wanted to eliminate the wait time of the employees and research scholars at the gates and was convinced that RFID could be the right solution for this problem.

Gemini Traze RFID Pvt. Ltd, a leading RFID product manufacturer and system integrator in India was chosen to do the implementation. The solution: It was suggested that RFID gate antennas and readers be installed at the four gates. The staffs were given RFID-enabled identity cards. These gates would continuously track movement of people. The RFID cards have unique identification numbers that identifies each employee as they walk through the antennas. The system is backed by a power supply in case of power failure and web enabled software tracks and records movement of people across the gates. In addition to tracking people movement across the gates, the system was also used to record employee attendance by taking in the entry and exit time.

Benefits

- Employees do not wait for verification at entry gate, saving time and avoiding queues at the gates in the morning and evening.
- The system is also used for employee attendance.

Source: www.traze.in

- For manufacturers: RFID enables detailed, automated monitoring of parts as they move through a facility, and quickly identifies the origin of defective components or products, even after they have been sold.
- For distributors: RFID manages inventories and fleets so effectively that manual task can be eliminated, processes can be dramatically accelerated, and shipping errors can be reduced.
- For retailers: RFID ensures appropriate stocking levels, tracks the origin and history of products; prevent theft or misplacement of goods and speeds up checkout lines.

8.14.8 Business Benefits

A technology generates a lot of benefits for users. Radio frequency identification is one such technology which has projecting a lot of business benefits to all concerned.

- Low-frequency tags: These tags are typically used for access control and security, manufacturing processes, harsh environments, and animal identification applications in a variety of industries which require short read ranges. Passive tags are used for a wide array of applications,

including building-access cards, mass transit tickets, and, increasingly, tracking consumer products through the supply chain.
- High frequency tags: Popular applications include: library tracking and identification, healthcare patient identification, access control, laundry identification, item-level tracking, etc.
- UHF tags: These tags are targeted towards supply chain tracking. UHF tags have the ability to identify large numbers of objects as they are moving through a facility and later through the supply chain. There are large numbers of additional markets with demand for UHF RFID technology such as transportation, healthcare, aerospace, etc.
- Microwave tags: They are mostly used in active RFID systems. Offering long-range and high-data transfer speeds at significantly higher cost per tag making them suitable for railroad car tracking, container tracking, and automated toll collection type applications as a reusable asset.

SUMMARY

Ideally, all layers of defences should be integrated and should work in tandem from central management, authentication and log consolidation, and correlation. Deploying a layered approach to internal security can protect your valuable corporate resources from malicious intrusions and intruders.

Threat sources are now more difficult to confront and reflect more unorthodox dangers, threat styles encompass the challenging combination of asymmetric commitment and deadly transfers, and threat targets face greater vulnerability and an undermining of public confidence in government protection capabilities. All these trends occur within an international system characterized by unpredictable dangers and clashing forces of anarchy, sovereignty, and globalization. Traditional deterrence policies have failed to eradicate threat, and in the wake of 9/11, the unstable combination of high threat attentiveness and low-security expectations seems common in the West. The net effect of these changes in threat is to make it extremely difficult to orient one's defense policies around a single unified and coherent notion of incoming dangers, pointing the way toward a more target-centred than initiator centered approach.

Adaptive threat defense is the need now where an end point security system can dynamically generate an attack signature and push it to the other end points and to the perimeter IPS devices to stop it from propagating to rest of the infrastructure. Similarly, security operations teams have challenges to fine-tune false positives, they struggle by logging into various devices to understand which logs are resulting into what logs; the need of the hour is that security operations team should be able to easily see the logs and link the same to configurations which are resulting into these logs.

The cost of implementing a security solution is varied depending on the policy requirements. But the feeling is that the cost needs to be relative to the implication of loss of information, which could even entail break of business operations. The potential loss from such a break can always justify the cost of a security solution. The cost of maintaining and upgrading these infrastructures continues to grow, driven by the need for pervasiveness, competition, innovation, and cutting-edge services. It also means, need to maintain the security, privacy, and reliable performance of communications across the growing network.

Key Terms

Access control Access control ensures that resources are only granted to those users who are entitled to them.

Access control list (ACL) A mechanism that implements access control for a system resource by listing the identities of the system entities that are permitted to access the resource.

Activity logging Electronic recordkeeping of such system or network actions as applications accessed, commands executed, files accessed, and trafic generated from a system.

Active RFID tags Tags containing their own power source.

Backdoor A backdoor is a tool installed after a compromise to give an attacker easier access to the compromised system around any security mechanisms that are in place.

Bastion host A bastion host has been hardened in anticipation of vulnerabilities that have not been discovered yet.

Covert channels Covert channels are the means by which information can be communicated between two parties in a covert fashion using normal system operations. For example, by changing the amount of hard drive space that is available on a file server can be used to communicate information.

Demilitarized zone (DMZ) In computer security, in general, a demilitarized zone (DMZ) or perimeter network is a network area (a subnetwork) that sits between an organization's internal network and an external network, usually the Internet. DMZs help to enable the layered security model in that they provide subnetwork segmentation based on security requirements or policy.

Firewall A logical or physical discontinuity in a network to prevent unauthorized access to data or resources.

Information warfare Information warfare is the competition between offensive and defensive players over information resources.

Intrusion detection A security management system for computers and networks. An IDS gathers and analyses information from various areas within a computer or a network to identify possible security breaches, which include both intrusions (attacks from outside the organization) and misuse (attacks from within the organization).

RFID Radio frequency identification (RFID) is a method of transmitting information using radio waves. RFID systems typically consist of a tag and a reader. The tag contains information identifying an item. A reader communicates with the tag and reads the information programmed in the tag's memory.

Worm A computer program that can run independently, can propagate a complete working version of itself onto other hosts on a network, and may consume computer resources destructively.

Review Questions

8.1 What is a firewall? Differentiate among the types of firewall.

8.2 Differentiate between a firewall and IDS.

8.3 Differentiate between perimeter security and endpoint. security.

8.4 Why does vulnerabilities emerge in systems?

8.5 Differentiate between vulnerability and threat.

8.6 Why is patch management essential from a security perspective?

8.7 Why according to you is vulnerability management essential from an organizational perspective?

Projects

8.1 Visit the websites of firewall vendors and identify the functionalities available in firewalls of different vendors. For the purpose of this exercise, make your comparison with at-least three different vendors.

8.2 Visit an IT firm and identify the various IT gadgets available in the firm to run its operations. Prepare an excel sheet of the gadgets. With your understanding of RISK classify the risk levels of the IT gadgets based on criticality of the usage with respect to the IT company. Remember that no two companies may have the same criticality factor. Provide a risk mitigation plan to reduce the risk levels.

8.3 New York-based ABC Unlimited was founded in 1991 by an Indian named Pooja as a venture capital attempt with an initial investment of about 1 million dollars as a family business. As you are aware, ABC has set up an excellent Data centre facility to take care of its Data consolidation. (Refer Chapter 1 for details of the case)

CIO Nayak was returning after an exhaustive meeting with the FBI to understand the implications of volatile IT data. Of late, he had suspicions of data leakage. He recalled the newspaper stories he had read about stolen laptops with customer records stored on them and about hackers trying to penetrate ecommerce sites. ABC unlimited was a supermarket chain with a good presence across the world.

Nayak wanted to understand data theft and data leakage to ensure fullest protection for ABC. Running his hand through his full head of dark black hair, he was contemplating to understand the impact of data leak with in his supermarket chain. He recalled the recent approval and took some comfort in his company having spent considerable time and money becoming compliant with the new payment card industry (PCI) standards for data protection. Though there were minor incidents within the organization, he was sure that there were no serious issues with respect to data-protection.

Discuss the technical and management plans which enabled ABC to carry out the security management of customer data customer particulars, credit card details with good data protection. Also help the organization to identify a suitable layered security architecture to protect the data, revenue, and brand of ABC.

Note: AS/400 is a mid-range server.

REFERENCES

Clark, David Leon 2002, *Enterprise Security: The Manager's Defense Guide*, Addison Wesley.

Finkenzeller, Klaus 2003, *RFID Handbook: Fundamentals and Applications in Contactless Smart Cards and Identification* John Wiley.

Gupta, Ajay and Scott Laliberte 2004, *Defend I.T.: Security by Example*, Addison Wesley.

Glover, Bill, 2006, *RFID Essentials*, O'Reilly.

Hunt, V. Daniel, Albert Puglia and Mike Puglia 2007, *RFID-A Guide to Radio Frequency Identification*, Wiley-Interscience.

IBM 2005, *RFID Sourcebook*, IBM Press.

Kovalick, Al 2006, *Video Systems in an IT Environment: The Essentials of Professional Networked Media*, Elsevier.

Miles, Stephen B., Sanjay E. Sarma and John R. Williams 2008, *RFID Technology and Applications*, Cambridge University Press.

Noonan, Wes and Ido Dubrawsky 2006, *Firewall Fundamentals*, Cisco Press.

Roussos, George 2008, *Networked RFID: Systems, Software and Services*, Springer.

Syme, Matthew and Philip Goldie 2003, *Optimizing Network Performance with Content Switching: Server, Firewall and Cache Load Balancing*, Prentice Hall.

Thornton, Frank, et al. 2006, *RFID Security*, Syngress.

Welch-Abernathy, Dameon D. 2004, *Essential Check Point FireWall-1 NG: An Installation, Configuration, and Troubleshooting Guide*, Addison Wesley.

CHAPTER

9 Office Tools

There's no such thing as a free lunch.
–Milton Friedman
Nobel prize-winning economist

Learning Objectives

After reading this chapter, you should be able to understand:

- the need for office tools
- Microsoft office suite and the productivity tools
- the equivalent open source office tools

9.1 Introduction

Software consists of digital bits that are downloaded onto the storage devices of a computer. Software is also a general name for audio and video discs that can be inserted into audio or video players connected to stereo systems or television sets. Software consists of software packages, music, or movie titles that are designed to perform different tasks. Software packages are generally produced by a large number of software firms that are independent of the hardware producers. For this reason, software packages are regarded as supporting services for the hardware. Thus, a larger variety of software supporting a certain hardware increases the value of a specific hardware machine.

The almost universal and ubiquitous office tool is the Microsoft (MS) Office suite, though in deference to other tools in the space. It allows the most non-technical individual to create, edit, change, review, and produce an extensive array of documents, from simple letters to quite complex and lengthy reports, passing through contracts and even marketing materials. Today's office tools could be described as write once, search often, and cut and paste even more.

The average MBA graduates and managers are not just knowledge workers. They are capable of being highly networked internal entrepreneurs and innovators. They are in an era of web office or virtual office. Web office solutions are using this philosophical approach to redefine how knowledge workers share information. With enterprise blogs and enterprise Wikis, knowledge workers will now have the ability to efficiently communicate with a large audience. Throughout web office, information will become efficiently reusable. Today, many knowledge workers feel overloaded because they are forced to react to a constant stream of e-mail, phone calls, and instant messages. E-mail, the phone, and instant messaging have one thing in common—they are all push workflows. In other words, they interrupt what you are doing. Theoretically, people can ignore all three, but generally, socially, it is difficult to get away with ignoring all three when you are in the office.

9.2 Microsoft and Office Automation

Microsoft was founded in 1975 by high school friends Bill Gates and Paul Allen. While students at Harvard, the two developed the programming language called BASIC, which was used in the first commercial microcomputer. The company continued to grow because they adapted their BASIC program to work on other computers. In 1980, International Business Machines (IBM) chose Microsoft to write its operating system (software that commands a computer's standard functions) for its new personal computer (PC). Microsoft purchased rights to quick and dirty operating system (QDOS), from Seattle programmer Tim Paterson. They renamed it as the Microsoft Disk Operating System or MS-DOS.

Microsoft introduced its first Microsoft Windows operating system in the mid-1980s, starting an avalanche of demand for PCs in the workplace and, finally at home. Windows opened up the world of computers to people with little technical background and allowed the PC industry to flourish in the late 1980s and early 1990s. By the mid-1990s, Windows was the world's leading operating system. Microsoft's products and services are aimed at empowering people and organizations by providing them with an easy way of finding and using information. In order to compete successfully in the software industry, Bill Gates believes that Microsoft must pioneer and orchestrate evolving mass markets.

9.3 Office Productivity

Microsoft Corporation is one of the world's leading independent software companies. Microsoft is known for its operating systems, including MS-DOS, Windows, Windows 95, Windows 98, and Windows NT; application software such as Microsoft Access (database products), Microsoft Excel (spreadsheets), Microsoft Word (word processing), and Microsoft Money (personal finance software); CDROMS including Encarta and Flight Simulator; an online service, Microsoft Network; online publishing, including *Slate* magazine; and a unique cable and Internet news service, MSNBC, in conjunction with NBC News.

Microsoft's competitors within the computer industry speak of their will to dominate markets with concern, fear, and admiration. Michael Kapar, Chief executive officer (CEO) of Lotus Development Corporation, argues, 'The question of what to do about Microsoft is going to be a central public policy issue for the next twenty years.' Lawrence Ellison, CEO of Oracle, another Microsoft competitor argues that Bill Gates and Microsoft do not just want to compete in the computer industry. Rather, they seek to dominate every aspect of the market, to eliminate their competition.

9.3.1 MS Office Suite

Microsoft Office has been around for more than a decade. Some of the individual programs that are part of the Office family date back to the 1980s. In Office 2003 and earlier versions, the default file formats are the familiar .doc (Word document), .xls (Excel workbook), and .ppt (PowerPoint presentation). By contrast, the default file formats in Office 2007 are based on extensible markup language (XML). To denote the change in format, the filename extensions associated with each format have changed, adding an 'x' at the end of each one; Word's new default format is .docx instead of .doc.

Microsoft's product Windows 98 was introduced in the summer of 1998 with little fanfare compared to the international hoopla that heralded the launch of Windows 95. Windows 98 improves upon Windows 95. It is more intuitive and allows users seamless access to the Internet with little effort. Before the introduction of Windows 98, the company unveiled Microsoft Office 97, a suite of programs that included word processing, scheduling, and database applications, as well as networking programs that allowed users to forward messages via the Internet and join in on the development of documents through networking applications.

In the previous versions of Office, there were menus and toolbars. Some of these have survived the Office 2007 world, but everyone is talking about the new Ribbon. The Ribbon is the toolbar and menu successor. Each Ribbon has a set of tools specific to that Ribbon's functionality that you cannot change. Each application has Ribbons that relate to the work for that particular application. For example, the Excel 2007 set of Ribbons will contain formulas and data Ribbons, whereas PowerPoint 2007 will have Ribbons that relate to animations and slide show. The purpose of the Ribbon is to organize different command options in a way that makes it easier for people to work because all the commands for a particular subject are together.

Microsoft packages the individual programs that make up Office 2007 into eight separate editions, which in turn are sold through a variety of channels.

9.3.2 MS Word

The oldest and the most mature of the Office programs, Word is also the most popular. Microsoft Word is a word processing software package. You can use it to type letters, reports, and other documents. It gives you the ablilty to use your home computer as well as your business computer for desktop publishing. It is an extremely versatile tool—ideal for creating short documents, such as letters and memos, with enough layout and graphics-handling capabilities to make it suitable for sophisticated publishing.

Essentials of Word

This section introduces you to the terminology and the basic Word skills you use in the program. In Word, files are called documents. Each new document is similar to a blank page. As you type and add additional text and other objects, your document gets longer.

Ribbon interface Word 2007 benefits from the addition of the Ribbon interface. It replaces menus, toolbars, and most of the task panes found in

Word 2003. The Ribbon is located at the top of the document window and is comprised of tabs that are organized by task or objects. The controls on each tab are organized into groups, or subtasks. The controls or command buttons in each group execute a command, or display a menu of commands or a dropdown gallery. Controls in each group provide a visual way to quickly make document changes.

Word provides three types of tabs on the Ribbon (Figure 9.1). The first type is called a standard tab, such as Home, Insert, Review, View, and Add-Ins that you see whenever you start Word. The second type is called a contextual tab, such as Picture Tools, Drawing, or Table, that appear only when they are needed based on the type of task you are doing. Word recognizes what you are doing and provides the right set of tabs and tools to use when you need them. The third type is called a program tab, such as Print Preview, that replaces the standard set of tabs when you switch to certain views or modes.

The two quickest ways to start the Microsoft Office Word 2007 program are to select it on the Start menu or double-click a shortcut icon on the desktop. By providing different ways to start a program, Office lets you work the way you like and start programs with a click of a button.

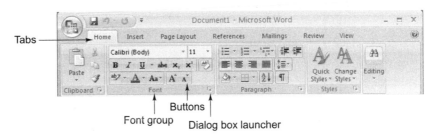

FIGURE 9.1 Word 2007—Ribbon

Unlike looking at a piece of paper, Word provides many views, such as the Reading Layout view, that helps you see the document in the best possible way for the task at hand. In Word 2007, you can display your document in one of five views: Draft, Web Layout, Print Layout, Full Screen Reading, or Online Layout (Figure 9.2).

- Draft View is the most frequently used view. You use draft view to quickly edit your document.
- Web Layout view enables you to see your document as it would appear in a browser such as Internet Explorer.

- Print Layout view shows the document as it will look when it is printed.
- Reading Layout view formats your screen to make reading your document more comfortable.
- Outline view displays the document in outline form. You can display headings without the text. If you move a heading, the accompanying text moves with it.

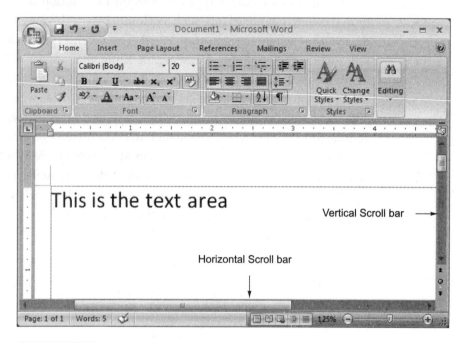

FIGURE 9.2 Word 2007—Text Area

Essentials of Interface

Dialog box A dialog box is a window that opens when you click on a Dialog Box Launcher. A dialog box allows you to supply more information before the program carries out the command you selected. After you enter information or make selections in a dialog box, click the OK button to complete the command. Click the Cancel button to close the dialog box without issuing the command. In many dialog boxes, you can also click an Apply button to apply your changes without closing the dialog box.

Status bar The status bar appears across the bottom of your screen and displays document information such as word count, page numbers, language, and current display zoom percentage and some Word program controls such as view shortcut buttons, zoom slider, and Fit To Window button.

Task panes Task panes are separate windows that appear when you need them such as Document Recovery, or when you click on a Dialog Box Launcher icon. A task pane displays various options that relate to the current task. Window panes are sections of a window, such as a split window. If you need a larger work area, you can use the Close button in the upper-right corner of the pane to close a task or window pane.

New document When you start Word, the program window opens with a new document so that you can begin working in it. You can also start a new document whenever Word is running, and you can start as many new documents as you want.

Word provides a collection of professionally designed templates that you can use to create documents. Start with a template when you have a good idea of your content, but want to take advantage of a template's professional look. A template is a document file that provides you with a unified document design, which includes themes so you only need to add text and graphics. In the new dialog box, you can choose a template from those already installed with Office Word or from the Microsoft Office online website, an online content library.

As your document gets longer, some of your work shifts out of sight. You can easily move any part of a document back into view. Scrolling moves the document line by line. Paging moves the document page by page. Browsing moves you through your document by the item you specify, such as to the next word, comment, picture, table, or heading.

You can copy and move text or data from one location to another on any Office document. When you copy data, a duplicate of the selected information is placed on the Clipboard. When you move text, the selected information is removed and placed on the Clipboard. To complete the copy or move, you must paste the data stored on the Clipboard in another location. With Microsoft Office, you can use the Clipboard to store multiple pieces of information from several different sources in one storage area shared by all the Office programs. You can paste these pieces of information into any Office program, either individually or all at once.

Once you type a document and get the content the way you want it, the finishing touches can sometimes be the most important. One of the first elements you can change is your font attributes. You can change the kerning, i.e., the amount of space between each individual character, for a special effect on a title or other parts of text. You can also apply a dropped capital letter to introduce a body of text, add a shade or border on your document.

The best place to understand why things work the way they do in Office Word 2007 is the three levels of formatting. Word organizes most of the formatting you can apply in your document into these three levels: font, paragraph, and section with font being the simplest of the three and section being the most complex.

Page margin Page margins are the blank spaces around the edges of the page. In general, you insert text and graphics in the printable area between the margins (Figure 9.3).

Add margins for binding: Use a gutter margin to add extra space to the side or top margin of a document that you plan to bind. A gutter margin helps ensure that text is not obscured by the binding.

Set margins for facing pages: Use mirror margins to set up facing pages for double-sided documents such as books or magazines. In this case, the margins of the left page are a mirror image of those of the right page.

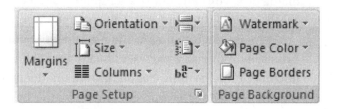

FIGURE 9.3 Word—Page Margin

- On the Page Layout tab, in the Page Setup group, click Margins.
- Click the margin type that you want. For the most common margin width, click Normal. When you click the margin type that you want, your entire document automatically changes to the margin type that you have selected.
- You can also specify your own margin settings. Click Margins, click Custom Margins, and then in the Top, Bottom, Left, and Right boxes, enter new values for the margins.

Essentials of Fonts and Formatting

Studies show that reading comprehension is greater with proper fonts with good formatting.

Fonts A font is a collection of alphanumeric characters that share the same typeface or design, and have similar characteristics. You can format text and numbers with font attributes, such as bold, italics, or underline, to enhance data to catch the reader's attention. The main formats you apply to text are available on the Home tab in the Font group or in the Font dialog box.

Kerning is the amount of space between each individual character that you type. Sometimes the space between two characters is larger than others, which makes the word look uneven. You can use the Font dialog box to change the kerning setting for selected characters. Kerning works only with True Type or Adobe Type Manager fonts.

Font formatting Font formatting can be applied to as little as one character, such as font face (for example, Times New Roman, Arial, Calibri) or font size. Any formatting available from the Font group on a Ribbon tab or the Font dialog box, as well as language setting, text borders, and text shading, are all types of font formatting. When you apply font formatting in your document, it is stored directly in the character to which you apply it.

Paragraph formatting Paragraph formatting can be applied to as little as one paragraph such as paragraph alignment or indents. Any formatting available from the paragraph group on a Ribbon tab (such as bullets, numbering, and paragraph borders and shading), or from the Paragraph dialog box are types of paragraph formatting. When you apply paragraph formatting in your document, it is stored in the paragraph mark at the end of the paragraph where the formatting appears.

Section formatting Section formatting is often thought of as page set-up formatting. Any formatting that can be applied from the Page Setup group on the Page Layout tab or from the Page Setup dialog box, as well as text columns, page, and footnote number formatting, and information about which header and footer appears on a page are all types of section formatting. To change any type of section formatting for just a part of the document, a section break is required. That is because all section formatting is stored in the section break at the end of the section. Note that for single-section documents, section formatting is stored in the last paragraph mark in the document.

Reveal formatting The Reveal formatting task pane is a great tool for troubleshooting document formatting because you can see all the formatting for a selection at a glance. But, reveal formatting can also help to bring into focus the concept of the three levels of formatting and how Word organizes formatting in a document.

Text alignment Text starts out positioned evenly along the left margin, and uneven, or ragged, at the right margin. Left-aligned text works well for body paragraphs in most cases, but other alignments vary the look of a document and help lead the reader through the text. Right-aligned text, which is even along the right margin and ragged at the left margin, is good for adding a date to a letter. Justified text spreads text evenly between the margins, creating a clean, professional look, often used in newspapers and magazines. Centred text is best for titles and headings.

The lines in all Word documents are single-spaced by default, which is appropriate for letters and most documents. But you can easily change your document line spacing to double or 1.5 lines to allow extra space between every line. This is useful when you want to make notes on a printed document.

To enhance the appearance of the text in a paragraph, you can format it using the buttons in the Paragraph group on the Home tab. You can quickly add a border and shade to the selected text.

Saving the document When you create a Word document, save it as a file on your computer so you can work with it later. When you save a document for the first time or if you want to save a copy of a file, use the Save As command. When you want to save an open document, use the Save button on the Quick Access Toolbar. When you save a document, Word 2007 saves 97-2003 files in an older format using compatibility mode and new 2007 files in an XML-based file format. The XML format significantly reduces file sizes, provides enhanced file recovery, and allows for increased compatibility, sharing, reuse, and transportability. The way that you save a document and the format that you save it in depends on how you plan to use the document. Save a document for the first time in the following ways.

- On the Quick Access Toolbar, click Save Button image, or press CTRL+S.
- Type a name for the document, and then click Save.

Word saves the document in a default location. To save the document in a different location, select another folder in the favorites link if your computer is running Windows Vista, or in the Save in list if your computer is running Microsoft Windows XP.

Essential Keyboard Shortcuts

- Find feature key: [Ctrl]-[F] Word's Find feature on the Find tab of the Find and Replace dialog box offers a wide variety of options. You can search simply for text; or for text with specific formatting (which you specify by using the Format drop-down menu); or for formatting without text (for example, you might search for the next instance of a specific style). You can constrain the search to match case or to find only whole words rather than matches inside other words.
- Replace feature key: [Ctrl]-[H] Word's Replace Feature offers similar functions to the Find feature: you can replace text, text with specific formatting, or just formatting. Use the Find Next button to find the next instance of the search item, and the Replace button to replace the current instance.
- Next Occurrence feature key: [Shift]-[F4], [Ctrl]-[Alt]-[Y] These shortcuts enable you to repeat your last search without displaying the Find and Replace dialog box. Alternatively, you can click the Next Find/Go To button at the bottom of the vertical scroll bar.
- Return to the previous editing point key: [Shift]-[F5], [Ctrl]-[Alt]-[Z] Word tracks the locations of the last three edits you made to a document. You can return to the last edit by pressing these shortcuts once, the second-last edit by pressing them twice, or the third-last edit by pressing them three times
- Repeat the previous action key: [Ctrl]-[Y], [F4], [Alt]-[Enter] These shortcuts all do the same thing: make Word repeat the previous editing action.
- Undo the previous action key: [Ctrl]-[Z], [Alt]-[Backspace] [Ctrl]-[Z] This is perhaps the most used keyboard shortcut for Windows applications. In Word, [Alt]-[Backspace] performs the same function and is more comfortable for some users.
- Select All key: [Ctrl]-[A], [Ctrl]-[5] A Select All command selects all the contents of the current object.
- Toggle All Caps key: [Ctrl]-[Shift]-[A] All caps capitalizes all letters in the selection but has no effect on non-letter keys.

- Toggle subscript key: [Ctrl]-[=] Subscript decreases the font size of the selected text and lowers it below the base line of the other characters.
- Toggle superscript key: [Ctrl]-[+] Superscript decreases the font size of the selected text and raises it above the baseline of the other characters.
- Display the Font dialog box key: [Ctrl]-[D] This is the easiest way of displaying the Font dialog box.
- Select the Font Size drop-down list key: [Ctrl]-[Shift]-[P] Use this shortcut to quickly activate the Font Size drop-down list without taking your hands off the keyboard.
- Increase the font size in jumps key: [Ctrl]-[>] With this shortcut and the [Ctrl]-[<] shortcut, Word uses the font sizes listed in the Font Size drop-down list and the Font dialog box.
- Decrease the font size in jumps key: [Ctrl]-[<] Use this shortcut to quickly decrease the font size to the next size that Word lists.
- Increase the font size by one point key: [Ctrl]-] Use this shortcut to increase the font size gradually.
- Decrease the font size by one point key: [Ctrl]-[Use this shortcut to decrease the font size gradually rather than in jumps.

Know Your Firm—Adobe

Adobe's selected products include: Acrobat (electronic document management software), Adobe Acrobat (document formatting software), Adobe ArtExplorer (painting and drawing software for children), Adobe Fetch (cataloging software), Adobe Gallery Effects (special-effects software), Adobe Illustrator (graphics software), Adobe PageMill (Web-page creation software), Adobe Persuasion (presentation software), Adobe PhotoDeluxe (personalized photo software), Adobe Photoshop (photographic image software), Adobe Premiere (film and video editing software), Adobe SiteMill (Internet link repair software), FrameMaker (document authoring software), PageMaker (page layout software), and PostScript. PostScript is a high-level computer language that communicates precise descriptions of computer-generated graphics, photos, and text to any output device with a PostScript interpreter. Adobe reorganized its businesses into five independent operating units which allowed management to better focus on its printing and publishing products and also to concentrate on opportunities in the rapidly growing graphics and Internet markets.

- 1982: Founded
- 1985: Begins marketing PostScript
- 1986: Goes public
- 1987: Creates Adobe Systems Europe subsidiary

Contd

- 1993: Begins licensing Adobe software to printer manufacturers and marketing Adobe Acrobat
- 1994: Merged with Aldus Corp.
- 1995: Acrobat viewing is integrated into Netscape's Internet software
- 1998: Initiates work on new PGML Web language with IBM, Sun, and Netscape

Source: www.adobe.com

9.3.3 MS Excel

Electronic spreadsheet analysis has become a part of the everyday work of researchers in all areas of engineering and science. Microsoft Excel, as the industry standard spreadsheet, has a range of scientific functions that can be utilized for the modelling, analysis, and presentation of quantitative data. It is easy to claim that Microsoft Excel is the most widely used numeric analysis and financial software in the world. It has made business and financial analysis available to millions of business people. Excel 2007 is incredibly useful for tasks as simple as balancing a checkbook or as complex as modelling a hostile takeover of a Fortune 500 corporation.

Essential Elements of Excel

Most Microsoft Excel users learn only a small percentage of the program's features. They know they could get more out of Excel if they could just get a leg up on building formulas and using functions. Unfortunately, this side of Excel appears complex and intimidating to the uninitiated shrouded in the mysteries of mathematics, finance, and impenetrable spreadsheet jargon. When you first start Excel, you see the window that appears in Figure 9.4.

The Ribbon is the toolbar area you see along the top of Excel. Each tab on the Ribbon contains different categories called 'groups.'

The task pane is a vertical pane of information that appears on the right side of your Excel program window when you perform certain operations.

Tabs, which are found on the Ribbon, are designed to give you everything you need in Office programs without wading through layers of menus, dialog boxes, and task pane options.

The Formula bar is the long, open box that appears above the column letters in the worksheet.

The box at the left end, the Name box, shows either the address of the selected cell or the name of the selected cell or range.

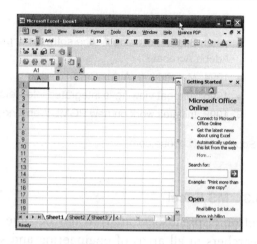

FIGURE 9.4 MS Excel—New Sheet

The Formula bar displays the underlying data in the active cell, whether the actual cell data is a formula or a value. The data displayed in the cell is the result of formatting the value in the cell and can be entirely different from the actual value in the cell. The status bar is the bar along the bottom of the Excel program window. Various components of Excel sheet are indicated in Figure 9.5.

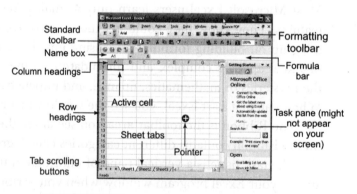

FIGURE 9.5 MS Excel—Components

Saving an Excel workbook is like saving any other file: Click the Save button on the toolbar (or press Ctrl+S). If the file has never been saved, the Save As dialog box opens. Type a filename in the File name dialog box, choose a folder in the Location dialog box, and click Save. To save a workbook as a template, choose Office. Save As and, in the Save As dialog box, click Template in the Save as type drop-down box. A workbook template file has the extension .xlt instead of .xls.

Essential Short Cuts

Some essential short cuts are as follows.

- Ctrl+C: Copies the current selection
- Ctrl+X: Cuts the current selection
- Ctrl+V: Pastes the most recent cut or copy
- Ctrl+Z: Undoes the last action; if pressed repeatedly, it undoes each preceding action
- Ctrl+': Copies the entry in the cell above
- Ctrl+;: Enters the current date
- Ctrl+Shift+;: Enters the current time
- Ctrl+D: Copies the cell at the top of a selection into all the selected cells below
- Ctrl+Enter: Enters the same entry in all selected cells
- Esc Clears: the currently open cell before you press Enter
- F4: Repeats the last action; if pressed repeatedly, it continues to repeat the same action

Essentials of Worksheets

The worksheet is the basic document in which a user can do work in Excel including entering, calculating, and organizing data. One or more worksheets are contained in each workbook. This is called an Excel file. Each worksheet will contain three worksheets by default. Select a single worksheet by clicking its sheet tab, found on the worksheet toolbar beneath the worksheet. The selected worksheet is called the active worksheet.

Worksheets are similar to pages within a book; you peruse through them like you flip the pages of a book. There are several ways to move and copy worksheets. Right click on the sheet tab and choose Move or Copy. Select a new position in the workbook for the worksheet or click the Create a copy checkbox and Excel will paste a copy of that worksheet in the workbook. The same shortcut menu for the sheet tab also gives you the option to insert, delete, or rename a worksheet.

The quickest way to move a worksheet is to click and drag the sheet tab to a new location among the other sheet tabs.

The quickest way to copy a worksheet is to drag the tab and press Ctrl when you drop it. With Excel, you can copy or transpose a row into a column and vice versa by using the Paste Special dialog box.

To rename a worksheet, double-click the sheet tab, type the new name, and then press Enter or click in the worksheet.

To add a new worksheet, right-click a sheet tab and click Insert. Moving or copying sheets to another workbook is almost as easy as moving or copying within a workbook: Open both workbooks and drag the sheet tab from one workbook to the other.

To hide a worksheet, click the sheet tab to make it the active worksheet, and then choose the Home tab on the Ribbon. In the Cells category, click Format – Hide and Unhide – Hide Sheet.

Excel work sheets (not applications) come in three fundamental forms: the Excel report, the Excel flat data file, and the Excel indexed list. Flat files are data repositories organized by row and column. Each row corresponds to a set of data elements called a record. Each column is called a field. A field corresponds to a unique data element in a record. Every data field has a column and every column corresponds to one and only one data element. Furthermore, there is no extra spacing and each row or record corresponds to a unique set of information. Flat files come with their own set of drawbacks. These are given below.

- Flat files often contain redundant data (that is, data that is duplicated multiple times). This naturally makes for unnecessarily large datasets.
- Flat files often contain irrelevant data columns. These columns are typically holdovers from another process that no one wants to delete.
- Flat files often contain blank or empty data elements. Because flat files are typically a mishmash of many subjects in one data table, it is not uncommon to have holes in the data.

Each worksheet is divided into a grid of rows and columns. A worksheet has a total of 16,384 columns and 1,048,578 rows, and cannot be expanded beyond those limits. The rows are identified by numbers down the left side of the worksheet, and the columns are identified by letters across the top of the worksheet. The grid units are called cells. Cells can contain numbers, text, comments, formats, and formulas.

Once we are clear with the cell, we need to understand the address of the cell. Each cell has an address. The address is a location on the worksheet, given by its row number and column letter. A cell reference is the cell's address on the worksheet in terms of its column letter and row number. One can identify the cell's reference by either looking at the row and column that intersect at the cell, or by selecting the cell and looking at the Name box.

To insert a row, right-click a row number and click Insert. To insert several rows, click and drag to select the number of rows you want to insert, and then right-click in the selected rows and click Insert. To insert a column, right-click the column letter and click Insert. To insert several columns, click and drag to select all the column letters, and then right-click in the selected columns and click Insert.

Essentials of Navigation

To navigate within a workbook, you use the arrow keys PageUp, PageDown, or the Ctrl key in combination with the arrow keys to make larger movements. The most direct means of navigation is with your mouse. Scroll bars are provided and work as they do in all Windows applications. Go ahead and try moving between cells in your newly opened Excel document with your mouse and then the PageUp and PageDown keys.

To move to other worksheets, you can click their tab with the mouse at the bottom of the screen (Sheet 1, Sheet 2, or Sheet 3) or use the Ctrl key with the Page Up and Page Down keys to move sequentially up or down through the worksheets. Go ahead and switch between your three sheets using the different methods described.

Essentials of Data Management

Entering data directly into cells in a worksheet is a simple task with possibilities and tweaks and techniques for doing things faster. Entering numbers is just like entering text, and when an entry consists of nothing but numbers, Excel assumes the entry is a calculable number. Leading zeroes are removed, the number is included in calculations, and number formatting is available. If, however, the entry contains any letters, Excel assumes the entry is text and ignores it in calculations. To enter data in a single cell, click the cell to select it, type the data, and press Enter.

To enter a fraction in a cell, leave a space between the whole number and the fraction, as in 8 1/5. When you press Enter, the fraction you typed appears in the cell and the decimal equivalent appears in the Formula bar. Data validation can help you avoid several kinds of incorrect data entry, especially helpful if someone else is entering data in your workbook.

One is permitted to use data entry rules to ensure that data entered has the correct format, and restrict the data entered to whole numbers, decimals, dates, times, or a specific text length. One can also specify whether the values need to be between, not between, equal to, not equal to, greater than, less than, greater than or equal to, or less than or equal to the values you specify.

Excel enables the owner to restrict the values a user can enter in a cell. By restricting values, you ensure that your worksheet entries are valid and that calculations based on them are valid as well. During data entry, a validation list forces anyone using your worksheet to select a value from a drop-down menu rather than typing it and potentially typing the wrong information.

Symbols and special characters serve many uses in Excel. Many financial applications, for example, call for currency symbols. Symbols and special characters are useful in column and row heads as part of the text describing column and row content. A smaller set of standard characters, called symbols, are always available for use which includes dashes, hyphens, and quotation marks.

The Excel report is a means of formatting and displaying data, often for managers or other users. A good Excel report makes judicious use of empty space for formatting, summarizes data where appropriate, and clearly marks data fields. A style is a named collection of formats you can share among users and apply across workbooks. Styles streamline the work of formatting so that you and others can apply a consistent set of formats to worksheet elements such as row heads, column heads, and data values. With styles, you maintain consistency in the way numbers, dates, times, borders, and text appear in cells. In Excel, the spreadsheet can be used both for data storage and for data reporting. Outlining was designed for use with structured information such as lists but can be used with any worksheet. Outlining a set of rows or columns creates a clickable button on the far left or top of your worksheet. The button displays either a minus sign or a plus sign, depending on what is displayed in the worksheet. Click the minus sign to hide rows or columns, and the plus sign to display them again.

Essential Calculation Tools

More than 300 functions that are built into Excel enable you to perform tasks of every kind, from adding numbers to calculating the internal rate of return for an investment. A formula consists of an equal sign, one or more functions, their arguments, operators such as the division and multiplication symbols, and any other values required to get your results. Many Excel functions do special-purpose financial, statistical, engineering, and mathematical calculations. Constructing formulas can be complicated, especially when you use several functions in the same formula or when multiple arguments are required in a single function. Excel's Function Wizard simplifies the use of functions. You can take advantage of the wizard for every one of Excel's functions, from the sum (SUM) function to the most complex statistical, mathematical, financial, and engineering function. A comment is a bit of descriptive text that enables you to document your work when you add text or create a formula. Comments in Excel do not appear until you choose to view them. Excel associates comments with individual cells and indicates their presence with a tiny red triangle in the cell's upper-right corner. View an individual comment by clicking in the cell or passing your cursor over it. View all comments in a worksheet by clicking the Review tab and then clicking Show All Comments.

AutoCalculate is a handy tool that we often use to calculate cells on the fly. AutoCalculate calculates cells in the worksheet temporarily but does not write formulas. AutoSum can write sums and average numbers, count all entries, and display the maximum and minimum numbers in a range. Sum is the function most people want to use with AutoSum so the default calculation is maintained as the sum.

A range reference is a rectangular range of cells identified by the references of the range's upper-left corner cell and lower-right corner cell. A colon is used to separate the references. An absolute cell reference is a fixed point and all references are with respect to the fixed point. A relative cell reference is a relative location with respect to the current location. The mixed cell reference is a mixture of absolute and relative locations.

All formulae can be used to calculate cells in the same worksheet. Simple formulae might consist of adding, subtracting, multiplying, and dividing cells on different worksheets, and even in different workbooks. In Excel, you can carry out calculations simple arithmetic, in three ways. One method is to use the plus (+), minus (-), multiplication (*), and division (/) signs. Start by typing an equal sign and the values to be added, subtracted, multiplied, or divided, each separated by an operator. Press Enter, and Excel does the calculations and displays the answer in the same cell.

- \+ (plus sign) Addition
- \- (minus sign) Subtraction
- * (asterisk) Multiplication
- / (forward slash) Division
- ^ (caret) Exponentiation

- () (parentheses) To group operations

A second method involves functions. Functions perform calculations on your information and make the results available to you. To use a function, type an equal sign followed by the function. As an example we can consider the function =SUM(). Place the numbers you want to add inside the parentheses, separating them with commas. If the numbers are on the worksheet, click on the cells. A third method is to use Excel's AutoSum feature, which offers a point-and-click interface for several functions, including SUM, AVERAGE, and COUNT.

It is possible to write formula that references cells in other worksheets or workbooks. The workbook with the formula is called a dependent workbook because it depends on input from other workbooks. The input workbooks are called source workbooks because they are the data source for the linking formulas. Cell names make formulas easier to read. The process of tracing formulas is called auditing, and there is a toolbar with buttons that do the work. A precedent cell is an input cell referenced in the formula. A dependent cell is a cell that uses the results of the formula. In Excel, you can name individual cells and groups of cells, called ranges. You can use named cells and ranges directly in formulas to refer to the values contained in them. When you move a named range to a new location, Excel automatically updates any formulas that refer to the named range. Excel evaluates all the values in the formula and returns the result. If Excel cannot calculate the formula, it displays an error in the formula's cell.

Excel includes more than 80 statistical functions. You can find these functions by using the Function Wizard in the Statistics category. Among the statistical functions, you will find more than a dozen types of descriptive statistics. With these statistics, you characterize both the central tendency of your data, such as mean, mode, and median, and the data's variability such as sample variance and standard deviation. Some of the prominent functions include internal rate of return (IRR) function, payment (PMT) function, etc. One can use Excel's IRR function to calculate the rate of return on an investment. When using the IRR function, the cash flows do not have to be equal, but they must occur at regular intervals. The LARGE function evaluates a series of numbers and determines the highest value, second highest, or Nth highest in the series, with N being a value's rank order. With a conditional formula, you can perform calculations on numbers that meet a certain condition like the highest score for a particular team from a list that consists of several teams. One can use conditional sums to identify and sum investments whose growth exceeds a certain rate. The SUMIF function combines the SUM and IF functions into one easy-to-use function. Using the PRODUCT function, you can multiply two or more numbers, and using the SQRT function, you can find the square root of a number. Excel formulas and functions permit calculations with dates and times. Excel's consolidate feature permits the user to generate a consolidated report. Excel provides a variety of functions that can be used to consolidate including SUM, COUNT, AVERAGE, MAX, MIN, and PRODUCT. One starts the consolidation process by selecting the location for your consolidated data. One may want to format the cells so that the incoming data displays properly. DCOUNT is a database function that counts the number of cells containing a number.

Sorting and filtering lists offers different views of data. Filtering works like a sieve through which the data is passed to display only data that meets the filter criteria. Sorting a list by one criterion, such as age, arranges your records for easy scanning. The Sort dialog box can be used to customize Options to specify a custom order.

Essentials of Pivot Table

Pivot tables are probably the most improved area of the Excel product, particularly from a business intelligence (BI) perspective. Most notable is the replacement of the old drag-and-drop interface with a new field well dialog box that enables users to simply check off which attributes they would like to see and Excel automatically provides a default layout. Pivot tables are one of the most commonly heard of, but least understood, mechanisms in spreadsheets because end-users do not realize how easily they can be created. The business purpose of adding a Pivot table and a Pivot chart to this spreadsheet is to improve the look and feel of the information being presented.

Pivot tables helps answer questions about different data sets. Pivot tables are based on lists. Lists are made up of rows and columns. The row and column labels of a Pivot table usually have discrete information, meaning the values fall into categories. The Pivot table layout consists of several elements: report filters, data, columns, and rows (Figure 9.6). Use the Pivot table Field List to organize the elements. You can use Pivot tables to compare and contrast the distribution of data across categories. You may need a variety of statistics to examine differences between categories. To aid you, Pivot tables can automatically calculate subtotals and grand totals for the columns and rows in your list. When calculating subtotals and grand totals, you have a choice of calculations from which to choose, including sum, average, count, standard deviation, minimum, and maximum.

322 Information Technology for Management

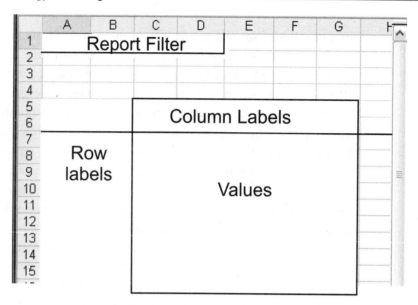

FIGURE 9.6 Elements of a Pivot Table

Within a Pivot table, you can create new fields, called calculated fields, which you base on the values in existing fields. Two Pivot table formatting options available in Excel are grouping and sorting. With grouping, you can hide detail so you can easily compare groups of data. When you group columns or rows, Excel totals the data, creates a field header, and creates a field with a drill-down button that displays either a plus or a minus sign. When you create a chart from a Pivot table, you can base your chart on summary statistics and you can adjust the row and column layout.

Essentials of What-if Analysis

What-if analysis or scenario analysis is sometimes also called deterministic simulation. It simulates the system to find possible effect of some possible scenarios. What-if analysis is the process of changing the values in cells to see how those changes will affect the outcome of formulas on the worksheet. What-if analysis is a systematic way of finding out how a change in one or more variables affects a result. Three kinds of what-if analysis tools come with Excel: scenarios, data tables, and goal seek.

Scenarios and data tables take sets of input values and determine possible results. A data table works only with one or two variables, but it can accept many different values for those variables. A scenario can have multiple variables, but it can accommodate only up to 32 values. Goal seek works differently from scenarios and data tables as it takes a result and determines possible input values that produce that result.

Scenario manager lets you vary one or more inputs to find out how the result changes. The advantage of the scenario manager is that it stores a series of values so you can create a single report or table showing how each value or combination of values influences the result.

The marketing function utilises a considerable amount of qualitative information which introduces uncertainty into the decision-making process. Qualitative decisions that must be taken by marketing managers include anticipating customer needs, forecasting product sales, and evaluating new sources of competition. Decision-makers often use spreadsheet models as a quantitative tool to obtain answers to various 'what-if' questions.

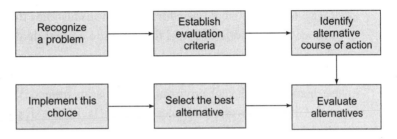

FIGURE 9.7 Decision Management Steps

When managers are able to see the immediate effect of changes to sensitive parameters such as sales volume or product price, they are better placed to evaluate feasible alternatives. A key objective of the marketing division is to produce sales forecasts. Many statistical models have been developed in an effort to improve forecasting accuracy, including regression-based models, the moving-average method, and exponential smoothing techniques. Business managers must choose the appropriate course of action which is most effective in attaining an organization's goals (Figure 9.7). The types of decisions that have to be taken vary considerably and depend upon the complexity of the problem to be solved. Usually, the level of complexity increases proportionally with the level of qualitative information required to solve the problem. Normally the what-if analysis is set up as a model in a spreadsheet. If a decision maker knows the exact values of all uncontrollable variables, then the model is said to be deterministic. The best decision is thus made under conditions of certainty. A payroll system is a typical example of a deterministic application because all inputs are already quantified. Excel provides some very powerful add-in tools such as Solver and the Analysis ToolPak, which contains a number of statistical and engineering functions. To activate the Analysis ToolPak, choose the Tools Data Analysis command.

Excel Macros

Macros provide the user with help on automating the tasks. A macro writer must have an understanding of Excel that is different from that of someone who utilizes it as a spreadsheet application. The programmer must have a fundamental understanding of how Excel interacts with its environment, processes information, and accesses information. Excel utilizes visual basic for applications (VBA) as its macro language. Visual basic for applications has a broader functionality than just programming macros in Excel. It has much of the same functionality as visual basic (VB), the language from which it was derived. Visual basic for applications also has the ability to interact with applications outside of Excel that provide a great deal of flexibility. At the top of the Excel hierarchy is the Object.

Loosely speaking, objects can be thought of as 'things', meaning, if it can be seen in the Excel environment, it is an object. Workbooks, Worksheets, and Charts are all examples of objects in the Excel VBA environment. Visual basic for applications utilizes objects to access information in Excel and perform manipulations on elements in the Excel environment. The most common objects utilized when working with Excel VBA are the Workbook, the Worksheet, and the Range. A fourth object which is slightly more abstract is the Application. An application is simply an instance of Excel that is running.

9.3.4 MS PowerPoint

This section provides the information you need to decide whether you should learn to use PowerPoint. Can you really use PowerPoint effectively, or will it get in the way of your doing what you already do so well? When properly used, PowerPoint can help you communicate, persuade, inspire, motivate, convince, and educate.

PowerPoint has always been an effective way to create PC-based slideshows for presentations to large audiences. Recent versions add the capability to create effective web-based presentations as well. PowerPoint was originally created to make it easier to print overhead transparencies. With the advent of the data projector, it also became possible to project images directly from the computer, and thus PowerPoint evolved to take advantage of a whole new world of presentation possibilities. PowerPoint, as a presentation tool, has evolved primarily in support of business presentations. Although PowerPoint is often identified with business, it has also found a home in schools, universities, and training centres. Indeed, PowerPoint can be utilized in education settings in ways probably never envisioned by its designers.

> Apple Computer designs, builds, and markets computers, peripheral equipment, and software for homes, businesses, educational institutions, and the government. Apple also provides customers with multimedia and connectivity solutions. Steve Jobs and Stephen Wozniak, two college dropouts, built the first Apple computer in Jobs' garage in 1976. This microcomputer, the Apple I, was a barebones creation with no monitor, casing, or keyboard. These were later added, as were software and ports for third-party peripherals.
>
> - 1976: Steve Jobs and Stephen Wozniak build the first Apple computer in Jobs' garage
> - 1977: Apple Computer, Inc. is incorporated.
> - 1980: Goes public at $22 per share
> - 1984: Introduces the Macintosh personal computer
> - 1985: Jobs leaves Apple and creates a new computer company, NeXT Incorporated
> - 1988: Apple reorganizes into four operating divisions
> - 1991: PowerBook notebook is introduced, gaining 21 percent market share in less than six months
> - 1996: Apple brings in new CEO Gilbert Amelio
> - 1998: NeXT is purchased, bringing Jobs back to Apple as a consultant

Essentials of MS PowerPoint

A presentation is any kind of interaction between a speaker and audience, but it usually involves one or more of the following visual aids: 35mm slides, overhead transparencies, computer-based slides, hard-copy handouts, and speaker notes. PowerPoint can create all of these types of visual aids, plus many other types that you learn about as we go along.

With PowerPoint you can

- use bulleted or numbered lists of text to summarize important points,
- use charts and diagrams to visually represent facts and figures,
- use graphics or pictures to illustrate things that are otherwise difficult to describe,

- use animations to illustrate processes or concepts or to gradually add information (such as bullets or figures) to a slide,
- add sound or video to give lifelike realism to slides, and
- quickly publish information to the Web.

PowerPoint is a communication and a visual tool. An effective presentation is one that communicates ideas, teaches concepts, or convinces or motivates listeners. PowerPoint helps in preparing such presentations and that is why it is one of the most popularly used tools. Some key elements for any engaging presentation are as follows.

- Know what the purpose for your presentation is: Do you want to entertain or inform, motivate or convince, provide facts or stimulate creative thinking?
- Organize yourself: Creating a plan or roadmap is crucial if you want to reach a target destination. Using an outline can help you show the big picture (major topics) and make sure the small picture (sub-topics) is covered as well.
- Practice your presentation: Know where you will stand and when and how you will move, what visual materials you'll use, and how you will interact with the audience.
- Capture the audience's interest: What can you use to grab people's attention and get them thinking along with you? How can you add fuel when the fire of interest is flickering? Can you use a story, an illustration, or some humour to keep them with you?
  Even the most important information can go unlearned or unnoticed if you don't get the audience's attention and keep it.
- Keep it relevant: Talking about critical success factors to third graders works no better than using exploding spiders in the boardroom to illustrate the need to lay-off employees.
- Try to assess whether the presentation has worked: Use overviews at the beginning and summaries at the end so that it is clear to the audience members what they should have learned.

Styles

Styles can automate the formatting of individual objects, but you can also apply overall themes to the entire presentation to change all of the formatting at once. A theme is a set of formatting specifications that are applied to objects and text consistently throughout the presentation. There are three elements to a theme: the colours, the fonts, and the effects. Colours are applied via a set of placeholders, as they were in PowerPoint 2003, but now you can apply tints or shades of a colour easily. Whenever you open a list or menu that contains a colour picker, you select from a palette. The top row contains swatches for the colours in the current theme, and beneath them are various tints (lighter versions) and shades (darker versions) of the colours. By applying theme colours instead of fixed colours, you enable objects to change colour automatically when you switch to a different theme.

PowerPoint's interface is typical of any Windows program in many ways, but as you learned earlier in the chapter it also has some special Office 2007 specific features as well. These are as follows.

- Title bar: Identifies the program running (PowerPoint) and the name of the active presentation. If the window is not maximized, you can move the window by dragging the title bar.
- Ribbon: Functions as a combination of menu bar and toolbar, offering tabbed 'pages' of buttons, lists, and commands.
- Office button: Opens the Office menu from which you can open, save, print, and start new presentations.
- Quick Access Toolbar: Contains short cuts for some of the most common commands. You can add your own favorites here as well.
- Minimize button: Shrinks the application window to a bar on the taskbar; you click its button on the taskbar to reopen it.
- Maximize/Restore button: If the window is maximized (full screen), it changes to windowed (not full screen). If the window is not maximized, clicking here maximizes it.
- Close button: Closes the application. You may be prompted to save your changes, if you made any.
- Work area: Where active PowerPoint slide(s) appear. Different views are available that make the work area appear differently.
- Status bar: Reports information about the presentation and provides shortcuts for changing the view and the zoom.

Essentials of Views

A view is a way of displaying your presentation on-screen. PowerPoint comes with several views because at different times during the creation process, it is helpful to look at the presentation in different ways. For example, when you add a graphic to a slide, you need to work closely with that slide, but when you rearrange the slide order, you need to see the presentation as a whole.

PowerPoint offers the following views.

- Normal: A combination of several resizable panes, so you can see the presentation in multiple ways at once. Normal is the default view.
- Slide Sorter: A light-table-type overhead view of all the slides in your presentation, laid out in rows, suitable for big-picture rearranging.
- Notes Page: A view with the slide at the top of the page and a text box below it for typed notes. (You can print these notes pages to use during your speech.)
- Slide Show: The view you use to show the presentation on-screen. Each slide fills the entire screen in its turn.

There are two ways to change a view: click a button on the View tab, or click one of the view buttons in the bottom-right corner of the screen.

Essential Benefits

In the current scenario, PowerPoint presentations in many forms, styles, and custom layouts are used by working professionals, academicians, students, medical practitioners, politicos, and others to showcase material, information, and data. Slideshows and complex presentations comprising of text, images, 3-D graphics audio clips, sound effects, and flashy animations are used creatively to enliven discussions, add value to lectures, make compelling speeches, and enrich studies and reports.

- Create powerful, dynamic SmartArt diagrams: It helps to easily create high-impact and dynamic workflow, relationship, or hierarchy diagrams from within Office PowerPoint 2007. Conversion of a bulleted list into a SmartArt diagram, or modify and update existing diagrams is possible. It is easy for users to take advantage of rich formatting options with new contextual diagramming menus.
- Easily reuse content with Office PowerPoint 2007 Slide Libraries: Wish there was a better way to reuse content from one presentation to another? With PowerPoint Slide Libraries, you can store presentations as individual slides on a site supported by Microsoft Office SharePoint Server 2007 and easily repurpose the content later within Office PowerPoint 2007. Not only does this cut down the time you spend creating presentations, but any

slides you insert can remain synchronized with the server version, so your content is always up-to-date.

- Flexibility and customization: Introduction of newer tools, custom templates and the ability to be used in many platforms makes PowerPoint a flexible and highly customizable tool. User ingenuity can be meshed with these flexibility and customization characteristics to present information and material to suit specific needs and audiences. Document themes enable you to change the look and feel of your entire presentation with just one click. Changing the theme of your presentation not only changes the background colour, but also the colours, styles, and fonts of the diagrams, tables, charts, shapes, and text within your presentation. By applying a theme, you can be confident that your entire presentation has a professional and consistent look and feel.
- Summarize detailed reports: PowerPoint presentations capture the essence and summarize key points of highly detailed reports, surveys, and studies on any subject or domain. Slides with statistical highlights and important facts are also used to present critical information.
- Educational value: PowerPoint presentations are used by professors and teachers to add value to their lectures and notes. Students use PowerPoint slides and presentations to enrich their assignments, projects, and other coursework submissions.
- Sales presentations: Sales professionals regularly use compelling PowerPoint presentations comprising of data and informative content to inform, educate, and enthuse different user audiences about products and services.
- Effective visual tool: The visual imagery made possible by juggling various PowerPoint combinations, design templates, and other tools is used by speakers to add value to speeches and presentations and engage audiences via projector screens during conferences, seminars, and symposiums.
- Add more security to your PowerPoint presentations: You can now add a digital signature to your PowerPoint presentations to help ensure their contents are not changed after they leave your hands, or you can mark a presentation as 'final' to prevent inadvertent changes. Using content controls, you can create and deploy structured PowerPoint templates that guide users into entering the correct information, while helping to protect and preserve the information in the presentation that should not be changed.

9.4 Lotus Development Corporation

Lotus Development Corporation is a pioneer spreadsheet developer and the world's number six software vendor. The core product line of Lotus Development Corporation is its spreadsheet program, known as Lotus 1-2-3. Spreadsheets assist in analysing financial data by performing difficult calculations instantly, such as those involving interest rates. The company was founded in 1982 when Mitchell D. Kapor, a software programmer, won 500,000 dollars in royalties for a software program that combined charts, statistics, and other challenging calculations. Kapor founded Lotus Development with these royalties. The company's ground-breaking Lotus 1-2-3 spreadsheet program took the still-young personal computer market by storm.

IBM's purchase of lotus in July 1995 signalled a new era for the company. Industry observers in *PC Week* suggested that IBM intended to maintain a hands-off approach, and let Lotus, along with its highly popular Notes program, influence the mother company, rather than the other way around. *Business Week* also held that the success of the merger would be tied closely to the success of Lotus Notes in the coming years. Black Enterprise called the merger a sign that IBM was gearing up to enter a groupware war with office-suite competitor Microsoft. On the heels of the merger, long-time Lotus chief Jim Manzi stepped down. According to *PC Week*, it appeared that IBM and Lotus's existing managers could not see eye-to-eye. Two of the biggest trends in Lotus' business during the late 1990s were office suites and the Net. Lotus entered the office suite market with SmartSuite in 1991. The next year, it entered into an agreement with Digital Equipment, to include a copy of SmartSuite for Windows with every computer the company sold. Lotus established a significant presence in the e-mail and Internet markets with its products Lotus Notes and cc:Mail.

9.4.1 Open Office

OpenOffice.org (OOo) is an open source office suite. It runs under Microsoft Windows, Sun Solaris, Linux, and Macintosh's OS X. Open source software is often referred to as 'Free Software'. OpenOffice.org started its life as a package called StarOffice, created by a German company named StarDivision. In April 1999, Sun Microsystems acquired StarDivision and, with it, StarOffice. In July 2000, Sun announced it would release the source code for StarOffice as open source and would sponsor development of the product as an open source office solution. A website, also called OpenOffice.org, was created to provide a home for the product and for those working on it.

Like Microsoft Office, Open office has a word processor (Writer), a spreadsheet program (Calc), and a presentation package (Impress). It also includes an HTML Editor (a first cousin to Writer), a drawing package (Draw), and a formula editor (Math). The easiest way to get OpenOffice.org is by downloading from www.openoffice.org. The home page of the site includes a Downloads link. OpenOffice.org is quite well-behaved in terms of what folders and files it creates and where it puts them. It places the bulk of what it creates in the folder specified for installation and subfolders created within that folder.

OpenOffice.org has its own file formats, but it also has the ability to read and write files in a variety of other formats. One of the appealing things about graphical user interfaces (GUIs) is the idea that knowledge is transferable from one application to another. For example, most Windows users expect the first three items on an application's menu to be File, Edit, and View. They further expect the File menu to contain items for creating, opening, and closing whatever objects the application deals with and the Edit menu to include Cut, Copy, and Paste items. Macros allow you to save a sequence of operations with a single name so you can do the same thing repeatedly.

Open office provides two ways to set up macros. The first is to record a macro, where the user performs the desired action with open office "watching." The second choice is to write it using the OpenOffice.org Basic language. OpenOffice.org uses a different version of Basic than Microsoft Office, so macros created in Office will not work in open office. Open office uses a multi-level structure for storing macros. There are some macros available throughout Open office. In addition, any document or template can have macros stored with it; such macros are available only when that document or template is open. Macros are organized into modules and libraries. A module is a group of macros, presumably with related functionality. A library is a group of modules, again presumably related by their function.

Thus it is seen that open office competes with MS Office in all aspects.

9.5 Collaboration Tools

A collaboration tool is something that helps people collaborate. Microsoft Exchange and Zimbra are projects that most corporations and large firms would use for collaborative calendaring. Tools such as Exchange and Zimbra offer many of the collaborative tools in one package.

9.5.1 Essentials of E-mail

The messaging market was more fragmented in 1996 than it is in 2010. The administrator who set out to deploy Exchange 4.0 had to cope with a plethora of competing standards, connections, and clients. Electronic mail, or e-mail, is probably the most common type of traffic to travel the Internet and other networks. When it was first developed, e-mail was designed to emulate the office paper memo. Messages could be sent to other individuals and they could reply to the messages. Since then, e-mail implementations have evolved to applications and, in some cases, products where even the hardware can send e-mails. In some systems, users can send messages or e-mail to applications, and the application will respond in the form of an e-mail. The ease of connectivity established by simple mail transfer protocol (SMTP), its extensions (ESMTP), and the easy access that we now enjoy to the Internet has revolutionized e-mail. This is true for corporate users and personal users.

A mailbox is usually a designated area of disc storage on a server that can only be accessed by the mailbox owner. In most implementations of e-mail, mailboxes are associated with user accounts so on each system where a user has an account, he or she may also have a mailbox. In order for a mail to be sent properly from one mailbox to another, each mailbox must have a unique e-mail address. The e-mail address is a multi-part hierarchical address. The first portion of the address indicates the user or mailbox account. The latter portion of the e-mail address indicates the location of the mailbox.

The structure of an e-mail message is very simple and consists of two parts that are separated by a blank line (Figure 9.8). The first part of the e-mail message is a header that contains information about who is sending the message who (is) are the recipient(s), the data, and the format of the message content.

The original intent of e-mail was to send only text messages, but people quickly realized that it would be useful to send other types of information and file formats. In order to provide compatible systems, the Internet Engineering Task Force (IETF) developed the multipurpose Internet mail extensions (MIME). The MIME specifications do not define just one technique for encoding binary data. It allows the sender and recipient to choose an encoding format that they both understand and that is readily available. The user's e-mail application will display the text portion of the message that the sender included with the attachment. The e-mail program will then either display the attachment or prompt the user for a location to save the file.

Outlook is not the only client that you can connect to Exchange. Ever since Microsoft realized that they had to support the Internet after the famous memo written by Bill Gates galvanized Microsoft's engineering groups in 1996, Exchange has been able to support other client protocols. Today, Exchange 2007 supports a broad range of Internet protocols from POP3 and IMAP4 on the client side to SMTP as the basis for messaging connectivity and transport, to hypertext transfer protocol (HTTP) for Web access, plus extensions that provide better security and functionality, like ESMTP and hypertext transfer protocol secure (HTTPS).

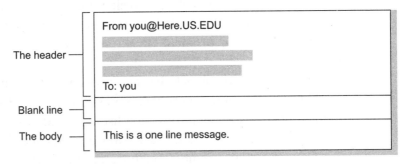

FIGURE 9.8 Structure of a Mail

The connection between the sending mail transfer program and the receiving mail transfer program is accomplished through transmisson control protocol (TCP) and another protocol called SMTP. Simple mail transfer protocol handles the connection establishment, transfer of information, and connection closure. Sendmail is an example of mail transfer program. The sendmail program needs to transport mail between a wide variety of machines. Consequently, its configuration file is designed to be very flexible. This concept allows a single binary to be distributed to many machines, where the configuration file can be customized to suit particular needs. The sendmail program is actually composed of several parts, including programs, files, directories, and the services it provides. Its foundation is a configuration file that defines the location and behaviour of these other parts and contains rules for rewriting addresses. A queue directory holds the mail until it can be delivered. An aliases file allows alternative names for users and creation of mailing lists. The configuration file contains relevant information for sendmail to do its job. Not all mail messages can be delivered immediately. When delivery is delayed, sendmail must be able to save it for later transmission.

The sendmail queue is a directory that holds the mail until it can be delivered. Aliases allow mail that is sent to one address to be redirected to another address.

Qmail is another mail transfer agent similar to sendmail. Qmail's design is rather different from its best-known predecessor, sendmail. Since its release in 1998, qmail has quietly become one of the most widely used applications on the Internet. It is powerful enough to handle mail for systems with millions of users. People who are familiar with sendmail often have trouble recasting their problems and solutions in qmail terms.

E-mail clients have evolved into elaborate suites that fulfil the user's every wish. Any measure of a mail client is its support for popular mail protocols, such as post office protocol (POP) and interactive mail access protocol (IMAP).

The POP3 protocol includes features such as requiring users to authenticate before they can remotely access a mailbox. It also includes the ability to perform queries such as retrieving a list of mailbox contents. A nice feature of POP3 is that a persistent connection between the POP3 client, the user requesting access to a remote mailbox, and the POP server does not have to be maintained. This is particularly useful for dial-up or occasional network access. A more sophisticated remote mailbox access protocol is available with IMAP. With IMAP, the e-mail maintains a central location for the mail. Some of the common mail user tools include Outlook, evolution, etc.

9.5.2 MS Outlook

Businesses of all kinds share common issues revolving around employee relations, planning, and training. Using Outlook as a dashboard or control panel, you can improve the way you confront and manage these issues (Figure 9.9).

Outlook 2007 might come as a part of the Microsoft Office suite, but it has its own publicized set of system requirements pertinent to running its myriad features.

When you first set up Outlook, a new 'profile' was automatically set up for you. An Outlook profile is what Outlook uses to remember the e-mail accounts and the settings that tell Outlook where your e-mail is stored. Your profile tells Outlook the following two basic things.

1) What e-mail account information to use
2) Where the data is delivered and stored

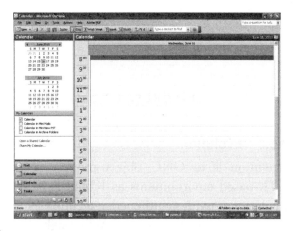

FIGURE 9.9 Microsoft Outlook—View of a Calendar

You probably want your work e-mails kept in a separate place from your personal e-mail, and also separate from your spouse's work e-mail and personal e-mails (Figure 9.10).

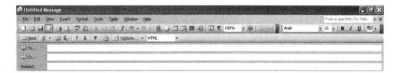

FIGURE 9.10 Microsoft Outlook—View of E-mail Screen

Sending an e-mail is probably one of the easiest and most useful tasks you can perform with Outlook and probably the task most widely associated with it. New e-mails appear in your inbox, displayed in bold text format until you read them. To open an e-mail, simply doubleclick it, and it will open in its own window. You can also see the contents of an e-mail in the Reading pane (by default this is on the right side of the screen, but you can adjust it). If you look beneath your Inbox folder in the Navigation Pane on the left side of the screen, you will see the Junk E-Mail folder where Outlook's junk e-mail filter will dispose of e-mails it deems to be spam. You can look in here for an e-mail that might have been mistakenly dumped as spam; right-click it, select Junk E-mail, and then choose Add Sender to Safe Senders List. The features of Outlook have been represented on the Word Ribbon bar (Figure 9.11).

FIGURE 9.11 Microsoft Outlook—View of the Ribbon Bar

9.5.3 Evolution

Evolution is to GNU object model environment (GNOME) and other GTK-based environments as KMail is to K Desktop Environment (KDE). Evolution provides cutting edge technology for integrated mail, address book and calendaring functionality. At first sight, Evolution looks like a pretty simple and easy to use e-mail client which is just what most users need from one. The truth is that it is a little more, and could easily satisfy the needs of almost anyone. Evolution is among those that currently support Exchange and GroupWise servers. Evolution comes with a data server that facilitates the integration with GNOME. It allows using applets in GNOME panel for showing appointments and tasks and also to add buddies in Gaim from evolution contacts, when Gaim is patched. The addressbook is Lightweight Directory Access Protocol (LDAP) compatible.

Evolution integrates seamlessly with GNOME, but it lacks a fast control via an icon in the kicker. Incoming mail is announced by a beep or a sound file of your choice, and you will need to install an add-on such as a GNU Krell Monitors (GK rell M) plugin to count the number of incoming mail messages without shifting the focus to the program window. What Evolution can do is collaborate with other popular GNOME applications, such as Pidgin or Planner, for advanced contact management.

Evolution also integrates SpamAssassin as a plugin. Filter rules are easy to create via the dropdown menu for a message. Evolution uses Bayesian spam filtering which is starting to be a standard in this area. This type of spam filtering is based on Bayes' theorem which, in the context of spam, says that the probability that an e-mail is spam, given that it has certain words in it, is equal to the probability of finding those certain words in spam e-mail, times the probability that any e-mail is spam, divided by the probability of finding those words in any e-mail.

9.5.4 Thunderbird

The Thunderbird mail client, which is developed by the Mozilla project, feels at home on various Unix platforms and also supports the Microsoft

Windows Operating System, unlike the other candidates. This would allow the administration to handle e-mail communications in a standardized way on Linux and Windows. Thunderbird allows users to set up multiple identities, and the program stores all settings, such as the use of signatures and certificates, or storage paths, separately for each account. Thunderbird's filtering options are many and more granular than those offered by Evolution. Thunderbird is highly configurable.

SUMMARY

In the office, we rely on technology and tools in order to work and communicate. Indeed, it is hard to imagine modern offices without technology. Having the right tools is key to getting the job done. Office productivity tools are applications that allow for the viewing, creating, and modifying of general office documents, for example, spreadsheets, memos, presentations, letters, personal database, form generation, image editing, etc. Office productivity tools also include applications for managing employee tasks. Office 2007 has added a Review tab for quick access to commands needed to review and revise existing documents. The various e-mail clients were outlined as a part of this chapter.

Key Terms

Alignment Alignment refers to the position of lines in a paragraph in relation to the document's left and right margins; i.e. whether they are left-aligned, centered, right-aligned, or justified (evenly spaced).

Arrange All Use the Arrange All command on the Window menu to see all of the available windows at one time. The windows are arranged horizontally, which is helpful if you are working with two or more documents at the same time.

Attachments These are files which you can send along with your e-mail, for example, graphics files, word processing documents, spreadsheets, etc. Any file you want to send along with your e-mail. Most free e-mail providers have a maximum size limit for attachments.

Bayesian filter An anti-spam program that evaluates header and content of incoming e-mail messages to determine the probability that it is spam. Bayesian filters assign point values to items that appear frequently in spam, such as the words 'money-back guarantee' or 'free'.

Current document The current or active document is the document you see on the Word screen.

Data source A data source contains the information from which a merged document is created. The data source is merged with a main document which specifies the kind of output required.

Document A document is any data file that is created by a specific application.

E-mail client A program used to read and send e-mail messages. As opposed to the e-mail server, which transports mail, an e-mail client is what the user interacts with.

Filters Filters automatically move incoming e-mails into separate folders according to criteria that you specify. These criteria may be based on who the e-mail is from, the priority, the subject, the main message, etc. This is a useful feature if you have a lot of incoming e-mails that need some way of being automatically organized.

Outline view Outline View provides a structured view of the document, arranging its contents according to heading levels and opening the Outlining toolbar for modifying the document's organization.

Paragraph formatting Paragraph formatting includes formatting options such as text alignment, indents, tabs, margins, and justification.

POP3 This is a computing standard that enables e-mails to be retrieved from a remote mailbox. That is, it allows you to collect e-mails from an account that you have on another computer (called a server, or host) to your own e-mail software, such as Outlook, Netscape Messenger, Eudora, Exchange, etc.

SMTP This is a computing standard that enables e-mails to be sent through another computer (called a server) from your own e-mail software, such as Outlook, Netscape Messenger, Eudora, Exchange, etc. It is complementary to POP3.

REVIEW QUESTIONS

9.1 What are the methods in which a document can be viewed in Windows 2007?

9.2 Outline the Ribbon interface of Word 2007 and compare the features with the earlier version of Word.

9.3 List any five short cut keys and their uses.

9.4 Carry out an exercise to compare the features of MS Office and Open Office suites.

9.5 What is SMTP used for?

9.6 What is MIME used for?

9.7 What is meant by an electronic mailbox?

9.8 What are the two required keywords in an e-mail header?

9.9 Briefly describe MS PowerPoint and views as applicable to PowerPoint.

9.10 Differentiate between MS Outlook and Evolution.

Projects

9.1 The office of a small independent realty company has four real estate agents, an executive secretary, and two assistant secretaries. They agreed to keep all common documents in a central location to which everyone has access. They had a lot of inconsistency in the various office communication formats. Of particular interest were the different varieties of letter formats which were being used by the employees to address different customers. The management wanted to normalize the formats, so that the external and internal communication formats are standardized. On careful study, it was found out that a few office correspondence were addressed to the vendors, a few office correspondence were addressed to the various customers, a few formats were purely internal. It is expected that in the near future the office will employ fifty more agents and three assistant secretaries apart from a couple of senior management personnel. Design a document template to capture the date, time, author, subject, and a few more document properties for all the cases discussed above. Also design a excel template to calculate the daily sales report to the top management, apart from a presentation template for all external customer presentations.

9.2 Visit the nearest IT firm and understand the various office activities that are being carried out with the help of Office suite, be it MS Office or Open Office. Prepare a gap report to address the areas of improvement.

References

Bisssett, Brian D. 2007, *Automated Data Analysis using Excel*, CRC, USA.

Barlow John F. 2005, *Excel Models for Business and Operations Management*, John Wiley, Great Britain.

Bluttman, Ken 2003, *Developing Microsoft Office Solutions: Answers for Office 2003, Office XP, Office 2000, and Office 97*, Addison Wesley.

Campbell, Tony and Jonathan Hassell, 2007, *Outlook 2007*, Apress, USA.

Cortese, Amy 1995, *Can IBM Keep Lotus Blooming?* Business Week, 30 October.

Dalgleish, Debra 2007, *Beginning Pivot Tables in Excel 2007*, Apress, USA.

Dellecave, Tom, Jr. 1996, 'Lotus Makes Play For SPA', Sales & Marketing Management, March.

Farber, Dan 1995, *How Will IBM Handle Lotus Marriage?* PC Week, 16 October.

Gibson, Stan 1995 'After a Season of Gloom, Can Lotus Blossom Again?' PC Week, 1 May.

Granor, Tamar E. 2003, *OOo Switch: 501 Things You Wanted to Know About Switching to OpenOffice.org from Microsoft Office*, Hentzenwerke Publishing, USA.

IBM 1998, *IBM: Lotus Development Corp. to Bring Real-time Dimension to Messaging and Groupware*, PressWire, 20 May.

Miller, Michael 2005, *Microsoft Windows XP for Home Users Service Pack 2 Edition*, Peachpit Press.

Morrison, Connie 2007, *Word 2007*, Apress, 2007, USA.

Prish, Shahar 2007, *Professional Excel Services*, Wiley, Canada.

Raymond, Eric S. 1996, *IBM Takes Notes Message to Heart*, PC Week, 15 January.

CHAPTER

10 Data Management Tools

Deception-perception. The two are inseparable aspects of human living. Finally, then, here are some principles of a deception-perception approach to systems... The systems approach begins when first you see the world through the eyes of another.
—C. West Churchman

Learning Objectives

After reading this chapter, you should be able to understand:

- the complexity of data management
- the need for DBMS
- the differences between the common DBMS offerings
- data mining
- and differentiate content management systems

10.1 Data Management

Data management systems play the most crucial role in building large application systems. Modern applications are no longer single monolithic software blocks but highly flexible and configurable collections of cooperative services. Therefore, within recent years, data management systems have faced a tremendous shift from the central management of individual records in a transactional way to a platform for data integration, federation, search services, and data analysis.

10.2 Database

Database researchers have been organizing workshops and/or panel discussions once in every three to five years, fully devoted to the question of whether there is anything interesting left in the database arena, since the mid-1980s. Database has proved to be exceptionally productive and of great economic impact. In fact, today, the database market exceeds 100 billion. Databases have become a first-order strategic product as the basis of information systems (IS), and support management and decision-making. Modern database technology has its roots in business data processing. The textbook on data organization, which was published in the early 1970s by Hartmut Wedekind, describes databases as an emerging technology, characterizes databases as the core of management information systems designed to accommodate different structural and operational schemes of a company and to provide a query interface. In the initial stages of computing, data were stored in file systems. The problems (redundancy, maintenance, security, the great dependence between data and applications, and, mainly, rigidity) associated with the use of such systems gave rise to new technology for the management of stored data, i.e, databases.

Paul Larson, senior researcher at Microsoft quoted, 'Database systems are as interesting as the household plumbing to most people. But if they don't work right, that's when you notice them.' There is a growing number of applications where databases are used to create a near real-time image of some critical section of the environment. These are connected together and operated on by the workflow software systems.

A database is a collection of interrelated data items that are managed as a single unit. This definition is deliberately broad because there is so much variety across the various software vendors that provide database systems. A database is a persistent, logically coherent collection of inherently meaningful data, relevant to some aspects of the real world. A database is a collection of information that is organized so that it can easily be accessed, managed, and updated. According to one view, databases can be classified according to types of content. Databases are designed to offer an organized mechanism for storing, managing, and retrieving information. They do so through the use of tables.

A database object is a named data structure that is stored in a database. The specific types of database objects supported in a database vary from vendor to vendor and from one database model to another. A file is a collection of related records that are stored as a single unit by an operating system.

Databases have a tenacious habit of just not going away. This is true for the real databases, discs, tapes, software, etc. that are run by big applications; those databases, through the sheer mass of data accumulated have an inertia that reliably protects them from being 'moved', technologically, platform-wise or in any other sense.

10.2.1 Database Management System

A database is a collection of data, typically describing the activities of one or more related organizations. A record is a representation of some physical or conceptual object. Each record has multiple attributes such as name, address, and telephone number. Individual names, addresses, and so on are the data. A database consists of both data and metadata. Metadata is the data that describes the data's structure within a database. The database stores metadata in an area called the data dictionary which describes the tables, columns, indexes, constraints, and other items that make up the database. Databases come in all sizes, from simple collections of a few records to mammoth systems holding millions of records.

A file is a collection of related records that are stored as a single unit by an operating system. An instance is a copy of the database software running in memory.

A database management system (DBMS) is a set of programs used to define, administer, and process databases and their associated applications. It is a piece of software that manages a database, a repository of interrelated data items that are often central to the business of an enterprise or institution. A DBMS that runs on platforms of multiple classes, large and small, is called scalable. A database is generally used by many diverse applications and multiple users, each of which may need only a fraction of the data. A DBMS provides a high-level data model and a related query and data manipulation language to applications.

Data model represents the data definitions and business logic rules that determine the relationships between items and groups of data. Data models are built from the following three main component types.

- Entities
- Attributes
- Relationships

A transaction is any one execution of a user program in a DBMS. This is the basic unit of change as seen by the DBMS. A lock is a mechanism used to control access to database objects. Two kinds of locks are commonly supported by a DBMS: shared locks on an object can be held by two different transactions at the same time, but an exclusive lock on an object ensures that no other transactions hold any lock on this object.

10.2.2 Data Models

The schema for a database is a plan for storing data in a collection of tables. The schema is consistent with our design and may look quite similar to it. A database's schema provides an overall organization to the tables. This structure is sometimes also called the complete logical view of the database. The database description consists of a schema at each of these three levels of abstraction: the conceptual, physical, and external schemas.

A data definition language (DDL) is used to define the external and conceptual schemas. The conceptual schema (sometimes called the logical schema) describes the stored data in terms of the data model of the DBMS. The physical schema specifies additional storage details. Essentially, the physical schema summarizes how the relations described in the conceptual schema are actually stored on secondary storage devices such as disks and tapes. External schemas, which usually are also in terms of the data model of the DBMS, allow data access to be customized (and authorized) at the level of individual users or groups of users. Any given database has exactly one conceptual schema and one physical schema because it has just one set of stored relations, but it may have several external schemas, each tailored to a particular group of users. Each external schema consists of a collection of one or more views and relations from the conceptual schema. A view is conceptually a relation, but the records in a view are not stored in the DBMS.

The data model is a logical view of how data are organized, which is generally very different from the way data are actually laid out on physical storage media. A DBMS allows a user to define the data to be stored in terms of a data model. A data model is essentially the architecture that the DBMS uses to store objects within the database and relate them to one another. These models contain a wide variety of constructs that help describe a real application scenario. A lot of models were proposed to manage data. When data models with more expressive power were born, DBMSs were capable of incorporating more semantics, and physical and logical designs started distinguishing one from the other as well. In the mid-1980s, one of those attempts was design automatization through the use of computer-aided software system engineering (CASE) tools. The CASE tools contributed to spreading the applications of conceptual modelling and re-launching DB design methodologies. Some of the data management models include hierarchical model, network model, relational model, and object-oriented data model.

Entities and Relationships

An entity is an object in the real world that is distinguishable from other objects. An entity set is a collection of similar entities. It is described using a set of attributes. All entities in a given entity set have the same attributes. For each attribute associated with an entity set, we must identify a domain of possible values. Each entity set requires a key. A key is a minimal set of attributes whose values uniquely identify an entity in the set. There could be more than one candidate key; if so, we designate one of them as the primary key. A relationship is an association among two or more entities. A relationship can also have descriptive attributes. Descriptive attributes are used to record information about the relationship. An instance of a relationship set is a set of relationships taken together.

Constraints are an important, although often overlooked, component of a database. Constraints are rules that determine what values the table attributes can assume. A column's domain is the set of all values that the column can contain. A constraint is a restriction on what a column may contain.

Database design has six steps: requirements analysis, conceptual database design, logical database design, schema refinement, physical database design, and security design.

Relational Data Model

Relational technology took the market by storm in the 1980s. The relational model provides the ability to relate records as needed rather than predefined when the records are first stored in the database. Moreover, the relational model is constructed such that queries work with sets of data.

Dr E. F. Codd of IBM first formulated the relational database model in 1970. The model evolved from the file systems that the databases replaced, with records arranged in a hierarchy much like an organization chart. A relational database is a collection of relations with distinct relation names. Relational databases have been acting as the electronic filing cabinets for the voluminous and complex data storage needs of large companies for over two decades. A relational database, simply defined, is a database that is made up of tables and columns that relate to one another. These relationships are based on a key value that is contained in a column. The relational database schema is the collection of schemas for the relations in the database. The schema specifies the relation's name, the name of each field (or column, or attribute), and the domain of each field. A domain is referred to in a relation schema by the domain name and has a set of associated values.

Relational databases have replaced databases built according to earlier models because the relational type has valuable attributes that distinguish relational databases from other database types. Probably, the most important of these attributes is that in a relational database one can change the database structure without requiring changes to applications that were based on the old structures. Relational databases offer structural flexibility; applications written for those databases are easier to maintain than similar applications written for hierarchical or network databases. The relational model is based on the notion that any preconceived path through a data structure is too restrictive a solution, especially in the light of ever-increasing demands to support ad hoc requests for information.

Codd invented a methodology of organizing the data content to reduce duplication and called the technique,'normalization'. Normalization helps to organize data better and enables reuse of data. It connects multiple systems of data together and adds complexity to the system. According to Codd, a relational database is made up of one or more relations. A relation is a two-dimensional array of rows and columns, containing single valued entries and no duplicate rows. Each cell in the array can have only one value, and no two rows may be identical.

Network Data Model

The network database model evolved around the same time as the hierarchical database model.

Object-relational DBMS

The origins of object-oriented databases trace their beginnings to the emergence of object-oriented programming in the 1970s. Relational database systems support a small, fixed collection of data types which has proven adequate for traditional application domains such as administrative data processing. Object-oriented concepts have strongly influenced efforts to enhance database support for complex data and have led to the development of object-database systems.

An object is a logical grouping of related data and program logic that represents a real-world, such as a customer, employee, order, or product. In the 1990s, object-relational systems emerged with Informix being the first to market, followed relatively quickly by Oracle and IBM.

Object-oriented database systems are proposed as an alternative to relational systems and are aimed at application domains where complex objects play a central role. The approach is heavily influenced by object-oriented programming languages and can be understood as an attempt to add DBMS functionality to a programming language environment. The object-oriented (OO) model actually had its beginnings in the 1970s, but it did not see significant commercial use until the 1990s. An object model is a description of content and metadata broken down into component parts that are joined using the data model as a starting point. These correspond to properties in the object-oriented world. An object-oriented database management system (OODBMS) can be defined as a DBMS that directly supports a model based on the object-oriented paradigm. Like any DBMS, it must provide persistent storage for objects and their descriptors (schema). The system must also provide a language for schema definition and for manipulation of objects and their schema. In addition to those basic characteristics, an OODBMS usually includes a query language and the necessary database mechanisms for access optimization, such as indexing and clustering, concurrency control and authorization mechanisms for multi-user accesses, and recovery.

Object-relational DBMSs extend the relational model to include support for object-oriented data modelling necessary to support a broader class of applications and, in many ways, provide a bridge between the relational and object-oriented paradigms. Object-oriented features have been added to the international structured query language (SQL) standard, allowing relational DBMS vendors to transform their products into object-relational DBMSs, while retaining compatibility with the standard.

Benefits of DBMS

A DBMS has many advantages. These are as follows.

- Data independence: Application programs should be as independent as possible from details of data representation and storage. The DBMS

provides an abstract view of the data to insulate application code from details.
- Efficient data access: A DBMS utilizes a variety of sophisticated techniques to store and retrieve data efficiently.
- Data integrity and security: If data is always accessed through the DBMS, the DBMS can enforce integrity constraints on the data.
- Data administration: When several users share the data, centralizing the administration of data can offer significant improvements.
- Concurrent access and crash recovery: A DBMS schedules concurrent accesses to the data in such a manner that users can think of the data as being accessed by only one user at a time.
- Reduced application development time: Clearly, the DBMS supports many important functions that are common to many applications accessing data stored in the DBMS.

10.2.3 Microsoft Access

Microsoft (MS) Access places the entire database in a single data file, so an Access database can be defined as the file that contains the data items. Oracle Corporation defines their database as a collection of physical files that are managed by an 'instance' of their database software product. An instance is a copy of the database software running in memory. The DBMS is software provided by the database vendor. Software products such as Microsoft Access, Oracle, Microsoft SQL Server, Sybase, DB2, INGRES, and MySQL are all DBMSs.

At its most basic level, you can use Access to develop simple personal database-management systems. Microsoft Access is a powerful program to create and manage databases. It is an application used to create small and midsize computer desktop databases for the MS Windows family of operating systems. An MS Access database is primarily a Windows file. Access is an excellent platform for developing an application that can run a small business. Its wizards let developers quickly and easily build the application's foundation.

As many as 10 data types can be defined in Access. You will probably use just a few of them. However, this section includes a brief description of all the types in case you are relatively new to Access. The 10 data types are: Text, Memo, Number, Date/Time, Currency, AutoNumber, Yes/No, OLE Object, Hyperlink, and Attachment.

Text Text is the most common data type you will use in Access. Technically, any combination of letters, numbers, characters, and spaces is text. However, any number stored as text cannot be used in a calculation. Examples of numbers commonly stored as the Text data type are customer numbers, product (SKUs), or serial numbers. Obviously, you would never perform any calculations on these types of numbers. The Text data type is limited to a maximum of 255 characters.

Memo The Memo field allows you to store text data that exceeds the 255-character limit of the text field.

Number The Number field is a numeric data type that is actually several data types under one heading. Use this data type with fields that might be summed or otherwise modified through arithmetic operations. After selecting the Number data type in the Design view of the table, go to the Field Size field at the top of the Field Properties menu. Selecting this menu will give you the following choices: Byte, Integer, Long Integer, Single, Double, Replication ID, and Decimal. The most common field sizes of the Number data type are Long Integer and Double. Long Integer should be selected if the numbers are whole numbers (no decimals). Double should be selected if decimal numbers need to be stored in that field.

Date/Time The Date/Time data type is used to record the exact time or date that certain events occurred. The posting date of a transaction and the exact time a service call was placed are perfect examples of fields where the Date/Time data type is most useful.

Currency The double field size of the Number data type can also be used for currency fields, but the Currency data type is ideal to store all data that represents amounts of money.

AutoNumber This data type is a long integer that is automatically created for each new record added to a table, so you will never enter data into this field. The AutoNumber can be one mechanism by which you can uniquely identify each individual record in a table, but it is best practice to use a unique record identifier that already exists in your data set.

Yes/No There are situations where the data that needs to be represented is in a simple Yes/No format. Although you could use the Text data type for creating a True/False field, it is much more intuitive to use Access's native data type for this purpose.

Object linking and embedding(OLE) object This data type is not encountered very often in data analysis. It is used when the field must store a binary file such as a picture or sound file.

Hyperlink When you need to store an address to a website, this is the preferred data type.

Attachment You can use attachments to store several files, and even different types of files, in a single field. The Attachment field is new for Access 2007 and stores data files more efficiently than using other fields like the OLE Object field.

> **Know your Firm—Brief Overview of Oracle**
>
> California-based Oracle Corporation is the world's largest enterprise software company, providing enterprise solutions to major companies worldwide. Three decades ago, Larry Ellison and his co-founders, Bob Miner and Ed Oates, understanding the tremendous business potential in the relational database model saw an opportunity other companies missed when they came across a description of a working prototype for a relational database and discovered that no company had committed to commercializing the technology. They never realized that they would change the face of business computing forever. Oracle has proved throughout its history that it can build for the future on the foundation of years of innovation, intimate knowledge of its customers' challenges and successes, and the best technical and business minds in the world. Oracle's first commercially available database software defied prevailing conventional wisdom that technology would never scale to large amounts of data or extensive numbers of users. After 30 years, Oracle remains the gold standard for database technology and applications in enterprises throughout the world. The company is the world's leading supplier of software for information management, and the world's second largest independent software company.
>
> Oracle database is a scalable and full-featured database, known for its cross-platform support of all popular operating system environments. Oracle's award-winning products and support enable customers to better use and apply the technology as a competitive advantage in their own businesses. Oracle's product line includes Oracle Database, Oracle Application Server, Oracle E-Business Suite, and Oracle Collaboration Suite. For the cost-effective Intel architecture platform, Oracle offers its products on Windows- and Linux-based systems.
>
> *Source*: www.oracle.com

10.2.4 DBMS and RDBMS

In business environments today, the demand for information access is increasing while the volume and complexity of data grows rapidly along with the variety of applications.

MySQL

MySQL is an open source, enterprise-level, multi-threaded, relational database management system. It was developed by a consulting firm in Sweden called TcX. The firm needed a database system that was extremely fast and flexible. It created MySQL which is loosely based on another database management system called mSQL. The product they created is fast, reliable, and extremely flexible. It is used in many places throughout the world. Universities, Internet service providers, and non-profit organizations are the main users of MySQL.

The reason for the growth of MySQL's popularity is the advent of the Open Source Movement in the computer industry. The Open Source Movement is the result of several computer software vendors providing not only a product but the source code as well. This allows consumers to see how their program operates and modify it where they see fit. This, and the popularity of Linux, has given rise the use of open source products in the business world. Because of Linux's skyrocketing popularity, users are looking for products that will run on this platform. MySQL is one of those products. It is more than just a database. It is a program that manages databases, much like Microsoft's Excel manages spreadsheets. Structured query language is a programming language that is used by MySQL to accomplish tasks within a database. It is a system that manages databases. It controls who can use them and how they are manipulated. It logs actions and runs continuously in the background.

TcX decided to make MySQL a multi-threaded database engine. A multi-threaded application performs many tasks at the same time as if multiple instances of that application were running simultaneously. Fortunately, multi-threaded applications do not pay the very expensive cost of starting up new processes. A separate thread handles each incoming connection with an extra thread that is always running to manage the connections. Multiple clients can perform read operations simultaneously without impacting one another. But write operations, to a degree that depends on the type of table in use, only hold up other clients that need access to the data being updated. While any thread is writing to a table, all other threads requesting access to that table simply wait until the table is free. Access to a MySQL database can be determined from the remote machine that can control which user can view a table. The database can be locked down even further by having the operating system play a role in security as well.

DB2

DB2 is a web-enabled relational database management system that supports Java. It can be scaled from single processors to symmetric multiprocessors to massively parallel processors and is multimedia-capable with image, audio, video, and text support. DB2 Universal Database is a powerful database management system that helps the user to manage data in complex and rapidly changing environments. This relational database management system (RDBMS) allows users to create, update, and control relational databases by using SQL. Data managed by DB2 database servers can be accessed and manipulated by applications on personal computer workstations running popular operating systems such as OS/2, Windows 95, Windows NT, Windows 3.1, and Macintosh.

MSSQL

Structured Query Language Server 2005 represents the relational database management software offering from Microsoft. This version has a wide array of features, capabilities, and functions. Several releases of the SQL Server software, called editions, are available to meet the various needs. Each edition has different features and costs. Some run on all current Microsoft operating systems; others require a particular platform. Microsoft may make changes to some editions with service packs. In addition to the production editions of SQL Server 2005, there is a Microsoft Developer Network (MSDN) Edition (also called the Developer Edition), which has the same features as the highest version of SQL Server. The Express Edition represents Microsoft's answer to free database offerings such as MySQL and replaces a previous version called Microsoft Server Desktop Engine (MSDE). Microsoft made this edition more capable than the older version, increasing the size of available databases and usable random access memory (RAM). It can participate in replication, serve as a witness machine in clustering (more on that later), contains the common language runtime (CLR) enhancement, integrates with visual studio, and has the same database engine that the other editions have. The Workgroup Edition is the next step up from the Express Edition. It has all the same features as Express and increases the available RAM to a maximum of 3GB, can run two Central processing units (CPUs), and has no database size limit. It also includes the full SQL Server Management Studio, the SQL Server Agent, and has full-text support. The Enterprise Edition adds high-end features such as online indexing, online page and file restore, fast redo, integration services advanced transforms, oracle replication, scale out report servers, full data mining, and greater text-search capabilities. Because SQL Server 2005 runs only on the Microsoft platform, it has a distinct advantage in being able to rely on operating system consistency and its programming interfaces. Many of these are implemented by programs rather than commands, but some commands can show activity that has an effect on SQL Server.

10.3 Transaction Processing Systems

Distributed computing often gives rise to complex concurrent and interacting activities. In some cases, several concurrent activities may be working together to solve a given problem. In other cases, the activities may be completely independent or essentially independent but need to compete for shared common system resources. In practice, different kinds of concurrency might coexist in a complex application which thus will require a general supporting mechanism for controlling and coordinating its various concurrent activities. Hence, there is a need for transaction management system.

A transaction is a transformation from one state to another. In the business world, a transaction is defined as an agreement between a buyer and a seller to exchange an asset for payment. While in the database world, the real state of the outside world is abstracted from and modelled by a database where the transformation of the state is reflected by an update of the database. From this perspective, a transaction can be defined as a group of operations executed to perform some specific functions by accessing and/or updating a database. These operations, are, in fact, a kind of program designed to consistently interact with a database system.

The database operations can be gathered together to form a unit of execution programme called a transaction. In other words, a transaction is a logical execution unit of database operations. A transaction transforms the database from one consistent state to another consistent state. A transaction is supposed to be complete if and only if all the related databases that are affected by this transaction are updated and all the updates are completed. It is important to stress the need of completing the updating of all databases before the transaction is taken as complete. Any deviation from this would lead to major disasters as incomplete information may have serious consequences. In a transaction processing system, very large numbers of transactions may read or cause changes to an enterprise's data store. It is important that transactions read valid data and that their updates are correctly recorded. Ensuring this is called preserving data integrity.

Each transaction must ensure that it always preserves the consistency of the database system. In order to retain and to protect the consistency of the database system, transactions will have the following atomicity, consistency, isolation, and durability (ACID) properties.

- Atomicity: Either all database operations of a transaction program are successfully and completely executed, or none of the database operation of this transaction program is executed. If a transaction cannot be completed in its entirety, all its effects must be undone. This requires a rollback facility in the transaction processing system. The atomicity of a state transition affected by a transaction is a general, unconditional property. It holds whether the transaction, the entire application, the operating system, or other components function normally, function abnormally, or crash.
- Consistency: A transaction must always preserve and protect the consistency of the database, i.e., it transforms the database from one consistent state to another. In other words, the result of a transaction that is committed fulfil the constraints of the database system. The results of a transaction must be reproducible and predictable. The transaction must always give the same results in the same conditions.
- Isolation: An on-going transaction must not interfere with other concurrent transactions, or be able to view intermediate results of other concurrent transactions. In other words, a transaction is executed as if it is the only existing execution program on the database system at any given time.
- Durability: The result of a transaction that has successfully committed is permanent in the database. The consistent state of the database is always survived despite any type of failures.

The ACID properties of a transaction ensure that

- a transaction always keeps the database in a consistent state,
- a transaction does not disturb other transactions during their concurrent execution processes, and
- the consistent state of the database system that is established by a committed transaction withstands software or hardware failures.

Transaction is the fundamental abstraction underlying database system concurrency and failure recovery. Transaction services ensure that sequences of database updates have been accurately and reliably committed as a single complete unit of work or that, in the event of failure, the database information is recovered. Transactions refer to all discreet tasks that must be performed in unison to accomplish a goal. Transactions may involve tasks that are done by one or more participant. When more than one participant is involved, it must be ensured that all the participants perform their tasks as promised. To guarantee this, there is some kind of transaction manager that all the involved participants trust, that controls the actions taken by various participants, and ensures that either all the participants do what they promise or none of the participants do anything. Transactions are short lived.

A transaction program starts from an initial consistent state of the database by invoking a Begin transaction method call. After that, one or a set of database operations of the transaction program are executed. When these database operations are completed, a new consistent database state is established as designed. The transaction program saves this new consistent state into the database by calling the Commit transaction method. The Commit transaction call ensures that all the database operations of the transaction program are successfully executed and the results of the transaction are safely saved in the database. If there is any error during the execution of the transaction program, the initial consistent state of the database is re-established by the Abort transaction call. The Abort transaction call indicates that the execution of the transaction program has failed and this execution does not have any effect on the initial consistent state of the database. The transaction is said to be committed if it has successfully executed the Commit transaction call, otherwise it is aborted. A transaction is called a read-only transaction if all of its database operations do not alter any database state.

A transaction processing application is a collection of transaction programs designed to carry out the functions necessary to automate a given business activity. The first online transaction processing application to receive widespread use was an airline reservation system—the SABRE system developed in the early 1960s as a joint venture between IBM and American Airlines. It was one of the biggest computer system efforts undertaken by anyone at that time, and still is a very large transaction processing system. Transaction processing systems of today, such as ticket booking systems, product ordering systems, or trouble ticketing systems, are real-time systems, i.e., correct functioning of the system is not only determined by whether the transaction was executed correctly, but whether it was completed within a certain amount of time. That is, the transaction response time is an important performance metric. This metric is also the most important user-perceived metric, and often the one by which the user judges the quality of the transaction processing system.

Now, as electronic commerce is handled at 'transaction aggregation points' on global networks, transaction workloads are moving increasingly to mid-range and high-end systems using client-server, distributed computing architectures. Online transaction processing provides the critical infrastructure that enables reliable and cost-efficient financial, commercial, travel, medical, and governmental operations.

Those transactions which are internal to the company and are related with the internal working of any organization are called internal transactions. The transactions which are external to the organization and are related with the external sources are regarded as external transaction. A transaction processing system supports different tasks by imposing a set of rules and guidelines that specify how to record, process, and store a given transaction.

Vendors of transaction processing systems quote transaction per second (TPS) rates for their systems. The performance of a transaction processing system depends heavily on the system input/output architecture, data communications architecture, and even more importantly on the efficiency of the system software.

10.3.1 Flat Transactions

The ACID properties are the basis for the classic form of transactions known as traditional transactions or flat transactions. Flat transactions represent the simplest type of transaction and for almost all existing systems, it is supported at the application programming level. A flat transaction is the basic building block for organizing an application into atomic actions. It can contain an arbitrary number of simple actions. Putting a flat transaction as an execution unit around a sequence of either protected or real actions makes the whole appear as indivisible as far as the application is concerned. These transactions are called flat because there is only one layer of control by the application. the transaction will either survive together with everything else (Commit), or it will be rolled back with everything else (Abort). The major restriction of flat transactions is that there is no way of either committing or aborting parts of such transactions, or committing results in several steps, and so forth.

While the interaction is in the process of changing system state, any number of events can interrupt the interaction leaving the state change incomplete and the system state in an inconsistent, undesirable form. Any change to system state within a transaction boundary, therefore, has to ensure that the change leaves the system in a stable and consistent state. Two concurrently running transactions can update different records within a single data store. The data store manager ensures and prevents two transactions from updating the same record. These functions are called concurrency control. The most commonly used method is locking. A lock is a mechanism by which use of a resource is restricted to the holder of the lock. Locking of data can be at file level, block level, or logical record level. Committing a change is the process of making it permanent or irreversible. A single transaction can update more than one database or file managed by the same data store manager. When two or more data store managers are involved in a single transaction, the transaction processing system coordinates the commitment control process and this process is called a two-stage commit.

A transaction processing system generally consists of a transaction processing monitor, which is a system program providing a middleware solution to manage transactions and control their access to a database management system. Transaction monitors were the first kind of middleware components for supporting distributed transaction processing and thus also the first kind of application servers. Atomicity and durability are guaranteed by the mechanism of recovery that is usually implemented by maintaining a log of update operations so that 'redo' and 'undo' actions can be performed when required. Isolation is guaranteed by the mechanism of concurrency control, which is implemented by using locks during the transaction process.

Several enhancements to the traditional flat transaction model have been proposed. One enhancement, permitting a finer control over recovery and concurrency, is to permit nesting of transactions. The partial rollback is supported by the mechanism of the save point. At each save point, special entries are stored containing the state of the database context in use by the transaction, and the identity of the lock acquired most recently. When a transaction fails, it can recover to the recorded save point, where it restores the corresponding context and releases locks acquired after this save point. Although the save point mechanism can be used in combination with flat transactions, it gives more hints to the later development of advanced transaction models that have been proposed since the mid 1980s, i.e., distributed transactions, nested transactions, chained transactions, etc. The outermost transaction of such a hierarchy is typically referred to as the top-level transaction. Nested transactions allow the construction of modular applications—the builder of an object can use transactions within its methods if those methods need to be transactional, and if they are subsequently invoked from within another transaction they will simply be nested. Thus, nested transactions could be concurrent. Distributed transactions consist of sub-transactions that may access multiple local database systems. In contrast to the distributed transactions, nested transactions adopt a top-down method to decompose a complex transaction into sub-transactions or child transactions according to their functionalities.

It is observed that there are two strategies adopted to extend the simple ACID transactions. One is to modularize a complex transaction with hierarchies. By this means, a big transaction is divided into smaller components, which can in turn be decomposed. This strategy has been applied in various transactions including distributed transactions, nested transactions, multilevel transactions, and open nested transactions. Another strategy is applied in chained transactions etc through decomposing a long-lasting transaction into shorter sub-transactions. By means of splitting the long processing time, each transaction can be divided into a sequential series of smaller components that are operated in a shorter time, thus minimizing the work lost during a clash.

10.4 Data Warehousing

Data provides no real value to business users if it is located on several different systems, in different formats and structures, and redundantly stored. Many IT organizations are increasingly adopting data warehousing as a way of improving their data management capabilities and relationships with corporate users. Proponents of data warehousing technology claim that the technology will contribute immensely to a company's strategic advantage. The origins of the data warehouse as a subject-oriented collection of data that supports managerial decision making was demand driven. A data warehouse provides a single, a more quickly accessible, and a more accurately consolidated image of business reality. It lets organizational decision-makers monitor and compare current and past operations, rationally forecast future operations, and devise new business processes. A data warehouse (DW) is best understood through its distinguishing characteristics of subject orientation, integration, time variance, and non-volatility. This is where the data warehouse comes into play as a source of consolidated and cleansed data that facilitates analysis.

10.5 Business Intelligence and Data Mining

Business intelligence (BI) systems are a range of new information technology (IT) tools that lets the ordinary office workers and managers—rather than IT professionals—access their companies' databases. The use of business intelligence is rapidly extending to other areas of business including e-commerce. For example, websites generate gigabytes of data a day that describe every action made by every visitor to the site. No bricks and mortar retailer carries the same level of detail about how visitors browse the offerings, the routes they take, and even where they abandon transactions. For example, a manufacturer of computer hard disk drives with annual sales of more than 3 billion uses BI to better manage its inventory, supply chains, product lifecycles, and customer relationships. Business intelligence enabled the company to reduce operating costs by 50 per cent.

Knowledge is the intellectual capital that an organization possesses. It is much more than data, as it includes the experience and expertise found within an organization. Information is generally objective, whereas knowledge includes elements of interpretation and understanding. Technological developments have prompted an explosion in the scope and depth of knowledge to which decision-makers have access. However, there is now so much information and knowledge available that what sets successful organizations apart is their ability to develop and use them creatively. Knowledge and information have to be collected, protected, and effectively and intelligently managed if they are to be valuable resources that guide and inform every stage of decision-making.

The development of DW and decision support systems (DSS) allowed the manipulation of data coming from heterogeneous sources and supported multiple-level dynamic and summarizing data analysis. Data mining is closely related to knowledge discovery in databases (KDD) and quite often these two processes are considered equivalent. Data mining is the process of exploration and analysis, by automatic or semi-automatic means, of large quantities of data in order to discover meaningful patterns and rules. Data mining is often defined as a set of mathematical models and data manipulation techniques that perform functions aiming at the discovery of new knowledge in databases. The functions, or tasks, performed by these techniques can be classified in terms of either the analytical function they entail or their implementation focus. In essence, data mining is the application of statistical methodologies to analysing data for addressing business problems. While marketing research allows for opportunities to be identified at a macro level, data mining enables us to discover granular opportunities that are not immediately obvious and can only be detected through statistical techniques and advanced analytics. Businesses, on the other hand, define data mining in terms of the application. For example, in banking and finance one speaks about credit scoring and risk analysis, and in marketing applications, data mining is described as the tool for modelling customer behaviour, churn analysis, customer acquisition, and cross-selling.

Knowledge discovery in databases is the process of extracting interesting, non-trivial, implicit, previously unknown and potentially useful information or patterns from data in large databases [Fayyad et al. 1996], where a pattern is considered to be interesting if it is easily understood by humans, valid on new or test data with some degree of certainty, potentially useful, novel, or if it validates some hypothesis that a user seeks to confirm.

Data mining can be applied to the vast majority of data organization schemes, most popular of which are the relational databases and data warehouses, as well as various transactional databases. Data mining is also used on a variety of advanced databases and information repositories, such as the object-oriented databases, spatial and temporal databases, text and multimedia databases, heterogeneous and legacy databases, and, finally, genomics databases and the World Wide Web (www) databases. For each database category, appropriate data mining techniques and algorithms have been developed to ensure an optimal result. To facilitate building market and customer intelligence, it is necessary to have integrated database systems that link together data from sales, marketing, customer, research, operations, and finance.

For example, metrics specific skills are also called measurement skills, which in the marketing consulting world refer to the identification and tracking of marketing campaign performance. Metrics skills include hands-on experience in tracking and measuring performance of a wide array of marketing communication channels.

10.5.1 Data Mining Tools

Early computers were designed, mainly, for number crunching. As memory became more affordable, we started collecting data at increasing rates. Data manipulation produced information through an astonishing variety of intelligent systems and applications. As data continued to amass and yield more information, another level of distillation was added to produce knowledge.

Data mining has evolved into a mainstream technology as a data analysis tool. The application of data mining on a dataset may lead to the discovery of a great number of patterns. Different methodologies of data mining attempt to mold the activities the analyst performs in a typical data mining engagement into a set of logical steps or tasks. To date, two major methodologies dominate the practice of data mining: CRISP and SEMMA. CRISP, which stands for cross industry standard process for data mining, is an initiative by a consortium of software vendors and industry users of data mining technology to standardize the data mining process. The original CRISP documents can be found on http://www.crisp-dm.org/. On the other hand, SEMMA, which stands for sample, explore, modify, model, assess, has been championed by SAS Institute. SAS has launched a data mining software platform (SAS Enterprise Miner) that implements SEMMA.

Data mining is often defined as a set of mathematical models and data manipulation techniques that perform functions aiming at the discovery of new knowledge in databases. The functions or tasks performed by these techniques can be classified in terms of either the analytical function they entail or their implementation focus. Businesses, define data mining in terms of the application. For example, in banking and finance one speaks about credit scoring and risk analysis, and in marketing applications, data mining is described as the tool for modelling customer behaviour, churn analysis, customer acquisition, and cross-selling.

A successful implementation of data mining modelling requires competencies in the following three areas.

- Understanding the problem domain necessitates the full understanding of the objectives of the project, the value added by the engagement, and the expected return on investment (ROI), and how the business processes are being impacted by the implementation of the data mining technology. For example, in credit card risk scoring applications, the analyst must understand the basics of credit card risk management strategies and the basics of the legal as well as the business procedures involved.
- Understanding the data is not limited to the names and descriptions of fields in the database or data warehouse, but also concerns the content of the fields, the meaning of each category, the meaning of outliers, missing values, any preprocessing that has been done on the data, and the sources of the data.

- Data mining modelling methods and software area of competency cover the methodology of data mining, the strengths and limitations of each data mining technique, and the modelling software. The analyst should know which technique to use with which dataset, and when.

Regression Models

Regression models are the oldest and most used models in data mining. Most data mining software packages include regression capabilities. They are also the most understood models in terms of performance, mathematics, and diagnostic measures for model quality and goodness of fit. Regression models have been applied to a very wide range of problems in finance, engineering, environment, medicine, agriculture, economics, military, and marketing, to mention a few.

Cluster Analysis

Cluster analysis is concerned with finding subgroups of the population that 'belong' together. In other words, we are looking for islands of simplicity in the data. Clustering techniques work by calculating a multivariate distance measure between observations. Observations that are close to each other are then grouped in one cluster. Clustering algorithms belong to three broad categories: agglomeration methods, divisive methods, and k-clustering. Because all cluster analysis methods rely on the calculation of some distance measure, all variables must be numeric.

A full discussion of data and web mining is beyond the scope of this book, and the reader is referred to references at the end of this chapter for further information.

10.6 Content Management

Enterprise content management (ECM) is the fastest growing category of enterprise software. Customers who are implementing or upgrading ECM systems are facing issues such as vendor lock-in, high maintenance costs, and lack of standardization. Open-source technologies and open standards are becoming powerful alternatives to commercial closed-source ECM software.

A content repository is a server or a set of services used to store, search, access, and control content. The content repository provides these services to specialist content applications such as document management, web content management systems, image storage and retrieval systems, records management, or other applications that require the storage and retrieval of large amounts of content. What distinguishes content management from other typical database applications is the level of control exercised over individual content objects and the ability to search content.

Content management systems (CMSs) systems will continue to grow and evolve as more commercial products become available at costs that organizations can afford, and as more open-source systems gain adherents. There are numerous CMSs existing in the market today, each offering its own specialized features. Documentum, Broadvision, Ektron products, Vignette Content products, and Interwoven product suite are some of the available content management systems in the market today.

10.6.1 Need for Content Management

A CMS offers a way to manage large amounts of web-based information that escapes the burden of coding all of the information into each page in HTML by hand. It is used to collect, manage, and publish content, storing the content either as components or whole documents, while maintaining dynamic links between components. A good CMS allows the content authors to create content in the form of articles through some predefined templates.

Organizations have spent billions on technology but have neglected the necessary strategy for identifying, organizing, and accessing needed information. A CMS allows the author to be concerned only about the core content and not about its look-and-feel and formatting, thus saving loads of time and pain. Some content management systems also optionally require the author to enter metadata for content, for example creator name, keywords, etc. so that these can be associated with the content and be used for indexing and searching the website. Content management systems are important to libraries because the sheer mass of their public Web presence has reached the point where maintenance is a problem. In many ways, content management is the Achilles' heel of the IT practitioner.

The author simply specifies the sequence of approvers to get the article approved and the automatic workflow does the rest of the work. It ensures that the content does not get published to the website until and unless the sequence of editors and approvers approve it via the automated workflow.

10.6.2 Portals

A portal is a website that aggregates contextually relevant information and services. In short, a portal distills knowledge from data. The right portal transforms how effectively a corporation conducts its business. Portals allow multiple layers of security. Content resources are abstracted from page markup. User roles restrict general access. Portals provide a mechanism that supports personalization. Personalization is blended from a mix of user interface (UI) preferences and programmatic rules. Portals facilitate application integration by interconnecting systems through data sharing and automated transactions. Portals allow information workers to create content once, reuse that content, and gather content from disparate sources to display within a single interface. In the infrastructure portals, users can create a unified interface to work with contents.

A portal page is made of a set of portlets. The portal page such as Welcome contains portlets such as Navigation, Language, and so on. Loosely speaking, portlets are fragments of an hypertext markup language (HTML) page, that is pieces of markup such as HTML, XHTML, wireless markup language (WML), and so on. The content of a portlet is normally aggregated with the content of other portlets to form the portal page. The life cycle of a portlet is managed by the portlet container. The content generated by a portlet may vary from one user to another, depending on the user configuration for the portlet.

Italian Car Maker's Need for a State-of-the-Art Website

Since 1947, Ferrari S.p.A., an Italian company based in Maranello, Italy, has manufactured race cars and street-legal vehicles that redefine performance and luxury. The company has also long participated in Formula One racing, with consistent success. Ferrari felt the need for on the fly updation of web content.

Ferrari opted for Share point 2007 after evaluating alternatives. Office SharePoint Server 2007 provided Ferrari the flexibility to publish a variety of content, from text-based pages to highly interactive experiences. It was found that with the new site, users can go on a virtual tour of a Ferrari plant or gain exposure to specific racing experiences. 'A fan can compare the lap times in a circuit 20 years ago with the times of the drivers of today, how a specific corner was raced over time, braking time, G-forces, and more,' says Castello. Ferrari is also encouraging fan interaction through discussion forums on its site, which are built using the blog capabilities of Office SharePoint Server 2007.

Using Office SharePoint Server 2007, Ferrari has enabled business users to manage site content independently. The company is reaching a broader audience with a more compelling Web experience. Statistics show a significant increase in site traffic since deploying the new site.

Source: http://www.microsoft.com/casestudies/, accessed in June 2009.

10.6.3 Open Source Content Management System

When the World Wide Web was born, creating even the simplest web page required learning the language of the Web—HTML. Since then, great strides in the power of web authoring software have been made with the availability of professional web editors such as Adobe Dreamweaver and Microsoft FrontPage. These types of editors have made the creation and maintenance of a website much easier by providing a graphical user interface for web construction and minimizing the amount of HTML coding required by the webmaster.

Infrastructure portals integrate all possible functions stated above. It covers collaboration and information sharing within a company, collaborative tools, content management and web publishing.

Alfresco

Alfresco, a new player in this market, is gaining a lot of momentum by providing content management solutions to enterprises using state-of-the-art open-source technologies and open standards.

Alfresco was founded in 2005 by John Newton, co-founder of Documentum, and John Powell, former chief operating officer (COO) of Business Objects. Its investors include the leading investment firms Accel Partners and Mayfield Fund. Alfresco is the leading open-source alternative for enterprise content management. It couples the innovation of open source with the stability of a true enterprise-class platform. The open-source model allows Alfresco to use the best-of-breed open-source technologies and contributions from the open-source community to get higher quality software produced more quickly and at a much lower cost.

The architecture of Alfresco is based on open standards, hence the applications built using Alfresco can be deployed on any environment such as Windows, Linux, Mac, etc., can use any relational database such as MySQL, Oracle, etc., can run on various application servers such as JBoss Application Server, Apache Tomcat, etc., can work with any Browser such as Mozilla Firefox, Microsoft Internet Explorer, etc., and can integrate with any portal such as JBoss Portal, Liferay Portal, etc.

Alfresco's document life cycle management features ensure that people in various company departments, divisions, and regions can work together to support all processes relating to a contract throughout its life cycle. Alfresco records management features provide a secure, auditable environment for creating, declaring, classifying, retaining, and destroying records. Organizations can ensure compliance by defining and enforcing policies for records use, storage, and disposition, with a legally defensible audit trail. The life cycle determines the disposition of the record including when the records will be cut off or grouped together, how long the records will be held, and what happens to the records after the hold period expires.

Alfresco provides an end-to-end solution by collecting paper documents and forms; transforming them into accurate, retrievable information; and delivering the content into an organization's business applications. The information is then full-text searchable and goes through various life cycles based on the organization's defined business process management. E-mails are considered as records in some organizations. Alfresco enables you to drag and drop emails from Microsoft Outlook into the file plan space. The system will extract the meta data from e-mail files and populate information such as whom the e-mail is from, who the recipients are, and the subject of the e-mail. E-mail content is stored in a secure and scalable repository and is full-text searchable. The Alfresco repository provides a set of reusable cross-cutting content management services such as content storage, query, versioning, and transformation, which may be utilized by one or more applications.

Mambo

Mambo began life as an internal CMS product created by engineers at the Miro Corporation of Australia. In April 2001, Mambo was initially released to the open source community. For its time, Mambo was an amazingly advanced CMS application to be freely available with full source code.

10.6.4 Commercial Tools

The Documentum product suite is an immensely vast sea and describing the complete set of offerings from Documentum within a single section would be unreasonable. Documentum provides ECM solutions enabling diversified organizations to integrate their distributed content and related business processes on a single platform, thus uniting teams to collaboratively create, manage, process, and deliver their unstructured content. Documentum's clientele includes several big organizations that are successfully utilizing its widespread capabilities in expanding their core business by reducing their operating costs, deriving better ROIs, and achieving increased customer satisfaction by delivering just in time.

The Content Server forms the heart of Documentum, providing essential services to create, version, manage, and archive content and objects in the Documentum system. Content Server (earlier known as 'eContent Server) houses a repository, which Documentum terms a 'Docbase', to store the various content and its associated properties (metadata). Whenever a client wants to make a connection with the server, an object called as DocBroker acts as a bridge or an intermediary. DocBrokers are termed connection brokers in Documentum release 5.3.

Folders within which documents are stored are objects, documents created are themselves objects, workflows used to get the documents reviewed are objects, and in fact the users creating the documents are also objects. Like any good CMS, Documentum internally manages multiple versions of the same document and maintains a history of all updates that have gone in since the initial creation of the document. Versioning is an automatic feature provided by the Content Server through version labels.

Content Server provides life cycles to automate the various stages in the life of a document. Simply speaking, a lifecycle is a sequence of states that describe the various stages in the life of an object. Documentum stores lifecycles in the form of a policy object (dm_policy) in the Docbase. Workflows consist of numerous activities, each comprising various tasks to be performed. Users or designated automated scripts carry out the tasks and pass over the document in question to the subsequent activity. What a workflow does, in essence, is routing the content through the various stages of its lifecycle via different users. Each user receives the designated task in one's Inbox and may also receive an e-mail notification for the same. Documentum stores workflow definitions in the form of business process objects (dm_process) in the Docbase.

Web Development Kit (WDK) is a framework provided by Documentum, on which one can build a web application for talking to the Content Server. The WDK architecture consists of the following.

- Presentation model: Incorporates JSP tag libraries to separate the UI from business logic.
- Component model: Incorporates server-side components that extend the Content Server functionality

Web Publisher is a browser-based web application provided by Documentum for content creation, review, and finally publishing to websites. In the Documentum architecture, the Web Publisher utilizes the content management services from Content Server, publishing services. The Web

Publisher defines its own specific groups in order to divide the responsibilities of the participating entities.

10.6.5 Microsoft Content Management Server

Microsoft Content Management Server (MCMS) is a part of Microsoft's integrated portal technologies, that is, products providing a comprehensive Web Services framework. It enables companies to rapidly develop, deploy, and maintain content-rich, highly volatile websites. It provides tools to implement and administer both the production and development environments. The MCMS content-management strategy hinges upon empowering a community of workers to author content, schedule updates, and administer a site, on its own—all while maintaining consistent quality and accessibility. Microsoft Content Management Server provides tools for organizing and automating dynamic content delivery.

A full discussion of content management systems is beyond the scope of this book, and the reader is referred to references at the end of this chapter for further information.

SUMMARY

Data management is a broad field of study, but essentially is the process of managing data as a resource that is valuable to an organization or business. Data modelling means first creating a structure for the data that you collect and use and then organizing this data in a way that is easily accessible and efficient to store and pull the data for reports and analysis. Transaction processing has been an important software technology for 40 years. Large enterprises in transportation, finance, retail, telecommunications, manufacturing, government, and the military are utterly dependent on transaction processing applications for electronic reservations, banking, stock exchanges, order processing, music and video services, shipment tracking, government services, telephone switching, inventory control, and command and control.

In this chapter we discussed what CMSs are, along with their benefits and drawbacks. We saw how content management systems can serve as a boon to the organization by reducing dependence on IT staff for maintaining business websites and by providing a secured, streamlined, and automated mechanism for entering and publishing data to the websites. On the other hand, content management systems do entail training overheads to get the business users familiar with using the automated systems.

Alfresco is the leading open-source alternative for ECM. It couples the innovation of open source with the stability of a true enterprise-class platform. The open-source model allows Alfresco to use best-of-breed open-source technologies and contributions from the open-source community to get higher quality software produced more quickly at much lower cost. Alfresco provides key features for a scalable, robust, and secure CMS to deliver trusted and relevant content to your customers, suppliers, and employees.

Documentum is an enterprise content management system that helps organizations integrate their unstructured content on a single platform. A Docbase can be thought of a logical repository, consisting of content files and their associated metadata. A DocApp, on the other hand is a packaging unit of a Docbase, consisting of objects from the Docbase. Web Development Kit is a framework provided by Documentum, on which one can build a web application for talking to the Content Server.

Key Terms

Access control It is the system of privileges and permissions that secures content and identifies who can read, create, modify, and delete content.

Access point device It is the device that bridges wireless networking components and a wired network.

Content management It describes a process that allows people to easily create and update content, especially on their websites

Content It is the intellectual capital of an organization. It is information separated from its presentation.

Database A database is a collection of records, or pieces of information. A database contains multiple tables. Tables contain records (rows) and fields (columns). For example, in a table containing personal information on people, each row might represent a person, while columns would be used to store the person's name, address, phone number, etc.

Document management systems (DMS) These are designed to assist organizations to manage the creation and flow of documents through the provision of a centralized repository, and workflow that encapsulates business rules and metadata.

Knowledge management It describes an organization's strategy to get the knowledge of its workers out of their heads and into an information storage and retrieval system where it can be used and reused.

ODBC ODBC (pronounced as separate letters) is the acronym for open database connectivity. ODBC is a standard database access method, which was developed by the SQL Access Group in 1992.

Review Questions

10.1 What is the difference between database and database management system.

10.2 Write a short note on MS Access as a DBMS package.

10.3 Differentiate between DBMS and RDBMS.

10.4 What is the need for content management in an organization?

10.5 Differentiate between open source content management and proprietary content management.

10.6 List a few open source content management solutions available in the open source community.

10.7 Compare any two open source CMS solutions available in the open source community.

10.8 Discuss a commercial CMS solution.

Projects

10.1 Visit the nearest IT organization and understand the CMS solutions deployed within the organization.

10.2 The office of a small independent realty company has four real estate agents, an executive secretary, and two assistant secretaries. They agreed to keep all common documents in a central location to which everyone has access. They also need a place to store their individual listings so that only the agent who has the listing should have access to. The executive secretary has considerable network administration knowledge, but everyone else does not. It is expected that in the near future the office will employ 100 more agents and three assistant secretaries and is planning to open branches across the country to enable better business. The organization is interested to manage a centralized content repository. Why would an organization need data management and what kind of a data management solution should the orgazation opt for. Advice the organization between open source and commercial content management solutions.

References

Briand, François-Paul and Michael Wirsching 2005, *Microsoft Content Management Server Field Guide*, Apress.

Elmasri, Ramez Shamkant Navathe 1999, *Fundamentals of Database Systems*, Addison Wesley, Delhi.

Kathuria, Gaurav 2006, *Documentum Set up, Design, Develop, and Deploy Documentum Applications*, Packt Publishing, UK.

Raisinghani, Mahesh 2004, *Business Intelligence in Digital Economy*, Idea Group Publishing, USA.

SAS 2006, *SAS Intelligence Platform: Overview, Second Edition*, SAS Press, USA.

Sharif, Munwar 2006, *Alfresco Enterprise Content Management Implementation*, Packt Publishing, UK.

CHAPTER

11 Web Tools

The dissemination of information is one of the cornerstones of modern civilization.
–John F. Budd

> **Learning Objectives**
>
> After reading this chapter, you should be able to understand:
>
> - the concept and need of web browsers
> - the differences between the types of browsers
> - the web authoring tools

11.1 Web Servers and Browsers

Web-based systems and applications now deliver a complex array of varied content and functionality to a large number of heterogeneous users. The interaction between a web system and its back-end information systems has also become more tight and complex. As we now increasingly depend on web-based systems and applications, their performance, reliability, and quality have become of paramount importance, and the expectations of and demands placed on web applications have increased significantly over the years. As a result, the design, development, deployment, and maintenance of web-based systems have become more complex and difficult to manage. web engineering uses scientific, engineering, and management principles and systematic approaches to successfully develop, deploy, and maintain high-quality Web systems and applications. The essence of web engineering is to successfully manage the diversity and complexity of web application development, and hence, avoid potential failures that could have serious implications.

The scope and complexity of Web applications vary widely—from small-scale, short-lived (a few weeks) applications to large-scale enterprise applications distributed across the Internet, as well as via corporate intranets and extranets. Web applications now offer vastly varied functionality and have different characteristics and requirements.

11.1.1 Deceptive Website

Web applications are not just web pages, as they may seem to a casual user. The complexity of many web-based systems is often deceptive and is not often recognized by many stakeholders. Many consider web development primarily an authoring work (content/page creation and presentation), rather than application development. Several attributes of quality web-based systems such as usability, navigation, accessibility, scalability, maintainability, compatibility and interoperability, and security and reliability often are not given the due consideration they deserve during development. Many web applications also fail to address cultural or regional considerations, privacy, moral and legal obligations and requirements. Most web systems also lack proper testing, evaluation, and documentation.

Web applications constantly evolve. In many cases, it is not possible to fully specify what a website should or will contain at the start of the development process, because its structure and functionality evolve over time, especially after the system is put into use. Web applications are inherently different from software. The content, which may include text, graphics, images, audio, and/or video, is integrated with procedural processing.

11.1.2 Web Servers

Web server is a software application that listens for client connections on a specific network port. When a connection is made, the web server then waits for a request from the client application. The client is usually a web browser, but it could also be a website indexing utility, or perhaps an interactive telnet session. The resource request, usually a request to send the contents of a file stored on the server, is always phrased in some version of the hypertext transfer protocol (HTTP).

In the early days of the Web, the National Center for Supercomputing Applications (NCSA) created a web server that became the number one web server in early 1995. However, the primary developer of the NCSA web server left NCSA about the same time, and the server project began to stall. In the meantime, people who were using the NCSA web server began to exchange their own patches for the server and soon realized that a forum to manage the patches was necessary. Thus, the Apache Group was born. The group used the NCSA web server code and gave birth to a new web server called Apache. The Apache server was assembled and released in early 1995 as a set of patches to the NCSA httpd 1.3 web server. Numerous individual programmers, loosely bound into a consortium initially called the Apache Group, contributed the original source code patches that made up the first Apache server. Less than a year after the Apache group was formed, the Apache server replaced NCSA Hypertext transfer protocol daemon (httpd) as the most-used web server.

The Apache Group has expanded and incorporated as a non-profit group. The group operates entirely via the Internet. However, the development of the Apache server is not limited in any way by the group. Anyone who has the know-how to participate in the development of the server or its component modules is welcome to do so, although the group is the final authority on what gets included in the standard distribution of the Apache Web server. This allows literally thousands of developers around the world to come up with new features, bug fixes, ports to new platforms, and more. When a new code is submitted to the Apache Group, the group members investigate the details, perform tests, and do quality control checks. If they are satisfied, the code is integrated into the main Apache distribution.

Apache offers many other features including fancy directory indexing; directory aliasing; content negotiations; configurable HTTP error reporting. The Apache server does not come with a graphical user interface for administrators. It comes with single primary configuration file called httpd.conf that you can use to configure Apache to your liking. All you need is your favorite text editor. However, it is flexible enough to allow you to spread out your virtual host configuration in multiple files so that a single httpd.conf does not become too cumbersome to manage with many virtual server configurations.

The first major change in Apache 2.0 is the introduction of multiprocessing modules (MPMs). Each MPM is responsible for starting the server processes and for servicing requests via child processes or threads depending on the MPM implementation. The main server configuration applies to the default website Apache serves. This is the site that will come up when you run Apache and use the server's Internet protocol (IP) address or host name on a web browser.

The official Apache website is www.apache.org. This site contains the latest stable version of Apache, the latest release version of Apache, patches, contributed Apache modules, and so on. Apache runs on just about any platform in use today, including Linux, FreeBSD, OpenBSD, NetBSD, BSDI, Amiga OS 3.x, Mac OS X, SunOS, Solaris, IRIX, HPUX, Digital Unix, UnixWare, AIX, SCO, ReliantUNIX, DGUX, OpenStep/Mach, DYNIX/ptx, BeOS, and Windows.

11.1.3 Web Browsers

A browser is an application program that provides a way to look at and interact with all the information on the World Wide Web. A web browser is a software application for retrieving, presenting, and traversing information resources on the World Wide Web. Technically, a web browser is a client program that uses HTTP to make requests of web servers throughout the Internet on behalf of the browser user. An information resource is identified by a Uniform Resource Identifier (URI) and may be a web page, image, video, or other piece of content. The first web browser, called the World Wide Web, was created in 1990.

Browsing hypertext has not changed much since the Web became mainstream in the early 1990s. Essentially you scroll through a page and click on a link and jump to a new page.

Microsoft released its web-browsing software, Internet Explorer, as a part of Windows 95, and its upgrade with Windows 98. The fact that Internet Explorer was offered free to purchasers of Windows products made it difficult for Netscape and other browser producers to sell their products.

11.1.4 Internet Explorer

The original IE 1.0 browser code was licensed from Spyglass. Windows Internet Explorer (formerly Microsoft Internet Explorer, abbreviated to MSIE or, more commonly, IE), is a series of graphical web browsers developed by Microsoft and included as a part of the Microsoft Windows line of operating systems starting in 1995. Most businesses have two main uses for the web browser: as the tool customers use to browse websites and as a piece of software running on PCs.

11.1.5 FireFox

Mozilla Firefox (henceforth just Firefox) began life as an experimental modification of the Mozilla Navigator component. The Mozilla Foundation's intention is to make the Suite obsolete, replacing it with Firefox and its flagship e-mail client, Mozilla Thunderbird. Firefox is a web browser that is smaller and faster. Firefox gives users a cleaner interface and faster download speeds. Firefox includes most of the features with which users of other browsers are familiar.

Tabbed browsing has been called the best thing since sliced bread and the biggest fundamental improvement in web browsing in years. It has also been criticized, rejected, and labelled a useless non-innovation. Tabbed browsing is not that new an idea; in fact, it was not invented by Firefox. Probably, NetCaptor, a third-party program that provides an alternative tabbed browsing interface for IE, pioneered this approach. The Mozilla Application Suite followed hot on the heels of that tool, and more recently, Opera2 has done the same. Firefox, being a derivative of the Mozilla Application Suite, naturally inherited its tabbed browsing capability. Currently, most graphical browsers support tabbed browsing. A quick (and incomplete) list of tab-enabled browsers reads: Firefox, Opera, Mozilla, Safari, Konqueror, Netscape, OmniWeb, and Camino. Tabbed browsing does not prevent the use of multiple windows; in fact, you can opt to use a browser that supports tabbed browsing in the same way as you use IE. Tabbed browsing suffers the same page title truncation problem as the taskbar, although a few more letters show in the given tab space. But, with the same number of windows open, it is still hard to work out which page is the one a user wants. Another redeeming feature of tabbed browsing is that this system allows you to juggle more web pages than would be possible using multiple separate windows. Tabbed browsing reduces those multiple windows into individual tabs in a single window. Smart Keywords are a great way to perform search queries from the location bar. Firefox comes with several Smart Keywords automatically installed.

Firefox comes with a download manager that displays all of your downloads in one place. Instead of displaying a download dialog for every single download, as IE does, Firefox gathers your downloads together in a single location where you can track their progress without having to contend with multiple windows. Finally, the Clean Up button is used to remove completed and cancelled entries from the download manager, which helps to keep the list of entries in the download manager to a same limit.

11.2 Web Authoring Tools

Authoring tools can enable, encourage, and assist users (authors) in the creation of accessible Web content through prompts, alerts, checking and repair functions, help files and automated tools. It is important for all people to have access to content. The tools used to create this information must, therefore, be accessible themselves. Adoption of these guidelines will contribute to the proliferation of Web content that can be read by a broader range of readers and authoring tools that can be used by a broader range of authors.

11.2.1 SGML

Standard generalized markup language (SGML) is one of the first standardized markup languages. It is governed by the International Organization for Standardization (ISO) 8879, developed in 1986, and it is widely used in industry and commerce for large documentation projects. Most of today's markup languages, including extensible markup language (XML), have descended from SGML. Markups are notations in a document that are not content. Markup languages provide a set of conventions that can be used for encoding texts. Markups for electronic documents perform a similar function by providing information about content, format, printing, and processing of a document. Markups can be of two types: procedural and descriptive. Procedural markups specify how to process the text and primarily deal with the formatting and presentation of the document. Since procedural markups do not contain information regarding content and are proprietary, they cannot be communicated between software packages and operating systems. Descriptive markups, on the other hand, contain information about the logical structure of text and content in the document. The basic premise behind descriptive markups is to keep content separate from the style of the document. Descriptive markups can help identify elements of the document structure, such as chapter, section, or a table of contents. These markups can also be used for presentation of content in different data formats such as HTML, portable document format (PDF), relational data tables, etc.

Authoring tool refers to the wide range of software used for creating Web content, including

- editing tools specifically designed to produce Web content,
- tools that offer the option of saving material in a Web format,
- tools that transform documents into Web formats,
- tools that produce multimedia, especially where it is intended for use on the Web,
- tools for site management or site publication, including tools that automatically generate websites dynamically from a database, on-the-fly conversion tools, and website publishing tools, and
- tools for management of layout.

11.2.2 Hypertext Markup Language

Hypertext markup language (HTML) was originally designed to be a content description language. Appearance and layout were secondary concerns, especially in the early days when text-based browsers, such as Lynx, were the only way to access the Web. Hypertext markup language specifies the rules for communication between browsers and web servers. The HTTP requests are sent as ASCII text and there are several keywords that permit different types of actions. The HTTP protocol defines the behaviour expected of the client (browser) and server components of an HTTP connection. A browser can be written only if it knows what to expect from the servers it connects to, and that behaviour is defined by the protocol specification (HTTP).

The HTML language is like other programming languages in that there are special words or phrases that are used to specify actions and variables. It uses tags to provide structure for the document and to indicate actions. Tags that specify the start of formatting instructions or an action appear as a name surrounded by the less than and greater than symbols. With the development of graphical browsers, however, HTML became increasingly cluttered with page layout and descriptive tags that did nothing to describe content. At the top of the document, the head tags identify the header portion of the HTML code. The body tags identify the section of the HTML document that is the body.

A simple HTML page is as shown below.

```
<html> <head>

<title>Sample HTML Document</title> </head>

<body> <p> \\
```

```
This is sample text on a sample page. Text can be
<b>bold</b> or <i>italic</i> or plain. </p> </body>

</html>
```

- The <html> tag tells the browser what kind of document it has encountered.
- The <head> tags contain basic information about the page. For example, the document title appears here. (The text of the title appears between its own <title> tags so a browser can find and display it.) If this were a more complex page, the head might also include some style information or even a script that animates text or pictures.
- The <body> tags surround the star of the show—the content of your page. Everything between these tags is what viewers will see in their browsers.
- <p> indicates the beginning of a paragraph and </p> the end of the paragraph.

The HTML codes, also referred to as HTML tags, are enclosed by the lesser than (<) and greater than (>) brackets and may be written in capital or lower case letters. The opening bracket is followed by an element, which is a browser command, and ends with the closing bracket.

Attributes are only contained in the opening HTML tags to the right of the element and are separated by a space and followed by an equal (=) sign. The value follows the equal sign and is enclosed in quotes <font size=2>.

At the heart of the design of the Web is the concept of the hyperlink. The clickable links on a Web page can point to resources located anywhere in the world. hypertext markup language uses concepts of hyperlinks, hypertexts, and tags to browse files on the Internet. A link is a special word or phrase in a web page that 'points' to another web page. When you click one of these links, your browser transports you immediately to the other page, no questions asked. Because these hypertext links are really the distinguishing feature of the World Wide Web, web pages are often known as hypertext documents. A hyperlink embedded in HTML-formatted page is only one way to use a uniform resource locator (URL); but it is the hyperlink that gave rise to the Web. A URL is a means of identifying a resource that is accessible through the Internet. Although the distinction is academic, a URL is a special case of a URI that is understood by web servers. Each URL is composed of three parts, a mechanism (or protocol) for retrieving the resource, the hostname of a server that can provide the resource, and a name for the resource. Markup is a detailed stylistic instructions written on a manuscript that is to be typeset and published on the World Wide Web. You can set up three kinds of links: links to your other web pages, links to a different location in the same web page, and links to any page anywhere on the Web.

The HTML can only be used to define and deliver simple report-style documents—lists, tables, headings, and some hypertext and multimedia. This limitation is due to the fixed and predefined set of markups or tags used by HTML. These tags are used for primarily formatting documents, and support fixed and simple document structure. The HTML provides static definitions of these documents and does not provide means to identify data, resulting in limited reuse and interchange of HTML documents.

What You See Is What You Get editors (WYSIWGs) provides the ability to edit web content, including Blog content. Less technical persons can use WYSIWGs without sifting through complex code. It is easier than ever to create a website with an HTML editor, as software developers continue to add tools that let you develop advanced features with style. Today's web authoring tools can provide the power to build an interactive, animated, state-of-the-art website suitable for anything from a personal web page to a midsize business site. New web designers do not need to know HTML to create discussion groups, pop-up windows, navigation bars, animated page transitions, dynamic HTML, or a dozen other advanced features in order to integrate them into a site with an elegant and consistent design.

Know Your Firm—America Online

America Online (AOL) was originally started as a service provided by a company called Quantum in 1989. Quantum changed its name in 1991 to America Online, offering PC users Internet access, e-mail, and an array of information and services. By 1995 America Online tapped into the Internet, allowing its users access to unlimited information as well. That same year the company introduced Global Network Navigator, a service solely geared toward the Internet. America Online also signed an exclusive marketing deal with Intuit, inventor of Quicken Financial Software. In May

Contd

> 1998, America Online announced it would acquire NetChannel, Inc. a Web-enhanced television company. The company claimed the acquisition would further its 'AOL Anywhere' strategy of making the AOL brand available on all emerging interactive platforms. Other attempts by America Online to increase customers and revenue have included joint efforts with big name rivals like Microsoft and AT&T. America Online agreed to boost Microsoft's Internet browser software in exchange for packaging America Online's software with every copy of Windows 95.
>
> - 1989: Founded as Quantum
> - 1991: Changes name to America Online
> - 1992: Stephen Case becomes CEO
> - 1995: Introduces Global Network Navigator
> - 1996: Reduces rates and gains new subscribers
> - 1998: Acquires NetChannel
>
> *Source*: www.aol.com

11.2.3 Cookies

Connections made using HTTP are called 'stateless', which means that after the user's computer receives the content of a requested page, the connection between the computer and the faraway web server is closed. The benefit of a stateless connection is quite simple. It enables one machine to serve a much higher volume of data. The downside to a stateless connection is that on occasion it might be helpful for a server to remember who you are. Cookies contain a piece of data that allows the remote Web server to recognize a unique connection as having a relationship to another unique connection. In short, the cookie makes sure that the server can remember a visitor through many steps in a visit or even when time has passed between visits. As a basic security measure, it should be noted that cookies are designed to be read only by a server within the same domain that created it.

Cookies enable myriad helpful features, such as the ability to personalize a website with the user's choice of colours, or language, or stock symbols on a stock ticker. It also enables features such as shopping carts on e-commerce websites, permitting the user to select multiple items over the course of a long visit and have them queued for purchase at the end of a visit. A special type of cookie, called a session cookie, is set to be automatically deleted after a relatively short period of time, usually within about 10 minutes after a user leaves a site. This type of cookie is typically used for remembering information over a short duration, such as what you may have stored in a shopping cart.

11.2.4 XML

Extensible markup language (XML) is a way of organizing and managing information, a constellation of supplementary technologies, and a paradigm for information handling for the Internet age. It is not a language; it is actually a set of syntax rules for creating semantically rich markup languages in a particular domain. The XML document is a basic unit of information and consists of content and markups, and follows the rules of XML. Extensible markup language elements include everything from the beginning tag to the ending tag and things in between. It requires strict adherence to the syntax.

Thus, a markup language is a set of words, or marks, that surround or tag, a portion of a document's content in order to attach additional meaning to the tagged content. The mechanism invented to mark content was to enclose each word of the language's vocabulary in a less-than sign and a greater-than sign. Containing content is achieved by wrapping the target content with a start and end tag. Thus, each vocabulary word in our markup language can be expressed in one of three ways: a start tag, an end tag, or an empty tag. The start and end tags are used to demarcate the start and end of the tagged content, respectively. The empty tag is used to embed semantic information that does not surround content. The XML specification defined two levels of conformance for XML documents: well-formed and valid. Well-formedness is mandatory, while validity is optional. A well-formed XML document complies with all the W3C syntax rules of XML like naming, nesting, and attribute quoting. A valid XML document references and satisfies a schema. A schema is a separate document whose purpose is to define the legal elements, attributes, and structure of an XML instance document. An XML document that conforms to these rules is called well formed.

Extensible markup language documents can be validated. The two main tools provided by XML to validate documents are called document type definition (DTD) and XML Schema. The XML documents with DTDs generally consist of two parts: the first part contains tags and content, and the second part formally describes syntax of the document. The XML Schema is a definition language that enables you to constrain conforming XML documents to a specific vocabulary and a specific hierarchical structure. The XML Schema is analogous to a database schema, which defines the column names and data types in database tables. Thus there are two types of documents: a schema document and multiple instance documents that conform to the schema. A good analogy to remember the difference between these two types of documents is that a schema is a blueprint (or template) of a type and each instance is an incarnation of that template.

A markup language's primary concern is how to add semantic information about the raw content in a document; thus, the vocabulary of a markup language is the external marks to be attached or embedded in a document. This concept of adding marks, or semantic instructions, to a document has been done manually in the text publishing industry for years.

The combination of elements and attributes makes XML well-suited to model both relational and object-oriented data.

11.2.5 Front Page Editor

Microsoft FrontPage is a web authoring tool that gives you everything you need to create and manage the site you want, whether you're creating a personal web page or a corporate Internet or intranet site. It lets you see a graphical map of your site, and you can easily add pages to it from there. It lets you use any combination of text- and graphics-based layouts and navigation tools. Graphic 'themes' optionally apply a consistent look to sites' banners, buttons, text, and background.

11.2.6 Dreamweaver

Dreamweaver is an excellent coding and development tool for new and experienced users alike. It has quickly become the preferred website creation and management program, providing a creative environment for designers (Figure 11.1). Adobe Dreamweaver CS4 allows a user to design, develop, and maintain web pages and websites. Designers and developers both use Dreamweaver CS4, which lets them create and edit content using either a visual layout or a coding environment. Dreamweaver CS4 also provides tight integration with other Adobe products such as Photoshop CS4 and Flash CS4 Professional. Dreamweaver is a comprehensive tool used for site design, layout, and management. Dreamweaver's many icon-driven menus and detailed panels make it easy to insert and format text, images, and media. Dreamweaver does not create graphics from scratch; instead, it is fully integrated with Adobe Photoshop CS4.

Dreamweaver includes site-management tools such as File Check In/Out and Design Notes. Dreamweaver has everything required for complete site management, including built-in file transfer protocol (FTP) capabilities between a server and your local machine, reusable objects (such as page templates and library items), and several safety mechanisms (such as link checkers and site reports) so that you can ensure that your site works well and looks good. Its tools include standard HTML objects such as tables and frames, plus pre-built scripts and behaviours, timeline-based activities, XML support, Cascading Style Sheet (CSS) support, and a JavaScript debugger. For webmasters, if you are designing your pages with CSS, the Browser Compatibility Check and CSS Advisor features will help you to locate and troubleshoot any potential display issues that may occur across different Web browsers. A view if the tool bar is provided.

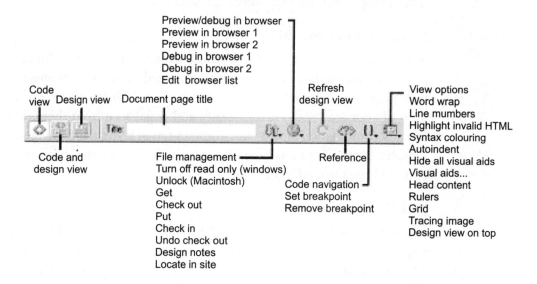

FIGURE 11.1 Dream Weaver—Toolbar

Dreamweaver's popularity is a result of its diversity. Its ability to take a site from conception through to launch—and maintenance afterward—makes it a preferred tool among industry professionals, businesses, and educational institutions. However, it remains easy and accessible enough for novice designers to get up and running quickly.

11.3 Web System

A typical web-based system will include several pages or screens in support of each user interaction or user function. Each of these pages will contain a variety of user interface elements such as hyperlinks, pull-down menus, fill-in forms, checkboxes, and radio buttons. The pages will be arranged in a particular hyperlinked structure, starting from the home page of the application. If possible, navigation should be transparent and effortless, so that the user can focus on the task at hand without being distracted by poorly designed navigation, or content that is difficult to scan visually. Content must simultaneously satisfy two basic criteria: it should be consistently presented and it must be easy for the user to scan and understand. The page designer determines how to partition and present information in a manner that is consistent with the overall visual design (look and feel) of the site.

Websites are highly interactive systems, used by many different kinds of end users. High usability is, therefore, an essential characteristic of websites. Website designers should pay attention to the ease of navigation, information content, download speed, and the availability. The following aspects are important to understand the ease of navigation.

- Provide meaningful hyperlinks, which make it easy for users to navigate around different parts of the website.
- Provide contextual links, for example, back and forward arrows that help users understand where they have previously been and where they can go from their present positions.
- Provide a summary of the website structure, in terms of the various Web pages and their linkages so that users can get an idea of the overall organization of the website.

To take care of the information content:

- Ensure that the information is current and relevant: implement an appropriate website content update strategy.
- Use graphics and multimedia elements appropriately, so as to add to the meaningfulness of the information, not distract the browser.
- Eliminate dead links.

To take care of the download speed:

- Have appropriate infrastructure design and upgrades including database connectivity, adequate wide area network capabilities, and an efficient overall distributed computing infrastructure.
- Use web caching and other algorithms to improve the rate at which the website pages download and the rate of information display between pages.

- Appropriately design the application logic so as to interface efficiently with back end databases.
- Efficiently design the database and information retrieval activities.

Today, more than 1.6 million commercial sites operate in the Web, all in fierce competition to grab the attention of potential buyers. Etailers are potential Web shops. Etailers are finding that it takes enormous marketing efforts to set themselves out from the crowd, inspire web shoppers to visit their sites, and get them to actually make a purchase. Prominent examples include CDnow [Hoffman 2000].

According to Timmers, a value network is a multi-enterprise network of relationships focused on integration of information flows to exploit information and knowledge in the network for strategic business objectives. The second type of value network is called relational value networks which are built through social and spatial proximity and which are based on long-term contracting relationship between firms.

SUMMARY

This chapter presented an overview on the web servers which are used for serving data.

A discussion on web servers with a focus on Apache was presented. Web authoring tools like FrontPage and Dreamweaver have been discussed. The basic building block of the Web is HTML. Hypertext markup language is the predominant markup language for web pages. It provides a means to create structured documents by denoting structural semantics for text such as headings, paragraphs, lists, etc., as well as for links, quotes, and other items.

Website usability is concerned with how easy and intuitive it is for individuals to learn to use and interact with a website, in order to quickly and easily accomplish their tasks on it. Since WWW is a hyperlink-based nonlinear medium and has a wide variety of users with varying levels of comfort with the technology, navigation has emerged as an important factor influencing usability. Information content and organization are two other important characteristics that determine usability. Download delay and website availability are primarily dependent on the underlying technical infrastructure of the website, which includes database design, web design, database connectivity, wide area network speeds, and the overall distributed computing infrastructure.

Key Terms

Browser It is software that will load and display a web page. A browser interprets the HTML or XML code from the web page files, executes embedded scripts and programs, provides encryption/decryption for security where needed, displays graphics (except text-only browsers), plays music and video, and provides links to related pages.

Extensible markup language (XML) It is an open standard for describing and defining data elements on a web page and business-to-business documents. It uses a similar tag structure as HTML. HTML defines how elements are displayed, while XML defines what those elements contain. The HTML uses predefined tags, but XML allows tags to be defined by the developer of the page.

Review Questions

11.1 You need to transfer a sound file between two different systems using file transfer protocol (FTP). Which file type would you specify for the file transfer?

11.2 What is HTTP and what is its purpose?

11.3 What are cookies and what is the need for cookies?

11.4 Handcode a welcome page using HTML standard tags.

11.5 Differentiate between the features of web editors.

11.6 What is XML and what is the need for XML?

11.7 What is a web server? Write a brief note on the Apache web server.

Projects

11.1 Analyse two web stores and submit a report on the use of screen, the colour schemes, the applicability of navigational interfaces, and the various services as available.

11.2 As you are aware, ABC unlimited is a supermarket chain. As a part of automation, the CIO has been asked to evaluate the applicability of a web front end for the store. The web front end will be used as the customer gate way to access the various services offered in the store. The web front end will also support interfaces to talk to the various back end services. The web system should remember the customer and the navigational interface should be well planned. This apart the web front end would go with a pleasing colour scheme.

You are expected to help the CIO build the web front design documentation with projected screen shots. The document should describe the navigational structure and the site map. You should also help the CIO with two attractive colour schemes.

References

Holzschlag, Molly E. 2005, *Spring Into HTML and CSS*.

Hoffman, Donna L. and Thomas.P. Novak 2000, *How to Acquire Customers on the Web*, HBR.

Lowe, D., *Web System Requirements: An Overview*, Requirements Engineering, 8, 102-113.

Menasce, D.A and Almeida, V.A.F. 2002, *Capacity Planning for Web Services: Metrics, Models, and Methods*, Prentice Hall, Upper Saddle River, New Jersey, USA.

Murugesan, S. 1998, 'Web Engineering', Presentation at the First Workshop on Web Engineering, World Wide Web Conference (WWW7), Brisbane.

Murugesan, S. 1999, 'Web Engineering: A New Discipline for Development of Web-based Systems', In Proceedings of the First ICSE Workshop on Web Engineering, Los Angeles (pp. 1-9).

Murugesan, S. and Deshpande, Y. (eds), 2001, 'Web Engineering: Managing Diversity and Complexity of Web Application Development', Lecture Notes in Computer Science—Hot Topics, 2016. Berlin: Springer Verlag.

Offutt, J. 2002, *Quality Attributes of Web Software Applications*, IEEE Software, Special Issue on Software Engineering of Internet Software, 19(2), 25-32.

Oppenheimer, D. and Patterson, D.A. 2002, *Architecture and Dependability of Large-scale Internet Services*, IEEE Internet Computing, September-October, 41-49.

Powers, David 2007, *The Essential Guide to Dreamweaver CS3 with CSS, Ajax, and PHP*, Apress.

Pressman, R.S. 2001, *What a Tangled Web We Weave*, IEEE Software, 18(1), 18-21.

Pressman, R.S. 2004, Applying Web Engineering, Part 3, *Software Engineering: A Practitioner's Perspective*, Sixth Edition, McGraw-Hill, New York.

Reifer, D.J. 2000, *Web Development: Estimating Quick-to-Market Software*, IEEE Software, 17(6), 57-64.2000

Woojong Suh 2005, *Web Engineering Principles and Techniques*, Idea Group.

PART III

Internet and Network Protocols

- **Chapter 12** Network Management Tools
- **Chapter 13** Protocols and Global Connectivity

CHAPTER

12 Network Management Tools

After 20 years of talking, this so-called convergence of computing and communications is happening.
–Craig R. Barrett
CEO of Intel

Learning Objectives

After reading this chapter, you should be able to understand:

- the basics of network management
- and comprehend simple network management protocol
- the differences between the various network management tools

12.1 Basics of IT Management

The networking community took the standards approach by first standardizing on management protocols, the structure of management information, and a subset of the management information. Then they developed products based on those standards. These products for network management have matured over the last several years. The Internet community has standardized on the simple network management protocol (SNMP) and associated structure of management information (SMI) and definitions of management information. Open management is now a widely held principle in the internetworking market. The telecommunications community has standardized the common management information service/protocol (CMIS/CMIP) and the associated open systems interconnection (OSI) management SMI and management definitions.

12.1.1 Protocols

As we have seen earlier, a protocol is a set of rules which is used by computers to communicate with each other across a network. A protocol is a convention or standard that controls or enables the connection, communication, and data transfer between computing endpoints. In its simplest form, a protocol can be defined as the rules governing the syntax, semantics, and synchronization of communication.

12.1.2 Simple Network Management Protocol (SNMP)

The simple network management protocol (SNMP) is a network management standard widely used in networks that support the transmission control protocol/Internet protocol (TCP/IP). Simple network management protocol is one of the standard operations and maintenance protocols for the Internet. It has been a key technology that enabled the Internet's phenomenal growth. Since its creation in 1988 as a short-term solution to manage elements in the growing Internet and other attached networks, SNMP has achieved widespread acceptance. It provides a method of managing network hosts such as workstation or server computers, routers, bridges, and hubs from a centrally located computer running network management software. Simple network management protocol performs management services using management systems and agents. It was derived from its predecessor SGMP (simple gateway management protocol). Simple network management protocol standards are defined in a series of documents called request for comments or requests for comments RFCs, proposed by the Internet Engineering Task Force (IETF).

The Internet Engineering Task Force is responsible for defining the standard protocols that govern Internet traffic, including SNMP. The IETF publishes RFCs, which are specifications for many protocols that exist in the IP realm. Documents enter the standards track first as proposed standards, then move to draft status. When a final draft is eventually approved, the RFC is given standard status.

The following list includes the current SNMP versions.

- SNMP Version 1 (SNMPv1) is the current standard version of the SNMP protocol. It is defined in RFC 1157 and is a full IETF standard. SNMPv1's security is based on communities, which are nothing more than passwords—plain-text strings that allow any SNMP-based application that knows the strings to gain access to a device's management information. There are typically three communities in SNMPv1: read-only, read-write, and trap. Read-only allows us to read data values but we cannot modify the data. The read-write community allows us to read and also modify the data. Trap allows us to receive traps.
- SNMP Version 2 (SNMPv2) is often referred to as community string-based SNMPv2. This version of SNMP is technically called SNMPv2c. It is defined in RFC 1905, RFC 1906, and RFC 1907, and is an experimental IETF. Even though it is experimental, some vendors have started supporting it in practice.
- SNMP Version 3 (SNMPv3) will be the next version of the protocol to reach full IETF status. It is currently a proposed standard, defined in RFC 1905, RFC 1906, RFC 1907, RFC 2571, RFC 2572, RFC 2573, RFC 2574, and RFC 2575. It adds support for strong authentication and private communication between managed entities.

Manager

In the world of SNMP, there are two kind of entities: managers and agents. A manager is a server running some kind of software system that can handle management tasks for a network. Managers are often referred to as network management stations (NMSs). An NMS is responsible for polling and receiving traps from agents in the network.

Simple network management protocol is a fairly simple request/response protocol where the manager polls managed devices periodically for updated information. A poll, in the context of network management, is the act of querying an agent for some piece of information. The polling frequency can be set by the network administrator. There are three types of polling:

- Monitor polling: To check that devices are available and to trigger an alarm when one is not
- Threshold polling: To detect when conditions deviate from a baseline number by a percentage greater than allowed (usually plus or minus 10 per cent to 20 per cent) and to notify the manager for review

- Performance polling: To measure the ongoing network performance over longer periods and to analyse the data for long-term trends and patterns

Agent

The second entity, the agent, is a piece of software that runs on the network devices being managed. It can be a separate program, or it can be incorporated into the operating system. The agent provides management information to the NMS by keeping track of various operational aspects of the device. The NMS can query the status of each interface on a router, and take appropriate action if any of them is down. Agents do not originate messages. However, an agent initiates traps, that is, alarm triggering events such as unauthorized system reboot and illegal access to the network. A trap is a way for the agent to tell the NMS that an unexpected event has occurred. When the agent notices that something bad has happened, it can send a trap to the NMS. This trap originates from the agent and is sent to the NMS, where it is handled appropriately.

Agents reside on the so-called managed devices. In other words, a managed device is a piece of equipment with an SNMP agent (software) built into it. The managed devices can be configured to collect specific pieces of information on device operations. Most of the information consists of totals, such as total bytes, total packets, total errors, and the like.

Management Information Base

A management information base (MIB) is a database of managed objects accessed by network management protocols. The SNMP contains two standard MIBs. The first, MIB I, established in RFC 1156, was defined to manage the TCP/IP-based Internet. MIB II, defined in RFC 1213, is basically an update to MIB I. The SMI provides a way to define managed objects and their behaviour. An agent has in its possession a list of the objects that it tracks. The MIB can be thought of as a database of managed objects that the agent tracks. Any sort of status or statistical information that can be accessed by the NMS is defined in an MIB. The SMI provides a way to define managed objects, while the MIB is the definition of the objects themselves.

Management information bases are categorized in accordance to the job they perform. They are referred to as MIB objects. Basic MIBs usually come packaged inside the network device operating system. They are programmed by vendors using a standard referred to as the Abstract Systems Notation One (ASN.1). Therefore, any MIB software package that is written in accordance with the SMI specifications, could be used on IBM MIBs, Cisco MIBs, HP MIBs, and others.

The SNMP manager and agent use an SNMP MIB and a relatively small set of commands to exchange information. The SNMP MIB is organized in a tree structure with individual variables, such as point status or description, being represented as leaves on the branches. A long numeric tag or object identifier (OID) is used to distinguish each variable uniquely in the MIB and in SNMP messages.

SNMP Messages

The SNMP uses five basic messages (GET, GET-NEXT, GET-RESPONSE, SET, and TRAP) to communicate between the SNMP manager and the SNMP agent. The GET and GET-NEXT messages allow the manager to request information for a specific variable.

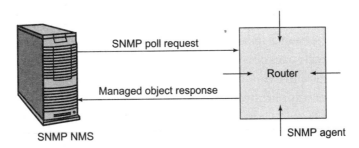

FIGURE 12.1 SNMP Agent

SNMP (Figure 12.1) can be used to do the following.

- Configure remote devices: Configuration information can be sent to each networked host from the management system. For instance, we can use SNMP to disconnect an interface on our router or check the speed at which a network adapter is operating.
- Monitor network performance: We can track the speed of processing and network throughput, and collect information about the success of data transmissions. For example, we could check the temperature of our power supply inside our system and shut it down if the temperature exceeds a predetermined value.
- Detect network faults or inappropriate access: We can configure trigger alarms on network devices when certain events occur. When an alarm is triggered, the device forwards an event message to the management system. Common types of alarms include a device being shut down and

restarted, a link failure being detected on a router, and inappropriate access.
- Audit network usage: We can monitor both overall network usage to identify user or group access, and types of usage for network devices and services.

12.1.3 RMON

Remote monitoring (RMON) is a bright example on the feasibility of extensive monitoring on routing elements. Even though the best vantage points for the collection of network measurements are the routing elements themselves, cost and performance considerations would probably deem any exhaustive solution infeasible. Remote monitoring Version 1 (RMONv1, or RMON) is defined in RFC 2819; an enhanced version of the standard, called RMON Version 2 (RMONv2), is defined in RFC 2021. RMONv1 provides the NMS with packet-level statistics about an entire local area network (LAN) or wide area network (WAN). RMONv2 builds on RMONv1 by providing network- and application-level statistics. These statistics can be gathered in several ways. One way is to place an RMON probe on every network segment you want to monitor.

In the RMON context, a single agent is capable of monitoring a complete shared LAN. The RMON agent is endowed with local intelligence and memory and can compute higher-level statistics and buffer these statistics in case of outage. Alarm conditions are defined, as well as actions that should be taken in response, in the spirit of the SNMP traps. Lastly, filtering conditions and the corresponding actions are defined so that the content of the filtered packets can be captured and buffered. Remote Monitoring offers great flexibility in combining the above primitives into sophisticated agent monitoring functions. However, this flexibility makes a full RMON agent implementation costly. Thus, RMON has been only implemented for LAN router interfaces, which are relatively low-speed. Implementations for high-speed backbone interfaces have proved to be infeasible or prohibitively expensive.

12.2 Network Management

The need for network management is apparent today throughout any organization deploying networks. The enterprise network in use by many organizations includes both local and wide area networking technologies and an assortment of network devices from different vendors, and is managed by a team of professionals in different geographic areas.

Companies are now using information management to gain a competitive advantage by delivering goods and services faster and more efficiently. While the benefits of network management are most apparent during a crisis, they are no less important in day-to-day network operation to control network faults, configurations, performance, or security. An integrated network management system can save time and money by reducing downtime and performance problems. The process of checking networks, logging, and notifying is generally accomplished using Ping and SNMP. There are proprietary products and clever ways to determine whether a service is up and running, like running telnet into a device to see if it is still there, but the basics shared by almost every network-management tool are these widely adopted and available methods.

Network management should be thought of as information systems for networks. It allows management to examine the way an important capital resource is being used, and provides the necessary information for adjusting the operation of that resource so that it more closely meets the needs of the organization. It involves the usage of information technology (IT) to measure and manage the effectiveness of the organization's networks. The overall goal of network management is to maximize network availability, performance, and benefits to network users.

Network management is accomplished through five basic management activities, each of which must be provided in an effective enterprise management system. These activities are discussed below.

Fault management is the discipline of detecting, diagnosing, bypassing, repairing, and reporting on network equipment and service failures.

Configuration and name management is concerned with maintaining an accurate inventory of hardware, software, and circuits in use throughout the enterprise, and provides the ability to change that inventory in a reliable and efficient manner in response to changing service requirements. Configuration and name management ensures the consistency and validity of operating parameters, addressing tables, software images, and hardware configurations.

Performance management is concerned with tracking and planning for the best utilization of network and computing resources and ensuring that the required resources are available to meet user service-level expectations.

Security management controls access to both the network and the network management systems. It protects the network and its management systems from unauthorized access or modification. Security management tells the manager who is using the network.

Accounting management measures network usage and computes charges for that usage. It tells the manager when the network is being used and about the cost of the resources consumed.

The problem of managing the enterprise network can be divided into three components: the users of network management, the objects to be managed, and the functional needs of network administrators. Each component influences network management architecture by imposing requirements for effective management.

Fault tolerance is the ability of a computer or an operating system to respond to a catastrophic event or fault, such as a power outage or a hardware failure, in a way that ensures that no data is lost and any work in progress is not corrupted. This can be accomplished with a battery-backed power supply, backup hardware, provisions in the operating system, or any combination of these. In a fault-tolerant network, the system has the ability either to continue the system's operation without loss of data or to shut the system down and restart it, recovering all processing that was in progress when the fault occurred.

12.3 Remote Management Tools

The approaches to integrated management of a multi-vendor computing and communications environment have been evolving in the marketplace over the past several years. Intelligent software brings multiple benefits to the management mix. It provides automation, provisioning, and standardization across all vendor systems. Combined with best practices (Information Technology Infrastructure Library processes, for example), it can address the business-driven part of IT management, and it enables the management of end-to-end services across diverse hardware and software components. With network management no longer a silo activity, and network metrics key to achieving a services-driven IT architecture, industry participants agree that intelligent hardware is key to the future of network management. Benefits of intelligent devices include faster reaction and resilience to hardware-related issues through self-management, savings on network resources, and optimization of hardware capability and performance. Sankara Nethralaya realized that it was imperative to have an IT-enabled infrastructure. The hospital opted for a network operating centre (NOC) based solution. Apart from providing the hospital with a 360-degree view of its infrastructure, the hospital's biggest benefit was the elimination of downtime. With NOC, the hospital, which had a myopic view of its network infrastructure, has broadened its vision and has gone beyond just a corrective surgery.

One of the biggest issues in managing today's IT infrastructure is ensuring the availability and performance of critical business applications, by proactively identifying problems within the underlying networks, systems, databases, and applications. One aspect of this issue is that, before Intel AMT, it was not possible to access detailed hardware information such as events, traps, and component states for systems that had crashed, hung, or were powered down. Also, there was no secure and reliable mechanism to remotely power up, power down, reboot, or access hardware information for devices.

12.3.1 Dashboards

Dashboards have become popular in recent years as uniquely powerful tools for communicating important information at a glance. Although dashboards are potentially powerful, the potential is rarely realized. A dashboard's success as a medium of communication is a product of design, a result of a display that speaks clearly and immediately. Dashboards can tap into the tremendous power of visual perception to communicate, but only if those who implement them understand visual perception and apply that understanding through design principles and practices that are aligned with the way people see and think.

A dashboard is a visual display of the most important information needed to achieve one or more objectives; consolidated and arranged on a single screen so the information can be monitored at a glance. For example, the online CRM suites such as Zoho, SFA provide customizable dashboards.

The fundamental challenge of dashboard design is the need to squeeze a great deal of information into a small amount of space, resulting in a display that is easily and immediately understandable. Some of the prominent IT management tools are as discussed below. All the tools provide customizable dashboards.

12.3.2 IBM—Tivoli

IBM realizes customers need to be able to measure how systems use applications and how well transactions perform across the infrastructure to enable autonomic and utility computing. IBM has come out with a management framework. IBM Tivoli Management Framework (TMF) is a system's management platform from IBM. IBM Tivoli Monitoring represents an ongoing effort to harness enterprise-level monitoring into a unified framework (Tivoli Management Framework). Tivoli Monitoring is actually a line of products within a large suite of management software that IBM sells under the Tivoli brand. IBM Tivoli Monitoring for Transaction Performance (TMTP) monitors application traffic as it flows through a network to provide detailed response time information on application transactions. The product uses server and agent software installed across an infrastructure on the Web, application and database servers, as well as on end-user clients, to determine application response times. The software collects the data, correlates it, compares it against preset desired response times, and alerts staff when transaction thresholds are missed.

There is also Tivoli Monitoring for Network Performance. It is important to bear in mind that Tivoli NetView, a product not in the Tivoli Monitoring line, provides in-depth network monitoring, and several of the Tivoli Monitoring products require Tivoli Framework. IBM Tivoli NetView is distributed network management software that helps keep your important business systems available. NetView discovers TCP/IP networks, displays topologies, monitors network health, and gathers performance data so that you can quickly identify the root cause of network failures.

12.3.3 CA-Unicenter

CA-Unicenter is a software that helps prioritize management, reduce costs, and increase efficiencies. These solutions map IT infrastructure to business processes and automatically detect, diagnose, repair, and recover complex problems across the entire technology stack supporting critical business applications and services.

The first step in ensuring the availability of critical business applications is to discover the applications, databases, systems, and networks that make up the underlying IT infrastructure. Unicenter NSM automatically discovers, classifies, and creates a baseline for performance for all the infrastructure elements that are on the network. Intel AMT enhances this ability by enabling Unicenter NSM to discover, classify, and monitor systems regardless of their power or operating system state.

12.3.4 HP OpenView

The HP OpenView network management architecture (NMA) is rooted in international and de facto industry standards. Network management is a distributed activity in that the user interfaces, management applications, and management services can be located in different systems throughout a network. The HP OpenView NMA communications infrastructure provides the facilities for establishing and maintaining communication between these components.

OpenView Internet Services (OVIS) provides probes that measure a wide range of services, for example, DNS, DHCP, FTP, ODBC, HTTP and many more. It measures, monitors, and reports against service-level objectives. OpenView Internet Services offers end-user emulation of major business critical applications as well as a single integrated view of the complete Internet infrastructure. The dashboard is the starting point for the user's view of OVIS monitoring and reporting. The main page gives access to many different reports and monitors, and provides a summary of each configured customer. A customer may be another company to whom the organization is providing IT services, an internal department within the company, or a subset of services that is chosen to configure for reporting and alarming purposes. The Alarms tab will display any alarms that have been generated due to failures to meet objectives. These alarms may also have been forwarded to other OpenView applications, such as OpenView Operations (OVO) or Network Node Manager. The integration of OVO with service desk also allows action by a help desk to solve user incidents that may be reported due to the alarm condition. An important element of the dashboard is the resource pane. This serves as a focus for display elements of the health workspace pane.

The OVIS service hierarchy provides a means of organizing the services on which you want to receive reports and problem notification. At the top of the services hierarchy is the customer, which could be the name of a company, Internet service provider, or any entity within a company. Below the customer is the service group. One customer may have one or more service groups; each service group may only contain services of the same type.

12.4 Open Source—IT Management

Many organizations want to keep track of machines on their networks; if a service fails, the relevant people should be informed quickly. There is a wide range of commercial software available to do this—HP OpenView, Patrol, and Netcool/Omnibus are among the better-known packages. However, these are all very expensive, and tend to require highly-specified machines as dedicated monitoring servers. One of the core challenges and criticisms associated with proprietary IT management platforms is functionality overkill. The four platforms that now dominate the market—from BMC, Computer Associates, HP, and IBM—all were designed for the upper echelon of the Fortune 100. The result is an overload of capabilities and features that the majority of companies do not want or need. Getting a product such as HP OpenView or IBM Tivoli configured and deployed often takes months, even years in many cases. Once the system is installed, users face rigid vendor lock-in scenarios.

The free-software community has produced a number of programs which serve the same task. While they tend not to have such slick graphical interfaces as the commercial offerings, they are generally much less demanding in terms of system resources, typically not even requiring a dedicated server but able to share a machine with other software. Open-source IT management solutions have three core characteristics that make them well-suited to the task of monitoring and managing heterogeneous IT environments: they provide open interfaces, they are built on component architectures that are highly configurable, and the open-source code is transparent and designed to be modifiable. This combination makes an open-source IT management solution an ideal 'manager of managers.'

Although the abilities to integrate easily and play the manager of managers role set open-source solutions apart from proprietary offerings, the fact remains that lowering costs is the primary driver in the growth of open-source IT management. Open-source solutions lower both the upfront cost and the long-term total cost of ownership (TCO) of IT monitoring and management in several ways.

- No licensing fees: With open source, software is essentially free. Customers pay only for enhancements, services, and support.
- Lower deployment costs: Because open-source solutions do not install unnecessary features, deployments are completed quickly and easily. Companies save money by paying only for the features they need. Whereas a proprietary solution may run in six figures and offer 100 different features, a company needing only 20 per cent of that functionality can pay only for those requirements if it goes with an open-source product.
- Low system administration overhead: This is because you manage only what you have installed. It is also because open-source solutions are efficient without offering gratuitous features. Open-source IT management solutions do not require expensive vendor-specific consulting and training.

- Low hardware costs: Open-source products typically run on inexpensive, industry-standard boxes, further lowering overall costs.

A number of open-source IT management products are gaining traction, including Nagios, for availability monitoring: Multi-Router Traffic Grapher (MRTG) for network device statistics, Nmap for network scanning and discovery, Ntop for network traffic analysis, SyslogNG for log file analysis, and Cacti for SNMP analysis and performance graphing. These products provide strong core functionality for an enterprise-class monitoring solution.

> **AutoTradeCenter Gets Results by Going Open Source**
>
> AutoTradeCenter, Inc. (ATC) manages customized, private-label websites for auto manufacturers and financial institutions, including companies such as Ford Credit, American Honda Finance, and DaimlerChrysler Financial Services. Given the nature of its business and its focus on hosted Internet solutions, ATC needs a reliable and highly available network to deliver quality service to its growing list of global customers.
>
> In 2001, ATC did not have a solution in place to monitor its network, servers, or applications. Instead, the IT team addressed network issues only after they became a problem.
>
> To compound this problem, ATC was enjoying a rapidly expanding customer base. As the company added more customers, the IT infrastructure was taxed with a growing number of buyers, sellers, searches and orders. According to Jorge Borbolla, ATC's CIO, a single new customer can mean 1,000 incremental network activities each day. ATC's IT team lacked adequate visibility into the network infrastructure; they couldn't track usage trends and therefore couldn't plan well for anticipated growth. In early 2002, Borbolla determined that the company needed an effective network monitoring solution that would satisfy two objectives. First, it must provide notification of any problem before it affects the customer, and second, the system had to monitor service level agreements that stipulated 99% uptime. After testing several open source solutions, the IT team turned to Nagios and was impressed with its escalation and notification features. Today, GroundWork Monitor, which extends Nagios with several open-source components and advanced features, evaluates everything from network connectivity, CPU loads, database availability and the status of each custom website.
>
> Source http://www.linuxjournal.com/article/8196

12.4.1 Nagios

Nagios is an open-source, Unix-based enterprise monitoring package with a web-based front-end or console. Nagios can monitor assets such as servers, network devices, and applications, essentially any device or service that has an address and can be contacted via transmission control protocol/Internet protocol (TCP/IP). It can monitor hosts running Microsoft Windows, Unix/Linux, Novell NetWare, and other operating systems. It can be configured to work through firewalls, virtual private network (VPN) tunnels, across secure shell (SSH) tunnels, and via the Internet [Turnbull 2006].

Nagios is designed to primarily run on the Linux operating system. There is no particular Linux distribution recommended as an operating system platform for Nagios, and it should have no issues running on your preferred flavour of Linux. Nagios can monitor a variety of attributes on organizational assets. These can range from operating system attributes such as central processing unit (CPU), disk, and memory usage to the status of applications, files, and databases. You can use a variety of network protocols, including HTTP, SNMP, and SSH, to conduct this monitoring. Nagios can also receive SNMP traps, and you can build and easily integrate your own custom monitoring checks using a variety of languages, including C, Perl, and shell scripts.

Sizing a Nagios server is greatly dependent on the environment intended to monitor. This is because the number of hosts and services intended to monitor and check has a material impact on the size of the host selected as a Nagios server. There are few ground rules available to determine exactly how many services and hosts you can monitor with a particular hardware configuration. The Nagios tool is capable of being deployed in a distributed model with multiple servers, collecting data about your assets, and reporting them to a central server

12.4.2 Cacti

Cacti is an open source, network monitoring and graphing tool written in PHP/MySQL. It uses the RRDTool (round-robin database tool) engine to store data and generate graphics, and collects periodical data through Net-SNMP. Cacti operation is divided into three different tasks. These are as follows.

- Data retrieval: Cacti retrieves data through poller. It is an application executed at a constant time interval as a schedule service under different operating systems.
- Data storage: There are a lot of options to do this task, such as SQL database and flat file database. Cacti uses RRDTool to store data. Round-robin database is a system to store and show time series data collected from different SNMP-capable devices. It consolidates historical

data based on consolidation functions such as AVERAGE, MINIMUM, MAXIMUM, and so on to keep the storage size minimum.

- Data presentation: Cacti uses this built-in graphing function to deploy customized graphing reports based on time series data collected from different SNMP-capable devices. This built-in graphing function supports auto-scaling and logarithmic y-axis.

RRDTool was developed by Tobi Oeticker, also known for his famous creation, MRTG. RRDTool is a high-performance data logging and graphing system, designed to handle time series data like network bandwidth, room temperature, CPU load, server load, and to monitor devices such as routers, uninterruptible power supply (UPS), etc. It is also known as the round-robin database tool, an industry standard, open source solution. It lets the administrator log and analyse data collected from all kinds of data sources (DS), which are capable of answering SNMP queries.

Cacti is extensively used in the industry for network monitoring activities.

12.4.3 Big Brother

Big Brother is one of the most established and popular web-based console monitoring packages available. It gives the user a console or dashboard look-and-feel with typical green, yellow, and red dots indicating system status. Big Brother Professional Edition (BBPE) is a simple way to measure the health of your heterogeneous IT environment at a glance. It is an easy-to-implement, affordable, web-based solution for IT infrastructure monitoring and diagnostics. Big Brother can monitor information such as connectivity (Ping), DNS, FTP, and HTTP, to name a few. A commercial version, Big Brother Professional Edition (PE), offers encryption and compiled versions for certain platforms is also available.

SUMMARY

In this chapter, discussions on the requirements of a monitoring infrastructure that forms part of the design, management, and operation of an IP backbone network were presented. This chapter has provided an overview of the basic network monitoring functionality and protocol SNMP. Simple network management protocol is especially well-suited for monitoring private services and indicators that are not accessible over a network. Central processing unit usage, swap utilization, random access memory (RAM) usage, and disk partition utilization are common examples. Simple network management protocol is also very useful for monitoring process activity on a device; agents that support the host-MIB allow you to remotely query the complete process table of a device, which includes the memory and CPU utilization each process is using, and the command and arguments of the command line used to start the process. Using just the process information we can monitor and alert on useful measures of device and service health such as critical processes stopping, too much memory, or CPU consumed over time for a process or group of processes and unusual numbers of processes running. We have discussed the need for a monitoring infrastructure comprising a coarse-grained component for continuous network-wide monitoring and a fine-grained component for on-demand monitoring.

This chapter provided the basic configuration of Nagios server to provide monitoring for hosts and services. The chapter outlined the basics of Nagios. Unlike many open-source projects and commercial software projects that are unwilling to change core features once a feature is stable, even if the reasons are valid and useful, Nagios dropped a piece of core functionality with version 2.x that many people found useful. With the advent of Nagios 3, we see this framework becoming even more stable and we see a growing number of Nagios Event Broker (NEB) modules available for Nagios.

Current market conditions dictate optimal use of the network resources and performance within the bounds defined in rather competitive service level agreements (SLAs). Network measurements can provide the necessary knowledge that could allow the formalization of different network provisioning and planning tasks. Moreover, the development of sound methodological approaches toward the estimation of different performance metrics can serve as a framework, according to which Internet service providers (ISPs), and customers can evaluate the performance offered by packet-switched networks such as the Internet.

Key Terms

ASN.1 Abstract Syntax Notation One is a formal language for the abstract (platform-independent) description of messages exchanged between machines. It is used to encode and decode messages in a wide range of applications, including SNMP. Objects such as integers are encoded in a manner called tag-length-value (TLV) that is independent of any processor architecture, such as big or little endian. The tag indicates the object type, the length is the object size, and the value is the encoded object. Abstract Syntax Notation One also allows structured (or nested) definitions

Agent A hardware device or software program that reports to an SNMP manager. In network alarm management, an SNMP agent is typically an RTU, but other network devices such as switches, routers, and hubs can also act as SNMP agents.

Community string It is an SNMP security password. There are three kinds of community strings: Read Community (which allows an SNMP manager to issue Get and GetNext messages), Write Community (which allows an SNMP manager to issue Set messages), and Trap Community (which allows an SNMP agent to issue Trap messages).

Event In SNMP terms, any change of status in a managed object in the network. SNMP equipment can generate traps for many different kinds of events, not all of which are important for telemetry. The ability to filter unimportant events is essential for high-quality SNMP alarm management

FCAPS Fault, configuration, accounting, performance, and security are the OSI functional areas of network management. In the Fault area, network problems are found and corrected. Root cause analysis may be used to give an exact reason for a given fault. In the Configuration area, network operation is monitored and controlled. Hardware and NE software changes are recorded along with an inventory of deployed equipment and firmware. In the Accounting area, resources are shared out fairly among network users. This area ensures that end users are billed appropriately. The Performance area is involved with managing the overall performance of the network. The Security area is used to protect the network against hackers, unauthorized users, and physical or electronic tampering.

Management information base (MIB) The MIB is a data structure that describes SNMP network elements as a list of data objects. To monitor SNMP devices, your SNMP manager must compile the MIB file for each equipment type in your network.

Trap An SNMP message issued by an SNMP agent that reports an event.

REVIEW QUESTIONS

12.1 What is SNMP and why is it required?
12.2 What is MIB and why is it required?
12.3 What is the need for RMON. What are the various versions of RMON?
12.4 What are commercial network monitoring software and why are they used?
12.5 Discuss HP OpenView in detail.
12.6 Discuss CA—Unicenter and compare it with HP OpenView.
12.7 How does Trivoli compare with OpenView and Unicenter?
12.8 What are the open source equivalent network monitoring tools?

Projects

12.1 Refer to Chapter 1 and understand the ABC Unlimited Case. The CIO of ABC is interested to set up a good network monitoring solution to monitor the various servers and IT infrastructure present in the data centre. As you are aware you have designed the data centre for ABC unlimited. Discuss the IP plans for implementing a suitable network monitoring solution for ABC. Also build a technical comparison and a commercial comparison of the various network monitoring solutions available for ABC Unlimited.

12.2 Visit the nearest IT firm to understand the deployment of network monitoring solution. Prepare a report on the same.

References

Barret, Craig R. 2004, *Business Week Online*, 8 March.

Turnbull, James 2006, *Pro Nagios 2.0*, Apress, USA.

CHAPTER

13 Protocols and Global Connectivity

Every truth passes through three stages before it is recognized. In the first it is ridiculed, in the second it is opposed, in the third it is regarded as self-evident.
—Arthur Schopenhauer
German philosopher

Learning Objectives

After reading this chapter, you should be able to understand:

- the protocols to appreciate the global connectivity
- Internet and its impact in business
- the differences between IP and TCP

13.1 Internet

Technological change is usually associated with progress. One sector in which the rate of technological change has easily surpassed the conventional growth pattern by orders of magnitude is the computing industry. Well, to start with, around 80 per cent of employees in the average company in the developed world have a computer on their desks today, as opposed to less than 10 per cent in the early 1980s. They also use these computers for the most part of their working day, as opposed to only 1–2 hours previously.

The fundamental technology that makes the Internet work is called packet switching, a data network in which all components (i.e., hosts and switches) operate independently, eliminating single point-of-failure problems. In addition, network communication resources appear to be dedicated to individual users but, in fact, statistical multiplexing and an upper limit on the size of a transmitted entity result in fast, economical networks. The modern Internet began as a US Department of Defense (DoD) funded experiment to interconnect DoD-funded research sites in the USA.

The Internet has no single owner, yet everyone owns (a portion of) the Internet. The Internet has no central operator, yet everyone operates (a portion of) the Internet. The Internet has been compared to anarchy, but some claim that it is not nearly that well organized.

Some central authority is required for the Internet, however, to manage those things that can only be managed centrally, such as addressing, naming, protocol development, standardization, etc. Among the significant Internet authorities are the following.

- The Internet Society (ISOC), chartered in 1992, is a non-governmental international organization providing coordination for the Internet, and its internetworking technologies and applications. The ISOC also provides oversight and communications for the Internet Activities Board (IAB).
- The Internet Activities Board (IAB) governs administrative and technical activities on the Internet.
- The Internet Engineering Task Force (IETF) is one of the two primary bodies of the IAB. The IETF's working groups have primary responsibility for the technical activities of the Internet, including writing specifications and protocols. The impact of these specifications is significant enough that ISO accredited the IETF as an international standards body at the end of 1994. RFC's 2028 and 2031 describe the organizations involved in the IETF standards process and the relationship between the IETF and ISOC, respectively, while RFC 2418 describes the IETF working group guidelines and procedures. The background and history of the IETF and the Internet standards process can be found in 'IETF–History, Background, and Role in Today's Internet.'

- The Internet Engineering Steering Group (IESG) is the other body of the IAB. The IESG provides direction to the IETF.
- The Internet Research Task Force (IRTF) comprises a number of long-term reassert groups, promoting research of importance to the evolution of the future Internet.
- The Internet Engineering Planning Group (IEPG) coordinates worldwide Internet operations. This group also assists Internet service providers (ISPs) to interoperate within the global Internet.
- The Forum of Incident Response and Security Teams is the coordinator of a number of Computer Emergency Response Teams (CERTs) representing many countries, governmental agencies, and ISPs throughout the world. Internet network security is greatly enhanced and facilitated by the FIRST member organizations.
- The World Wide Web Consortium (W3C) is not an Internet administrative body, per se, but since October 1994 has taken a lead role in developing common protocols for the World Wide Web to promote its evolution and ensure its interoperability. The W3C has more than 400 member organizations internationally. The W3C, then, is leading the technical evolution of the Web, having already developed more than 20 technical specifications for the Web's infrastructure.

13.1.1 ARPANET and DARPANET

The Advanced Research Projects Agency (ARPA) was formed with an emphasis towards research, and thus was not oriented only to a military product. From its inception ARPA significantly funded many US university research labs, and as early as 1968 had a close relationship with Carnegie-Mellon University, Harvard University, MIT, Stanford University, UCB, UCLA, UCSB, University of Illinois, and the University of Utah, as well as leading industry labs including Bolt Beranek and Newman, Computer Corporation of America, Rand, SRI, and Systems Development Corporation. Most of these labs were connected to the ARPANET soon after it was developed in order to enable cross-fertilization of research activity. The ARPA's Program Plan for the ARPANET was titled 'Resource Sharing Computer Networks'. It was submitted on 3 June 1968, and approved by the Director on 21 June 1968.

In the early 1970's the word 'Defense' was prefixed to the name, and ARPA became known as DARPA. By the late 1990's, DARPA reported to the Director for Defense Research and Engineering and had about 250 staff and a budget of USD 2 billion. A typical project was funded with between

10 and 40 million dollars over a period of four years, and drew support from several consultants and one or two universities.

DARPA's mission has been to assure that the USA maintains a lead in applying state-of-the-art technology for military capabilities and to prevent technological surprise from its adversaries. The DARPA organization was as unique as its role, reporting directly to the Secretary of Defense and operating in coordination with but completely independent of the military research and development (R&D) establishment. DARPA program managers have always had complete control over program funding, unprecedented flexibility in management capabilities, and direct responsibility for making their program a success.

The ARPANET was the first wide area packet switching network, the 'Eve' network of what has evolved into the Internet we know and love today. The ARPANET was developed by the Information Processing Techniques Office (IPTO) under the sponsorship of DARPA, and conceived and planned by Lick Licklider, Lawrence Roberts, and others.

13.1.2 Current Systems

While the Internet today is recognized as a network that is fundamentally changing social, political, and economic structures, and in many ways obviating geographic boundaries, this potential is merely the realization of predictions that went back nearly 40 years.

The hierarchical structure of domain names is best-understood if the domain name is read from right-to-left. Internet hosts names end with a top-level domain name. World-wide generic top-level domains (TLDs) include:

- .com: Commercial organizations
- .edu: Educational institutions; largely limited to four-year colleges and universities
- .net: Network providers; laregely limited to hosts actually part of an operational network from about 1994 to 2001 but now open to anyone.
- .org: Non-profit organizations
- .int: Organizations established by international treaty
- .gov: Government organizations
- .mil: US military (managed by the US Department of Defense Network Information Center)

In November 2000, the first new set of TLDs were approved by ICANN, The seven new TLDs are as listed below.

- .aero: Aviation industry
- .biz: Businesses
- .coop: Business cooperatives
- .info: General use
- .museum: Museums
- .name: Individuals
- .pro: Professionals

13.2 Protocols

A protocol is the standard procedure for regulating data transmission between computers. Protocol is an agreed-upon format for transmitting data between two devices. It determines the type of error checking to be used. The definition of a common technical language has been a major catalyst to the standardization of communications protocols and the functions of a protocol layer. The protocol generally accepted for standardizing overall computer communications is a seven-layer set of hardware and software guidelines known as the open systems interconnection (OSI) model.

- Data compression method, if any
- How the sending device will indicate that it has finished sending a message
- How the receiving device will indicate that it has received a message

The widespread use and expansion of communications protocols is both a prerequisite for the Internet, and a major contributor to its power and success. The pair of Internet protocol (IP) and transmission control protocol (TCP) are the most important of these, and the term TCP/IP refers to a collection (or protocol suite) of its most used protocols.

Some common protocols are internetwork packet exchange (IPX), TCP/IP, and net basic input/output (BIOS) extended user interface (Net-BEUI). Protocols can either be mandated by one company or organization, or created, used, and maintained by the entire networking industry.

13.2.1 Seven Layer OSI Stack

The OSI management model is a layered architecture (plan) that standardizes levels of service and types of interaction for computers exchanging information through a communications network. The open systems interconnection reference model (OSI reference model or OSI model) was originally created as the basis for designing a universal set of protocols called the OSI protocol suite. The protocol stack denotes a specific combination of protocols that work together, a reference model is a software architecture that lists each layer and the services each should offer. The OSI reference model is commonly used to describe, in an abstract manner, the functions involved in data communication. This model, originally conceived in the International Organization for Standardization (ISO), defines data communications functions in terms of layers.

In the OSI reference model, each layer is responsible for certain basic functions such as getting data from one device to another or from one application on a computer to another. The functions at each layer depend and build on the functions called services, provided by the layers below it. Communication between peer entities at a given layer is done via one or more protocols; this communication is invoked via the interface with the layer below.

The OSI model defines a networking framework for implementing protocols in seven layers. Control is passed from one layer to the next, starting at the application layer in one station, proceeding to the bottom layer, over the channel to the next station and back up the hierarchy.

The OSI model separates computer-to-computer communications into seven layers, or levels, each building upon the standards contained in the levels below it. The lowest of the seven layers deals solely with hardware links; the highest deals with software interactions at the application-program level. When data is sent from a source device down the OSI management model, each layer attaches its own header to that information

Application Layer

This layer supports application and end-user processes. Communication partners are identified, quality of service is identified, user authentication and privacy are considered, and any constraints on data syntax are identified. Everything at this layer is application-specific. This layer provides application services for file transfers, e-mail, and other network software services. Telnet and file transfer protocol (FTP) are applications that exist entirely in the application level. Tiered application architectures are a part of this layer.

Presentation Layer

This layer provides independence from differences in data representation (for example, encryption) by translating from application to network format, and vice versa. The presentation layer works to transform data into the form that the application layer can accept. This layer formats and encrypts data to be sent across a network, providing freedom from compatibility problems. It is sometimes called the syntax layer. It provides independence to application processes from differences in data representation (that is, in syntax); syntax selection and conversion by allowing the user to select a 'presentation context' with conversion between alternative contexts.

Session Layer

This layer establishes, manages, and terminates connections between applications. The session layer sets up, coordinates, and terminates conversations, exchanges, and dialogues between the applications at each end. It deals with session and connection coordination. It provides mechanisms for organizing and structuring dialogues between application processes; mechanisms allow for two-way simultaneous or two-way alternate operation, establishment of major and minor synchronization points, and techniques for structuring data exchanges.

Transport Layer

This layer provides transparent transfer of data between end systems, or hosts, and is responsible for end-to-end error recovery and flow control. It ensures complete data transfer. The services include establishing and terminating node connections, message flow control, dialogue control, and end-to-end data control.

Network Layer

This layer provides switching and routing technologies, creating logical paths, known as virtual circuits, for transmitting data from node to node. Routing and forwarding are functions of this layer, as well as addressing, internetworking, error-handling, congestion control, and, packet sequencing. It provides independence from data transfer technology and relaying and routing considerations; masks peculiarities of data transfer medium from higher layers. This layer provides the Internet protocol. The Internet protocol includes an addressing scheme that identifies the source and destination address of the packet being transported. Additional protocols available in this layer include the address resolution protocol (ARP) and the Internet control message protocol (ICMP).

Data Link Layer

At this layer, data packets are encoded and decoded into bits. It furnishes transmission protocol knowledge and management and handles errors in the physical layer, flow control, and frame synchronization. It is also responsible for defining data formats to include the entity by which information is transported, error-control procedures, and other link control procedures. The data link layer takes the raw stream of bits of the physical layer and provides the functionality of sending and receiving a meaningful message unit called a frame and also provides error-detection functions. The data link layer is divided into two sub-layers: the media access control (MAC) layer and the logical link control (LLC) layer. The MAC sub-layer controls how a computer on the network gains access to the data and permission to transmit it. The LLC layer controls frame synchronization, flow control, and error checking.

Physical Layer

This layer conveys the bit stream electrical impulse, light or radio signal through the network at the electrical and mechanical level. It provides the hardware means of sending and receiving data on a carrier, including defining cables, cards, and physical aspects. At this layer, cable connections and the electrical rules necessary to transfer data between devices are specified. Examples of physical layer standards include RS-232, V.24, and the V.35 interface. Fast Ethernet, RS232, and ATM are protocols with physical layer components. It provides electrical, functional, and procedural characteristics to activate, maintain, and deactivate physical links that transparently send the bit stream; it only recognises individual bits, not characters or multi-character frames

Computers use protocols to talk to each other, and when information travels between computers, it moves from device to device, or layer to layer as defined by the OSI management model. Each layer of the model has different protocols that define how information travels. The layered functionality of the different protocols in the OSI model is called a protocol stack. In other words, protocol stacks are sets of protocols that work together on different levels to enable communication on a network. For example, the protocol stack on the Internet, incorporates more than 100 standards, including FTP, IP, simple mail transfer protocol (SMTP), TCP, and telnet protocol.

13.2.2 TCP/IP Stack

Internet began as a research effort to link networks operated on packet switched methods. It emerged as a layered design, built on the principles of encapsulation to carry end-to-end Internet packets. It is important to understand that the Internet is not a new kind of physical network. It is a collection of interconnecting networks. Internet as a technology has grown with increased complexity built on TCP/IP architecture. Transmisson control protocol/Internet protocol is a result of protocol research and development conducted on the experimental packet switched network, ARPANET, funded by DARPA. This consists of a large collection of protocols that have been proposed as standards by the IAB. Transmisson control protocol/Internet protocol presented a radical departure from the traditional computer networking services during its development and emerged as a modular family of protocols providing a wide range of highly segmented functions. The Internet and TCP/IP are so closely related in their history that it is difficult to discuss one without also talking about the other. They were developed together, with TCP/IP providing the mechanism for implementing the Internet. Transmisson control protocol/Internet protocol has over the years continued to evolve to meet the needs of the Internet and also smaller, private networks that use the technology.

The TCP/IP suite of protocols is the set of protocols used to communicate across the Internet. It is also widely used on many organizational networks due to its flexibility and wide array of functionality provided. Transmisson control protocol/Internet protocol, in fact, consists of dozens of different protocols, but only a few are the 'main' protocols that define the core operation of the suite. The Internet is the primary OSI network layer (layer three) protocol that provides addressing, datagram routing, and other functions in an internetwork. The TCP is the primary transport layer (layer four) protocol, and is responsible for connection establishment and management and reliable data transport between software processes on devices.

- Internet protocol: It is responsible for moving packet of data from node to node. It forwards each packet based on a four byte destination address (the IP number). The Internet authorities assign ranges of numbers to different organizations. The organizations assign groups of their numbers to departments. Internet protocol operates on gateway machines that move data from department to organization to region and then around the world.

- Transmisson control protocol: It is responsible for verifying the correct delivery of data from client to server. Data can be lost in the intermediate network. Transmisson control protocol adds support to detect errors or lost data and to trigger retransmission until the data is correctly and completely received.

Transmisson control protocol/Internet protocol is most often studied in terms of its layer-based architecture and the protocols that it provides at those different layers.

13.2.3 Internet Protocol

Internet protocol is a protocol situated at the network layer of the OSI model. It is the basis of the worldwide network commonly known as the Internet. The protocol provides a formal definition of the layout of a datagram and the formation of a header composed of information about the datagram. It is the protocol that hides the underlying physical network by creating a virtual network view. It is designed to interconnect packet switched communication networks to form an Internet. It transmits blocks of data called datagrams received from the IP's upper layer software to and from the source and destination hosts. The ability of IP to run over anything is one of its important characteristics.

Internet protocol does not guarantee either delivery or in-order delivery of datagram, i.e., it does not guarantee the integrity of data in the datagram. The process of dividing the datagram into small pieces is called fragmentation. An IP datagram can be fragmented and forwarded across any available route.

Functions of IP

The IP provides the following four main functions.

- It provides the basic unit of measure for data transfer.
- It provides addressing mechanism for movement of data in a network.
- It provides for routing of data packets.
- It takes care of datagram fragmentation.

13.3 IP Addressing Mechanism

The IP's service model is made up of two separate parts: an addressing part and a datagram data delivery part. The IP's addressing scheme provides a way to uniquely identify the hosts in a network. Internet protocol is responsible for the routing of a datagram, determining where it will be sent, and devising alternate routes in case of problems.

The purpose of a network is to allow multiple participating devices (stations) to exchange information. By definition, a network comprises multiple devices. It allows information exchange among these multiple devices. An important feature of the IP is the ability to transparently use a wide variety of underlying network architectures to transport IP packets. This is achieved by encapsulating IP packets in whatever packet or frame structure the underlying network uses.

To achieve this task with full transparency, the IP needed an addressing structure, which developed as a two-level hierarchy in both addressing and routing. One part of the address, the network part, identifies the particular network a host is connected to, while the other part, the local part, identifies the particular endsystem on that network. Address is a means used to uniquely identify each network device as a sender or receiver of information. The most important characteristic of an address is its uniqueness. A unique address enables the delivery of the packets to the correct destination. This introduces the need for addressing mechanism.

There are several registries responsible for blocks of IP addresses and domain-naming policies around the globe.

13.3.1 IP V4-Address System

Internet protocol address is an identifier for a computer or device on a TCP/IP network. Networks using the TCP/IP route messages based on the IP address of the destination. The format of an IP address is a 32-bit numeric address written as four numbers separated by periods. These addresses are unique. Each address will identify one and only one network element in the Internet. The IP defines a set of unique address spaces. An address space is the total number of addresses space used by the protocol.

A typical IP address looks like this:
216.27.61.137

To make it easier for us humans to remember, IP addresses are normally expressed in decimal format as a 'dotted decimal number' like the one above.

13.3.2 Limitations of IPV4

IP version 4 (IPv4) has proven to be robust, easily implemented, and interoperable. It has stood up to the test of scaling an internetwork to a global utility the size of today's Internet. This is a tribute to its initial design. However, the initial design of IPv4 did not anticipate the following.

The recent exponential growth of the Internet and the impending exhaustion of the IPv4 address space Although the 32-bit address space of IPv4 allows for 4,294,967,296 addresses, previous and current allocation practices limit the number of public IPv4 addresses to a few hundred million. As a result, public IPv4 addresses have become relatively scarce, forcing many users and some organizations to use a network address translation (NAT) to map a single public IPv4 address to multiple private IPv4 addresses. Although NATs promote reuse of the private address space, they violate the fundamental design principle of the original Internet that all nodes have a unique, globally reachable address, preventing true end-to-end connectivity for all types of networking applications. Additionally, the rising prominence of Internet-connected devices and appliances ensures that the public IPv4 address space will eventually be depleted.

The need for simpler configuration Most current IPv4 implementations must be either manually configured or use a stateful address configuration—protocol such as dynamic host configuration protocol (DHCP). With more computers and devices using IP, there is a need for a simpler and more automatic configuration of addresses and other configuration settings that do not rely on the administration of a DHCP infrastructure.

The requirement for security at the Internet layer private communication over a public medium such as the Internet requires cryptographic services that protect the data being sent from being viewed or modified in transit. Although a standard now exists for providing security for IPv4 packets (known as Internet protocol security, or IPsec), this standard is optional for IPv4 and additional security solutions, some of which are proprietary, are prevalent.

13.3.3 IPV6 Address system

The IPv6 protocol includes the following features.

- New standardized header format
- Larger address space
- Multicast and anycast
- Stateless address configuration
- Built-in security
- Better support for quality of service (QoS)

- Extensibility

To address the issues of IPv4 and other concerns, the IETF has developed a suite of protocols and standards known as IP version 6 (IPv6). This new version, previously called IP-The Next Generation (IPng), incorporates the concepts of many proposed methods for updating the IPv4 protocol.

13.3.4 Transmission Control Protocol

The TCP, documented in RFC 793, is a widely used protocol employed by many applications for reliable exchange of data between two hosts. It is a stateful protocol. This means, one must set up a TCP connection, exchange data, and terminate the connection. Transmission control protocol guarantees orderly delivery of data and includes checks to guarantee the integrity of data received, relieving the higher-level applications of this burden.

Transmission control protocol is a connection-oriented protocol and is said to be 'reliable', although this word is used in a data communications context. It establishes a session between two machines before data is transmitted. Because a connection is set up beforehand, it is possible to verify that all packets are received on the other end and to arrange retransmission in case of lost packets. Because of all these built-in functions, TCP involves significant additional overhead in terms of processing time and header size.

Transmission control protocol fragments large chunks of data into smaller segments if necessary, reconstructs the data stream from packets received, issues acknowledgments of data received, provides socket services for multiple connections to ports on remote hosts, performs packet verification and error control, and performs flow control.

13.3.5 User Datagram Protocol

User datagram protocol (UDP) is a 'connectionless' protocol and does not require a connection to be established between two machines prior to data transmission. It is, therefore, said to be .unreliable. The word unreliable, used here as opposed to 'reliable' in the case of TCP and should not be interpreted against its everyday context.

Sending a UDP datagram involves very little overhead in that there are no synchronization parameters, no priority options, no sequence numbers, no timers, and no retransmission of packets. The header is small, the protocol is streamlined functionally. The only major drawback is that delivery is not guaranteed. User datagram protocol is, therefore, used for communications that involve broadcasts, for general network announcements, or for real-time data.

13.3.6 Multi-protocol Label Switching

Multi-protocol label switching (MPLS) was originally presented as a way of improving the forwarding speed of routers, but is now emerging as a crucial standard technology that offers new capabilities for large-scale IP networks. The essence of MPLS is the generation of a short fixed-length label that acts as a shorthand representation of an IP packet's header. The MPLS service protocol management solutions focuses on the traffic engineering capabilities built into the operating systems of core router network devices which are essential in managing telecom and service provider backbone networks.

13.4 Telephone System

A network is a composition of communication devices and links that connect at least two nodes that consist of hardware and software. These connected communications devices and links perform interactions between nodes using exact prescriptions, including protocols.

Telecommunication network consists of national networks and the international network [Broek 2000]. An international network crosses international borders. International networks are often characterized as no different from a national network—particularly tempting in a world where country boundaries are disappearing every day. An international network is defined as a network that operates in at least two different countries, with a management organization (coordinating body) in each of the countries responsible for the management of that part of the network that lies in that country. A national network is a combination of public and private networks. Public networks are for general use; private networks can be used only by employees of the organization. The fixed public network is called the public switched telecommunication network (PSTN). It is the global collection of interconnects originally designed to support circuit-switched voice communication system. It is a circuit-switched network that is used primarily for voice communications worldwide. The PSTN provides the traditional plain old telephone service (POTS) to residences and many other establishments. The term PSTN used to refer to the public communication system that provides local, extended local and long distance telephone service. The PSTN is composed of telephone exchanges networked together to form a nationwide and worldwide telephone communications system. It is public in the sense the system is available to anyone who can afford the service. All calls are switched calls. A caller's conversation is broken into pieces and these pieces are sent simultaneously over many connections to reach a receiver at the other end. The individual pieces are 'switched' from one telephone device to another until they reach their final destination at the receiving end.

There are two types of telephone networks: circuit-switched networks and packet-switched networks. Telecommunication network consist of exchanges, trunks, and subscriber lines. Trunks are circuits between exchanges, and the group of trunks between a pair of exchanges is known as a trunk group. A telephone exchange is often also called a telephone switch. Originally, the telephone exchange was created as a means of a provider receiving and inbound phone signal, interacting with a subscriber, and then switching the signal to whomever the subscriber wished to speak with. The setup and release of connections in telecommunication networks are triggered by signals. Starting and ending a call involve signaling between the subscribers and their local exchanges and, for inter-exchange calls, signaling between the exchanges along the connection. This was referred to early on in the history of telephony as exchanging a call. An exchange made up of thousands of local lines grouped together into a single switched grouping connected to and switched from a central office. This physical grouping relates to how phone numbers are grouped together.

13.5 Voice Over Internet

Digital networking for telephones was invented in the 1920s, but the first digital networks would not leave the laboratory until much later, in 1964. When digital networks were implemented, the telephone carrier companies began using a technique that permitted them to accept analog telephone calls coming into their switching facilities and convert those signals into digital form for transmission. Today, most phone companies have updated their equipment to include digital service. Over time, the POTS network gave way to the PSTN, or public switched telephone network (Figure 13.1).

The Internet protocols are the basis of IP networking, which supports corporate, private, public, cable, and even wireless networks. Voice over Internet protocol (VoIP) is a general term for a family of transmission technologies for delivery of voice communications over IP networks such as the Internet or other packet-switched networks. Voice over Internet Protocol, means voice transmitted over a computer network. It was developed in 1995 and is gradually replacing the PSTN. It has particular appeal to those who want to use their computer network to carry their telephone calls, thereby saving the expense of running different networks for each.

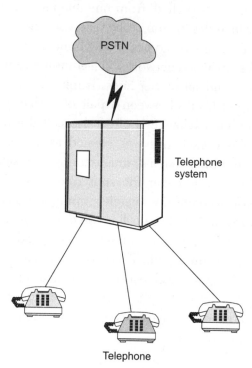

FIGURE 13.1 Telephone Network

Voice over Internet protocol requires a connection to the Internet through an Internet service provider (ISP), a VoIP service to extend the reach to traditional land lines, and VoIP software to actually place calls. Voice over Internet protocol systems employ session control protocols to control the set-up and tear-down of calls as well as audio codecs which encode speech allowing transmission over an IP network as digital audio via an audio stream. Voice over Internet protocol is often referred to as IP telephony (IPT) because it uses IPs to make enhanced voice communications possible. Voice over Internet protocol telephony services are inter-operable, meaning that they work well over all kinds of networks. They are also highly portable, which means they will work with any IP-enabled device such as an IP telephone, a computer, or even a personal digital assistant (PDA). Voice over Internet protocol technology enables traditional telephony services to operate over computer networks using packet-switched protocols. Packet-switched VoIP puts voice signals into packets, similar to an electronic envelope. Along with the voice signals, the VoIP packet includes both the caller's and the receiver's network addresses.

Voice over Internet protocol is used over hand-held phones and soft phones. A soft phone is a software that works on a laptop computer or pocket PC and provides most of the functionality of a traditional desk phone. If a user can connect to a network, the soft phone provides a way to reap the benefits of IP telephony regardless of location. For example, Nortel has introduced a range of soft phones. With VoIP, phone numbers are no longer tied to specific wires and switches. Voice over Internet protocol routes calls based on network addresses, and phone numbers are simply used because that is what people are familiar with. Voice over Internet protocol technology has enabled telephony signals to run over dedicated networks using packet-switched protocols. One of the preferred methods of running VoIP in the corporate sector is to use dedicated lines. Instead of being primarily dependent on the PSTN for its telephone service requirements, companies using VoIP protocols can send and receive telephone calls over their private computer networks.

Voice over Internet Protocol is a truly disruptive technology and one of the early challenges it will pose is to the existing regulatory frameworks and rules. Since telecommunications is a (tightly) regulated industry and the Internet is totally unregulated, the key issue is whether VoIP should be regulated. The voice element suggests that it should, while the IP element suggests that it should not.

BT is already offering VoIP solutions for all businesses from small and medium enterprises (SMEs) to the corporate enterprise. BT Business Information Systems has selected Cisco Systems and Nortel Systems to provide 'new wave' products to complement its existing portfolio. Cisco's IP private branch exchange (PBX) provides customers with a revolutionary option for their communications, bringing IP telephony to the desktop over a single network infrastructure, one that would normally be used for just data applications. The highly scalable IP PBX will be built around Cisco's Call Manager and IP telephone sets from Cisco's AVVID range. Nortel Network's IP enabled Meridian 1 and the Business Communications Manager (BCM—previously known as enterprise edge), are also available to provide existing customers with an evolutionary path into an IP future.

SUMMARY

In this chapter we briefly discussed some protocols, in order to establish their relationship with the OSI model. Protocol suites are the real-world implementation of the OSI model. The OSI model has seven layers, each of which outlines tasks that allow different devices on the network to communicate. Software applications running on computer devices use Internet protocols to establish and manage information flow in support of applications carried over the Internet. Much as a common set of standard protocols lies at the core of the Internet, common standards and a common body of software are features of many applications at the edge, the most common being those that make up the World Wide Web (WWW or Web).

The physical layer specifies transmission of bits across the network media. The data link layer packages data into frames and provides for reliable transmission of data. This layer contains the media access control and logical link control sublayers. The media access control sublayer is responsible for access to the network media. Access can be provided using a contention or deterministic system. In a contention-based system, any device can transmit when it needs to. Deterministic systems require that a device first possess the right to transmit. Ethernet is a contention system, while Token Ring is deterministic. The logical link control sublayer establishes and maintains network connections and performs flow control and error checking. The Network layer is responsible for routing data across the network. Data is converted to datagrams at this layer, which are sent using connectionless transmission to a specific network address. The Transport layer provides end-to-end reliability using connection-oriented transmissions. Data at the Transport layer is packaged in segments and sent using connection-oriented transmissions, in which an acknowledgment is sent after data is received. If the sender receives no acknowledgment, the data is then re-sent.

The Session layer allows users to establish communications between devices using easily remembered computer names. Dialogue between devices is managed at this layer. At the Session layer, data is packaged as packets. The Presentation layer negotiates and establishes the format for data exchange. Data compression and translation are handled at this layer. The Application layer is the final layer of the OSI model. At this layer all of the interaction between the user's application and the network is handled.

An international network crosses borders and operates in various countries, aspects of the countries, such as the language, telecommunications operators, the offering of telecommunications services, and the telecommunications regulations may influence cost-effective management of the network.

Key Terms

ARPANET The network created by the Advanced Research Projects Agency (ARPA) that became the Internet.

ARP Address resolution protocol (ARP) maps an Internet protocol address (IP address) to a media access control address (MAC address) that is recognized on the local network. The media access control address (MAC address) is the physical machine address, typically that of the network interface card (NIC).

Binary Binary means the use of only two values, zero and one, in encoding data. All digital computers primarily use some form of binary encoding, such as 8 or 16 or 32 binary digits at a time. Characters that you see on screen or type with your keyboard are normally encoded with 8 binary digits.

DHCP DHCP (dynamic host configuration protocol) is a protocol that enables Network Administrators to manage Internet Protocol (IP) addresses from a central location. Using TCP/IP, each machine that connects to the network requires a unique IP address.

DNS Domain names system (DNS) is used to locate and translate Internet domain names into Internet Protocol addresses (IP addresses). A domain name is a meaningful and easy-to-remember name for an Internet address.

Internet protocol This is a protocol of the network layer. This is useful for network addressing requirements.

Protocol The rules of order by which a communications network is operated.

Review Questions

13.1 List a few significant Internet authorities.
13.2 Discuss the TCP/IP stack.
13.3 List the layers of OSI and explain the dependencies between the layers.
13.4 List the various functions of the Internet protocol.
13.5 Differentiate between the IP addressing mechanisms present among the different versions of IP protocol.

Projects

13.1 Prepare a presentation on the various layers of OSI describing the functions of the layers.
13.2 As a network administrator, you are preparing for a presentation to explain the functions of each of the seven layers of the OSI model. You have decided to use an example of how to communicate with a user in a different network than yours. You may use a brief sentence in explaining the function of each layer.

References

Broek, Floris Vanden 2000, *International Networks Cost-Effective Strategies for the New Telecom Regulations and Services*, CRC, USA.

Duck, Michael and Richard Read 2003, *Data Communications and Computer Networks: For Computer Scientists and Engineers*, Prentice Hall, Great Britain.

Kularatna, Nihal and Dileeka Dias 2004, *Essentials of Modern Telecommunications Systems*, Artech House, USA.

Muthukumaran, B. 2005, *Introduction to High Performance Networks*, Vijay Nicole, Chennai, India.

Peterson, Larry L. 2003, *Computer Networks: A Systems Approach*, Morgan Kaufmann, USA.

Prasad, K.V. 2004, *Principles of Digital Communication System and Computer Networks*, Charles River Media.

Tanenbaum, Andrew S. 2002, *Computer Networks*, Prentice Hall.

PART IV

IT Management

- **Chapter 14** E-business Highway—Business Automation Platform
- **Chapter 15** Infrastructure Management
- **Chapter 16** Security Management
- **Chapter 17** Information Management
- **Chapter 18** Audit

CHAPTER

14 E-business Highway—Business Automation Platform

Today, the principal use of the World Wide Web is for interactive access to documents and applications. In almost all cases, such access is by human users, typically working through web browsers, audio players, or other interactive front-end systems. The Web can grow significantly in power and scope if it is extended to support communication between applications, from one program to another.
—From the W3C XML Protocol Working Group Charter

Learning Objectives

After reading this chapter, you should be able to understand:

- intranet and its requirement within an organization
- the differences between intranet, extranet, and Internet
- the various intranet–Internet services

14.1 Intranet

Twenty years ago, most executives looked down on computers as proletarian tools—glorified typewriters and calculators—best relegated to low-level employees such as secretaries, analysts, and technicians. It was the rare executive who would let his/her fingers touch a keyboard, much less incorporate information technology (IT) into his/her strategic thinking. Today, that has changed completely. Chief executives now routinely talk about the strategic value of IT, about how they can use IT to gain a competitive edge, about the 'digitization' of their business models. Behind the change in thinking lies a simple assumption—as IT's potency and ubiquity have increased, so has its strategic value.

Internet services (or network services) are a form of distributed applications in which software components running on networked hosts coordinate via ubiquitous communication protocols on the Internet. The components that accept and handle requests are often referred to as servers and the components that issue requests to and receive responses from servers are called clients. The hypertext transfer protocol (HTTP) is the most popular standard protocol endorsed by the WWW Consortium (W3C) for communication between web servers and clients like Microsoft Internet Explorer (IE) and Mozilla Firefox. A defining characteristic of Internet services is openness. An open system means its components can be extended or reimplemented without affecting its original system functions. It is in contrast to closed systems in which developers must have complete system-level knowledge and full control over the disposition of all system components. A key requirement for the openness is well-defined and stable interfaces between the system components.

The surge in interest and media coverage of intranets recalls the early and heady days of the World Wide Web's burst on the corporate mindscape. Intranet is the new corporate hero, dazzling the covers of business magazines and leading panel discussions at business conferences. Internet and Internet-based technologies enable an expectation of seamless and real-time movement of ideas, expertise, and information, regardless of our location or physical state. This is further defined by the devices we choose to use, as well as the format in which we contribute to or interact with information and knowledge.

An intranet is a private network. A network based on transmisson control protocol/Internet protocol (TCP/IP) protocols (an Internet) belonging to an organization, usually a corporation, accessible only by the organization's members, employees, or others with authorization is classified as intranet. Typically, an intranet includes connections through one or more gateway computers to the outside Internet. Many companies have implemented some form of an internal IP network, known as an intranet. The corporate intranet has been hailed as the most important business tool since the typewriter. The web browser is the universal interface for the intranet. This universal interface will decrease the learning curve of new applications, decrease the need for costly proprietary software upgrades at the desktop, and provide a common look and feel to all applications.

The first corporate intranets were nothing more than hypertext markup language (HTML) versions of paper documents that were updated about as often as their paper counterparts. These pages are difficult to maintain and rarely keep the interest of the end-user. Intranets that rely on static HTML documents to deliver information to employees quickly become stale. Linking the web-intranet interface to internal databases changed the way the intranets were viewed. This new data-driven intranet offers the user up-to-date information and point-and-click access to information otherwise unavailable.

According to Gartner Group, Forrester, and other technology forecasters, corporations spent nearly USD 64 billion on intranet hardware, software, and related services worldwide in 2000.

A website is the product of a scientific effort with highly managed content, studied and calculated audience targeting, and dependencies on relationships with services such as Google, Yahoo!, and other Internet aggregation mechanisms that expose sites and their related content. Websites and intranets are not lifeless, static constructs. Rather, there is a dynamic, organic nature to both the information systems and the broader environments in which they exist. This is not the old world of yellowing cards in a library card catalogue. Here we are talking about complex, adaptive systems with emergent qualities; rich streams of information flowing within and beyond the borders of departments, business units, institutions, and countries; and messiness and mistakes, trial and error, and survival of the fittest. We use the concept of 'information ecology' composed of users, content, and context to address the complex dependencies that exist. In short, we need to understand the business goals behind the web site and the resources available for design and implementation. We need to be aware of the nature and volume of content that exists today and how it might change a year from now. We must also learn about the needs and information seeking behaviours of our major audiences. Good information architecture design is informed by all three areas.

14.1.1 Purpose of Intranet

Today, most industries and individuals are information dependent. Early adopters of information, intensive work, including information workers are found in service-based organizations such as healthcare and education, manufacturing, and energy. Employees want better information faster. Companies are looking for an open-architecture, cost-effective solution for distributing information throughout their organization. Information creation and distribution are carried out by information professionals. An information professional is defined as any individual with training in and responsibility for facilitating the exchange of ideas, expertise, and information between people or between people and documented knowledge. The subject of their work is the user, provider, or contributor of the information as well as the ways in which the information and its exchange is organized and made available in usable forms and formats. An informational professional will co-opt and work on intranet to manage the knowledge flow. The experience of the information user, provider, or contributor is dependent on the interaction that the information professional creates and enables between himself or herself and the information, including the context in which the interaction takes place. Information professionals are essential to the design, implementation, and continued improvement of the experience people have when engaged with information, ideas, and expertise in making sense of a given situation, when creating new knowledge, or when making a decision. Today, in the knowledge economy where the collective knowledge of an organization is its greatest asset, there is a growing acceptance that most information workers live in a world of multitasking and continuous partial attention. We know their needs for information forms and formats vary and they are usually situation-specific, limitless in terms of natural boundaries such as language and physical place, and more sophisticated in terms of expectations around personalization.

In today's very competitive environment, having fast access to accurate information can be crucial for the sales and marketing staff. It can be the difference between making a sale or giving someone time to look elsewhere. Intranets offer tremendous potential as a communication, collaboration, and knowledge building tool that will create new, more efficient ways of doing business. The purpose of an intranet in an organization is primarily to share company information and computing resources among employees. Intranets are very effective with assisting the executive in addressing the roles and activities discussed addressing communication with internal employees and external partners. Intranets are part of the organization that puts them in place, and, as such, they are subject to the organizational dynamics. Organizations exist in complex environments as well. Customers, partners, benefactors, economies, markets, and competitors, and even entire industries can and do impact the organization in predictable and unpredictable ways. Information professionals are faced with the challenge of aligning the functionality of an intranet-based offering of products and services in the context of highly dynamic environments.

All websites and intranets exist within a particular business or organizational context. Whether explicit or implicit, each organization has a mission, goals, strategy, staff, processes and procedures, physical and technology infrastructure, budget, and culture. This collective mix of capabilities, aspirations, and resources is unique to each organization. Information architectures must be uniquely matched to their context. The vocabulary and structure of website and intranet is becoming a major component of the evolving conversation a business has with customers and employees. It influences what they think about products and services. It tells them what to expect from a company in the future. It invites or limits interaction between customers and employees. The information architecture perhaps provides the most tangible snapshot of an organization's mission, vision, values, strategy, and culture.

An intranet is also used to facilitate working in groups and for web conferences. Building an employee intranet can be a key productivity tool for a small and medium business (SMB), particularly as companies grow beyond the efficacy of sneakernet and explore electronic methods of sharing and disseminating corporate knowledge. The 500 to 1,000-person companies are probably the ones getting the highest value from their intranets.

The intranet content is broadly defined to include the documents, applications, services, and metadata that people need to use or find on a site. To employ a technical term, it is the stuff that makes up the site. Originally, intranets were a series of static HTML pages. Today, they are dynamic and complex. Websites and intranets are becoming the unifying means of access to all digital formats within the organization. Oracle databases, product catalogs, Lotus Notes discussion archives, technical reports in Microsoft Word, annual reports in portable document format (PDF), office supply purchasing applications, and video clips of the chief executive officer (CEO) are just a few of the types of documents, databases, and applications.

Intranets are used for real-time conferencing, for expertise location and real-time conversations, as well as for team collaboration. Intranets, as enablers of a connected enterprise, bring together individual, team, unit, and organization-wide virtual spaces. The intranet provides metrics and support security, privacy, authentication, informed decision-making with respect to governance and content management. A significant challenge for intranets has been deciding what needs to be managed and what does not. The intranet is focused more and more on individual and group productivity, whether through deliberate top-down direction or through team choices. Intranets will be used more often to aid executive decisions by providing visualizations of cross-departmental data and cross-unit performance measurements. Dashboards are popular again, and so is the next generation of executive information systems. While technology continues to be a strong focal point when developing an intranet, more organizations realize that a deep understanding of information-user behaviour and organizational priority setting needs to drive decision-making. Intranet users can be very sophisticated, particularly those for whom the Internet is a part of their daily lives. Web-based technologies and new standards, such as simple object access protocol or service oriented architecture protocol (SOAP), have made it easier (not easy, but easier) to integrate web applications in a single user interface and harness information from across organizations.

There are different models of what happens when users look for information. Modelling users' needs and behaviours forces us to ask useful questions about what kind of information the user wants, how much information is enough, and how the user actually interacts with the architecture. Intranets are under-exploited in many organizations. While technology is likely to continue to enable increasingly sophisticated choices, managing expectations, and ensuring intranets are relevant, reliable, and accurate; it will rely on an intersection of technology, organizational need, and the capacity of the organization to meet the needs and to identify and prioritize them appropriately.

14.1.2 Intranet Benefits

Intranets are simply internal Internet that conveniently make information and applications more accessible to an organization without the expensive and time-consuming application rollouts. Corporations with an intranet have the ability to transparently access computing resources, applications, and information that may be located anywhere in the world. Some of the common benefits of intranet are as listed below.

Access to Information

The intranet provides a way for employees to gain better access to more time-sensitive information. Intranets also allow information to be rapidly and economically deployed to a dispersed group of employees. A marketing planner for a global pharmaceutical concern notes that before the deployment of an intranet, his division was spending a sizable amount of US dollars per month on information mailings to sales representatives. Information dissemination was obsolete by the time the information was received by global representatives and affiliates. The development costs of an intranet database were more than offset by savings in printing and mailing.

Collaboration

Intranets help employees collaborate on business processes, such as product development or order fulfilment, which create value for a company and its customers. Specifically, intranets centralize the business process in an easily accessible, platform-independent virtual space. Successful intranets allow employees from a variety of departments to contribute the different skills necessary to carry out a particular process.

Increased Productivity

Productivity increases from intranets arise from more rapid and easier access and exchange of information. Intranets also allow for flexibility in the time of delivery of information. Intranets break though departmental walls to help accomplish business processes more efficiently. For example, a customer complaint might involve people and information from the accounting, sales, and marketing department. Even though the employees necessary to resolve the complaint work in different departments, they are all involved in the process of customer service.

Process-based Corporate Communities

Management gurus are helping companies move away from vertical, hierarchical organizational lines towards horizontal, process-oriented groups that link cross-functional teams focused on the same set of business tasks. The problem is that this requires significant interaction between departments, functions, even countries. Intranet, emerged as the ideal vehicle for creating and empowering process-based corporate communities. The corporate intranet can help a company organize around 'communities of process' both on- and off-line. When Texas Instruments initiated a process-centred organization, oriented around collaborative work groups, software development time fell from 22 to eight months. The Texas Instruments intranet was established after this shift and was designed to reflect and enhance the new organization. Whether it precedes or follows the organizational shift, an intranet that encourages this type of collaborative work environment can provide a significant return on investment (ROI).

Typically, larger enterprises allow users within their intranet to access the public Internet through firewall servers that have the ability to screen messages in both directions so that company security is maintained. Decisions on how intranet development will be handled must be made from a business perspective. In many organizations, information systems (IS) is the controller of all web servers, and all requests must go through IS to set up anything on the network.

14.1.3 Weblog

A weblog is a website (equated to an intranet in most of the companies) that gets updated many times in a day with content presented in a reverse chronological order, that is, the latest content gets pasted on to the top of the page, then the second latest, and so on. Most weblogs or blogs have features that enable the user to update the content via Web. This means that for posting the content, a weblog owner does not need any knowledge on technical matters such as HTML or file transfer protocol (FTP). One reason for the phenomenal success of this web technology is the availability of numerous free tools that enable anyone on the Net/intranet to start a weblog quickly. Almost all the tools contain most of the features required by a weblog that include the facility to post via the Web, tools that let readers express their comments, calendar interface that allows visitors to directly access any day's postings, and the facility to publish new content information through an RSS file.

Case Study—Intranet Dashboard Boosts Australian Job Site

SEEK is Australia's number 1 job site and training provider. It has two customer-facing segments to its business—the SEEK.com.au job board which serves employers and jobseekers, and its online training and development business, SEEK Learning.

With the candidate and skills shortage biting hard in 2008, SEEK's phenomenal growth as Australia's number one online employment website was putting pressure on its customer service staff. As SEEK's product range broadened, its customer service staff increasingly needed quick and easy access to information in response to customer queries.

Contd

SEEK recognized that it needed a new intranet to act as an 'internal information hub' for both its customer service staff and 400-strong personnel generally. It implemented Intranet DASHBOARD (iD) in October, and overnight, the SEEK intranet has become the single point of information for staff at the customer service front line. It has also enhanced internal communications and seen ownership of the intranet pass from IT into the hands of the broader business.

Cisco Systems 'Not an Intranet Portal'

In a study entitled, 'Not an intranet portal,' Cisco Systems found that the intranet significantly reduced the cost of processing expense reports. The study stated that:

- In 1996, Cisco processed 54,000 reports. The amount of dollars processed was USD19 million.
- This required four staff, with a total processing time of four days.
- The cost per report was USD 50.69.
- In 1999, the process had been moved to the intranet. 145,000 expense reports were processed, amounting to USD 77 million.
- The number of staff required was three and the total processing time was four days.
- The cost per report was USD 1.90.

14.1.4 Return on Investment

On reviewing the existing return on investment studies or questioning company executives on their claims of multimillion dollar savings, one finds that calculating intranet return on investment (ROI) is more art than science and more guesstimate than calculation. An intranet can deliver ROI by either reducing the cost or expanding the ability to communicate. By shifting manual processes to the intranet, the cost of accessing and processing information is reduced.

The key is to link intranet ROI to bottom line issues that senior management cares about. These include cost savings, increased productivity, and gaining competitive advantage. Technology managers say their companies often earn back the money spent on intranet development costs in one to three years in savings from expense reductions in widely divergent areas, from printing and postage to employee productivity.

14.2 Extranet

An extranet is a private network that uses Internet protocols, network connectivity, and possibly the public telecommunication system to securely share part of an organization's information or operations with suppliers, vendors, partners, customers, or other businesses. An extranet can be viewed as part of a company's intranet that is extended to users outside the company (for example, normally over the Internet). An intranet is a use of Net and web technology that happens inside your own organization. An extranet is an intranet that people outside your formal organization have access to. For example, consider the initiatives of Maruti. Maruti achieved interconnectivity in the 1990s by leveraging the Internet and deploying an extensive extranet to reach its business partners. This information business to business (B2B) access-point on the Internet allowed suppliers and dealers to access information pertaining to the production plan, supply status of components and vehicles, and status of payments. One was the extranet itself that helped them interface with Maruti for all order placements, enquiries, vehicle schedules, warranty claims, fund reconciliation, and so on. The other was a small application running on the dealers' personal computers (PCs) to manage their daily activities.

Intranet/extranet form a major part of the information systems of many corporates. A lot of companies use intranets/extranets as a powerful tool to manage their resources and work in a systematic manner. One of the most valued services is helping companies enhance their business and communication channels by using intranet and extranet technology. Intranets and extranets make use of all the technology of the internet to supply new and novel ways through secure managed channels and helps to monitor and control business in a cost-effective way. If used properly, they can drastically speed up the pace with which one can communicate within a geographically dispersed community.

14.3 Benefits of Intranets and Extranets

A few benefits of intranets and extranets are as listed below.

14.3.1 Tangible Benefits

The tangible benefits of intranets and extanets are as follows.

- Inexpensive to implement, easy to use, just point and click
- Saves time and money, better information faster
- Connects across disparate platforms, scaleable, and flexible

- Puts users in control of their data
- Huge reduction in paper cost
- Reduction in training and orientation costs for the company; a definite advantage for the human resources (HR) department.

14.3.2 Intangible Benefits

It is not only the removal of paper that leads to organizational benefit. What is done with that information in this new web-enabled environment has a huge impact.

The tangible benefits of intranets and extranets are as follows.

- Improves decision-making
- Empowers users
- Builds a culture of sharing and collaboration
- Facilitates organizational learning
- Breaks down bureaucracy
- Improves quality of life at work
- Improves productivity

14.4 Internet Services

In the 1950s and early 1960s, prior to the widespread internetworking that led to the Internet, most communication networks were limited in that they only allowed communications between the stations on the network. The Internet began in the late 1960s as a research project sponsored by the US Defense Department's Advanced Research Projects Agency (ARPA). The original network, named ARPANET, was launched in October 1969 and included just two sites: the Stanford Research Institute (SRI) and the University of California, Los Angeles (UCLA). This spread of internetwork began to form into the idea of a global internetwork that would be called 'the Internet', and this began to quickly spread as existing networks were converted to become compatible with this. The name 'Internet' was first used in December 1974, and over time the Internet expanded to include other government agencies, universities, research labs, and businesses. This spread quickly across the advanced telecommunication networks of the western world, and then began to penetrate into the rest of the world as it became the de facto international standard and global network. The Internet is named after the Internet protocol, the standard communications protocol used by every computer on the Internet.

Most Internet data travels along a collection of telephone lines and fibre-optic cables that span the world. This collection of lines and cables makes up the so-called backbone of the Internet. Data travels along this backbone at nearly the speed of light, so that one can usually access data on the other side of the world in seconds.

14.5 World Wide Web

In the late 1990s, low-cost personal computers and an extensive, relatively easy-to-use Internet helped computers spread to the majority of households in many developed countries. Many of the activities for which people use the Internet are long-standing and well-rooted in our social system. For instance, one can maintain contacts with friends and family though telephone calls, visits, and letters or can meet new people by joining formal organizations. One can turn to the newspaper for news or weather updates, go to the library for research on a variety of topics, look at advertisements and buy consumer magazines for product information, or visit the bank to conduct financial transactions. By definition, this penetration of the Internet and mobile telecommunications into the way we achieve fundamental goals of connecting to other people, finding information, or entertaining ourselves is changing how we live our lives.

The Web is a system of information distribution using the Internet. It is a global hypertext system of linked documents written based on a set of rules called HTML. These documents can contain both text and non-text (graphics, audio, video) material. Embedded links allow jumps, both within a single document and between documents on the same computer or on other computers anywhere on the Net and computers linked to the Net exchange documents based on the HTTP. World Wide Web browsers are clients which can retrieve and display WWW documents based on the rules laid down by HTML.

14.5.1 Telnet

Telnet allows a user to remotely log in to another computer and run applications, provided the user has a password for the distant computer or the distant computer provides publicly available files.

This is a set of procedures that enables a user of one computer on the Internet to log on to any other computer on the Internet, provided the user has a password for the distant computer or the distant computer provides publicly available files. Telnet is also the name of a computer program that uses those rules to make connections between computers on the Internet.

Many computers that provide large electronic databases, such as library catalogs, often allow users to telnet in to search the databases. Many resources that were once available only through telnet have now become available on the easier-to-use the World Wide Web.

14.5.2 FTP

FTP is an acronym for file transfer protocol. It is a network protocol used to exchange and manipulate files. It is used to exchange files between computer accounts, to transfer files between an account and a desktop computer, or to access software archives on the Internet.

File transfer protocol is a member of the TCP/IP protocol suite that is designed to transfer files between a server and a client computer. This protocol allows bidirectional (download and upload) transfer of binary and American Standard Code for Information Interchange (ASCII) files. The user needs an FTP client to transfer files to and from the remote system, which must have an FTP server. Generally, the user also needs to establish an account on the remote system to FTP files, although many FTP sites permit the use of anonymous FTP. An FTP uses the port numbers 20 and 21. One of these port numbers is for the request, and the other for the download.

Many universities, government agencies, companies, and private individuals have set up publicly accessible archives on the Internet. There are thousands of these sites that contain a myriad of programs, data files, and informational text. At these sites, public directories and files that may be read by the rest of the world via FTP are set aside. These directories are usually named /pub. You can usually find specific directions and information about the site in greeting messages or in files with names such as README.

14.5.3 Search Engine and Data Delivery

The term 'web search engine' refers to a service available on the public Internet that helps users find and retrieve content or information from the publicly accessible Internet. It is a program that searches documents for specified keywords and returns a list of the documents where the keywords were found. A web search engine is a tool designed to search for information on the World Wide Web. Search engines are becoming the most important gateway used to find content.

A search engine is basically composed of three essential technical components: the crawlers or spiders, the index or database of information gathered by the spiders, and the query algorithm that is the 'soul' of the search engine. This algorithm has two parts: the first part defines the matching process between the user's query and the content of the index and the second (related) part of this algorithm sorts and ranks the various hits. The process of searching can roughly be broken down into four basic information processes, or exchanges of information. These are:

- information gathering,
- user querying,
- information provision, and
- user information access.

The web search process of gathering information is driven primarily by automated software agents called robots, spiders, or crawlers that have become central to successful search engines. Once the crawler has downloaded a page and stored it on the search engine's own server, a second programme, known as the indexer, extracts various bits of information regarding the page. Numerous search engines are available on the Internet. Novices will feel most comfortable in a friendly, easy-to-manipulate environment such as Yahoo!, Google, AltaVista, and HotBot, the engines that store the most sites. Most of the directories and smaller search engines provide links to the behemoths such as AltaVista, Google, or Excite. Not so long ago, at the beginning of this century, a lot of search engines were active, and it was the general assumption that competition between search engines would discipline the market. It was also assumed that both information providers and users would be able to benefit from this. Although the number of search engines is still significant, this cannot be said about their market shares.

Common user queries follow a 'pull'-type scheme. The search engines react to keywords introduced by the user and then submit potentially relevant content. Web search engines work by storing information about many web pages, which they retrieve from the WWW itself. Typically, a search engine works by sending out a spider to fetch as many documents as possible. Another program, called an indexer, then reads these documents and creates an index based on the words contained in each document. Each search engine uses a proprietary algorithm to create its indices such that, ideally, only meaningful results are returned for each query. Over the years, one has witnessed a steady transformation. Storage, bandwidth, and processing power have increased dramatically, and automation has become more efficient. Search engines have gradually shifted from a reactive response to the user (pull) to proactively proposing options to the user (push).

While text-based search is efficient for text-only files, this technology and methodology for retrieving digital information has important drawbacks when faced with other formats than text. For example, images that are very relevant for the subject of enquiry will not be listed by the search engine if the file is not accompanied with the relevant tags or textual clues. For instance, although a video may contain a mountain coloured brown and the search engine will not retrieve this video when a user inserts the words 'mountain coloured brown' in his/her search box. The same is true for any other information that is produced in formats other than text.

Search engines generate income mainly from one source, i.e., advertising. Let us take Google as an example. Google generates almost all of its income from advertising. This income is generated mainly by 'Google AdWords'. AdWords enables advertisers to create their own advertisements and state how much money they are willing to spend. They are then charged on the basis of the number of times that the advertisement is clicked on. The second source of income consists of placing the advertisements on third parties' websites. Some search engines offer the opportunity of 'buying' a high position on the list of search results. There are different variations of this. The simplest method involves literally selling the position. Other search engines priority-index the pages of paying parties, so that they rank higher in the list of search results.

14.5.4 Web Services

Web services are rapidly becoming the de facto distributed enterprise computing technology that is widely used for enabling collaborative business processes. By now, most enterprises have gained some initial experience in deploying predominantly internal business applications by consuming web services developed either in-house or offered by third parties. Web services are modular, self-contained applications or application logic developed per a set of open standards. A web service is any service that is available over the Internet, uses a standardized extensible markup language (XML) messaging system, and is not tied to any one operating system or programming language. Web services depend on remote procedure invocation mechanism.

Web services operate by interchanging data that is in the form of XML. The reliance on XML-based data is the fundamental premise of web services.

To be even more precise, it should be noted that the data interchange is done using XML documents. Thus, the input parameters to a web service are in the form of an XML document. The output of a web service will also always be an XML document. Extensible markup language enables all types of information to be exchanged across disparate systems in an easier and better manner than was possible in the past. It provides data from disparate applications and platforms with a standardized 'interchange' format. There are several alternatives for XML messaging. For example, you could use XML remote procedure calls (XML-RPC) or SOAP, both of which are described later in this chapter. Alternatively, you could just use HTTP GET/POST and pass arbitrary XML documents. Any of these options can work. A complete Web service is, therefore, any service that:

- is available over the Internet or private (intranet) networks
- uses a standardized XML messaging system
- is not tied to any one operating system or programming language
- is self-describing via a common XML grammar
- is discoverable via a simple find mechanism

The SOAP is an XML-based protocol for exchanging information between computers. Although SOAP can be used in a variety of messaging systems and can be delivered via a variety of transport protocols, the initial focus of SOAP is remote procedure calls transported via HTTP. It, therefore, enables client applications to easily connect to remote services and invoke remote methods. Other frameworks, including CORBA, DCOM, and Java Remote method invocation (RMI), provide similar functionality to SOAP, but SOAP messages are written entirely in XML and are therefore uniquely platform- and language-independent. For example, a SOAP Java client running on Linux or a Perl client running on Solaris can connect to a Microsoft SOAP server running on Windows 2000. It, therefore, represents a cornerstone of the web service architecture, enabling diverse applications to easily exchange services and data. The SOAP specification defines three major parts: SOAP envelope specification, data encoding rules, and RPC conventions.

The SOAP includes a built-in set of rules for encoding data types. This enables the SOAP message to indicate specific data types, such as integers, floats, doubles, or arrays. Most of the time, the encoding rules are implemented directly by the SOAP toolkit you choose, and are therefore hidden from you. It is nonetheless useful to understand the basics of SOAP encoding, particularly if you are intercepting SOAP messages and trying to debug an application.

With web services, we move from a human-centric Web to an application-centric Web. This does not mean that human beings are entirely out of the picture. It just means that conversations can take place directly between applications as easily as between web browsers and servers.

Most enterprises achieve this by web service enabling legacy applications and enterprise information systems. Implementing a thin SOAP layer on top of existing applications or software components that implement the web services is by now widely practiced by the software industry.

The lifecycle of the Web services development methodology comprises five distinct phases: planning, analysis and design (A&D), realization, execution, and deployment that may be traversed iteratively.

The planning phase is a preparatory step (Phase-1) that serves to streamline and organize consequent phases in the methodology. The analysis and design phase is the next phase (Phase-2) that specifies web services and business processes in a step-wise manner. Service realization (Phase-3) transforms specifications from the analysis and design phase into implementations. The development phase (Phase-4) aims at deploying the service and process realizations and publishing interfaces in a repository. The final phase entails execution (Phase-5), which supports the actual binding and run-time invocation of the deployed services.

Service coupling refers to the degree of interdependence between business processes or web services. Representational coupling of business processes should not depend on specific representational or implementation details and assumptions underlying business processes. Identity coupling of connection channels between services should be unaware of who is providing the service. It is not desirable to keep track of the targets (recipients) of service messages, especially when they are likely to change. Communication protocol coupling of the number of messages exchanged between a sender and addressee in order to accomplish a certain goal should be minimal, given the communication model. Service cohesion defines the degree of the strength of functional relatedness between operations in a service or business process. The service aggregator should strive to offer cohesive (autonomous) processes whose services and service operations are strongly and genuinely related to one another.

Napster's introduction of digital distribution of music in the late 1990s was a rule-breaking strategic action that created chaotic conditions for the content companies. Established music companies, such as Bertelsmann, EMI, and Universal Music, were frantically searching to invent new business models for distributing music but were shackled by strong corporate strategic inertia. Movie companies, similarly, were considering defensive and offensive moves to deal with the radically increased opportunities for piracy associated with digital distribution. Open-source software, especially the emergence of Linux during the 1990s as a viable alternative to Microsoft in the enterprise computing market segment, also represented rule-breaking strategic action that forced proprietary software companies to reconsider their strategies and business models in the early twenty first century.

14.5.5 Surfing

Surfing the Internet is usually seen as fun, dangerous, or a tremendous waste of time, depending on who you ask. All of the above is true, but what is also true is that the Internet is a necessity for today's computer users. The online surfer behaves according to the motto 'I will just take a look'. Often online surfers surf aimlessly on the web and jump from one website to the next. In order to attract online surfers, the address of the website must be known or be attainable via different links. The surfer develops into an online consumer if the company succeeds in establishing goal-directed and repeated contact with the customer. The online consumer keeps the web address in his/her collection of bookmarks and clicks on it for certain services and information. He/she reads the desired information, prints it out, or stores it locally in his/her personal computer for further use. Among the activities performed by an online consumer is the provision of simple feedback, such as the ordering of brochures or the downloading of visual material. He/she demonstrates a greater interest than a passive consumer (surfer) through the retrieval of goal-directed information. The company, therefore, has the chance to encourage and fulfil a need of the online consumer.

14.6 The Global Village

The question of whether or not the world is really becoming a 'global village' is a complicated one, and inevitably involves the consideration of a wide variety of different circumstances.

With the growing demand for digital information, manufacturers of digital devices such as personal computers, Internet appliances, cell phones, cameras, and personal digital assistants, must design products that can communicate with one another. These devices communicate using special languages called protocols. A communication protocol is a set of detailed rules that govern the behaviour of networks of devices connected through copper wire, fibre-optic cables (FOC) or wireless technologies. The types and number of new devices and their complexity are growing rapidly, leading to the evolution of new, and increasingly complex, connectivity protocols. With the explosive growth in Internet connections worldwide, networked communication has the potential to shrink geographic distances and facilitate information exchange among people of various backgrounds.

New information and communication technologies make it possible for developing countries not only to integrate into and compete in the global economy, but also to gain access, and add their contributions to a growing body of global information and knowledge resources on development issues. Empowered by information technology such as search engines and automatic filters, IT users are spending more of their waking hours plugged into the Internet, choosing to interact with information sources customized to their individual interests.

One point clearly supporting the argument that the world is, in fact, moving towards or has already become a global village is the explosion of the Internet over the last two decades as a medium of mass communication and global connectivity.

14.7 Bandwidth

The Internet, in the most simplest of terms, is a group of millions of computers connected by networks. These connections within the Internet can be large or small depending upon the cabling and equipment that is used at a particular Internet location. Bandwidth demands for the Internet are growing rapidly with the increasing popularity of web services. Although high-speed network upgrading has never stopped, clients are experiencing web access delays more often than ever. There is a great demand for latency tolerance technologies for fast delivery of media-rich web contents. It is the size of each network connection that determines how much bandwidth is available. In computer networking and computer science, digital bandwidth, network bandwidth, or just bandwidth is a measure of available or consumed data communication resources expressed in bit/s or multiples of it. Traffic is simply the number of bits that are transferred on network connections. Bandwidth, therefore, is measured in bits (a single 0 or 1). Bits are grouped in bytes which form words, text, and other information that is transferred between your computer and the Internet.

Most applications transmit data using TCP over the wide area network (WAN). File transfer protocol is one example of an application that uses TCP; others include web browsing, e-mail, iTunes, and CRM. In each case performance and throughput are affected by the following.

- The capacity of the network link—such as 155Mb/s
- The congestion on that link—the number of people or applications sharing the link
- The distance the data needs to travel as measured as latency and round-trip time—the time it takes for data to travel over the network and then acknowledge the receipt of that data.

14.7.1 Cost of Bandwidth

The growth, in terms of users adapting to the Internet and bandwidth, outperforms even Moore's law that was defined for the growth of the semi-conductor industry. Internet traffic is doubling every four months and e-commerce revenue is expected to be 600 billion by 2002. Approximately, 150 million people around the world are connected to the Internet as compared to a single-digit number just few years ago. More and more people have become dependent on the Internet for their daily communication and information exchange.

One of the largest components of any monthly IT budget is the cost for network connectivity—either Internet access or a private WAN. The 'bandwidth' cost varies depending on the type of network, the speed, and your time commitment. Also, these prices vary around the world based on the service provider and country providing them.

When purchasing network capacity, many people take the cost/month and divide that by the theoretical network capacity to get the price/capacity.

14.7.2 Internet Service Providers

Enterprises have several options to consider before they begin their search for an Internet service provider (ISP). The requirement and the extent of their dependence on the service will determine the options they can opt for. The requirement of an enterprise could be a broadband or a narrowband connection. For a majority of enterprise users, broadband is the most practical option. Broadband empowers the Internet. It is changing, and will continue to change, the types of content (or media) that are available on the Internet. It will bring rich media, which were previously the preserve of professional environments, into the home. Our living rooms will be transformed into concert halls, movie theaters, and games arcades. So, where the Internet transformed the definition of entertainment, broadband is in the process of reshaping our concept of the home. Between them, they are redefining the popular perception of home entertainment.

14.7.3 Types of Bandwidth

The pricing of access services has emerged as an important policy issue with the liberalization of telecommunications markets. The long-distance service providers who offer either clear bandwidth or IP services must do proper capacity planning to accommodate the demand and reliability aspects. The main services provided on the logical layer of Internet service provision are Internet traffic services and Internet access services. These are provided on top of local communications infrastructure and serve to transmit Internet traffic between the end-users premises and a point of presence of an ISP's network and Internet backbone services, which are provided over long-distance communications infrastructure and serve to transmit data within an ISP's networks and between ISPs' networks. The communication lines over which Internet traffic is transmitted are part of the physical layer of Internet service provision. Cost of bandwidth is inversely proportional to volume. As the volume increases, the cost per megabit will drop substantially. Without enough bandwidth, this new industry will not take off. With a few exceptions, both in theory and practice, access payments are treated as per-unit charges on the amount of the volume of access services purchased by new entrants. In most jurisdictions, regulated access prices are treated in a cost-oriented manner.

But before deciding in favour of a particular ISP, an enterprise should ponder over the parameters listed below and weigh all options

- Estimate the right bandwidth
- Evaluate the ISP's backbone
- Extent of coverage
- Points of presence

Dial-up

The simplest and most common connection over the public switched telephone network (PSTN) is the dial-up connection. We use this connection each time we call up our Internet provider. Using conventional telephone lines and a typical modem, we can attain speeds up to 56 Kbps, but often we are connected at lower speeds.

Leased Line

This is a cost-effective Internet access solution offered by a majority of ISPs. Corporates can receive high-speed Internet connectivity and pay for the bandwidth according to the usage. It is considered to be a reliable network, and corporates can get a leased line from basic service providers. Leased lines come in 64 kbps, 124 kbps, 256 kbps, and 512 kbps. Leased lines are offered by some of the service providers such as Videsh Sanchar Nigam Limited (VSNL) and Bharat Sanchar Nigam Limited (BSNL).

Dedicated leased lines are either analog or digital. Presently, most are digital. Because of digital transmission, digital leased lines are faster than analog lines, and they are less susceptible to interference. Digital data service (DDS) is the class of service offered by telecommunications companies for transmitting digital data as opposed to voice. Digital data service transmission requires a device called channel service unit/data service unit (CSU/DSU). The CSU is used to terminate a DS1 or DS0 (56/64 Kbps) digital circuit. It performs line conditioning, protection, loop-back, and timing functions.

VSAT

Satellites can be used for accessing the Internet one-way or two-way. To access the Internet, the corporate user has to put up a very small aperture terminal (VSAT), which comprises an antenna and an adapter card, at the subscriber's PC connected to the external outdoor antenna by a cable. The user sends the request by the normal telephone line and downloads through a satellite broadcast that is received by the VSAT and comes to the PC. In the two-way mode, both the request and the download happen through the satellite. This offers significant benefits to consumers, including an 'always-on' connection that saves time when dialing-up to the Internet and eliminates the need for a second telephone line. It is provided by many operators such as HECL, Bharti Broadband, HCL Comnet, and Comsat Max.

SUMMARY

This chapter outlined the need for business automation platform. Today, browsing through a vast indescribable amount of information and entertainment data at anytime and anywhere has become a reality. The intranet, in many instances, grew out of grass-roots initiatives with little thought to connecting the intellectual assets of the enterprise. The primary purpose has been to communicate with employees and to share broad-based policy documents. Going forward, there are several opportunities centred on collaboration and leveraging ideas. Search machines, such as Google, bring a certain amount of order to this knowledge chaos and it was this particular discovery by two Stanford students that revolutionized global knowledge within a very quick time. Accessing data quickly assumes that the data can be read and is available somewhere in the World Wide Web. Its readability is based on file formats, classifications, and index or metadata which must be defined before being accessed by the search machine.

Business today depends on information and access to information. Information access is enabled via the intranet, the extranet, and the Internet. The various incarnations of the network also provide the user with a variety of services. Some of the common services include services such as FTP, telnet, etc.

Key Terms

Portal A type of website that provides access and interface to multiple sources of information. The sources of information may be autonomous or tightly integrated.

Search engine Software that identifies web content, organizes and indexes it, searches it, and displays results based on relevance.

Web page A defined set of information (also known as content) that is presented to a web user. A web page is usually served by a web server. Web pages may be part of a website.

Review Questions

14.1 What is the need for intranet in an organization?

14.2 What are the benefits of intranet?

14.3 What is an extranet and how does it differ from the intranet?

14.4 What are the tangible benefits of intranet and extranet?

14.5 List the various services available for use within an organization.

Projects

14.1 You are the network administrator for your company and you have acquired a new Unix server that uses the entire TCP/IP protocol suite. Your client computers must be able to retrieve the data that is on the server and perform file transfer operations. Will you need additional protocols to satisfy the needs of your client computers? Explain.

14.2 You are the network administrator for your small network that uses Microsoft Windows-based client computers using NetBEUI. Your company will soon acquire a new NetWare-based server. Your client computers must be able to retrieve the data that is on the server and perform file transfer operations. Will you need additional protocols to satisfy the needs of your client computers? Explain.

REFERENCES

Kangas, Kalle 2003, *Business Strategies for Information Technology Management*, Idea Group, USA.

CHAPTER

15 Infrastructure Management

A journey of a thousand miles begins with a single step.
—Confucious

Learning Objectives

After reading this chapter, you should be able to understand:
- the need for infrastructure management
- the basic concepts of ITIL
- the levels of support
- the differences between the various levels of support

15.1 Introduction

Achieving strategic alignment between information systems (IS) organization and business has been a top IS management issue for more than a decade. In the past, achieving strategic information technology (IT) alignment was expected to result primarily from a periodic IT planning process. Today, however, the emphasis is on a continuous assessment of the alignment of IT investments in not only systems, but also IT people and IT processes. The foundation of successfully managing IT is to have an appropriate IT organization in place. Information technology is a general organization capability. Similar to functions like finance or human resources or manufacturing, IT function has processes and data elements it needs to manage. It manages the definition of process, data, and system architectures. It manages the creation and operation of physical data and software artifacts implementing them, and hardware computing platforms supporting those artifacts. It manages process concepts such as change and incident tickets, work orders, services, and systems as cooperatively defined with the client, and more. It also manages the human and financial resources necessary to support the IT capability.

The turbulence in which companies operate today has reached peak levels. Since 1995, mergers and acquisitions have increased in absolute numbers and size. Mergers permit economies of scale, which translates to more and larger workforce reductions. Despite drivers toward re-centralization, the forces that originally led to diffusion of IT management responsibility still exist. The need to respond quickly to competitive thrusts continues to increase the value of independent IT decision-making by line managers.

The development and management of IT infrastructure is clearly a centralized IS responsibility. However, it only solves half of the IT management and planning problem. The IT organizations can then move on to managing selected aspects of IT. The detailed structure of IT organizations is as varied as the range of businesses carried out by companies, and yet with the exception of companies where IT forms an integral part of the end product, the essence of IT service requirements remains very similar. The variety observed in IT organizations delivering this service appears to be primarily a product of the rapid growth of IT over time, all the while exposed to pushing and pulling from various parts of the business and IT itself.

At the most fundamental level, organizations invest in technology in compliance with mandated legal and accounting requirements and the like. At the next level, an enterprise expends resources to maintain its existing base of IT assets, including hardware and software maintenance, system licenses and upgrades, security services, desktop, storage, and printer expansions and replacements. These investments are meant to keep the system up and therefore, are not discretionary; nor are these costs stagnant. They go up with inflation and keep growing as new workers are added or as the network and related IT infrastructures grow. Because none of these IT products and services run on their own, IS must also provide significant and cost-effective end-user operations, and production support and troubleshooting. Later, since neither the requirements of IS customers nor the evolution of IT itself are static, there is a constant need to enhance existing IT products and services and to invest strategically in new IT capabilities.

The services delivered by information systems to its customers across the enterprise have evolved over time and are in a constant state of flux as both the business needs of the organization and its underlying enabling technologies evolve. Information systems management will provide its customers with a comprehensive understanding of the ongoing services delivered to them by the IS organization. Furthermore, service-level management establishes a routine for the capture of new service requirements, for the measurement and assessment of current service delivery, and for alerting the customer to emerging IT-enabled business opportunities. Thus, IS service delivery management will ensure both the IS resources focused on delivering the highest value to the customer and that the customer appreciates the benefits of the products and services so delivered. Taken together, these various layers of IT investment establish the boundaries of the IS organization's internal economy.

Viewed from a business perspective, the whole idea is that the business should be supported in exploiting IT to its advantage. This implies that IT delivers the following three core services that are normally required irrespective of the size or geographic spread of a company.

- Run applications: Business applications, such as enterprise resource planning, customer relationship management or supply chain management systems, need to be reliably operated and maintained. Business users have to be supported in using these applications.
- Run infrastructure: The whole IT infrastructure, on which business applications run and which ultimately provides users with a modern office environment, needs to be operated and maintained. The infrastructure itself covers personal computers (PCs), data centres, networks, and basic office tools such as e-mail and wordprocessing. Users need to be supported in using the infrastructure.
- Integrate new solutions: Information needs that remain open with deployed applications need to be resolved. This activity can range from development of simple end-user applications or decision-support queries to assistance in selection and deployment of entirely new systems.

15.2 IT Function

A business does not exist without a market. Apart from the business units, there is a unit that is responsible for support of the business unit which is called IT function, or IT. This IT function may be centralized in one department, or decentralized within the various business units. The IT functions within companies deliver products and services to the internal market. The demand for IT services is defined by the lines of business of the company. Balancing demand and supply is a very important characteristic of IT performance management.

The IT value of a company depends on the strategic focus of the company and the level of perception of the IT services that are delivered to that company. As business units become more aware of IT possibilities, and IT services become more mature, IT departments are more and more pressed to prove their added value to the business. The business value of IT is subjective. It is about the perceived added value of IT to a business, based on a price/performance ratio.

15.2.1 IT Management

Business processes are increasingly dependent on complex IT systems and the amount of information accumulated by systems is large. This opens up threats and opportunities for a firm and it is an implicit expectation that IT should manage both. There are two extremes of IT organizations that can in principle result from the forces between IT demand and supply in global firms: either totally decentralized or centralized.

In the former extreme, each unit has its own distinct IT organization which delivers all IT services to the respective units. The IT services themselves may or may not duplicate those provided in other units, but either way the responsibility and authority for IT implementation lie with the unit and the role played by the IT head office, if any, is that of coordination. The IT head office may not even have information on IT costs and head count on IT with respect to individual units.

By contrast, at the other end of the scale in the centralized extreme, a large, centrally controlled core organization provides harmonized IT services to satellite affiliates which themselves only carry out those IT tasks which have to be done on site.

Most IT organizations fall somewhere between the two extremes of centralization or decentralization, with one of the most common constellations being global management of wide area networks, regional data centres, and local help desks and support. The degree of freedom behind the variants above is the ability in a multinational firm to distribute or consolidate the IT organization geographically. However, the IT organization can also be distributed across business functions in any particular location. Wherever the served business functions or units are large and their requirements sufficiently specific, dedication of at least part of the IT organization to them may be warranted.

The common response to this challenge is for the IT organization to adopt a matrix structure that faithfully reflects the matrix structure of the business served. A matrix organization may be justified for a business as a whole, however, the IT organization is an order of magnitude smaller and runs the risk of becoming excessively fragmented in smaller locations.

Business processes require optimization, and to optimize them they must be measured. The economic efficiency of IT services can be assessed by examining the relationship between the IT service effect on cost and quality of business processes, and the IT service production cost. Of course, these production costs include the application system's costs of production and maintenance, and the costs of planning the IT services. This view of economic efficiency is necessary in order to make decisions regarding the provision of additional IT services. However, it can also be used as an element of control for examining the effects of IT services on business processes, in order to recognize the right time for terminating the provision of an IT service.

15.3 Overview of ITIL Framework

In today's competitive business climate, IT has moved from a support organization for the latest applications and technology to focus on business service delivery. Information technology executives are challenged to demonstrate increased value for each IT investment while striving for continuous service improvement and a secure environment. Companies rely on IT to conduct day-to-day business, fuel growth and differentiate themselves from their competition. Front-page issues, such as identity theft and Sarbanes-Oxley legislation, have made businesses a lightning rod for public scrutiny. Addressing issues such as these while simultaneously enabling business growth, has elevated IT from a support function to a true business driver. For IT to deliver maximum value, it cannot be managed as isolated islands of technology. Leveraging disparate technology silos into a unified platform is a well-proven strategy in the software industry. SAP combined manufacturing, financial, human resource (HR) software and defined enterprise resource planning (ERP), Siebel combined automated marketing, sales force automation, and customer service software and defined customer relationship management (CRM) are some of the major business support services maintained by IT.

Applying best practices across governance, business service management, security management, and their sub-disciplines is a new challenge that requires an enterprise approach to IT management. Because the IT value chain requirements are still emerging, any system intended to manage it will require flexibility, just as current ERP systems provide frameworks for adapting them. Such adaptations, however, limit the forward compatibility of the systems with subsequent vendor releases; techniques for managing this problem must be part of the automation strategy. In response, IT has turned to best practices such as IT Infrastructure Library (ITIL), COBIT, ISO17799, and ISO27001 to tackle service improvement, IT governance, and risk management.

The ITIL implementation is one of the hottest topics in IT today. 'Information Technology Infrastructure Library' was published in 1987 to modest critical acclaim and spent its formative years as the preserve of large government and corporate IT departments with equally large budgets. It is a compendium of best practices from many companies in many industries. It is a documented set of processes designed to define how a company's IT functions can operate. It contains a series of statements defining the procedures, controls, and resources that should be applied to a variety of IT-related processes. It represents the best thinking of thousands of people about how IT should be run, what impact IT can have on the business it supports, and how to gain the most value from IT investments. One of the stated goals of ITIL is to help decision-makers take better decisions by ensuring that adequate IT information is available to support those decisions. Information technology infrastructure library is now big business and is considered by some as the solution that will finally legitimize the IT function within the business world. The latest information on ITIL comes from the UK Office Government Commerce (OGC) through its website http://www.best-management-practice.com/.

It even appears that some large outsourcers and contract vendors may be resistant to establishing a system of record for IT dependencies, as some of their revenue stream derives from continually reanalysing dependencies or monopolizing their knowledge of them. Being the sole source of understanding for complex system dependencies gives any outsourcing partner a measure of power over an organization's operations that should be construed as significant risk. The ITIL insists complete access, ownership to management information at its source.

15.3.1 BS15000 and ISO20000

Standards work best when they describe something in quantifiable terms that can be independently tested and verified. Standards define a definite output in unambiguous language to prevent miscommunications and misunderstandings. With the publication of British Standards (BS) 15000 and its internationalized counterpart International Organization for Standardization ISO 20000, IT service management has taken a giant leap towards acceptance and legitimacy. Both standards are based upon the ITIL documentation library and outline a system for the management of the IT function geared towards the provision of IT services.

15.4 Management and Measurements

Metrics play an important role in performance management. It is evident that without proper measurement of the performance, IT managers will not be able to control the IT activities. The concept of metrics management is essential to process improvement frameworks such as Six Sigma. Processes are controlled by metrics. But what is a metric? A metric is a measurement. It is an information and not an activity. It is an information that drives activity.

A metric is a variable that can be measured and used to quantify the performance of a marketing effort. With respect to marketing efforts, metrics fall into the following categories: return metrics, investment cost metrics, operational metrics, and business impact metrics. Return metrics are often referred to as key performance indicators (KPI) or success metrics. The costs of marketing programs, goods sold, and capital are investment cost metrics that must be optimally related to metrics measuring investment returns. Business processes require metrics. The use of metrics to assess and guide process is called performance management or business performance management.

A hierarchical metrics structure is characteristic of performance management and the business intelligence methods supporting it. The hierarchy of metrics may progress from simple operational reporting to complex, derived leading indicators. Such approaches have become well- established in many types of business activities, and attention is now turning to measuring IT similarly.

Metrics are the quantifiable objects that management needs to analyse in order to get feedback on the execution of their plans. These plans can be strategic, tactical, or operational, resulting in strategic, tactical, and operational metrics.

Another widely used term for metrics is performance indicators. Examples of IT metrics are the availability of an application, number of help desk calls per month, total cost of ownership for a desktop, etc. [Wiggers et al. 2004]. For each metric or performance indicator, multiple measurements need to be defined, implemented, measured, and reported. These measurements can be automatically generated or may be gathered using extensive user surveys. Other IT performance metrics of importance are IT human capital effectiveness measures such as the desired inventory of internal IT skillsets, the optimal number of external employees as a percentage of the total IT workforce, and the ideal turnover rate for internal IT employees to ensure knowledge-retention as well as an infusion of new skills.

Information technology staff are geographically and temporally separated from the systems and applications they manage. All troubleshooting, monitoring, and management of systems and applications occur remotely. The systems and applications being managed are complex and hugely configurable. They are also made up of multiple components such as hardware, software, networking devices, networks, and supporting infrastructure such as environmental and electrical systems. In the event of a problem, many of these components need to be checked in order to eliminate them as a cause. The systematic planning and implementation of fault-tolerant architectures and the use of systems without single points of failure help to minimise the likelihood of major IT incidents and service outages. Information technology controls are nothing without the appropriate management system to monitor their implementation, validity and usage. Process and procedures are needed to oversee the IT environment as well as to continually assess the level of residual risk that is inherent in the complete system of IT including the infrastructure, application portfolio, organizational structure, procedures, practices, and people. Implementing IT service management processes can go some way to holistically managing the risks associated with IT.

The ITIL promotes the setting up of regular dialogues between service providers and service consumers to thrash out requirements and concerns regarding delivery quality, etc. The output of such conversations are intended to form service level agreements (SLAs) which clearly define what the business expects of the IT function and the performance metrics in place to measure compliance against the agreed terms.

15.4.1 IT Service as a Pproduct

A company's success today is integrally linked with its IT services. Services are predictable events, which are easily metered and with which IS personnel and their customers have considerable experience and a reasonably firm set of expectations. Any disruptions or degradations in IT service can cause serious damage to business. That means IT professionals are under considerable pressure to ensure that required IT service levels are delivered to users. Changes to the IT environment are planned and implemented in a structured manner to ensure that the desired result is achieved. Where multiple changes are required to be managed as part of a wider program, they are grouped together and managed as a release. The components of the IS service delivery management process include the comprehensive mapping of all IS services against the enterprise communities that consume those services. Information Systems management needs to segment its customer base and conceptually align IS services by customer. If the IS organization already works in a business environment where its services are billed out to recover costs, this task can be easily accomplished.

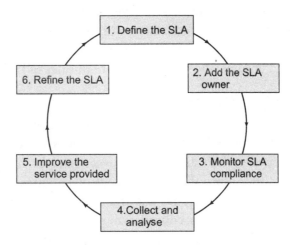

FIGURE 15.1 SLA Management Cycle

The purpose of the SLA is to identify the value that IS brings to the organization or departments being served. Service level agreement is a tool for communicating vital information to key constituents on how they can most effectively interact with the IS organization (refer to Figure 15.1). Typically, the document includes contact names, phone numbers, and e-mail addresses and the levels of services. Service level agreement also help shape customer expectations and identify customer responsibilities in dealing with IS. They define IS performance metrics for the resolution of problems and for responding to customer inquiries. Considering the complexity and dynamic nature of today's enterprise networks, it can be quite a challenge to achieve the high levels of service that users require and demand.

Inventory and asset management provide an accounting of the entities which comprise the enterprise infrastructure, and track each entity through its own life cycle. Configuration management encompasses both inventory and asset management. In addition, it includes establishing and maintaining the relationships between entities to provide a holistic view of the IT infrastructure to support all other IT service management processes. Configuration management database (CMDB) is a functional data repository that unifies and simplifies the management of configuration information. This information is leveraged by the other dependent processes. Configuration management database does not participate within the balanced process model as it is a repository rather than a process. It does naturally participate as a repository and point of reference for all processes that leverage it.

In total, the service-level management process will ensure the proper alignment between customer needs and expectations on one hand, and IS resources on the other. The process clearly defines roles and responsibilities, and ensures understanding on both sides of IS service delivery.

15.4.2 Service Life Cycle

In today's fast-paced communications environment, the ability to support a wide range of network architectures, service types, and industry standards may mean the difference between success and failure. It is anticipated that by 2015, more than 75 per cent of IT infrastructure will be purchased as a service from external and internal hosting providers. To understand the IT infrastructure as a collection of related entities, one considers an object model of the enterprise. Beginning with a traditional view of the IT infrastructure as a group of technologies, the model should at least encompass all the IT assets which comprise the enterprise IT infrastructure. These assets, from a company perspective, include the processes, roles, and services. Adopting the ITIL framework for service management, every distinct entity that is part of the infrastructure can be denoted as a configuration item (CI). The dynamic management of CI defines the service life cycle of IT infrastructure.

The service life cycle model was developed with this diverse set of services and networks in mind. Service life cycle management governs the creation and management of service instances for customers. The framework exploits a model-driven approach with information about a service instance captured in a series of model states and separable supplemental models that describe software and infrastructure provider information. Service life-cycle management comprises nine core elements: service sales and marketing, service contract management, customer service and support, installed-base management, warranty and claims management, field service, depot repair, service parts management, and financial management. These functionalities provide true service life-cycle management, which not only reduces costs and increases revenue, but also increases customer retention, a critical competitive differentiator in today's fast-changing marketplace.

A service is a collection of IT resources that is provided to a customer to enable them to operate their business on a day-to-day and long-term basis. Software providers provide software that is packaged as a service. Software vendors create software. Customers contract with an infrastructure provider or software provider to consume a service. A service implements business processes for customers. A service instance provides the service to a customer.

Service providers can offer services with greater process maturity and improved efficiency while amortizing infrastructure and labour costs across the many businesses that they serve. Rather than supporting solely circuit or packet-based networks, the service life cycle model is truly network convergent. It enables service providers to integrate the platform into multiple networks simultaneously, preserving investments in legacy infrastructure while deploying cutting-edge infrastructure to support new IP-based services. As the demand for higher value services increases, there will be a greater need for service customization, automation for the provisioning and management of services, and the ability to offer services that satisfy non-functional requirements.

A service catalogue identifies the services that can be provided. Given the context of supporting high-value enterprise services for a software vendor, such as SAP, each entry in the catalogue describes a service that is a collection of related business processes. Once a service has been selected by the customer, IT function creates an entry in the catalogue to track the service instance. The service life cycle model transitions between states as the tool-set operates on the service instance. The infrastructure design service uses configuration parameters and requirements information collected from the customer and software vendor in previous states. This helps to select an appropriate infrastructure design pattern from a collection of design alternatives for the service.

15.4.3 Interaction Management

Interaction management is a term some companies use to describe a class of web services that support online relationships (called communities) and transactions between employees, customers, partners, and suppliers. Today, service desk software is built on interaction management principles.

15.4.4 Request and Incident Management

An event is raw material. It is any operational signal emitted by any production CI. Information Technology Infrastructure Library implies that an event is equivalent to an automatically detected incident. Events are also indicators of capacity consumption and supports measurements like hardware utilization, memory, transactions, and so forth.

Information Technology Infrastructure Library defines incident as any event which is not part of the standard operation of a service and which causes, or may cause, an interruption to, or a reduction in, the quality of that service. Incident management is the restoration of service. The overall process of incident and problem starts with the identification of an interruption to established production services.

This can be through a user contact or through operational monitoring. The service is first restored to operational status and, if warranted, forwarded to problem management. This provides proactive root cause analysis and recommended resolutions, which may involve software or hardware changes, process changes, training changes, and so forth.

15.4.5 Problem Management

Problem management plays an important role in the detection of problems, and provides solutions to them and prevents their recurrence. Problem management is the root cause analysis and prevention of future outages, including proactive initiatives. The objective of problem management is to minimize the impact of problems on the organization.

Problem management begins at the start of a service life cycle and should be applied to all aspects of IT including application development, server building, desktop deployment, user training, and service operation. As more problems are discovered, recorded, researched, and resolved, IT will experience fewer failures. If problem management is performed during the period when a service is envisioned, planned, designed, built, and stabilized, the service will be deployed into productive use with fewer failures and higher customer satisfaction.

Problem control is concerned with identification of the underlying cause of incidents. Problems should be classified by category, impact, urgency, priority, and status. Classification is important because it makes the burden of management and reporting easier. A problem which will impact two users of a particular module of a program will have a far lower urgency than a problem that impacts higher number of users. The status of the problem will also affect the urgency, for example a problem with a usable work around will probably be less urgent than one where no work exists, regardless of the number of users affected.

The goal of proactive problem management is to prevent incidents by identifying weaknesses in the IT infrastructure, before any issues have been caused. Proactive problem management will involve trend analysis and targeted preventative action. Many organizations are performing reactive problem management to some degree; very few organizations are successfully undertaking the proactive approach to problem management simply because of the difficulties involved in implementation.

15.4.6 Change Management

A change is an authorization to alter the state of some configuration item (CI). The ITIL defines change as the addition, modification, or removal of approved, supported, or base-lined hardware, network, software, application, environment, system, desktop build, or associated documentation. It defines request for change (RFC) as form, or screen, used to record details of a request for a change to any CI within an infrastructure or to procedures and items associated with the infrastructure.

A change by definition affects CIs, and CIs are objects under change control. In the context of enterprise IT, an atomic change is a change that is successfully applied or is rolled back completely. A consistent change means that the change, when deployed, leaves the item in a stable state. If a change has some part that would be rolled back and another part would stay, it should be framed as two changes.

An isolated change means that it can go in without affecting other changes or item functionality, and is not affected by other concurrent changes. A durable change is one that, once executed, is stable and permanent. Changes may require a service request to implement. This will depend on the maturity of the IT organization.

15.4.7 Service Centre Organization

Service can generically be defined as a collection of capabilities, with each capability being a set of actions performed by an entity in response to a request issued by another entity. The given service definition points to several considerations. First, to be accessed a service must provide access points and interfaces. Second, some authentication mechanisms should be performed in order to preclude intruders from accessing the service offered by the entity of interest. Third, talking of constrained requests means the possibility of putting some requirements over the service capabilities.

Managing a service provisioning system at run-time necessitates a logical representation of the activities being carried by that system, especially service sessions. At operation time, the service dynamics is described by its behaviour and the on-going communication groups. An organization is said to be a service-centred organization when it satisfies the above requirements.

15.4.8 Immediate Response Model

Implementing the immediate response model means that the service flow will attempt to handle calls as and when they arrive. Customers will contact the service centre using the contact methods provided for them. These methods can include the telephone, a line to stand in, e-mail, fax, chat room, etc. Once customers have contacted the service centre, the help desk agent must have some mechanism for routing them to the next available agent or engineer or service representative. The routing mechanism can be as simple as one line for each agent, or it can be a sophisticated call tree that routes the customer to the most appropriate queue based on customer input. The customer then waits for the next available agent in that queue.

There are other routing options in between. For example, if a call distribution system is not available, one or more receptionists can take calls and then manually route them to the appropriate person or queue in the next tier. The pool of agents that provide immediate response are in Tier 1. Tier 1 can be one pool of agents or it can be multiple pools of agents. This model ignores complexity and variability because it assumes that each resource in the pool can handle any issue that comes in. The volume and complexity of requests play an equally important role in determining how to structure the service centre organization.

15.4.9 Managed Response Model

The managed response model is very similar, at least from an organizational perspective, to the immediate response model. The organization of the resources that handle the request in the managed response model is no different than the structure of the immediate response model. The same people can handle these requests. Just as in the immediate model, the calls are routed to the appropriate queue depends on the deployed tool set like service desk. Many help desk systems now have the ability to automatically log, prioritize, and route the request to a queue.

By definition, the managed response model delays the servicing of the request, which may not be acceptable to the customer. A more significant disadvantage is that it has the potential to actually generate more calls to the service centre than if the request had been initiated over the phone in the first place. This occurs because the agent who handles the request will often have to contact the customer to gather additional information.

15.4.10 Service Desk/Help Desk

The ITIL emphasizes that the service desk is a function, not a process, which is debatable. Service request management is repeatable and measurable, which qualifies it as a first-class process. Service request management most importantly requires a user interface (online, phone, fax, walk-up, etc.), and workflow, so that requests are tracked and routed for correct fulfilment. It requires a defined process for establishing new orderable services and assessing the performance of those currently offered. It requires a robust workflow capability with complex conditional routing of tasks and the ability to monitor and manage request queues for performance. It also requires knowledge management, in the sense of managing unstructured data, for the frequent use of scripts or documented interactions or activity sequences undertaken by the service desk staff.

15.4.11 Levels of Support

Level of support indicates a specific extent of technical assistance in the total range of assistance that is provided by an information technology product to its customers. The levels are designed to group support personnel based on their abilities and subsequently pay.

Level 1 support are the phone and e-mail ticket responders. They are the first responders and least skilled of the bunch. Level 2 support are expected to handle basic system administration tasks and coordinate with Level I support operators to resolve customer issues. Level 2 is also responsible for all change implementation. Level 3 is a spillover group that covers the staff trained to resolve the problems which cannot be handled by Level 2 support personnel. Staff in the Level 3 classification are generally expected to have a solid working knowledge of the company's systems and software and use that skill base in conjunction with innate problem solving skills. Level 3 support personnel are the last resort when confronting complex customer issues.

15.5 Information Technology Measurements

Information technology service management focuses on the planning, monitoring, and control of IT services. Nearly 80 per cent of costs of IT investment consists of costs of integration, operation, and exploitation of IT systems. These costs are referred to as IT service costs. Due to the various defined and undefined IT services, it is difficult to manage IT services in a formal and efficient way.

15.5.1 Measurement Metrics

Uptime is important. In high-value, high-transaction volume operations, it can cost a lot.

15.5.2 Uptime and Downtime

Uptime is a computer industry term for the time during which a computer is operational. Downtime is the time when it is not operational. Uptime is sometimes measured in terms of a percentile. For e-commerce sites, uptime and downtime is highly relevant because downtime could translate to lost revenues, and in this highly competitive world, it could very well mean your business' survival.

$$\text{uptime percentage} = \frac{u}{d} \times 100 \qquad (15.1)$$

where $u =$ uptime, $d =$ downtime

Uptime	Uptime Percentage	Maximum Downtime Per Year
Six nines	99.9999	31.5 seconds
Five nines	99.999	5 minutes 35 seconds
Four nines	99.99	52 minutes 33 seconds
Three nines	99.9	8 hours 46 minutes
Two nines	99.0	87 hours 36 minutes
One nine	90.0	36 days 12 hours

15.5.3 Availability

Availability management implies first the existence of a robust monitoring and event management architecture, typically one of the earlier internal tool acquisitions for most IT organizations. Availability management is concerned with ensuring that systems are available for their intended purpose (for example, user access), in keeping with any service level agreements (SLAs). The business service topology mappings built for availability purposes should be aligned with the central configuration management information store to prevent redundant dependency management.

15.5.4 Bit Error Rate

On the surface, bit error rate (BER) is a simple concept. Its definition is simply:

$$\mathrm{BER} = \frac{\mathrm{Errors}}{\mathrm{Total\ number\ of\ bits}} \qquad (15.2)$$

With a strong signal and an unperturbed signal path, this number is so small as to be insignificant. The bit error rate is the percentage of bits that have errors relative to the total number of bits received in a transmission.

15.5.5 Mean Time between Failure

Mean time between failure (MTBF) is a reliability term used loosely throughout many industries. It has been used for over 60 years as a basis for various decisions. Mean time between failure is largely based on assumptions and definition of failure and attention to these details are paramount to proper interpretation. It impacts both reliability and availability. Reliability is the ability of a system or component to perform its required functions under stated conditions for a specified period of time. Availability, on the other hand, is the degree to which a system or component is operational and accessible when required for use. Mean time between failure is a basic measure of a system's reliability. It is typically represented in units of hours.

15.5.6 Mean Time to Repair MTTR

Mean time to repair (or recover), (MTTR) is the expected time to recover a system from a failure. This may include the time it takes to diagnose the problem, the time it takes to get a repair technician onsite, and the time it takes to physically repair the system.

15.6 Outage Management

The word outage may have different meanings depending on who is analysing the event. IEEE 1159 refers to power interruptions lasting up to three seconds as momentary, interruptions lasting more than three seconds and less than one minute as temporary, and interruptions greater than one minute as sustained. Utilities may have their own definitions or classifications for an outage, which may or may not follow the guidelines outlined in IEEE 1159. For example, a utility may combine momentary and temporary outages into a broader momentary outage category and may determine outages lasting several minutes which are still to be classified as momentary. While an outage is typically caused by the clearing of a fault on the distribution system, it may also be caused by a fault or open circuit on the customer premise.

Planned outages or disruptions to services are the result of a planned action on the part of IT and communication services. Accordingly, the manner in which they can occur is for the most part predictable, and thus they can be managed in a controlled manner, which minimizes the potential disruption to the clients. Gartner initiated coverage of outage management systems (OMSs) in 2007 to address the needs of clients seeking solutions to improve their emergency network restoration process.

15.6.1 System and Network Outage

Outage is the unavailability of a equipment or a service. There are two types of outages namely, system outage and network outage. An outage can occur as a scheduled outage or an unscheduled outage. A scheduled outage is carried out as a planned activity with very low interruption to the customers. However, unscheduled outages are the most critical events which requires special attention. Dealing with unscheduled outages is the most critical operational activity on a distribution network. It requires a high level of coordination between the customer call centre, the control centre dispatcher, and the field crews. The amount of data to be absorbed can be very large and the speed of response has to be as fast as possible. Under these conditions, it is vital that the dispatcher have a set of tools that ensures that each and every outage is dealt with as efficiently as possible. Prioritization of outages must be done quickly. Conditions can change rapidly and the dispatcher as well as the rest of the organization must be kept up to date.

15.7 Service Level Agreements

Today's businesses demand a granulated measure of outsourcing project success, one that balances business objectives against the parameters of risk, business impact, and business continuity. Service level agreements enable this by focusing on the highest quality or timeliness of project completion delivered at the lowest cost, using measurements of defined metrics tied to vendor performance. Service level agreements validate expectations of the respective parties and set parameters for measuring project success. This important tool measures the status of an outsourcing project and empirically scores the performance of the vendor using measurable and enforceable results.

Service level management is the name given to the processes of planning, coordinating, drafting, agreeing, monitoring, and reporting on SLAs. They are fundamental to both providers and recipients of services. They define the terms of engagement and the rules that will govern the relationship. Service level agreements, a formal negotiated agreement between two parties, is sometimes called a service-level guarantee. It is a contract that exists between the service provider and the customer, designed to create a common understanding about services, priorities, responsibilities, etc. Service level agreements define the service provider vs. service customer relation in a formal way.

Structuring an SLA is an important, multiple-step process involving both the client and the vendor. In order to successfully meet business objectives, SLAs best practices dictate that the vendor and client collaborate to conduct a detailed assessment of the client's existing applications suite, new IT initiatives, internal processes, and currently delivered baseline service levels. This will enable a clear vision for IT to effectively strategize, manage, control, develop, and utilize the given resources while enhancing the quality of services to enterprise users. An effective SLA can help to ensure that the outsourcing vendor is helping the customer meet or surpass business and technology service levels, which in turn leads to increased productivity and flexibility, and improved standardization and capacity. The issues mentioned in an SLA defines levels that the service can meet. These levels of service are expressed in terms of parameters and metrics upon which penalties and liabilities are to be enforced on the violator party by an independent authority. Once IT and business objectives have been identified and agreed to, the IT staff or outsourcing vendor should collect baseline performance metrics for the applications that will be outsourced. By assessing existing service levels, both parties can agree to the appropriate measures of success.

According to the International Telecommunication Union (ITU), quality of service (QoS) is defined as 'the collective effect of service performances, which determine the degree of satisfaction of a user of the service. The quality of service is characterized by the combined aspects of service support performance, service operability performance, service integrity, and other factors specific to each service'.

Service level management is akin to the demand management area in so far as it implies customer relationship management. It requires close coordination with demand, portfolio, incident, request, and availability management, as these may all be sources for service level measurements. Service level management requires metrics and business intelligence capabilities. Information technology can manage the entire range of SLAs, from defining SLAs and monitoring compliance to collecting and analysing KPI data, addressing problem areas, and continually refining the services offered.

15.7.1 Components of SLA

A well-defined SLA, typically includes a number of components. These are discussed below.

A description of the provided service nature includes the type of the service, and a description of the technical issues associated with the service, such as, network connectivity, operation, and maintenance; in addition the server's and client's configurations.

The level of responsiveness and reliability of the service includes availability requirements, and how soon the service performs in a normal state.

Service problems reporting includes people who will be contacted in case of a certain problem, in addition to steps and formalities that must be followed to guarantee quick resolving of the problem.

Problem response time and resolution defines the time after which a reported problem will be solved.

Monitoring and service reporting defines how quality levels are monitored and reported, who is responsible of that, how, and how often.

Liabilities on the service provider if promises were not met describes a service customer getting extra credits, such as the customer may have the power to terminate the contract, or ask for refunding.

Extra conditions is a component in the SLA describing conditions upon which the SLA is not valid anymore. This can be true, for example, in case of natural causes. Also, such cases may involve the customer himself/herself trying to breach the security of the service provided or the network.

15.8 Outsourcing

As a business practice, outsourcing is flourishing in almost every conceivable domain. Organizations today outsource software development, innovation, and research and development efforts and even functional areas such as marketing, human resource administration, and finance and accounting. Sourcing refers to the act of transferring work, responsibilities, and decision rights to someone else. Sourcing is normally conducted with an external party. Outsourcing is the act of transferring the work to an external party. The organization transferring these is referred to as the client, the organization that conducts the work and makes decisions is the vendor, and the scope of the work is captured in a project [Linder 2004].

Outsourcing has moved from initiatives that were financially motivated to the current stage of being strategically motivated. Financially motivated outsourcing efforts have been around since the early days of commerce. Strategically driven outsourcing efforts are capability- and competency- intensive, and are initiated to tap into specialized expertise, knowledge, processes and capabilities found outside the organization, and to use these as inputs to help improve the effectiveness and efficiency of operations. Strategic outsourcing often involves partnerships between the client organization and multiple vendors. Outsourcing is subcontracting a process, such as product design or manufacturing, to a third-party company. Outsourcing can be defined as a contractual agreement that entails the procurement of goods and/or services from an external provider.

15.8.1 Types of Outsourcing

Outsourcing efforts can be segmented by location, ie where the work is done: on-site or off-site. On-site work involves having members of the vendor's team conduct work within the premises of the client organization. Off-site work is where the vendor conducts work at its location. Outsourcing efforts are also segmented by the way the activity is carried out. Insourcing and outsourcing are the two basic sourcing strategies. Often organizations have their own IS departments from where they insource their IS needs. The responsibility and delegation of tasks involving the firm's IS needs are handled internally. Hence, when the service provider to the client is a client-entity such as a subsidiary or the internal IS department, it is known as insourcing.

Due to various factors, organizations (clients) often need to outsource work to external entities. Hence, when the service provider is a non-client entity, such as a vendor/supplier it is known as outsourcing [Sahay et al. 2003]. The term 'outsourcing', although not specific to IS in that it reflects the use of external agents to perform one or more organizational activities. For instance, Dell Computers has opened operations in places such as Brazil, where they own the outfits. Dell routinely sends software work to these locations because of the availability of a skilled workforce and cost savings.

The outsourcing life cycle is made up of the following stages.

- Strategic assessment
- Needs analysis
- Vendor assessment
- Negotiation and contract management
- Project initiation and transition
- Relationship management
- Continuance, modification, or exit strategies

Outsourcing services fall in two broad categories: technology services outsourcing and business process outsourcing.

15.8.2 Technology Services Outsourcing

The fast emerging business world of today requires companies to use sophisticated and fast computer systems and software. These technologies and systems also need to be scalable and highly adaptive. The following are some of the different categories that come under technology services.

- E-commerce
- Network/infrastructure
- Software/applications development
- Telecom
- Web development and hosting

15.8.3 Business Process Outsourcing

The new global scenario requires that each company finds its own niche field that can add value to the world economy. Thus companies now try to focus their resources on areas that give maximum yield. Thus, the term business process outsourcing (BPO) came into being around 1995. The proliferation of the Internet and its emergence as a business tool helped to make BPO highly popular.

The sub-categories of services that come under BPO are as follows.

- Customer relations/customer contact management
- Finance/accounting processes
- Logistics
- Equipment management
- Security
- Supply chain/procurement management

15.8.4 Outsourcing Trends

Information technology outsourcing has expanded in recent years to include applications development, network management, and desktop management in addition to more traditional data processing and call centre services. Cost reduction has, typically, been the primary driver for outsourcing initiatives, but as the demand for round-the-clock service becomes more prevalent, outsourcing is increasingly being viewed as a means to improve performance over the long term.

Information systems outsourcing is an increasingly common business practice in which a company contracts all or part of its information systems operations to one or more outside information service suppliers.

SUMMARY

Many solution vendors are jumping on the ITIL band wagon to add weight to their claims about their products and services. The industry is already awash with people claiming to certify solutions and organizations against ITIL. Ultimately, ITIL will probably be a good thing for everyone involved in the IT industry. It will require all IT professionals to raise their games to a reasonable level and may even help eradicate some of the more technology-centric practices of the past. It is important to understand incident management and problem management and differentiate between them accordingly.

Service level agreements between users act as a bridge for service delivery. They are documents which understand the service requirement and describe them. Information technology measures the service delivered against the service level agreement. A typical SLA has a set of components. Service level agreements are managed by service level reports (SLRs). Service level reports are the primary tools to manage SLAs. The SLR should be produced regularly. The report should state expectation for the reporting period and how the organization meets these expectations.

Key Terms

Change management The practice of administering changes through the help of tested methods and techniques in order to avoid new errors and minimize the impact of changes being introduced is called change management. This includes careful control of activities such as planning, communicating, testing, coordinating, scheduling, and monitoring resources and services.

Change request Any hardware or software alteration that could impact any customer in the production environment.

Planned outage This is a planned change that has gone through the appropriate change management procedures. It is not an emergency or break/fix change. A planned outage will appear on the public-view calendar and inform customers of an upcoming change, that may or may not cause an actual outage.

Service level agreement (SLA) It is an agreement that supports the services to various units, customers against specific measurable service levels based on the type of SLA.

Review Questions

15.1 What is a business process
15.2 What is ITIL and what is the need for ITIL?
15.3 Differentiate between problem management and incident management.
15.4 What is a CMDB? Why is it so important in the ITIL framework?
15.5 What do you understand by service life cycle management?
15.6 What is the need for outage management?
15.7 Differentiate between immediate response model and managed response model.
15.8 Differentiate between MTTR and MTBF.
15.9 What is an service level agreement and why is it required?
15.10 What are the components of SLA?

Projects

15.1 Visit the nearest IT company and understand the implementation of CMDB within the organization. Prepare a summary report of CMDB implementation.
15.2 Visit the nearest manufacturing firm and review the service level agreement executed by them with the IT service provider. Prepare a summary report of the service level agreement as understood by you.

15.3 The office of a small independent realty company has four real estate agents, an executive secretary, and two assistant secretaries. They agreed to keep all common documents in a central location to which everyone has access. They also need a place to store their individual listings so that only the agent who has the listing should have access to. The executive secretary has considerable network administration knowledge, but everyone else does not. It is expected that in the near future the office will employ ten more agents and three assistant secretaries. You are planning implement an effective IT management. You plan to identify the IT management metrics and outsource IT management to a third party through a structured service level agreement.

Design appropriate IT management metrics and a suitable service level agreement for outsourcing the IT management.

REFERENCES

Addy, Rob 2007, *Effective IT Service Management: To ITIL and Beyond*, Springer, India.

Adelman, Sid, Larissa Moss, and Majid Abai, *Data Strategy*, Addison Wesley.

Al-Hakim, Latif 2007, *Challenges of Managing Information Quality in Service Organizations*, IGI Global, USA.

Bannister, Frank 2004, *Purchasing and Financial Management of Information Technology*, Elsevier, Great Britain.

Buchta, Dirk, Marcus Eul, and Helmut Schulte-Croonenberg 2007, *Strategic IT Management*, Gabler, Germany.

Click, Rick L, Thomas N Duening, 2005, *Business Process Outsourcing: The Competitive Advantage*, John Wiley, New Jersey.

Gottaschalk, Peter 2006, *Managing Successful IT Outsourcing Relationships*, IRM Press, USA.

Halvey, John K. 2005, *Information Technology Outsourcing Transactions Process, Strategies and Contracts*, John Wiley, New Jersey, USA.

Hann, Robert W. 2006, *Information Markets A New Way of Making Decisions*, AEI Press, USA.

Harris, Michael D., David E. Herron, and Stasia Iwanicki 2008, *The Business Value of IT: Managing Risks, Optimizing Performance, and Measuring Results*, CRC Press.

Heywood, J. Brian 2001, *The Outsourcing Dilemma the Search for Competitiveness*, Prentice Hall, Great Britain.

Klosterboer, Larry 2008, *Implementing ITIL Configuration Management*, IBM Press, USA.

Linder, Jane C., 2004, *Outsourcing for Radical Change: A Bold Approach to Enterprise Transformation*, AMACOM, USA.

Maizlish, Bryan, Robert Handler 2005, *IT Portfolio Management Step-by-Step*, John Wiley, New Jersey, USA.

Michael, Gentle 2007, *IT SUCCESS! Towards a New Model for Information Technology*, John Wiley, England.

Power, Mark J., Kevin C Desouza, and Carlo Bonifazi 2006, *The Outsourcing Handbook*, Kohen Page, Great Britain.

Pecht, Nash 1994, *Predicting the Reliability of Electronic Equipment*, Proceedings of the IEEE, Vol. 82, No. 7.

Sahay, Sundeep, Brian Nicholson, and S. Krishna 2003, *Global IT Outsourcing: Software Development Across Borders*, Cambridge University Press, UK.

Schniederjans, Marc J., *Outsourcing Management Information Systems*, Idea Group, USA.

Verma, D. C. 2004, *Service Level Agreements on IP Networks*, Proceedings of the IEEE, Vol. 92, No. 9, September.

Wiggers, Peter, Henk Kok, and Maritha de Boer-de Wit 2004, *IT Performance Management*, Elsevier.

CHAPTER

16 Security Management

War consists largely of acts that would be criminal if performed in time of peace—killing, wounding, kidnapping, destroying or carrying off other people's property. Such conduct is not regarded as criminal if it takes place in the course of war, because the state of war lays a blanket of immunity over the warriors. But the area of immunity is not unlimited, and its boundaries are marked by the laws of war.

—Telford Taylor

Learning Objectives

After reading this chapter, you should be able to understand:

- the CIA triangle
- the concepts of security controls
- password management
- the concepts of cyber war and cyber weapons

16.1 Introduction to Security

Security is now a part of the network rollout rather than being looked at as a separate activity. With more networks migrating to Internet Protocol and fibre becoming its backbone, new challenges are arising for the security managers. The dream for any network manager would be to have a self-defending and self-healing network. The scope of security has moved beyond the firewalls and intrusion-detection systems at the periphery level to one of end-to-end security at all levels.

Today, businesses need to extend their radar, both to protect existing assets and to build for the future by leveraging emerging technology as a growth engine. The benefits of converging IT physical security are beginning to be realized. The heavy investment made in IP-enabled enterprise networking, with its promise that virtually anything can be distributed across it, is making chief information officers (CIOs) challenge the long-standing presumption of using separate infrastructures for physical security systems.

The increasing complexity of software-intensive and telecommunication products, together with pressure from security and privacy legislation, are increasing the need for adequately validated security solutions. In order to obtain evidence of the information security performance of systems needed for the validation, services or products, systematic approaches to measuring security are needed. The field of defining security metrics systematically is very young. Because the current practice of security is still a highly diverse field, holistic and widely accepted measurement and metrics approaches are still missing.

16.1.1 CIA Triangle

The confidentiality, integrity and availability (CIA) triangle is the industry standard for computer security. It has existed since the development of the mainframe. It is still used today due to it's characteristics namely, confidentiality, integrity, and availability which are still just as important in today's society. However, it no longer addresses the full breadth of security concerns faced today, so it serves as a foundation for a more advanced system, known as the expanded CIA triangle. The CIA has expanded into a more comprehensive list of critical characteristics of information security management.

Confidentiality of information ensures that only those with sufficient privileges may access certain information. When unauthorized individuals or systems can access information, confidentiality is breached. Confidentiality requires that the proper techniques be in place to prevent unauthorized viewing of company assets. Simple shoulder spying, dumpster diving all the way up the malicious chain to corporate espionage, and social engineering fall into this realm. To protect the confidentiality of information, the following measures are used.

- Information classification
- Secure document storage
- Application of general security policies
- Education of information custodians and end users

Integrity is the quality or state of being whole, complete, and uncorrupted. The integrity of information is threatened when it is exposed to corruption, damage, destruction, or other disruption of its authentic state. Corruption can occur while information is being compiled, stored, or transmitted. Integrity stops information from being modified or untrusted. Mistakes at work should not be able to break the integrity of the corporate policies. Users are often involved in breaching integrity by speaking to unauthorized media sources, modifying databases on the fly, and generally not adhering to the policies put in place. The policies also have to be iron-clad so as to not put users in a position where they can make a mistake that would void the credibility of a document in the first place.

Availability is the characteristic of information that enables user access to information without interference or obstruction and in a required format. A user in this definition may be either a person or another computer system. Availability does not imply that the information is accessible to any user; rather, it means availability to authorized users. Availability is often the last bastion of the triangle as it is the one that is most directly influenced by resources beyond your control. Availability covers areas such as power failures, hard drive failures, and pretty much all system availability failures.

16.2 Two Views of Security

It seems logical that any business, whether a commercial enterprise or a not-for-profit business, would understand that building a secure organization is important to long-term success. When a business implements and maintains a strong security posture, it can take advantage of numerous benefits. An organization that can demonstrate an infrastructure protected by robust security mechanisms can potentially see a reduction in insurance premiums being paid. A secure organization can use its security program as a marketing tool, demonstrating to clients that it values their business so much that it takes a very aggressive stance on protecting their information.

Security, by its very nature, is inconvenient, and the more robust the security mechanisms, the more inconvenient the process becomes. Employees in an organization have a job to do; they want to get to work right away. Most security mechanisms, from passwords to multi-factor authentication, are seen as roadblocks to productivity. One of the current trends in security is to add disk encryption to laptop computers. Although this is a highly recommended security process, it adds a second login step before a computer user can actually start working. For most organizations, the cost of creating a strong security posture is seen as a necessary evil, similar to purchasing insurance. Organizations do not want to spend the money on it, but the risks of not making the purchase outweigh the costs. Because of this attitude, it is extremely challenging to create a secure organization. The attitude is enforced because requests for security tools are often supported by documents providing the average cost of a security incident instead of showing more concrete benefits of a strong security posture.

Whether the threat is internal or external is a subjective issue and is very vertical-dependent. A software company can have an offsite employee and still face a security breach during his/her usage of the corporate network. Similarly, if a company gives network access to its dealer, the dealer is now an internal user but the access is external.

16.2.1 Sources of Threats

The threat of attacks (threat vectors) on the business information systems has become such a persistent issue that the organizations have accepted it as part of doing business in the new digital age. 'Most security breaches are committed by individuals who possess intimate knowledge of the systems they are attacking.', said Norman Inkster, President of KPMG Investigation & Security Inc., Canada and Chair of KPMG's International Forensic Accounting Committee. If senior management understood that, they might handle their security issues very differently. The threat from inside an organization has historically accounted for majority of the loss suffered by businesses. According to the Webster's Dictionary, an insider is defined as an officer of a corporation or others who have access to private information about the corporation's operations, especially information relating to profitability.

The generic disgruntled employee covering current employees, ex-employees, contractors, and consultants can be the most common type of an inside attacker threat vector. The term 'hacker' refers to individuals internal to an organization who are sympathetic to the hacker mentality or ethos. This mentality is characterized by a disregard for convention and rules, loose ethical boundaries, ambiguous morality, disregard for private property, and an innate curiosity. Petty criminals take advantage of opportunities that present themselves at the workplace and do not usually join an organization with the intent to steal from it. Once employed, they take advantage of lax security and opportunities to conduct criminal activities.

Information preserved in uncorrupted form still face the threat vectors, such as misappropriated information, threatening its confidentiality and authorized usage. Some physical threats to information, such as fire and theft, predate the digital age. Physical damage to information asset is another obvious threat. Small, unattached items are the easiest to lose, either through theft or accident.

There are a host of threat vectors. Understanding the factors that may be directly or indirectly responsible for the threat should allow us to choose better mitigation strategies and, in some cases, be preventative and proactive rather than being solely reactive. The phenomena of an uncertain economy, poor corporate governance, downsizing, and cheap labour in foreign countries has contributed to the transient workforce that leads to threat vectors in today's business world.

> A considerable amount of forensic analysis of Windows systems today continues to centre around file system analysis; locating files in the active file system, or carving complete or partial files from unallocated space within the disk image. However, a great deal of extremely valuable information is missed if the Windows Registry is not thoroughly examined, as well.

16.3 Security Management Controls

Security is based on converged physical security on a design philosophy that includes a strategy for managing physical access to corporate resources. A design model balances the basic security elements of technology, monitoring, and response. Security measures must strike a balance between security and functionality. Part of the strength of that balance is in creating the awareness that physical security exists, so the attempt is to place the security measures strategically and make them conspicuous. By simply making people aware of monitoring devices and other physical security measures helps to deter theft or trespass.

Monitoring security systems from a remote location provides the ability to centralize the administration and response. One of the benefits of integrating physical security with information technology (IT) is the ability to use a smaller, centralized team of individuals to monitor and respond to events throughout an entire region. Event-based response and signal prioritization ensure that the most important events receive immediate attention, and they help facilitate continuity of response throughout the enterprise. Closely related to remote monitoring, the solution must provide for precision response.

Defense in depth provides for multiple layers of security at a facility that is appropriate to asset risk. Defence in depth for physical security begins with designing facilities with the strategy for physical security in mind, and it considers property boundaries, building approach, parking areas, ingress and egress points of a building, and flow of human traffic through the building. It also includes the physical security devices, such as access card readers that grant or prevent access and log activity at facility entry points, biometric authentication, camera systems, hardened construction, and other discreet sensors that monitor specific areas. All these functions combined provide a layered defence strategy in protection of corporate resources.

16.4 Physical Security

Physical security is the protection of personnel, hardware, programs, networks, and data from physical circumstances and events that could cause serious losses or damage to an enterprise, agency, or institution. This includes protection from fire, natural disasters, burglary, theft, vandalism, and terrorism. A more verbose definition of physical security is: 'The protection of building sites and equipment and all information and software contained therein from theft, vandalism, natural disaster, man-made catastrophes, and accidental damage. It requires solid building construction, suitable emergency preparedness, reliable power supplies, adequate climate control, and appropriate protection from intruders.'

A model security facility is one where all necessary systems are in place, tried and tested, to protect people, operations, interdependence and information without affecting day-to-day operations. It is one where everyone knows why the systems are in place and what they have to do.

Physical security involves a number of distinct security measures which form part of a layered or defence in depth approach to security, which must take account of the balance between prevention, protection, and response. Physical security measures, or products such as locks and doors, are categorized according to the level of protection offered.

The layered approach to physical security starts with the protection of the asset at source then precedes progressively outwards to include the building, estate, and perimeter of the establishment. Approach routes, parking areas, adjacent buildings and, utilities/services beyond the perimeter should also be considered. To ensure appropriate physical security controls, departments must consider the following factors.

- The impact of loss of the site or asset
- The level of threat
- The vulnerability
- The value, protective marking, or amount of material held
- The particular circumstances of the establishment, including considerations of environment, location, and whether occupancy is sole or shared

There are three main components to physical security. First, obstacles can be placed in the way of potential attackers and sites can be hardened against accidents and environmental disasters. Such measures can include multiple locks, fencing, walls, fireproof safes, and water sprinklers. Second, surveillance and notification systems can be put in place, such as lighting, heat sensors, smoke detectors, intrusion detectors, alarms, and cameras. Third, methods can be implemented to apprehend attackers and to recover quickly from accidents, fires, or natural disasters.

Physical security covers all the devices, technologies, and specialist materials for perimeter, external, and internal protection. This covers everything from sensors and closed-circuit television (CCTV) to barriers, lighting, and access controls. The goal of implementing an integrated physical security plan is in achieving sensible and sustainable security. Integrated physical security plans are by their very nature a compromise, a careful balancing act between what needs to be done and what can be done weighed against what is in the best interest of the facility and its normal day-to-day procedures. Integrated physical security planning should not be undertaken in isolation.

Manned guarding is a key element of integrated physical security. Guards provide deterrence against hostile activity and facilitate a rapid response to security incidents.

Guards may either be directly employed by an agency, or be employed by a commercial guard force. Guard duties and the need for, and frequency of, patrols should be decided by considering the level of threat and any other security systems or equipment that might already be in place.

16.4.1 Locks and Physical Security

Physical security plays a huge role in your network's overall security. Physical security starts with physical access to the resource. Resource centres are usually kept under lock. There are four basic types of locks that companies use on computer room doors: key locks, combination locks, keycard locks, and biometric locks. Each of these locks has advantages and disadvantages. Key locks are probably the most common type of lock. However, anyone with basic lock-picking skills can get past a key lock relatively easily.

Keycard locks is a favourite option for a number of reasons. First, there are several types of keycard locks, but many of them have the cardholder's identity encoded on the card's magnetic strip. The card readers are linked to a central computer that keeps track of card scans. Therefore, if someone who is not authorized to access the computer room door tries to use his/her card to get in anyway, the computer logs the fact that the person tried to get in. An e-mail alert can then be sent to the security coordinator and to the person's supervisor.

Biometric locks are probably a good solution, but they tend to be expensive. However, depreciating technology costs are bringing biometric devices to the reach of small and medium business (SMB) industries.

16.4.2 Asset Tagging and Asset Life Cycle Management

All types of electronic equipment ranging from computers to copiers to fax machines will meet an end-of-life point in their office journeys. Organizations usually have an inventory control function that is designed to track large capital assets. When organizations track IT assets, however, the techniques used to track other capital assets may fall short and not deliver all the possible value to the organization. In addition, many benefits over and above financial tracking are lost.

Asset management forms the basis for the operational systems. The first phase in the asset life cycle is procurement. When an asset is procured it enters the asset management system and begins to be managed. Ideally, the procurement system should feed the asset management system the data on the new asset as soon as the purchase order is completed. The receiving organization should acknowledge receipt in a way that confirms the asset in the asset management system and notifies the purchasing system so that payment can be made.

The second phase of the asset lifecycle is deployment. When an asset is deployed, the system should be updated with relevant data such as location, responsible party in the organization, configuration, vendor, warranty, and any other data that will be useful in managing the asset. The location may be a physical location or simply a link to some other asset that contains the asset being deployed.

The third phase of the asset management cycle is usage. Usage is not simply a static flag but could be periodically updated by operational software that measures asset usage so that valuable assets not being used can be redeployed.

When an asset is no longer being used, it is decommissioned. Decommissioned assets may still be useful to the organization, in which case they can be redeployed. If not, it is likely that they still have some salvage value and the asset management system should track them until salvage has been completed.

The total cost of ownership (TCO) is a measure of all aspects of owning and operating an asset. A properly functioning asset management system can reduce TCO by eliminating costs from duplication of assets or from wasting assets by not using them after they are purchased. An asset management system also reduces the effort required to track assets, reduces the risk of software license non-compliance, and facilitates better asset operations.

16.5 Access Control

The mechanism which defines user access is called access control. Access control refers to the practice of controlling and monitoring access to a property or asset. Access control relies on and coexists with other security services in a computer system. Access control is concerned with limiting the activity of legitimate users. It is enforced by a reference monitor which mediates every attempted access by a user to objects in the system. The reference monitor consults an authorization database in order to determine if the user attempting to do an operation is actually authorized to perform that operation.

It is important to make a clear distinction between authentication and access control. Correctly establishing the identity of the user is the responsibility of the authentication service. Access control assumes that authentication of the user has been successfully verified prior to enforcement of access control via a reference monitor. The effectiveness of the access control rests on a proper user identification and on the correctness of the authorizations governing the reference monitor. It is also important to understand that access control is not a complete solution for securing a system. It must be coupled with auditing.

Physical access control can be achieved through a combination of manned guarding, and mechanical or technical means. When deciding which access control measures to deploy, departments must ensure that they consider the security measures in an integrated manner, such as combining automated access control systems with photo passes and CCTV.

16.5.1 Access Control Principles and Objectives

In access control systems, a distinction is generally made between policies and mechanisms. Policies are high-level guidelines which determine how accesses are controlled and access decisions determined. Mechanisms are low-level software and hardware functions which can be configured to implement a policy.

The subject-object distinction is basic to access control. Subjects initiate actions or operations on objects. These actions are permitted or denied in accordance with the authorizations established in the system. Authorization is expressed in terms of access rights to objects.

The access matrix is a conceptual model which specifies the rights that each subject possesses for each object. There is a row in this matrix for each subject, and a column for each object. Each cell of the matrix specifies the access authorized for the subject in the row to the object in the column. The task of access control is to ensure that only those operations authorized by the access matrix actually get executed. This is achieved by means of a reference monitor, which is responsible for mediating all attempted operations by subjects on objects. A popular approach to implementing the access matrix is by means of access control lists (ACLs). Each object is associated with an ACL, indicating for each subject in the system the accesses the subject is authorized to execute on the object.

When the server receives a request, it uses the authentication information provided by the user in the bind operation and the access control instructions (ACIs) defined in the server to allow or deny access to directory information. The server can allow or deny permissions for actions on entries such as read, write, search, and compare. The permission level granted to a user may depend on the authentication information provided. Access control instructions are stored in the directory as attributes of entries. The ACI attribute is an operational attribute; it is available for use on every entry in the directory, regardless of whether it is defined for the object class of the entry.

16.5.2 Access Control Mechanisms

Today, the best known US computer security standard is the Trusted Computer System Evaluation Criteria (TCSEC). It contains security features and assurances, exclusively derived, engineered, and rationalized based on US Department of Defense (DoD) security policy, created to meet one major security objective—preventing the unauthorized observation of classified information. The TCSEC specifies two types of access controls: discretionary access control and mandatory access controls (MAC). Discretionary access controls requirements have been perceived as being technically correct for commercial and civilian government security needs, as well as for single-level military systems. As defined in the TCSEC and commonly implemented, discretionary access control (DAC) is an access control mechanism that permits system users to allow or disallow other users access to objects under their control. Discretionary access control as the name implies, permits the granting and revoking of access privileges to be left to the discretion of the individual users. A DAC mechanism allows users to grant or revoke access to any of the objects under their control without the intercession of a system administrator. Access control decisions are often determined by the roles individual users take on as part of an organization. This includes the specification of duties, responsibilities, and qualifications.

Access control works at a number of levels. The access control mechanisms, which the user sees at the application level, may express a very rich and complex security policy. The access controls provided with an operating system typically authenticate principals using some mechanism, such as passwords or Kerberos, then mediate their access to files, communications ports, and other system resources.

The primitive goal of an access control framework is to efficiently manage the entities (users and resources) under its control. A typical deployment of an access control framework is a combination of the following three logical components: an access control model, policies, and enforcement mechanisms. The access control model provides means to arrange, efficiently manage entities, and define relations among them. The policy languages are employed to provide properties that are difficult to achieve under the access control model alone, for example, context-sensitive access requests. And, the enforcement mechanisms are employed to enforce outcomes of access requests to a resource. They are also used in situations where certain requirements are contrary to the inherent properties of the underlying access control model. Their effect can often be modelled by a matrix of access permissions, with columns for files and rows for users.

Another way of simplifying access rights management is to store the access control matrix a column at a time, along with the resource to which the column refers. This is called an access control list or ACL. Access control lists have a number of advantages and disadvantages as a means of managing security state. These can be divided into general properties of ACLs and specific properties of particular implementations. Access control lists are widely used in environments where users manage their own file security, such as the Unix systems common in universities and science laboratories. Where access control policy is set centrally, they are suited to environments where protection is data oriented; they are less suited where the user population is large and constantly changing, or where users want to be able to delegate their authority to run a particular program to another user for some set period of time.

16.5.3 Logical Access Control

Information residing on a system that is accessed by many users, however, can also create problems. A significant concern is ensuring that users have access to information that they need but do not have inappropriate access to information that is sensitive. It is also important to ensure that certain items, though readable by many users, can only be changed by a few. Logical access controls are means of addressing these problems. Logical access controls are protection mechanisms that limit users' access to information and restrict their forms of access on the system to only what is appropriate for them. Logical access controls are often built into the operating system, or may be part of the 'logic' of applications programs or major utilities, such as database management systems (DBMS).

The concept of access modes is fundamental to logical access control. The effect of many types of logical access control is to permit or deny access by specific individuals to specific information resources in specific access modes. A few access modes are discussed below.

Read only: This provides users with the capability to view, copy, and usually print information but not to do anything to alter it, such as delete from, add to, or modify it in any way. Read-only accesses are probably the most widely allowed to data files on IT systems.

Read and Write: Users are allowed to view and print as well as add, delete, and modify information. Logical access control can further refine the read/write relationship such that a user has read-only ability for one field of information but the ability to write to a related field.

Execute: The most common activity performed by users in relation to applications programs on a system is to execute them. A user executes a program each time he/she uses a word processor, spreadsheet, database, etc. Users would not ordinarily be given read or write capabilities for an application, since it would appear in a format that is unintelligible to most users. It might be desirable, though, for software programming specialists to be able to read and write applications.

Successfully refining, implementing, and managing these different access modes have resulted in greatly improved information sharing, both for government and industry as well as for the general public.

Logical access control can enhance the security provided by physical access control by acting as an additional guard against unauthorized access to or use of the system's resources. It can also augment physical access control by providing added precision, since different users are able to perform different functions.

16.5.4 Physical Access Control

Physical access control was the main means of protecting information on an IT system. Access to information was controlled solely by controlling access to the system. Once logged onto a system, though, a user could generally access all of its data. In some environments, this is not a problem. Physical access control may be sufficient in environments where all users of a system need an access to all of the information on it and need to perform all of the same types of accesses in relation to it.

16.5.5 Passwords/Keys/Tokens

Passwords are probably the most common way of protecting information on an IT system in that they are the most frequently used means for users to be identified and authenticated on the system. Thus, they are often the first line of protection afforded by an IT system. In addition, passwords are also used to protect data and applications on many IT systems. Passwords are also used frequently in PC applications as a means of logical access control. For instance, an accounting application may require a password in order to access certain financial data or invoke a sensitive application.

The primary advantage of password-based logical access control is that it is provided by a large variety of PC applications and thus often does not have to be implemented as a new/separate feature on an operating system. The drawbacks of this approach centre on the difficulty for users to manage even moderate numbers of passwords. The security of a password-based system is significantly diminished when users write down their passwords. If users need to use more than a few different passwords in the course of their work, there will be a strong likelihood that they will write them down, thus exposing the IT resources the passwords were meant to protect. Also, if passwords are the same for several different applications, then a user who learns the password for one can gain access to the others.

16.5.6 Biometric Identification

The word biometric can be defined as life-measure. It is used in security and access control applications to mean measurable physical characteristics of a person that can be checked on an automated basis. Biometrics is a measurable biological and behavioural characteristic that can be used for automatic recognition. Security personnel look for biometric data that does not change over the course of life. Biometrics are automated methods of recognizing a person based on a physiological or behavioural characteristic. Among the features measured are face, fingerprints, hand geometry, handwriting, iris, retina, vein, and voice.

Biometric technologies are becoming the foundation of an extensive array of highly secure identification and personal verification solutions. As the level of security breaches and transaction fraud increases, the need for highly secure identification and personal verification technologies is becoming apparent. Biometric-based authentication applications include workstation, network, and domain access, single sign-on, application logon, data protection, remote access to resources, transaction security, and web security.

The following factors are needed to have a successful biometric identification method.

- The physical characteristic should not change over the course of the person's lifetime.
- The physical characteristic must identify the individual person uniquely.
- The physical characteristic needs to be easily scanned or read in the field, preferably with inexpensive equipment, with an immediate result.
- The data must be easily checked against the actual person in a simple, automated way.

Fingerprint Identification System

Among all the biometric techniques, fingerprint-based identification is the oldest method which has been successfully used in numerous applications. Everyone is known to have unique, immutable fingerprints. The best-known biometric systems are those that identify persons by using their fingerprints. The chance of two people having the same print is less than one in one billion. Biometric-based authentication applications include workstation, network, and domain access, single sign-on, application logon, data protection, remote access to resources, transaction security, and web security. Trust in these electronic transactions is essential to the healthy growth of the global economy. Utilized alone or integrated with other technologies such as smart cards, encryption keys, and digital signatures, biometrics are set to pervade nearly all aspects of the economy and our daily lives. Utilizing biometrics for personal authentication is becoming convenient and considerably more accurate than current methods (such as the utilization of passwords or personal identification numbers). This is because biometrics links the event to a particular individual (a password or token may be used by someone other than the authorized user), is convenient (nothing to carry or remember), accurate (it provides for positive authentication), can provide an audit trail, and is becoming socially acceptable and cost-effective.

A fingerprint is made of a series of ridges and furrows on the surface of the finger. The uniqueness of a fingerprint can be determined by the pattern of ridges and furrows as well as the minutiae points. Minutiae points are local ridge characteristics that occur at either a ridge bifurcation or a ridge ending.

Fingerprint matching techniques can be placed into two categories: minutae-based and correlation-based. Minutiae-based techniques first find minutiae points and then map their relative placement on the finger. Correlation-based techniques require the precise location of a registration point and are affected by image translation and rotation. A commercial fingerprint-based authentication system requires a very low false reject rate (FRR) for a given false accept rate (FAR).

A typical automated fingerprint identification system (AFIS) system requires a user to place a finger on the machine for as little as one-half a second to two seconds. Many devices analyse the position of the endpoints and junctions of print ridges (minutiae) of the fingerprint. Others count the number of ridges between points, while some approach the fingerprint from an image processing perspective.

The Biometric Consortium serves as a focal point for research, development, testing, evaluation, and application of biometric-based personal identification/verification technology. The Biometric Consortium organizes a premier biometrics conference every fall.

16.6 System Security

The IT systems available within the firm need to be made secure. Computer Emergency Response Team (CERT-In) has recommended a series of guidelines to ensure system security. The document covers general topics required for setting up a server in secure environment.

16.6.1 Hardening Systems

System hardening is the process of evaluating a company's security architecture and auditing the configuration of their systems in order to develop and deploy hardening procedures to secure their critical resources. System hardening, also called operating system hardening, helps minimize these security vulnerabilities. System hardening is a step-by-step process of securely configuring a system to protect it against unauthorized access, while also taking steps to make the system more reliable. Hardening systems is a defense strategy to protect against attacks by removing vulnerable and unnecessary services, patching security holes, and securing access controls.

The purpose of system hardening is to eliminate as many security risks as possible. This is typically done by removing all non-essential software programs and utilities from the computer. While these programs may offer useful features to the user, if they provide 'back-door' access to the system, they must be removed during system hardening.

16.7 Password Management

Passwords are a critical part of information and network security. Authentication of individuals as valid users, via the input of a valid password, is required to access any shared computer information system. Each user is accountable for the selection, confidentiality, and changing of passwords required for authentication purposes. Passwords serve to protect user accounts but a poorly chosen password, if compromised, could put the entire network at risk. Password is the personal key to a computer system. Passwords help to ensure that only authorized individuals access computer systems. Passwords also help to determine accountability for all transactions and other changes made to system resources, including data.

16.7.1 Password Guidelines

Even when passwords are encrypted, they can be guessed or 'cracked', especially when they match a dictionary word or permutation. Choosing the right password is something that many people find difficult, there are so many things that require passwords these days that remembering them all can be a real problem. Passwords should be treated as confidential information. No employee is to give, tell, or hint at their password to another person, including IT staff, administrators, superiors, other co-workers, friends, or family members, under any circumstances. Do not use the 'Remember Password' feature of applications and do not create a 'hot key' for password use. Hackers also have easy access to very powerful password-cracking tools incorporating extensive word and name dictionaries. Passwords should never be dictionary words or names. The cracking tools will also check for simple tricks like words spelled backwards or simple substitution of certain characters.

16.7.2 Secure Password

To an attacker, a strong password should appear to be a random string of characters. Each character that you add to your password increases the protection that it provides many times over. Your passwords should be 8 or more characters in length; 14 characters or longer is ideal.

16.7.3 Age and Length of Password

We regularly find attacks targeting user credentials (NetworkID and passwords). It is believed that the pathway to much of the Internet based fraud is through the compromise of user IDs and passwords on users, machines, either via installation of malware/trojans or via brute-force guessing of passwords. An article about the underground economy by the security research firm Team Cymru demonstrates that the there is a market value for a stolen user credential because of the potential benefits of access to bank accounts, corporate or university systems, or even administrator privileges to home computers.

The presence of this threat, combined with the condition of our dependency on the user ID and password combination, highlight the need to change network passwords on a frequent basis. The higher frequency of password changes reduces the probability that an unauthorized user can compromise your password via guessing. A general rule of thumb is to estimate the timeframe necessary to brute-force guess passwords (undetected) and set the maximum password age just below that threshold. This does not mitigate the keystroke logging trojan, but that is where one-time passwords can be of benefit.

Maximum password age setting determines the period of time (in days) that a password can be used before the system requires the user to change it. The user can set passwords to expire after a number of days between 1 and 999, or the user can specify that passwords never expire by setting the number of days to 0. It is a security best practice to have passwords expire every 30 to 90 days, depending on the environment.

16.7.4 Password Best Practices

Using a single password is the equivalent of using a single key for your car, house, mail box, and safety deposit box—if you lose the key, you give away access to everything. If your password is compromised on one system, using different passwords on different systems will help prevent intruders from gaining access to your accounts and data on other systems. For example, system managers should use different passwords for their personal account and their privileged account. If the personal account password is accidently revealed, the privileged account is still protected. Similarly, users should use different passwords for their pop email account and interactive logons. Some of the best practices in password management include the following.

- Avoid sequences or repeated characters as passsword.
- Avoid using only look-alike substitutions of numbers or symbols as password.
- Avoid your login name as password.
- Avoid dictionary words in any language as password.
- Use more than one password everywhere.
- Never provide your password over e-mail or based on an e-mail request.
- Do not type passwords on computers that you do not control.
- When receiving technical assistance, do not divulge your password to the IT specialist, but stay with your computer and enter the password yourself when required.
- Change your password regularly.
- Never store your password on a computer file.
- Never write down your password.
- If you believe your password may have been compromised, change it immediately.

Your password must

- be memorized, not written down
- be at least eight characters long
- contain changes in case
- contain three non-alphabetic characters

16.8 Communication Security

All communication over the Internet is made using the IP suite. The Internet protocol suite consists primarily of the IP and the transmission control protocol (TCP). It has become common to use the term TCP/IP to refer to the whole protocol family. The TCP/IP architecture divides data into packets and then independently routes each packet through the network. The Internet protocol suite provides no security at all.

The compound word COMSEC is prevalent in the US Department of Defense (DoD) culture with hundreds of secondary and tertiary words. COMSEC is used to protect both classified and unclassified traffic passed via tactical or strategic switched systems within the Department of Defence. The Internet is the world's largest interconnected network. However, many applications using the IP suite require or could benefit from a mechanism that provides a higher-level of security involving such aspects as confidentiality, data integrity, and authentication. Security protocols can be utilized on all layers in the protocol suite to protect data in different ways and to the extent needed.

A lot of security measures are need to be taken to ensure safety of data in transit. This includes device security, firewall security, switch security, router security, modem security, and encryption technologies.

16.9 Information Security

Information security is an issue because most of our core business processes incorporate IT, and technology has started to break down the stovepipes that used to protect corporate data. Information security metrics are used in various scenarios to measure the success of an information security architecture (ISA) framework, the implementation of new technology, processes and policies, and level of compliance with the security standard.

16.10 Risk Management and Business Continuity Planning

The last two decades have seen the exploitation of advances to develop extremely complex, tightly-coupled systems within the major infrastructures demanded by the current society. Risk analysis methodologies aim to inform management on the cost-effectiveness of security systems designed to reduce system risk to acceptable levels. This proved to be an ambitious target even in the early 1970's when computers performed limited tasks in secure environments. In the current environment of complex, interdependent systems and highly insecure environments it would appear that mere risk identification is in itself a daunting task.

For all types of infrastructure, the information systems (IS) that support their operations are vitally important. These ISs however introduce much of the system complexity, their networks increase the speed of interdependency interactions, and their ubiquity can often mask the low-level extent of the coupling.

16.10.1 Risk Analysis

A business impact analysis involves identifying the critical business functions within the organization and determining the impact of not performing the business function beyond the maximum acceptable outage. Types of criteria that can be used to evaluate the impact include: customer service, internal operations, legal/statutory, and financial. Most businesses depend heavily on technology and automated systems, and their disruption for even a few days could cause severe financial loss and threaten survival. The continued operations of an organization depend on the management's awareness of potential disasters, their ability to develop a plan to minimize disruptions of mission critical functions, and the capability to recover operations expediently and successfully. The risk analysis process provides the foundation for the entire recovery planning effort. Security begins with a thorough knowledge of the assets and protection of those assets begins with knowing what to protect. After a proper risk assessment, it can be seen which assets are the most valuable and which are the most vulnerable. Security then takes what resources are available and assigns them to protect the most important elements first.

16.10.2 Risk Analysis and Assessment

All facilities face a certain level of risk associated with various threats. These threats may be the result of natural events, accidents, or intentional acts to cause harm. Regardless of the nature of the threat, facility owners have a responsibility to limit or manage risks from these threats to the extent possible.

A risk assessment is a collaborative process that attempts to answer the following questions: what assets needs to be protected, who/what are the threats and vulnerabilities, what are the implications if the assets were damaged or lost, what is the value of the assets to the company, and what can be done to minimize exposure to the loss or damage. The core areas in a risk assessment are: scope definition, data collection, policies and procedures review, threat and risk analysis, vulnerability assessment, and development, implementation, and audit of recommendations. The first step in a risk management program is a threat assessment. A threat assessment considers the full spectrum of threats (i.e. natural, criminal, terrorist, accidental, etc.) for a given facility/location. The assessment should examine supporting information to evaluate the likelihood of occurrence for each threat. A risk assessment is also a continual process that should be reviewed regularly to ensure that the protection mechanisms currently in place still meet the required objectives. Risk assessment and analysis provides a comprehensive review of a company's overall network design and security policies to determine any vulnerability that may cause exposure to security risks. The service also identifies the areas where improvements should be made to enhance the security policies on an ongoing basis.

16.10.3 Business Continuity Planning

Business continuity (BC) is the ability of a business to continue its operations with minimal disruption or downtime in the advent of natural or intentional disasters. It begins with a plan that addresses all risks and secures systems that are vital to business operations. It is imperative for companies that are heavily dependent on IT infrastructure to design and implement a business continuity plan (BCP). A business continuity plan should provide an enterprise-wide risk-based approach, covering people, processes, technology and extended enterprise to ensure continuing availability of business support systems and minimize disruption risks.

Business continuity and disaster recovery (DR) are two terms that are used interchangeably. Disaster recovery is an older term that is used for recovery of IT systems. It also connotes a reactive approach to managing recovery. Business continuity is the current usage that connotes a broader scope then just IT. It also suggests a proactive approach in keeping the business available. Business continuity is a process and not a technology solution. It is not a one-time effort. A plan that will work for you is one that involves the right people, has processes in place, and has IT and infrastructure support to keep the business going.

The key to business continuity lies in understanding the business, determining which processes are critical to staying in that business, and identifying all the elements crucial to those processes.

Specialized skills and knowledge, physical facilities, training and employee satisfaction, as well as information technology must all be considered. It is by thoroughly analysing these elements that a company can accurately identify potential risks and decide to accept, mitigate, or transfer those risks. Chief information officers should use a BCP to make their organizations less susceptible to outages.

Large businesses and multinationals are now immensely serious about business continuity. Organizations, especially from the banking and finance and manufacturing sectors have started implementing BCP. Many organizations are heavily dependent on IT infrastructure. So if disaster strikes and these organizations cannot recover quickly enough, the consequences could affect business along the entire value chain. Business revenue drops, brand equity takes a beating; there is loss of customers (who choose alternatives), and permanent loss of shareholder value.

Disasters, both natural and intentional, are unpredictable. Natural disasters could be earthquakes, floods, hurricanes, or fire. Intentional disasters are caused by disgruntled humans and range from virus/hacker attacks to nuclear attacks. Then there are other causes for business disruption such as hardware and communications failure.

A BCP is insurance against such disasters and ensures that key (if not all) business functions continue.

Some of the elements of best practices include the following.

- Ensure sufficient latent capacity will be immediately available to assure rapid failover and recovery.
- Test capacity availability without disrupting ongoing operations.
- Install redundant network capacity dedicated to business continuity.
- House failover equipment in a separate location from main production equipment and provide further redundancies, such as sourcing electrical supplies, from different power grids.
- Establish and maintain relationships with vendors to assure quick delivery of replacement personal computers(PCs), network hardware, furniture, and telephones in the event of a facility-wide disaster.

- Secure adequate funding from end-user departments to implement and maintain adequate business continuity protection.
- Acquire, train, and retain skilled personnel to manage complex interdependencies and specialized elements of business continuity.
- Make adequate provisions for adding recovery support staff in the event of a regional or natural disaster.
- Using a technology solution provider for part of or for all these requirements is attractive for companies that prefer to focus on already scarce resources for driving revenue-growth.

16.10.4 Business Continuity in Distributed Environments

The pressures to implement business continuity software that can span the enterprise and recover application servers grow with each passing day. As per Gartner's recommendations, the best way to get started is to conduct a Business Impact Analysis (BIA). This will identify most crucial systems in the business and the effect of an outage on business. The greater the potential impact, the more money a company should spend to restore systems quickly.

In distributed environments, companies generally have multiple operating systems. Operating systems can be categorized by technology, ownership, licensing, working state, usage, and by many other characteristics. In practice, many of these groupings may overlap. Companies want the flexibility to recover applications running on any of these operating systems while using storage that they have available at the DR site to do the recovery.

A major reason that companies implement specific business continuity solutions for specific applications is due to how they manage high numbers of write inputs/outputs (I/Os). High performance (i.e., high write I/Os) applications put different demands on business continuity software than those that protect application servers with infrequent write I/Os.

Enterprise BC needs to monitor these wide area network (WAN) connections, provide logical starting and stopping points if the connection is interrupted, and resume replication without loosing data or negatively impacting the application which it is protecting.

Computer Associates's business continuity initiative is called crisis life cycle management. The initiative leverages Computer Associate's technologies and services to address all areas of business continuity best practices from needs assessment and planning to contingency command-and-control capabilities. It also incorporates new technology designed to provide a centralized common view of 'silos' of operational information, enabling organizations to proactively identify exposures and organizational threats and to create appropriate recovery plans. Crisis life cycle management also leverages the expertise and experience of computer associate services, which has assisted customers in safeguarding their operations and successfully recovering from a wide range of contingencies.

EMC's business continuity technologies and services have long enabled organizations to maintain real-time backups of their information at remote locations to protect against the unexpected.

> **Worm Targets Facebook**
>
> A new variant of the Koobface worm is making the rounds. Facebook users need to be aware of this and other attempts to use clever social engineering related to Facebook to trick them into installing malware. Like the earlier versions of Koobface, this one starts out as a Facebook message from a friend that includes a link to a website with something enticing like 'a video of you.' As with the old ones, in order to view the video you have to apply what purports to be an update to your Adobe Flash player but is in fact the malware.
>
> *Source*: Larry Seltzer, blogs.pcmag.com/...new_koobface_variant_preys_on.php

16.11 Security Standards and Assurance

With compliance emerging as one of today's most prevalent business issues, multiple corporate functions are beginning to converge in a federated approach to addressing quality, risk, and overall compliance management. As the visibility of compliance continues to rise, there is a concurrent increase in the importance placed on IT and the role of the CIO. As the regulatory environment continues to change with marked frequency and measurable complexity, so do the requirements for automated, repeatable controls, and processes around the classic information compliance drivers: internal controls over financial reporting, controls to protect and govern the use of personal information, protection of intellectual property, records management, and e-discovery rules.

16.11.1 ISO 27001

The scope of security consulting and audit for certifications like BS7799 (the only globally accepted security standard), drawing of request for proposal (RFP), security architecture design, etc. got a favourable response from customers. Also, what was good for network security service providers was the fact that there was a significant increase in the average value and scope of security consulting assignments.

16.11.2 OCTAVE

Operationally Critical Threat, Asset, and Vulnerability Evaluation (OCTAVE) is built on 10 volumes of material supporting the methodology, including background materials, guidance, worksheets, to manage information security risk. The OCTAVE-S and the OCTAVE Method are the two methods developed at the Software Engineering Institute (SEISM) consistent with the OCTAVE criteria, the essential requirements of an asset-based, strategic assessment of information security risk. OCTAVE is a risk-based strategic assessment and planning technique for security. It is self-directed and is managed by people from an organization. They assume responsibility for setting the organization's security strategy. OCTAVE is targeted at organizational risk and focused on strategic, practice-related issues. OCTAVE-S is a variation of the OCTAVE approach that was developed to meet the needs of small, less hierarchical organizations.

16.12 Information Infrastructure

Global proliferation in computer interconnectivity, most notably profound growth in the use of Internet, has revolutionized the way governments, societies, and the world at large communicate and conduct business. The information age has brought changes that challenge our ability to ensure the availability, integrity, and security of systems and information infrastructures.

For the purpose of our discussion, information infrastructure is the framework of interdependent networks and systems comprising identifiable industries, institutions, and distribution capabilities that provide a continuous flow of goods (information) and services essential to the defence and economic security of a country and the smooth functioning of government at all levels, and the society at large. Protecting these infrastructures against physical and electronic attacks and ensuring the availability of the infrastructures will be complicated.

Information resides on information infrastructure and travels on this information highway, enabling a nation to carry on its functions. Nations are expected to build information defence systems. The information defence system evolution need to exceed the continued technological advances, and meet emerging information needs with information infrastructure that can withstand offensive or exploitative threats.

National strategies to secure information infrastructure and cyberspace is a part of any country's overall effort to protect itself. It is an implemented component of the national strategy for national security and is complemented by a national strategy for the physical protection of critical infrastructures and key assets [Bush 2003].

Information infrastructure is vulnerable to attack. While this in itself poses a national security threat, the linkage between IS and traditional critical infrastructures has increased the scope and potential of the information warfare threat.

As information infrastructures become increasingly interdependent and complex, we also grow increasingly dependent on them. Massive computer networks provide multiple links between and among systems that, if not properly secured, can be operated from isolated locations to gain illicit access to data and operations in other systems. The use of such massive networks and cyber systems has brought not only advances in the quality of life, but also new threats to the international community. This technology- intensive information age brings with it opportunities of cyber crime and information warfare. Examples of cyber crimes include critical infrastructure attack, fraud, online money laundering, criminal uses of Internet communications, ID fraud, use of computers to further traditional crimes, and cyber extortions.

Safeguarding information infrastructures systems is becoming increasingly challenging as new technologies are evolving and attackers are inventing new technologies. The time for study of cyber and information warfare is ripe as security studies itself comes under more critical eye, coupled with the more continued attention in our deliberations about international relations.

16.12.1 Information Warfare

The beginnings of 1999 and 2000 saw a computer virus named Melissa wreak havoc upon computer systems around the world. When a Philippino hacker launched the 'Love Letter' virus in 2000, estimated loss of damage in the US was in the range of 4–15 billion dollars. But the US government could not do anything to prosecute the hacker or to recover the damages because at that time the Philippines had no laws prohibiting such crimes [Adams 2001].

For the purpose of this discussion, information warfare is defined as actions taken to achieve relatively greater understanding of the strengths, weaknesses, and centres of gravity of an adversary's military, political, social, and economic infrastructure in order to deny, exploit, influence, corrupt, or destroy those adversary information-based activities thorough command and control warfare and information attack.

Information warfare involves much more than computers and computer networks. It is comprised of operations directed against information in any form, transmitted over any media, including operations against information content, its supporting systems and software, the physical hardware device that stores the data or instructions, and also human practices and perceptions [Wilson 2004].

In 1991, Martin Van Creveld published *The Transformation of War*. Information warfare (IW) is motivated by opportunities that arise from an ever-increasing dependence upon vulnerable information systems. It also provides a new set of weapons different in many ways from those of conventional warfare. Information warfare realms of military, political, economic, social, and physical manifestations consists of a complex, interdependent infrastructure of systems and processes that are subject to attack and exploitation by a range of adversaries. Specific vulnerabilities to a realm occur throughout the information spectrum. Regardless of borders or geography, all digital information assets are at least potentially vulnerable to IW threats.

Strategic information warfare (SIW) is the deliberate electronic sabotage of a nation-state's national information infrastructure. This could take the form of crashing the financial markets of a nation. This could also lead to the deliberate shutting down of the power grid in the capital city of an adversary. In tactical situations, the goal is often to detect and destroy nodes, or to jam the traffic. Jamming can involve not just noise insertion but active deception. The increasing use of civilian infrastructure, and in particular the Internet, raises the question of whether systematic denial-of-service attacks might be used to jam the traffic.

Information warfare, in its wider sense, is daily used between individuals and corporations. Computer system penetrations are reported daily to emergency report teams that are in charge to take countermeasures. Often, the attackers argue that they do not commit a crime but improving the security of the system by pointing to its weaknesses. However, data disclosure and denial of service are a serious problem.

16.12.2 Electronic Warfare

Electronic warfare differs as well as overlaps information warfare. According to Schleher, the goal of electronic warfare is to control the electromagnetic spectrum. It is generally considered to consist of the following.

- Electronic attack such as jamming enemy communications or radar and disrupting enemy equipment using high-power microwaves.
- Electronic protection which ranges from designing systems resistant to jamming, through hardening of equipment to resist high-power microwave attack, to the destruction of enemy jammers using anti-radiation missiles.
- Electronic support which supplies the necessary intelligence and threat recognition to allow effective attack and protection. It allows commanders to search for, identify, and locate sources of intentional and unintentional electromagnetic energy.

Asymmetry created by information and communication technology (ICT) is among the six forms of asymmetry identified by [Metz and Johnson 2001]. Nations and organizations can exploit asymmetric advantages by strategically employing ICTs in war against enemies (for example, cyber attacks) as well as by using ICTs in facilitating other functions contributing to attack and defense such as communications, detection of threats from enemies, gathering intelligence, etc. Information and communication technology ICT deployments by terrorist groups, nations, and individuals involve some forms of positive and negative asymmetries. Positive asymmetry entails capitalizing on differences to gain an advantage. Negative asymmetry involves 'an opponent's threat to one's vulnerabilities' [Hollis 2007].

16.13 Information Operations and Cyber Weapons

Dominating the information spectrum is as critical to conflict as occupying the land or controlling the air. This is achieved through suitable information operations (IO). Responding to the possibilities (and vulnerabilities) inherent in the Internet's interconnectivity and the worldwide spread of new forms of communication, IO has emerged as a 'new category of warfare'. Information operations conceives information and information systems as both new tools and new objectives for military activities.

Information operations is conducted during time of crisis conflict to affect adversary information and information systems while defending one's own information and systems. Reported IO include operations security, computer network operations, and electronic warfare. Reported IO activities include:

- attempts to infiltrate networks,
- attempts to steal or sabotage information, and

- attempts to paralyse high-technology systems.

16.13.1 Cyber Weapons

In a 1999 study, the National Institute of Standards and Technology, (NIST) [Wilson 2004] found that many newer attack tools, available on the Internet, can now easily penetrate most networks, and many others are effective in penetrating firewalls and attacking Internet routers. Information operations provides a new weapon that can be deployed instantaneously and surreptitiously thousands of miles away from its target. Although its effects can certainly equate to those of kinetic force (for example, the death and destruction that would flow from unleashing a computer virus on a nuclear power plant's operating system).

An important question at this juncture is what cyber weaponry will do to conflict dynamics. The impact of cyber weaponry has been estimated to be higher. Perpetrator ambiguity is a problematic feature in this regard, with a clear potential for conflict aggravation through mistaken attribution of responsibility for attacks and retaliation against innocent parties perhaps as an intended effect by the real perpetrator.

A malicious computer code that attacks information systems may in theory be treated as a weapon of war within the scope of the laws of armed conflict, and attempts are now being made by some international organizations to classify and control malicious computer code.

Information operations activities are carried out with cyber tools or cyber weapons. Cyber weapons are becoming easier to obtain, easier to use, and more powerful. Some of the cyber tools include:

- offensive attack tools, such as viruses, Trojan horses, denial-of-service attack tools;
- dual use tools, such as port vulnerability scanners, and network monitoring tools; and
- defensive tools, such as encryption tools, Intruder detection system/intruder prevention system (IDS/IPS) and firewalls.

Information warfare is omnipresent in today's newspapers. Several times a week, you can follow reports concerning new, miniaturized weapon systems. Ironically, IW is not 'armed' in the traditional sense. Dramatic hypothetical accounts of IW abound and best serve to introduce this little-known realm.

16.13.2 Virtual Cyber Weapons

The IW weapons could more likely be used in the near future as terrorist weapons rather than on the battlefield by the regular armies. An intelligence operator hacks into a nation's telecommunications network, planting a computer code that destroys the software running that system. An operator hacks into a target nation's computer network coordinating air or rail traffic to re-program the systems to shut down without warning. These are some of the extreme hypothetical examples which would lead to drastic results [Rowe 2007].

16.13.3 Cyber Warrior

By the end of 2010, Army units hope to digitize the battlefield and link every soldier and weapon system with wireless links. A report on the twenty-first century land warrior, out of the Marine Officer training camp at Fort Quantico, shows the new battle gear for a so called cyber warrior: a lightweight helmet with mounted display, night vision sensors and flat video panel (all voice activated), integrated headgear, body armour with room for a computer in the lumbar area which gives friend or foe identification capability to the soldier, detects mines and chemicals, and has a built-in global positioning system (GPS).

In order to develop its IW capabilities still further, China has established the Academy of Military Sciences Military Strategy Research Centre, the PLA Academy of Electronic Technology, and the General Staff Department 3rd Sub-department to boost research for the study of IW weapons and strategy. It has also built an information warfare simulation centre for training its 'network warriors' [Anand, 2006] and established training programmes for its officers and commanders [Bolt and Brenner, 2004].

16.13.4 Information Exploitation

Information exploitation (IE) involves espionage that in the case of IO is usually performed through network tools that penetrate adversary systems to return information or copies of files that singly, or collectively, enable the military to gain an advantage over the adversary. Economic superiority is increasingly as important as military superiority. And the espionage industry is being re-tooled with this in mind. There is no one reason behind the increase in economic espionage and trade secret theft in recent years. Rather, a number of factors appear to have contributed to the problem: the end of the Cold

War, increased access to and use of computers and the Internet, it is profitable and often guilt-free, the lack of company resources to investigate and pursue such illicit activity, and the hesitancy to report such theft to the authorities. In 1994, the *New York Daily News* reported that American business executives were stunned in 1991 when the former chief of the French intelligence service revealed that his agency had routinely spied on US executives travelling abroad (and) that his agency had regularly bugged first-class seats on Air France so as to pick up conversations by travelling executives, (and) then (entered) their hotel rooms to rummage through attach cases. [Robbins 1994].

16.14 Operations and Information Dominance

Although information has always played a major role in military operations, the new emphasis on information dominance plays a major role in military operations.

This new emphasis greatly expands its role to that of a major realm for exerting control over the enemy. The 1991 Gulf War inspired widespread realization of the immense importance of information superiority in a modern conflict. [Erikson 1999]. Disruption of Iraqi's flow of information by destroying the command and control systems which provided critical tracking information for their fighters and surface to air missiles and disrupting their internal communication systems was an example of information warfare operations [Winn 1994].

In discussing cyber threats, it should first be made clear that the use of the Internet and its possible successors for propaganda, criminal activities, and for open-source intelligence collection has been proved. The issue here is the possibility of using digital information networks to do harm in more direct ways, be it to the Internet infrastructure itself or to other infrastructures increasingly dependent on it. In the past, a person had to be physically present at a key point to perform sabotage, as a trespasser, an insider, or a combination of the two. In networked society, these categories are translated to the logical domain. Obviously, increasing connectivity is a key enabler of cyber attack.

Technological monoculture benefits the cyber attacker because methods and resources of attack can be freely moved to and launched from anywhere to any target. One may conceive a piece of malicious software that affects some key function of the Internet and then also of every intranet and extranet where the same malicious code is successfully implanted.

16.14.1 Defensive Information Warfare

Defensive information warfare [Worden 2001, Panda 1999] is the practice of protecting an organization's computer systems against attack by hostile sources. The information security literature is replete with details of differing methods that a hostile attacker may choose to attempt to harm an organisation's information system. In early discussions about information warfare, Jensen pointed out that the goals of information warfare are centred on the concept of the precision strike that paralyzes the enemy without creating large numbers of casualties among either friendly or hostile forces [Jensen 1994].

A group of Russian computer hackers stole 10 million Dollar from Citibank by infiltrating its computer network. *The New York Times*, 1995. Developing systems capable of managing the volumes of data created by the network and processing it to generate relevant information to a wide variety of forces with different goals and requirements in a manageable form is one of the biggest challenges to achieving the goal of information dominance [Endsley 1997].

Because of the particular characteristics of cyberspace—in particular because of its dual use nature, the difficulty in differentiating a probe from an attack, the scale-free and unbounded nature of potential consequences, and the limited capabilities for definitive attribution - the traditional emphasis on imposing a cost or penalty on a specific adversary may not be alone sufficient.

16.14.2 Defensive Measures

The concept for defending the information infrastructure and the information components of other critical infrastructures includes the following principles [DoD 1996].

- Critical functions must be capable of being performed in the presence of information warfare attacks.
- Some minimum essential infrastructure capability must exist to support these critical functions.
- Point and layered defences are preferable to area defenses.
- The infrastructure must be designed to function in the presence of failed components, systems, and networks. The risk associated with failed components, systems, and networks must be managed since it cannot be avoided.
- The infrastructure control functions should not be dependent on normal operation of the infrastructure.

- The infrastructure must be capable of being repaired.

Tactical warning, damage control, attack assessment, and restoration ensures the continuance of these critical functions and activities in the presence of disruptions or attacks. The essence of tactical warning is monitoring, detection of incidents, and reporting of the incidents. Monitoring and detection of infrastructure disruptions, intrusions, and attacks are also an integral part of the defense against information warfare.

Future battles will incorporate diverse forces operating over large distance in a wide variety of missions. Operations will be characterized by the use of more widely distributed forces working with a network of distributed sensor and smart weapon systems. Technological monoculture benefits the cyber attacker because methods and resources of attack can be freely moved to and launched from anywhere to any target.

Information warfare is a double-edged sword. As the global information network becomes truly global, a disruption in one node of the system could have unknown consequences throughout the rest of the network. This would be called cyber collateral damage. leading to a scenario similar to digital Pearl Harbor. There are multiple countries, organizations, and individuals who have mastered the technical know-how to devise and conduct such an attack.

The greatest, and certainly to a Westphalian nation-state-centred universe most revolutionary, challenge for regulation is the increasing cooperation between national, regional, and international networks of regulators, to regulate the Global Information Society.

SUMMARY

This chapter introduced the need for security practices. Recent developments in information system's technologies have resulted in computerizing many applications in various business areas. Data has become a critical resource in many organizations; therefore, efficient access to data, sharing data, extracting information from data, and making use of information has become an urgent need. As a result, there have been many efforts not only on integrating the various data sources scattered across several sites but also on extracting information from these databases in the form of patterns and trends. These data sources may be databases managed by database management systems, or they could be data warehoused in a repository from multiple data sources. The emphasis was on security standards and cyber warfare.

The need for risk analysis and information security was discussed. The need for business continuity was discussed. The impact of BOTS on various industrial segments was discussed. Malicious BOTS marked the next major step in the criminalization of malicious code, a significant step up from Trojans. Malicious BOTS are often thought of as a combination of a remote access Trojan (RAT) and a worm, able to provide an attacker with remote access and the ability to spread like a worm

Key Terms

Assessment surveys and inspections An analysis of the vulnerabilities of an AIS. Information acquisition and review process designed to assist a customer to determine how best to use resources to protect information in systems.

Assurance A measure of confidence that the security features and architecture of an AIS accurately mediate and enforce the security policy.

Attack An attempt to bypass security controls on a computer. The attack may alter, release, or deny data. Whether an attack will succeed depends on the vulnerability of the computer system and the effectiveness of existing countermeasures.

Audit The independent examination of records and activities to ensure compliance with established controls, policy, and operational procedures, and to recommend any indicated changes in controls, policy, or procedures.

Audit trail In computer security systems, it is a chronological record of system resource usage. This includes user login, file access, other various activities, and whether any actual or attempted security violations occurred, legitimate, and unauthorized.

Computer fraud Computer-related crimes involving deliberate misrepresentation or alteration of data in order to obtain something of value.

Passive attack Attack which does not result in an unauthorized state change, such as an attack that only monitors and/or records data.

Passive threat The threat of unauthorized disclosure of information without changing the state of the system. A type of threat that involves the interception, not the alteration, of information.

Penetration The successful unauthorized access to an automated system.

Penetration signature The description of a situation or set of conditions in which a penetration could occur or of system events which in conjunction can indicate the occurrence of a penetration in progress.

Penetration testing The portion of security testing in which the evaluators attempt to circumvent the security features of a system. The evaluators may be assumed to use all system design and implementation documentation, that may include listings of system source code, manuals, and circuit diagrams. The evaluators work under the same constraints applied to ordinary users.

Port scan A port scan is a series of messages sent by someone attempting to break into a computer to learn which computer network services, each associated with a 'well-known', port number, the computer provides. Port scanning, a favorite approach of computer cracker, gives the assailant an idea where to probe for weaknesses.

Risk assessment A study of vulnerabilities, threats, likelihood, loss or impact, and theoretical effectiveness of security measures. The process of evaluating threats and vulnerabilities, known and postulated, to determine expected loss and establish the degree of acceptability to system operations.

Threat assessment Process of formally evaluating the degree of threat to an information system and describing the nature of the threat.

Review Questions

16.1 Differentiate between confidentiality, integrity and availability?

16.2 What are security controls and why do we need them.

16.3 What are the common security controls normally deployed in a firm.

16.4 What do we understand from security risk analysis?

16.5 What do we understand from business continuity?

16.6 How would a company benefit from business continuity practices? Justify your statement.

16.7 What is the need for business continuity standards in large business initiatives?

16.8 Differentiate between cyber war and cyber industrial espionage.

16.9 What are the various cyber weapons and what are the countering strategies?

16.10 What is your understanding of cyber robots?

Projects

16.1 Using the Internet, prepare a report on cyber robots and their industrial impact. What would be the hidden danger due to bots in industries

16.2 Prepare a detailed risk and security plan to safeguard ABC's IT systems from internal and external IT risks.

References

Adams, J. 2001, *Virtual Defense Foreign Affairs*, May/Jun, 98-112.

Anand, V. 2006, *Chinese Concepts and Capabilities of Information Warfare. Strategic Analysis*, 30(4): 781-797.

Alberts, Christopher Audrey 2002, *Managing Information Security Risks: The OCTAVE SM Approach*, Addison Wesley, USA.

Barnes, James C. 2001, *A Guide to Business Continuity Planning*, John Wiley.

Berkowitz, B. D. 1995, Warfare in the Information Age. Issues in Science & Technology, 11(1), 59-67.

Bolt, P.J. and Brenner, C.N. 2004, Information Warfare Across the Taiwan Strait, *Journal of Contemporary China*, 13(38), 129-150.

Bush, G. (Ed.), *The National Strategy to Secure Cyberspace*, Washington.

Clark, David Leon 2005, *Enterprise Security, The Manager's Defense Guide*, Addison Wesley Information Technology Series, USA.

Cobb, A. 1999, *Electronic Gallipoli? Australian Journal of International Affairs*, 53(2), 133- 150.

DoD 1996, Report of the Defense Science Board Task Force on Information Warfare—Defence, Office of the Undersecretary of Defense for Acquisition & Technology.

Endsley, Mica R and William M Jones, 2007, *Situation Awareness*, Information Dominance and Information Warefare., Tech Report 97-01

Eriksson, Anders 1999, *The Nonproliferation Review*, Spring-Summer.

Hollis, Duncan B. 2007, *New Tools, New Rules: International Law and Information Operations*, Research Paper No. 2007-15, Legal Studies Research Paper Series, Temple University.

Jensen, O. E. 1994, ' Information Warfare: Principles of Third-Wave War' Airpower Journal, 6(4).

Jaquith, Andrew 2007, *Security Metrics, Replacing Fear, Uncertainity, and Doubt*, Addison Wesley, USA.

Klevinsky, T. J., Laliberte Scott, and Ajay Gupta, 2002, *Hack I.T.: Security Through Penetration Testing*, First Edition, Addison Wesley .

Metz, S., Johnson, D. V. II. 2001 *Asymmetry and U.S. Military Strategy: Definition, Background, and Strategic Concepts*, Carlisle Barracks, PA.: US Army War College, Strategic Studies Institute, January.

Provos, Niels, Thorsten Holz 2007, *Virtual Honeypots: From Botnet Tracking to Intrusion Detection*, Addison Wesley Professional.

Panda, B. and Giordano, J. 1999, *Defensive Information Warfare*, Communications of the ACM, 42(7), 30-32, (1999).

Robbins 1994, *In the New World of Espionage, the Targets are Economic*, N. Y. DAILY NEWS, Sept. 5, available in Dow-Jones News (Publications Library).

Rowe, N.C 2007, 'War Crimes from Cyber-Weapons,' Journal of Information Warfare.

Shim, Jae K., Anique A. Qureshi, and Joel G. Siegel 2000, *International Handbook of Computer Security*, AMACOM.

Shulman, Mark Russell 1999, *Legal Constraints on Information Warfare*, Center for Strategy and Technology Air War College.

Szor, Peter 2005, *The Art Of Computer Virus Research And Defense*, Addison Wesley Professional, 2005.

The New York Times 1995, Russians Arrest 6 in Computer Thefts, Sept. 27.

Wilson, Clay 2004, *Information Warfare and Cyberwar: Capabilities and Related Policy Issues*.

Winn, Schwartau, W. 2000, *Asymmetrical Adversaries*. Orbis, 44(2), 197-206.

Winn, Schwartau 1994, *Information Warfare, Chaos on the Electronic Superhighway*, Thunder's Mouth Press.

Worden, S. P. and France, M. E. B. 2001, *Towards an Evolving Deterrence Strategy: Space and Information Dominance*, Comparative Strategy, 20(5), 453-466.

CHAPTER

17 Information Management

> [...]be aware of the challenges of organizing information on the Web. Language is ambiguous, content is heterogeneous, people have different perspectives, and politics can rear its ugly head.
> –Lou Rosenfeld and Peter Morvile

Learning Objectives

After reading this chapter, you should be able to understand:

- the need for information management in an organization
- the life cycle of information management
- and react on data quality problems

17.1 Information Architecture

As Peter Drucker noted, a distinguishing feature of the twentieth century was the emergence of knowledge as a resource to be exploited in and of itself, with emergent properties at scale requiring significant experience, specialization, and infrastructure for support. Treating information in this respect parallels the evolution of other enterprise resources. Virtually all organizations: small, medium, and large, across the world depend on technology to some degree. Software development and systems administration are skills sought worldwide as gateways to greater economic security and worldwide investment in information technology (IT) continues apace, into the hundreds of billions and trillions of dollars. With all this importance and investment, it is ironic that the activity of managing information technology itself continues to be one of the most troubled areas in today's large organizations. Information technology teams and their sponsors rely on rumour, impression, and educated guesses when faced with critical decisions.

Organizations often experience difficulties when they try to use information across business functions, system boundaries, or organization boundaries. They are frustrated when they believe they have the data to perform a business function but cannot do it. A firm may wish to perform a trend analysis or build closer relationships with its customers or partnering organizations. All too often, the information technology department cannot provide the integrated information requested by the consumer or cannot deliver the needed information within the time frame required by the consumer. These problems have confronted organizations for a long time.

Data represent real-world objects in a format that can be stored, retrieved, and elaborated by a software procedure and communicated through a network. The process of representing the real world by means of data can be applied to a large number of phenomena, such as measurements, events, characteristics of people, the environment, sounds, and smells. Data are extremely versatile in such representation. Data is a corporate asset and has to be consistent across the entire corporation, not just within the business function or division where it originated. The e-world has broken down corporate data glass walls. Data is distinguished, implicitly or explicitly into three types of data. They are as follows.

- Structured: When each data element has an associated fixed structure. Relational tables are the most popular type of structured data.
- Semi-structured: When data has a structure which has some degree of flexibility. Semi-structured data are also schema-less or self-describing. Extensible markup language (XML) is the markup language commonly used to represent semi-structured data. Some common characteristics are the following.
  - Data can contain fields not known at design time. For instance, an XML file does not have an associated XML schema file.

- The same kind of data may be represented in multiple ways. for example, a date might be represented by one field or by multiple fields, even within a single set of data.
- Among fields known at design time, many fields will not have values.
• Unstructured: When data are expressed in natural language and no specific structure or domain types are defined. Information is derived from data and information is an abstract entity. It has no separate existence on its own.

Information has special characteristics. It is easy to create but hard to trust. It is easy to spread but hard to control. It influences many decisions. The problem of how best to ensure data and data quality has been on the radar of enterprises for more than a decade now.

In recent years, increasing attention has been focused on the challenge of organizing information. Yet this challenge is not new. People have struggled with the difficulties of information organization for centuries. The field of librarianship has been largely devoted to the task of organizing and providing access to information. The phrase 'information architecture' appears to have been coined, or at least brought to wide attention, by Richard Saul Wurman, the author, editor, and/or publisher of numerous books that employ fine graphics in the presentation of information in a variety of fields.

An information architecture (IA) has been described in many ways, each tackling the question from often slightly different perspectives. Although no one definition offers a definitive answer, they all have a common thread. Some of the definitions are given below.

- Information architecture is the science of figuring out what an information user wants and constructing a blueprint of requirements before diving in for a development.
- Information architecture describes a specialized skill set which relates to the interpretation of information and expression of distinctions between signs and systems of signs.
- Information architecture is the frameworks, processes, projects, policies, and procedures to manage and use valuable enterprise information assets. This includes plans, policies, principles, models, standards, frameworks, technologies, organization, and processes that will ensures that integrated data delivers business value and aligns business priorities and technology.

The concept of an IA evolved from IBM's pioneering efforts in this area, resulting in the development of the Business Systems Planning (BSP) methodology. The concept was further developed by Dr Zachman. It facilitates the establishment of the underlying infrastructure for managing the information asset and is analogous to city/town planning—which allows the city to build on and thrive in a managed fashion.

Information architecture is the art of expressing a model or concept of information used in activities that require explicit details of complex systems. Among these activities are library systems, content management systems, web development, user interactions, database development, programming, technical writing, enterprise architecture, and critical system software design.

The architectures provide a map of these three components that form the foundation for information management planning efforts.

According to Wurman, an information architect is

1. The individual who organizes the patterns inherent in data, making the complex clear
2. A person who creates the structure or map of information which allows others to find their personal paths to knowledge
3. The emerging twenty first century professional occupation addressing the needs of the age focused upon clarity, human understanding, and the science of the organization of information

17.1.1 Information Architecture Components

A set of policies, tools, practices, and processes called information life cycle management (ILM) is used to align the business value of information with the most appropriate and cost-effective IT infrastructure. An information architecture provides easy storage, access, and retrieval of information. Understanding what is entailed in the architecture means examining each of the components.

The list of components include:

- Plan
- Policies
- Principles
- Models
- Technologies
- Organizations
- Processes

Plans are the initial part of the IA and is a decomposition of business plans. Plans are not technology plans. The business plan includes business drivers, goals, and objectives. The highest level of the organization makes statements that set the tone, direction, and most important, mandates for the management and use of information and are called policies. Information principles are one of the key core components—these express the beliefs and philosophies of the organization in terms of how it will view and treat information.

A variety of models act as the architects to present the abstractions that make up the IA—typically data models, process models, metrics/measures models, and occasionally object-based models. Standards are a list of what goes where, what means what, and who is allowed to do the thing that is to be done.

Technologies are a list of necessary technology, expressed by classes of tools, hardware, and software. The fact that IA cannot exist without a sustaining organization, is often overlooked. Therefore, the organization chart of the business, along with roles, responsibilities, and reporting lines is required. The origination accountable for information must execute policy and principles. The discreet process required for the day-to-day and strategic functions are also described and detailed.

17.1.2 Challenges of Organizing Information

In recent years, increasing attention has been focused on the challenge of organizing information. Yet this challenge is not new. The Internet is forcing the responsibility for organizing information on more of us each day. Till early 2007, data generated by Fidelity India's business units was stored and managed using discrete technologies and traditional Wintel servers. The business's number one priority was the availability of data and its integrity. Fidelity has a state-of-the-art data management system. Data centralization reduced the cost of storage. Not only that, it also improved data handling and security. De-duplication and compressions features have been used to optimize storage and efficiently utilize available space. The creation of fully redundant cluster of exchange servers with optimum load balancing has led to a highly efficient infrastructure.

Information must be organized in order to permit analysis, synthesis, understanding, and communication. Organizing information is important because it allows a person to:

- manage and retain the information more efficiently;
- communicate the information more effectively;

- recognize the need for further information and discard unneeded information;
- recognize trends, clusters, and other patterns in the information gathered; and
- synthesize disparate pieces of information into new knowledge.

17.1.3 Information Maps

Information maps are versatile tools to capture and organize information. They are diagrams that represent visually the way topics and concepts are related and organized. They are used to understand and handle complex information, to generate ideas, to plan, and to organize.

A concept map is a way of capturing understanding of a topic on a computer screen or on paper, in a way that shows how concepts are related and helps its users visualize knowledge. A concept map lets the person or group making it see whether they have understood all the parts of the topic being mapped, and the nature of the relationship between the parts.

17.2 Information Life Cycle Management

How do organizations make sure that the information they manage is valid and of quality and how do they effectively manage their information life cycle? Organizations need to improve the use of information structures and reuse of already existing information.

We readily recognize that information has a life cycle, and that its value changes over time. The e-mail messages, proposals, and contract drafts we exchange with our partners, for example, are crucial to cementing our business relationships. Information has a life cycle. It is created, changed, and is finally destroyed. Information life cycle management manages this lifecycle to optimize the use of resources, meet regulatory requirements, and ensure the integrity of the information. When a life cycle has been developed for a class of information, it can be expressed as a series of policies.

We store these information assets on our desktop hard drives and work group servers, and expect to have instant access to current and prior versions. An information life cycle is dependent on the needs of an organization and the nature of the information. How and where we store our information assets affects the day-to-day operations of our enterprise computing environment. We need to optimize our electronic storage costs and benefits across our information life cycle. Frequently accessed content needs to be managed for high availability. Infrequently accessed content needs to be managed for low availability.

Information life cycle management promises to align a company's information storage infrastructure with its business needs. It is the strategic means for an enterprise to gain maximum value from its information assets, at the lowest total costs, at every point in the information life cycle.

The business value of ILM, in short, depends on our capability to categorize content—accurately and succinctly. But traditionally, we can only identify stored information assets as discrete files, tagged by the sparse set of core operating system attributes—date last modified, date created, file name, and file size.

17.2.1 Unstructured and Structured Information

Information today comes in a wide variety of types, for example it could be an e-mail message, a photograph or an order in an online transaction processing system (OLTP).

Chief information officers (CIOs), directors of IT operations, and other line IT managers, by comparison, are concerned about information growth and cost. They realize that the volume of information requiring electronic storage is growing rapidly— 50 per cent per year in many organizations. They realize that more than 80 per cent of this new information is unstructured content, and that more than 95 per cent of this unstructured content is unmanaged. Yet despite the information explosion, IT budgets are flat and IT departments are continually struggling to do more with less.

17.2.2 Importance of Life Cycle Management

In today's highly complex technological environment with increasing compulsory regulations surrounding the storage of information and data, an ILM process needs to be a corporate policy.

17.2.3 Data Classification

Data classification is a key underlying activity that enables an economically feasible implementation of ILM. It is a process that defines, at a minimum, the performance, recovery, and retrieval characteristics of an enterprise's different sets of data, grouping them into logical categories to allow storage management activities to be mass-customized. Data classification is a process that defines the access, recovery, and discovery characteristics of an enterprise's different sets of data, grouping them into logical categories to facilitate business objectives.

522 Information Technology for Management

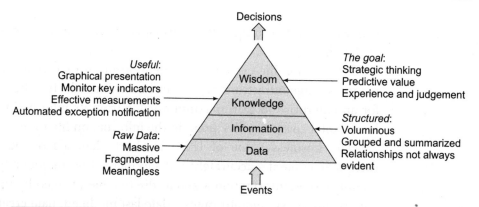

FIGURE 17.1 Knowledge and Decisions

An effective data classification initiative forces the alignment of storage management with business processes. Data classification is becoming a necessity in controlling the cost of storage and storage management. The data classification techniques define classes of data, each of which will be managed with a set of rules or policies (refer to Figure 17.1). Without a solid, business-based taxonomy, the benefits of information life cycle management cannot be achieved.

17.3 Information Economics

Computers are not useful without having monitors attached, or without having software installed. Compact disc (CD) players are not useful without CD titles, just as cameras are not useful without films. Stereo receivers are useless without speakers or headphones, and airline companies will not be able to sell tickets without joining a particular reservation system. All these examples demonstrate that, unlike bread which can be consumed without wine or other types of food, the markets we analyse in this book supply goods that must be consumed together with other products (software and hardware). In the literature of economics, such goods and services are called complements. Complementarity means that consumers in these markets are shopping for systems (for example, computers and software, cameras and film, music players and cassettes), rather than individual products. The fact that consumers are buying systems composed of hardware and software or complementary components allows firms to devise all sorts of strategies regarding competition with other firms. Complementarity turns to be a crucial factor in the markets for information goods.

Information has always been a valuable asset to those who possess it. Information economics is a theory in microeconomics that has developed simply because of the unique nature of information. An information economy is based upon the premise that information has economic value and requires an information marketplace in which such value can be exchanged. Consuming information does not prevent another person from consuming that information. All can share equally in the consumption, according to the information economics theory. However, this economic reality runs counter to the popular historic notion that information is free. Economic costs are associated with discovering, gathering, processing, manipulating, archiving, and even using information.

Software, or more generally any information has the highly noticeable production characteristic in which the production of the first copy involves a huge sunk cost (cost that cannot be recovered), whereas the second copy (third, fourth, and so on) costs almost nothing to reproduce. The cost of gathering the information for the *Britannica* encyclopedia involves more than one hundred years of research as well as the life-time work of a good number of authors. However, the cost of reproducing it on a set of CDs is less than five dollars. The cost of developing advanced software involves thousands of hours of programming time, however, the software can now be distributed without cost over the Internet. In economic terms, a very high fixed sunk cost, together with almost negligible marginal cost implies that the average cost function declines sharply with the number of copies sold out to consumers. This by itself means that a competitive equilibrium does not exist and that markets of this type will often be characterized by dominant leaders that capture most of the market.

Information economics therefore indicates that the monetary value of information must be presented in such a way as to create an opportunity for trade, if that is the end desire of the supplier. A genuine marketplace of information products cannot be sustained without the existence of a fairly sophisticated and mature legal system that guarantees proprietary rights to the producers and processors of information products and the providers of information services.

17.3.1 Copyright

Copyright law defines the property rights of the product being sold. Patent law defines the conditions that affect the incentives for, and constraints on, innovation in physical devices and, increasingly in software and business processes. When consumers place a high value on software variety, competition between the hardware firms intensifies simply because undercutting becomes more profitable. This is because undercutting increases the software variety of the winning brand. Consider a single monopoly software firm which supplies one piece of software to the entire economy. Some consumers gain an extra utility from the services and support provided by the software firms to those customers who pay for the software, whereas other consumers are 'support-independent' and do not gain from the support provided by the software firms. Each consumer in the economy has three options: the consumer can buy software, pirate software, or not use any software. In case of pirating, the consumer does not pay for the software and does not receive any support from the software firm. Software support (customers' service) is monopolized by the software writer, hence support-oriented consumers cannot steal the software and buy service from a third-party provider.

Daimler Chrysler–A Case Study in Enterprise Data Quality Success

Daimler Chrysler Corporation, a world leader in the automotive manufacturing industry with sales over 15 billion, relies on its customer database to enhance marketing effectiveness and reduce operating costs.

Challenges for the customer included improvement of data quality to enhance marketing effectiveness and reduce costs, apart from designing a single customer view within its network.

The following benefits were derived.

- **Enhanced direct marketing effectiveness:** Central database has helped Daimler Chrysler to disseminate well-targeted messages to its customers and owners. With accurate customer information, Daimler Chrysler carefully evaluates the results of its direct marketing efforts to reduce waste and determine which campaigns have been the most successful.
- **Reduced operating costs:** The firm has achieved savings by processing its own mailing lists using Group 1 Software products. For example, over 500 million pieces were validated in the last year. With over fifty million records in the customer database, a substantial improvement in data quality has delivered huge savings in mailing costs alone.
- **Increased use of data across the enterprise daimler:** Chrysler's business analyst teams are now able to access and generate marketing lists and analytic reports on customer information without IT staff assistance. The firm's

Contd

> advertising agency, BBDO Detroit, and its dealers have also benefited from improved data quality and availability of data.
>
> *Source*: http://www.g1.com/Resources/CS/Daimler-Chrysler-Data-Quality/ 20 April 2009

17.4 Data Quality Problem

Data is the driver of business decisions, actions, and transactions. Data is ubiquitous in an organization. The same piece of data is used multiple times for multiple purposes. Data is used to process transactions quickly (efficiency) and to fuel analytic packages to enable better decision making (effectiveness). We live in an era of unprecedented data abundance and aggregation. The sheer variety of new information available on the Internet, in databases, and from other sources has changed the way we conduct business, undertake research, and communicate. Most of the changes are positive. Yet, increased reliance networked data has also introduced new challenges. Accurate data is a fundamental requirement of good information systems. And yet most information systems contain significant amounts of inaccurate data, and most enterprises lack enough of a basic understanding of the concepts of information quality to recognize or change the situation. Accurate data does not come free. It requires careful attention to the design of systems, constant monitoring of data collection, and aggressive actions to correct problems that generate or propagate inaccurate data.

Data quality is a measure of the value of data in relation to potential data problems. A data quality domain is an application or use of data that imposes a set of data quality rules, each of which is associated with a degree-of-importance for the domain. One of the primary reasons data quality continues to erode has to do with organizations' attitudes toward new technology. A good data quality program begins with a comprehensive data quality assessment. The objective of the assessment is to systematically go through the data and identify all data problems and then measure their impact on various business processes.

One serious problem we need to address is that of dirty data or inaccurate information that resides in various data repositories. The reasons for bad data quality include fast-changing business dynamics, systems management gaps, uncontrolled application proliferation and people issues. The narrow definition of data quality is that it is about bad data—data that is missing or incorrect. A broader definition is that data quality is achieved when a business uses data that is comprehensive, consistent, relevant, and timely.

There are a number of theoretical frameworks for understanding data quality. Data quality is broadly defined as the totality of features and characteristics of a data set that bear on its ability to satisfy the needs that result from the intended use of the data. Data accuracy is one of the foundational features that contribute to data quality. In addition, data quality has two essential components: content and form. These definitions are important to keep in mind when considering ways to minimize data inaccuracies, as they illustrate why the task of fixing dirty data requires more than merely providing right information. The collection of data quality issues are collectively called data quality problems. Data quality problems cost organizations millions of dollars, waste vast amounts of time and resources, and deceive management into making very poor decisions. Data quality can be a huge expense to companies and most organizations do not seem to give process discipline enough attention, which can end up creating a lot of bad data. Many executives and managers dramatically underestimate the tremendous damage poor quality data inflicts on their organization each year. Over the next decade, more than 25 per cent of critical data in companies will continue to be flawed, that is, the information will be inaccurate, incomplete, or duplicated, according to research and advisory firm Gartner, Inc. Gartner expects that three-quarters of large enterprises will make little to no progress towards improving data quality until 2010.

17.4.1 Data Quality—Why is it Important?

Data quality investigations are all designed to surface problems with the data. This is true whether the problems come from stand-alone assessments or through data profiling services to projects. Choosing dimensions to measure the level of quality of data is the starting point of any data quality related activity. Though measuring the quality of ICT technologies, artifacts, processes, and services is not a new issue in research, for many years several standardization institutions have been operating.

Models are used in databases to represent data and data schemas. Techniques correspond to algorithms, heuristics, knowledge-based procedures, and learning processes that provide a solution to a specific data quality problem. Methodologies provide guidelines to choose, starting from available techniques and tools, the most effective data quality measurement and improvement process.

Companies today are under increasing pressure to make better business decisions in less time, with less risk, while producing higher quality results. The challenges are enormous, as are the many issues that can arise and potentially jeopardize success. Among the most pervasive problems companies face is the consistently poor quality of internal data that is used to draw conclusions and make decisions. The problem with data is that its quality quickly degenerates over time.

In any organization, it is critical that the importance of data quality be properly explained. From the chief executive officer (CEO) to the lower managerial levels, a sensible and understandable rationale must be given in order to obtain buy-in and motivate active participation in the data quality effort. A thorough cost/benefit analysis is critical to any data quality program. Early on, perhaps before, but certainly immediately after, any kickoff meetings and motivational exhortations, the executive team would require an economic justification, the value proposition underlying data quality and a data quality program. It is this economic rationale for the added value brought by good data quality or at the very least identification of opportunities and competitive advantages lost by poor data quality, that will support the other initiatives necessary to achieve awareness and to implement a viable data quality program. The CEO obtains the data requested and remains unaware of the cost of producing it. Consequently, executives are reluctant to commit scarce resources to an apparent non problem area of data quality unless there is a defensible cost/benefit justification to show the costs of neglecting the data quality problem and the economic benefits of initiating the data quality project. Senior executives may be further reluctant if the organization is performing adequately, particularly if calling attention to the problem would affect the corporate image. Among the commonly used rationales are as follows.

- Data of high quality is a valuable asset.
- Data of high quality can increase customer satisfaction.
- Data of high quality can improve revenues and profits.
- Data of high quality can be a strategic competitive advantage.

Beyond these broad rationales, more specific ones can be offered. To repeat an old aphorism, 'an ounce of prevention is worth a pound of cure.'

Clearly, data quality is not a one-time effort. A lot of time and effort is required to improve the data quality. The process of ensuring organizational data quality is broadly defined as data governance. The problem with data governance programs has been that most companies are taking a top-down approach, while more pressing short-term business demands are derailing efforts and distracting resources. As a result, only a very small percentage of companies have an active data governance program in place because in most cases a top-down approach does not work. What is most critical for companies that want to make better business decisions by leveraging higher quality data is to begin implementing a data governance process. A proactive approach to data governance and data quality enables the identification of the introduction of flawed data within the application framework, the flawed processes that are responsible for injecting unexpected data can be corrected, eliminating the source of the data problem.

Organizations are forced to validate their data quality for reliability and effective decision-making. This is achieved through organizational initiatives to ensure high levels of data quality. The ability to motivate data quality improvement as a driver of increasing business productivity demonstrates a level of organizational maturity that views information as an asset, and rewards proactive involvement in change management. Some of the organizational data quality initiatives are listed below.

- Organizations need a way to formalize data quality expectations as a means for measuring conformance of data to those expectations.
- Organizations must be able to baseline the levels of data quality and provide a mechanism to identify leakages as well as analyse root causes of data failures.
- Organizations must be able to effectively establish and communicate to the business client community the level of confidence they should have in their data, which necessitates a means for measuring, monitoring, and tracking data quality.

High-quality data has six key attributes: accuracy, reliability, credibility, timeliness, completeness, and appropriateness. The degree to which data meets each of these criteria determines its quality. Data quality refers to the degree of excellence exhibited by the data in relation to the portrayal of the actual phenomena.

> **Case Study–Irish Life & Permanent**
>
> Mergers and acquisitions frequently create data quality hassles, especially where two customer databases need to be consolidated.
>
> That was the case for financial services company Irish Life & Permanent formed from the 1999 merger of insurance company Irish Life and the Irish Permanent bank.
>
> The company has embarked on a massive clean-up operation, in preparation for implementing IBM's WebSphere Customer Center product for master data management (MDM), which will enable it to establish a single record of its entire customer base.
>
> A massive data clean-up was underway, using IBM Information Server to bring together matching records and consolidate them. When agreed thresholds for matches are not reached, the system marks the records as 'suspect' and flags them for human examination.
>
> Using that approach, the group has already identified 117,000 exact duplicates – where there are two or more records of the same customer – within the Irish Life database alone.
>
> Now that those duplicates have been eliminated and the rest of the data has been cleaned up, the group started loading that database into WebSphere Customer Center. The data held in the bank database, meanwhile, will also cleaned and loaded into the MDM system.
>
> Benefits include a single reliable data sheet for different customers.
>
> *Source*: http://www.ikmagazine.com/xq/asp/ txtSearch.CRM/exactphrase.1/sid.0/ articleid.17E21301- 5E70-4374-B646-E53ABCB7FD3E/ qx/ display.htm, downloaded on 20 April 2009

17.4.2 Manifestation of Data Quality Problems

One of the most frustrating issues associated with data quality improvement is not knowing how bad data really affects the organization. In some companies, the methods to address poor data quality may incorporate interim data corrections, nominal customer service adjustments, multiple copies of non-standard data, or other 'topical' solutions, all of which incur some 'correction' costs to the company but probably do not address the source of the problem. Gartner's research shows that poor-quality customer data leads to significant costs, such as higher customer turnover, excessive expenses from customer contact processes like mail-outs and missed sales opportunities. But companies are now discovering that data quality has a significant impact on their most strategic business initiatives, not only sales and marketing. Other back-office functions, such as budgeting, manufacturing and distribution, are also affected [Batini 2006].

We can divide the costs associated with poor data quality into the soft costs, which are clearly evident but yet hard to measure, and the hard impacts, whose effects can be estimated and measured.

- Hard Costs: Hard costs are those costs whose effects can be estimated and/or measured. These include:
  - Customer attrition
  - Error detection
  - Error rework soft costs
- Soft costs: These costs are those that are evident, clearly have an effect on productivity, yet are difficult to measure. These include:
  - Difficulty in decision-making
  - Time delays in operation
  - Organizational mistrust
  - Lowered ability to effectively compete
  - Data ownership conflicts
  - Lowered employee satisfaction
  - Error prevention
  - Customer service
  - Fixing customer problems
  - Delays in processing
  - Delayed or cancelled projects

17.4.3 Improved Data Quality Benefits

The cornerstone of sound business decision-making revolves around having access to accurate and timely information. Improved data quality can add to the company's bottom line, either through optimization in operational systems or by improving the value of knowledge generated through a business intelligence process. The following kinds of improvements are typically the result of improved information quality [Al-Hakim 2007]

- Improved throughput for volume processing
- Improved customer profiling
- Decreased resource requirements
- Predictability in project planning and completion

17.4.4 Data Quality Assurance

A data quality assurance program is an explicit combination of organization, methodologies, and activities that exist for the purpose of reaching and maintaining high levels of data quality. Quality assurance, quality control, inspection, and audit are terms applied to other activities that exist for the purpose of maintaining some aspect of the corporation's activities or products at a high level of excellence. The goal of a data quality assurance program is to reach high levels of data accuracy within the critical data stores of the corporation.

A straightforward approach to assess the economic trade-offs of undertaking a data quality program, one used often to evaluate information systems projects, is to quantify the benefits and costs and to apply the basic value relationship.

$$\text{Value} = \text{Benefits} - \text{Costs} \tag{17.1}$$

The challenge reduces to how well one can quantify costs and benefits, what assumptions are made in weighing the two, and how one incorporates intangible benefits into the analysis.

Improvement assumes that the current state of data quality is not where you want it to be. Much of the work is to investigate current databases and information processes to find and fix existing problems. This effort alone can take several years for a corporation that has not been investing in data quality assurance.

The data quality audit is a business rules-based approach that incorporates standard deviation to identify variability in sample test results.

17.4.5 Assessing Data Quality

Traditionally, data quality research and practice have revolved around describing and quantifying the intrinsic quality of individual data records or rows in a database table. Most current approaches to data quality are rooted in the evaluation of traditional data matching or duplicate detection techniques, such as precision and recall graphs [Bilenko and Mooney 2003]. However, these techniques are inadequate for modern knowledge-based entity resolution techniques where two records for the same entity may present entirely different representations and can only be related to each other through a priori assertions provided by an independent source of associative information.

Data are said to be reliable when they are:

- complete
- accurate

Data is expected to be reliable and consistent. Consistency refers to the need to obtain and use data that are clear and well-defined to yield similar results in similar analyses.

17.4.6 Data Quality Metrics

Data quality metrics assessment applies a process of semantic refinement that quantifies data quality performance to develop meaningful metrics associated with well-defined data quality dimensions. The refinement steps include the following.

- Identifying the key data assertions associated with business policies.
- Determining how those data assertions relate to quantifiable business impact.
- Evaluating how the identified data flaws are categorized within a set of data quality dimensions and specifying the data rules that measure their occurrence.
- Quantifying the contribution of each flaw to conformance with each business policy.
- Articulating and implementing the data rules within a drillable reporting framework.

Data quality investigations turn up facts. The primary job of the investigations is to identify inaccurate data. The data profiling process will produce inaccurate facts that in some cases identify specific instances of wrong values. Data profiling is defined as the application of data analysis techniques to existing data stores for the purpose of determining the actual content, structure, and quality of the data. Data profiling is a process that involves learning from the data. It cannot find all the inaccurate data. It can only find rule violations. This includes invalid values, structural violations, and data rule violations. Some inaccurate data can pass all the rule tests and yet still be wrong.

Data profiling is an indispensable tool for assessing data quality [Wootton 2007]. It is also very useful at periodic checking of data to determine if the corrective measures are effective or to monitor the health of the data over time. Data profiling uses two different approaches for examining data. One is discovery, whereby processes examine the data and discover characteristics from the data without the prompting of the analyst. In this regard, it is performing data mining for metadata. This is extremely important to do so because the data will take on a persona of itself and the analyst may be completely unaware of some of the characteristics. It is also helpful in addressing the problem that the metadata that normally exists for data is usually incorrect, incomplete, or both.

The second approach to data profiling is assertive testing. The analyst poses conditions he believes to be true about the data and then executes data rules against the data that check for these conditions to see if it conforms or not. This is also a useful technique for determining how much the data differs from the expected. Assertive testing is normally done after discovery.

The output of data profiling will be accurate metadata plus information about data quality problems

Dimensions of data quality are often categorized according to the contexts in which metrics associated with the business processes are to be measured, such as measuring the quality of data associated with data values, data models, data presentation, and conformance with governance policies.

There is a strong temptation for quality groups to generate metrics about the facts and to grade a data source according to:

- number of rows containing at least one wrong value,
- graph of errors found by data element,
- number of key violations (non-redundant primary keys, primary/foreign key orphans),
- graph of data rules executed and number of violations returned,
- breakdown of errors based on data entry locations, and
- breakdown of errors based on data creation date.

Metrics can also be useful to show improvements. If data is profiled before and after corrective actions, the metrics can show whether the quality has improved or not.

SUMMARY

In this chapter we have perceived that data quality is a multi-disciplinary area. This is not surprising, since data in a variety of formats and with a variety of media are used in every real-life or business activity, and deeply influence the quality of processes that use data. While data quality is a relatively new research area, other areas, such as statistical data analysis, have addressed in the past some aspects of the problems related to data quality. With statistical data analysis, also knowledge representation, data mining, management information systems, and data integration share some of the problems and issues characteristic of data quality, and, at the same time, provide paradigms and techniques that can be effectively used in data quality measurement and improvement activities.

Key Terms

Archive A term for inactive records which need to be kept permanently or for an extended period of time

Data Facts, measurements, or observations

Data management The terms 'Data Management', 'Information Management' and 'Content Management' are all used to encompass the policies, strategies, processes, and technologies used to manage an organization's information throughout the stages of its life cycle.

Information life cycle Activities involved in managing information throughout its life, e.g., information acquisition, creation, retention, storage, retrieval, communication, utilization, and destruction.

Knowledge management Systematic approaches to help information and knowledge emerge and flow to the right people, at the right time, in the right context, in the right amount, and at the right cost so they can act more efficiently and effectively.

Knowledge map A visual representation of the knowledge of an organization or the knowledge underlying a business process.

Metadata Descriptive and cataloguing data which systematically identifies various attributes of a class of items such as file/folder and documents.

Structured information Information that can be organized in the rows and columns of spreadsheets or relational databases (for example, a series of data samples).

Unstructured Information that is not as easy to organize in the rows and columns of spreadsheets or relational databases. Examples include audio or video files, images, e-mail messages, text documents, slide presentations, and Web pages. Unstructured information is typically more difficult to find, analyse, or interpret than structured information.

Review Questions

17.1 What are the types of data types documented?
17.2 What is information architecture?
17.3 What are the information architecture components. Briefly outline them?
17.4 What are the challenges of organizing information?
17.5 Why is data quality important?
17.6 Briefly describe the problem of data quality?
17.7 What is the correlation between data quality and business?
17.8 What are the manifestations of data quality problem?
17.9 What are data quality metrics?
17.10 What is data profiling and how is it useful for an industry?

Projects

17.1 Visit a company which is a few years old and try to catalogue the information life cycle practices of the company.

17.2 Visit a company which is a few years old and try to list the quality metrics practised by the company.

References

Al-Hakim, Latif 2007, *Information Quality Management: Theory and Applications*, Idea Group.

Barton, Robert 2003, *Global IT Management A Practical Approach*, John Wiley.

Batini, Carlo, Monica Scannapieca 2006, *Data Quality Concepts, Methodologies and Techniques*, Springer.

Bennet, P.Lientz and Lee Larssen 2007, *Manage IT as a Business Professional and Trade Series*, Elsevier.

Betz, Charles T. 2007, *Architecture and Patterns for IT Service Management, Resource Planning, and Governance*, Elsevier.

Davis, Jim, Gloria J. Miller, and Allan Russell 2006, *Information Revolution Using the Information Evolution Model to Grow Your Business*, John Wiley.

Härder, Theo and Wolfgang Lehner 2005, *Data Management in a Connected World*, Springer.

Little, David B., Skip Farmer, and Oussama El-Hilali 2007, *Digital Data Integrity The Evolution from Passive Protection to Active Management*, John Wiley.

Luenberger, David G. 2006, *Information Science*, PUT.

Petrocelli, Tom 2005, *Data Protection and Information Lifecycle Management*, Pearson.

Prahalad, C.K. and M.S. Krishnan, *The New Age of Innovation.*, Tata McGraw Hill.

Reid Roger and Gareth Fraser-King 2007, *Data Lifecycles Managing Data for Strategic Advantage*, John Wiley.

Rosenfeld, Lou and Peter Morvile 2002, *Information Architecture for the World Wide Web*, Second Edition, Chapter 5: Organization Systems, P. 74, O'Reilly.

Wootton, Cliff 2007, *Developing Quality Metadata*, Elsevier.

Yang W. Lee et al. 2006, *Journey to Data Quality*, MIT Press.

CHAPTER

18 Audit

Virtually everything in business today is an undifferentiated commodity, except how a company manages its information. How you manage information determines whether you win or lose.
—Bill Gates

> **Learning Objectives**
>
> After reading this chapter, you should be able to understand:
> - the concepts of audit
> - the need for controls and internal controls
> - and apply the principles of audit

18.1 Introduction

In this globally networked economy, information is more valuable than ever. A lot of it is highly confidential and sensitive. Information deals with a variety of issues and shoulders numerous responsibilities. Information is increasingly recognized as a valuable resource that needs to be managed. Today's business environment is rapidly and constantly changing. The adoption of audit, especially IT audit, is globally catching up in the enterprise agenda and is being taken seriously after the Enron and 7/11 disasters. Information technology auditing is the evaluation of information systems (IS), practices, and operations to assure the integrity of an entity's information. Such evaluation can include assessment of the efficiency, effectiveness, and economy of computer-based practices with computer as an audit tool. Technology is one of the key factors that are forcing auditors to reassess their approach to auditing. Other factors are the evolving regulations and audit standards calling for auditors to make better use of technology. These forces are creating a new audit environment, and audit professionals who understand how to evaluate and use the potential of emerging technologies can be invaluable to their organizations.

Chief information officers (CIOs) today play a very vital role in an organization. Whether it is financial services, health care, or the public sector, securing sensitive business data is a cause of concern. A key challenge CIOs face is to balance costs while keeping the enterprise abreast with the latest technologies to improve productivity. But even more challenging is maintaining a fine balance between information security and convenience and productivity. These are carried out through well-structured internal controls.

Many years ago, internal controls were purely concerned with bookkeeping. They were checks that bookkeepers did to detect errors and fraud. Indeed, the English word control derives from an ancient bookkeeping control that involved keeping two records of transactions and checking one against the other, the roll. The Latin term for this is *contra rotulus*. However, in recent decades the phrase internal control has become ambitious and now claims to include much more than bookkeeping.

Managers from different functions might join together to form a permanent network or team, formulating policies and processes to assist in specifically managing information behaviour. Auditing is an accepted management technique to carry out this requirement. Many different types of audits currently exist in the commercial world. Information technology audit is one among them. Today, IT auditing is a profession with conducts, aims, and qualities that are characterized by worldwide technical standards, an ethical set of rules (ISACA Code of Ethics), and a professional certification program (Certified Information Systems Auditor, CISA). It requires specialized knowledge and practicable ability, and often long and intensive academic preparation. IT auditing is an integral part of the audit function because it supports the auditor's judgement on the quality of the information processed by computer systems. It is needed to evaluate the adequacy of information systems to meet processing needs, to evaluate the adequacy of internal controls, and to ensure that assets controlled by those systems are adequately safeguarded.

This chapter is designed for the professional and those who wish to learn about the Information Systems Audit and Control community, as well as those aspiring to enter the profession.

18.2 Systems

Companies require systems, structures, and processes to operate globally. A system represents a set of dependent elements forming a single unitary entity. A system can be defined by the following elements: inputs, outputs, transformation process, system structure, and its state. A process is nothing more than a structured set of activities and decisions to do a certain job. Common to all the processes is that they consist of one or more activities. The activities require an effort and have a result, and they are implemented by adding some form of value to the effort.

The value of the activities may vary, but the purpose is to implement something that is planned and thoroughly considered. These are strong levers to affect behaviour, since they embody the norms and values of the culture of the organization. They are powerful catalysts for change or significant inhibitors to it. When systems, structures, and processes are not aligned with desired new values and behaviours, cultural transformation efforts are ultimately futile. Examples of these include the following.

- Systems: Management and measurement systems reward and recognition systems
- Structures: Hierarchical or team-based structures, functional or matrix structures
- Processes: Customer relationship management (CRM), integrated product development (IPD)

What is available here is an opportunity to create value. The rapidly merging fields of internal control and risk management often look established, standardized, and even dull (refer to earlier chapters of risk). Most people working in this territory are pioneers trying to get by with a handful of crude tools. It is no surprise that many people today view internal control and risk management as dreary chores of little real value, done only to placate regulators and auditors.

The risks in a computer-based system include both the risks that would be present in any manual processing system and those that are unique to an automated environment. In a manual system, errors are made individually, but in computer systems errors are made in quantity because automated systems apply rules (good and bad) consistently. Additionally, automated systems can contain errors that can trigger errors in an unrelated part of the system and so on. For example, when Equity Funding Corporation of America (EFCA) declared bankruptcy in 1973, the minimum direct impact and losses from illegal activity were reported to be as much as 200 million dollars. Further estimates from this major financial fraud escalated to as much as 2 billion dollars, with indirect costs such as legal fees and depreciation included. These losses were the result of a computer-assisted fraud in which a corporation falsified the records of its life insurance subsidiary to indicate the issuance of new policies. In addition to the insurance policies, other assets, such as receivables and marketable securities, were recorded falsely.

Risks associated with automated applications include the following.

- Weak security
- Unauthorized access to data
- Unauthorized remote access
- Inaccurate information
- Erroneous or falsified data input
- Misuse by authorized end users
- Incomplete processing
- Duplicate transactions
- Untimely processing
- Communications system failure
- Inadequate training
- Inadequate support

The greatest threats to the integrity and privacy of the information system come from inside the organization. These threats include:

- degradation of the validity, accuracy, and reliability of data resulting from errors produced by incompetence or carelessness;
- loss or destruction of assets by malicious acts; and
- deliberate disclosure of private or privileged information.

The best defence against these threats is a combination of actions to reduce the threats supplemented by actions, and risk mitigate the threats. Risk is the probability that an event or action will adversely affect the organization. The defence includes installation and maintenance of basic routine safeguards, such as password protection of computer access, and the use of access tables to authorize the kinds and extent of access that each individual is given to the information assets of the corporation (refer to earlier chapters on information security risk).

There is a lot of scope for getting greater business value from risk control. A risk management approach will often play an important role in identifying and establishing the needs of an organization in relation to creating and managing information resources. If we trace the development of the ideas behind risk management and internal control, some exciting possibilities emerge. There are opportunities to borrow ideas and techniques from competing approaches and fuse them into one. There are opportunities to combine specialist risk control teams in large organizations. And there are opportunities to combine and rationalize the thinking behind internal control and risk management to create one simpler perspective. Corporate-wide and complex enterprise workflows require the establishment of comprehensive internal control systems, as well as their integration in the financial reporting cycle and risk management. Further, a risk management system must be established that is capable of identifying and addressing business risks, and of mitigating, managing, and monitoring them through those appropriate controls. More often what people want is assurance, a feeling of comfort, some protection from losses or just from embarrassment.

System planners must ensure that provisions are made for the following.

- An adequate audit trail so that transactions can be traced forward and backward through the system
- Ensuring technology provided by different vendors are compatible and controlled
- Adequately designed and controlled databases to ensure that common definitions of data are used throughout the organization, that redundancy is eliminated or controlled, and that data existing in multiple databases is updated concurrently
- Handling exceptions to, and rejections from, the computer system
- Unit and integrated testing, with controls in place to determine whether the systems perform as planned and meet the business objectives

- Controls over changes to the computer system to determine whether the proper authorization has been given and documented
- Adequate controls between interconnected computer systems
- Adequate security procedures to protect the data and availability of data on demand
- Authorization procedures for system overrides and documentation of those processes
- Determining whether organization and government policies and procedures are adhered to in system implementation
- Backup and recovery procedures for the operation of the system and subsystems with assurance of business continuity
- The documentation and existence of controls over the accounting for all data (for example, transactions) entered into the system and controls to ensure the integrity of those transactions throughout the computerized segment of the system
- Training user personnel in the operation of the system

With computer-based financial reporting systems, new auditing procedures must be continually developed and improved. The electronic computations performed, such as additions or deletions of records or fields within records, and assurances that the transactions were authorized must be done through the computer in concert with the transaction flow. As one can experience, there are significant differences in the techniques of maintaining adequate internal control in computer-based processes.

18.3 Audits

Audits are a part of the overall control system of an organization and provide several important control functions. Auditing is defined as a systematic process of objectively obtaining and evaluating evidence regarding the current condition of an entity, area, process, financial account, or control and comparing it to predetermined, accepted criteria and communicating the results to the intended users. The purpose of auditing is diagnostic, i.e., to discover, check, verify, and control some or other process/resource in an organization.

Audits are conducted in many different forms in organizations today. In different countries, different national prerequisites apply to who is allowed to perform different types of audits. The various types of audits include the following.

- A quality system audit measures an organization's capability to meet the quality requirements.
- Management audits are carried out to validate the business strategic plan reflecting the business objectives.
- A process audit verifies the validity of process to deliver the expected output.
- System audits are carried out to ensure that a business management system is sufficiently comprehensive to control all of the activities within that business.
- Procedural audits verify the documented practices and its completeness to ensure the implementation of approved policies and are capable of controlling the organization's operations.
- The major purpose of an information audit is the identification of users' information needs, as well as how well these needs are met by the information services department. The information systems audit is used for its investigation of the way in which technological tools are used to manage information resources.
- The communication audit focuses on organizational information flow patterns.
- Information mapping focuses on the identification and use of organizational information resources.
- The knowledge audit as knowledge management (or strategic information management) is the 'highest'/last level of information management, and therefore logically follows on information management and information auditing.
- The intelligence audit is used because of its relationship with both information and knowledge management.

Audits may be needed in different phases of the life of digital objects. The general consensus is that measures for proper management and preservation have to be taken from the very beginning, that is the design of systems that create or produce them. For example, ISO9001:2000 defines audit to be of the following three types.

- First party audits of an organization, or parts of an organization by personnel employed by that organization. These audits are usually referred to as internal audits. Members of a business evaluate their own processes with established criteria, with respect to their organization. This

is the least effective form of auditing, as the auditors will find it difficult to criticize their own work.
- Second party audits are carried out by customers upon their suppliers and are completed by an organization independent of the organization being audited. These audits are usually referred to as external audits or vendor audits.
- Third party audits are carried out by personnel who are neither connected to the customer nor to the supplier. They are usually employees of certification bodies or registrars such as British Standard Institute (BSI), etc.

Computer-based auditing has traditionally been considered from two perspectives: a systems-based approach and a data-based approach.

- A systems-based approach can be used to test the application's controls to determine if the system is performing as intended. Approaches to internal control reviews are primarily based on a review of the application system in terms of input-output relationships and program reviews.
- Data-based approach view of computer-assisted auditing focuses on the data and is commonly called transaction or data-based auditing. This approach is primarily used during the conduct phase, providing the auditor with increasingly more detailed information about the audit entity. Often this technique is used to verify the accuracy, completeness, integrity, reasonableness, and timeliness of the data. It is also often used to address Sarbanes-Oxley compliance requirements.

The primary participants needed for conducting an audit are auditor, auditee, and client. The person conducting the audit is called the auditor, lead auditor, or audit team leader. The organization being audited or investigated is called the auditee. There is also a client, the person or organization that has requested the audit. At least three aspects will be involved in performing an audit: the object, the aspects, and the framework.

18.3.1 Information Audit

An information system is a particular type of work system that uses IT to capture, transmit, store, retrieve, manipulate, or display information, thereby supporting one or more other work systems. Information systems provide users in organizations with computer support to accomplish business tasks. The IT system is a component of the information system that is in charge of collecting, processing, transmitting, storing, and presenting the data by using computing systems. The functionality of an information system includes the allocation of resources such as data, communication services, or hardware devices to the users. Single-user tasks usually belong to (business) processes that are executed to realize a certain business objective. Typical examples of business processes include processing insurance claims, processing tax forms, order fulfilment, or recruitment of employees, etc.

The information system resides in the informational flows and circuits and all the methods and techniques used to process the data. It is said to be process-aware if it supports process enactment by scheduling the activities according to the specified rules of the respective process type.

Controls in an information system reflect the policies, procedures, practices, and organizational structures designed to provide reasonable assurance that objectives of information systems will be achieved. The controls in an information system ensure effectiveness and efficiency of operations, reliability of financial reporting, and compliance with the rules and regulations. An effective information system leads the organization to achieve its objectives. It uses minimum resources in achieving the required objectives.

Information technology audit is the process of collecting and evaluating evidence to determine whether an information system has been designed to maintain data integrity, safeguard assets, allow organizational goals to be achieved effectively, and use resources efficiently. Data integrity relates to the accuracy and completeness of information as well as to its validity in accordance with the norms. It analyses the systems and the networks with the view of measuring the efficiency of technical and procedural control in order to minimize the risks. It entails the systematic examination of the information resources, information use, information flows, and the management of these in an organization. It is an important element in the process of feedback. It is an instrument of evaluation and provides information that can be used to plan and implement corrective actions. It involves the identification of users' information needs and how effectively these are being met. Information Systems Audit and Control Association (ISACA) has developed a document with the title IS standards, guidelines, and procedures for auditing and control professionals.

One of the results of an information audit is knowledge of available information sources and where these are, i.e., information inventory. The information inventory is analysed in terms of the usefulness of the information sources and according to this information, decisions regarding archiving and/or disposing can be made. This can enhance the use of information. An information audit is an effective marketing tool in itself as it heightens information awareness. The comparison of the information inventory to the identified information needs will highlight the information gap. The information audit provides detailed and accurate information of the organizational information environment as well as an understanding of the way in which the organization functions.

18.3.2 Audit Teams

There are two necessary components for an audit to be successful. The first is an auditor with the right skills, education, and experience. The second is the audit process itself. The auditor evaluating today's complex systems must have technical skills to understand the evolving methods of information processing.

A group of auditors will form an audit team. Audit teams are composed with consideration to the type, content, and extent of the audit to be conducted. In accordance with the dual-control principle, each audit should be conducted by at least two internal audit team employees, one of whom has to act as the audit lead. The audit lead should be nominated early. In case of global audits, the end of the audit planning phase is a good time to appoint the audit lead. For all audits, the appointment of an audit lead should be made well ahead before the audit announcement is sent out. The audit manager is responsible for selecting the audit teams. When selecting the audit team members, consideration should be given to audit content, cultural group, and linguistic requirements, as well as personal aspects. Audit teams primarily should be able to meet the requirements of the audit and carry out the audit successfully. It is also important that the team members are good at socializing and are able to work together, especially in international assignments.

Each team member must understand the audit process before the beginning of the audit. The IT auditor must know the characteristics of the users of the information system and the decision-making environment in the auditee organization while evaluating the effectiveness of any system. It is therefore important to hold joint kick off meetings, where important aspects of the audit is highlighted. In addition, the team member responsible for taking minutes should be identified and the manner for presenting interim results should be clarified. Access to sensitive data should be discussed. Generally, information-based business environment, business professionals who are technically competent in IT or IT specialists who understand the accounting, commerce, and financial operations are in high demand for IT auditing careers.

Audit documentation is the written record that supports the auditor's representations and conclusions. Documentation serves as a basis for review and is used to plan and perform the engagement. It includes records of planning and performing the work, as well as a record of the procedures performed and evidence obtained. Audit documentation is sometimes referred to as workpapers or working papers. Audit documentation must be prepared for each engagement conducted according to relevant standards. It must be sufficiently detailed and clear. Audit documentation may be in the form of memoranda, confirmations, correspondence, schedules, audit programs, and letters of representation. Audit documentation may be kept on paper, in electronic files, or in other media. Audit documentation should show compliance with appropriate standards. For every relevant assertion, the documentation should support the basis for the auditor's conclusions.

External auditors seek to obtain sufficient evidence to support an opinion on overall fairness of the financial statements and accounting. Their perspective is therefore mainly backwards, what happened last year. The internal audit on the other hand is mainly interested in the situation now and in the future. External auditors also need to evaluate the system of internal control. Mainly external auditors are interested in internal controls only as far as they concern the financial statements and accounting. Also the main difference between internal and external auditors is their line of reporting and responsibility. Internal audit works for the top management of the organization whereas external auditors work for the stakeholders and to some extent to the authorities.

18.3.3 Audit Schedule

Auditing departments create annual audit schedules to gain agreement from the board on audit areas, communicate the audit areas with the functional departments, and create a project/resource plan for the year. Audit schedule should be linked to current business objectives and risks based on their relative cost in terms of potential loss of goodwill, loss of revenue, or non-compliance with laws and regulations. Annual schedule (refer to Table 18.1) creation is the process of determining the total audit hours available, then assigning universe items to fill the available time.

Table 18.1 Annual Audit Schedule

Fn	Jan	Feb	Mar	Apr	May	Jun	Jul	Aug	Sep	Oct	Nov	Dec
Fn1	X			X			X			X		
Fn2		X			X			X			X	
Fn3	X			X			X			X		
Fn4		X			X			X			X	
Fn5	X			X			X			X		

The audit performance record has to be consulted to ensure that the scheduled audits can in fact be conducted based on previous performance and time requirements. It is also important to schedule reserve capacities for unscheduled audits. This allows identifying noticeable capacity over or under utilization in time to make adjustments to the schedule. An adjustment of the assignment during the year may affect audits still to be conducted. Different time zones and different personal circumstances result in new challenges to audit teams. Once the available audit hours are determined, audit management can prepare the audit schedules.

Annual audit schedule comprises the creation of risk profiles, the compilation of the audit inventory, as well as the creation of the annual audit plan and of the regional team-based execution plans. The culmination of the annual audit scheduling involves lining up audits for a year slotted into the available weeks and months under consideration of personnel capacities.

Annual audit scheduling begins with the creation of risk profiles for all possibly relevant auditable entities. For all those entities a risk profile is created, and they are subsequently added to the audit inventory. The schedule is then compiled based on the topics in the audit inventory and their respective risk profiles. The schedule is derived from the inventory based on the priorities given by the risk assessment. This approach enables audit managers to continuously map their inventory to the schedule and, if required, address any unforeseeable events and audit needs.

18.3.4 Audit Plan

The audit process is well known in the government, financial, quality assurance, and systems environments, focusing on different aspects including information management. Audit lists allow the auditors to examine individual process steps or audit objects comprehensively and thoroughly. Audits are usually planned and initiated by the audit department in relation to the status and importance of the various activities of an organization. An important part of the process for managing an audit function involves planning. Planning covers both administration of the audit office as well as administration of the audit assignment. For successful audits, we need to know what we want to achieve (audit objectives), determine what procedures we should follow (audit methodology), and assign qualified staff to the audit (resource allocation). Thus an audit plan partitions the audit of IT into discrete segments. These segments describe a computer systems audit as a series of manageable audit engagements, identifying the associated risks and risk mitigation steps.

To maintain a clear focus on audit objective(s), the audit team typically performs an audit in six sequential tasks. During the initial audit, a meeting is scheduled between the auditor and all the key operating personnel to kick off the project. The meeting agenda focuses on: audit objectives and scope of work, facility rules and regulations, roles and responsibilities of project team members, and description of scheduled project activities.

It is important in the planning phase to first consider those audits that must be conducted, taking available capacities into account. Then other audits are added in line with their priority as identified during the risk assessment. Within each team, the audit engagements are sorted according to their priorities and appear either as fixed engagements or potential engagements depending on their risk rating.

The annual audit schedule is embedded into the audit performance record, which provides up-to-date annual statistics of completed audits and the status of audits not conducted. The audit performance record affords a quick overview of the activities of internal audit.

The next step after creating the annual audit schedule is the preparation of the actual execution plan. The following points are of importance at the activity-related level.

- The scheduled audits have to be assigned according to the number and skills of the auditors. Qualification, experience, availability, etc. are important for the composition of each audit team.
- The time planning and sequence of the various audits is a closely related issue.

18.3.5 Audit Preparation

Audit preparation is composed of all the work that is involved in initiating an audit. The functions include audit selection, definition of audit scope, initial contacts and communication with auditees, and audit team selection. In preparation for the audit, for example, the auditor should become familiar with prior audit reports on the financial accounting systems to be audited. Audit scope should clearly state the process areas, controls, geographic or functional areas, time period, and other specifics to delineate the area to be reviewed. Audit objectives are formal statements that describe the purposes of the audit. By defining appropriate objectives at the outset, management can ensure that the audit will verify the correct functioning and control of all key audit areas.

18.3.6 Audit Procedures

Management dictates how the organization will be divided into subgroups that control small portions of a company. Policies and procedures are only as good as the management structure which formed them and enforces the action taken. Each function in the organization, including internal audit and IT, needs complete, well-documented polices and procedures to describe the scope of the function of its activities and the interrelationships with other departments. The IT auditor should examine the corporate structure of the policies and procedures set by the management to understand the business objectives and functions of the organization.

The core of the audit process is analysing controls to determine if they are adequate or need improvement. An increasing degree of interaction between business processes requires an integrative design for auditing procedures, ultimately involving all levels and areas of management and all enterprise units. Auditing procedures are the activities that the auditor performs to obtain sufficient, competent evidence to ensure a reasonable basis for the audit opinion. Firstly, they are detective control mechanisms by which auditors identify and investigate variances or deviations from predetermined standards. Secondly, they are used as preventive control mechanisms because the expectation of an audit should deter individuals from engaging in fraudulent financial reporting or making careless errors.

18.3.7 Internal Audit

Contemporary systems carry risks such as non-compatible platforms, new methods to penetrate security through communication networks, and the rapid decentralization of information processing with the resulting loss of centralized controls. The management of internal audit has to deal with the challenge of recognizing this and taking appropriate actions. Internal control techniques of IT auditing find their genesis in the traditional auditing discipline. Basic principles of internal control as applied to the manual information processing systems are required to be applied to the information and communication technology supported information systems. These controls, if applied, lead information and communication technology supported information systems to maintain the data integrity.

The internal audit function is a control function with a company or organization. The primary purpose of the internal audit function is to assure that management authorized controls are being applied effectively. Internal audit is a part of the internal monitoring system of an organization. It is an integral part of the organization and functions under the policies established by the top management. Top management generally establishes the statement of purpose; authority and responsibility for the internal audit unit approved by the top management should be consistent. The audit system comprises all monitoring measures and precautions put in place within the company to secure assets and guarantee the accuracy and reliability of the accounting system. This task is managed with objective-based and compliance-focused comparisons between the existing condition and the accepted criteria, as required by all applicable policies, regulations, and laws. Internal audit examines and evaluates the adequacy of the organization's controls, risk assessments controls, and assess the adequacy of communication systems and evaluate monitoring activities.

An internal audit is generally conducted by a team of auditors. As internal audits vary in size and content, the size of the internal audit teams working on each audit also fluctuate. One of the auditors acts as the team lead who is responsible for planning and overseeing the audit, as well as communicating with the auditees, while other audit team members execute the audit activities. As per the standards, the internal auditors have to have free access to records, personnel and physical properties of the employer. The objective of internal audit is to assist members of the organization in the effective discharge of their responsibilities. Audit independence is a very critical component if a business wishes to have an audit function that can add value to the organization. The audit report and opinion must be free of any bias or influence if the integrity of the audit process is to be valued and recognized for its contribution to the organization's goals and objectives. Therefore, internal audit furnishes them with analysis, appraisals, recommendations, counsel, and information concerning the activities reviewed.

Since the three components of audit risk can offset or reinforce each other, audit risk is determined by multiplying its components.

$$\text{Audit risk} = \text{inherent risk} \times \text{control risk} \times \text{detection risk} \qquad (18.1)$$

Once the overall audit risk acceptable for the audit has been defined and the inherent and control risks have been determined, the tolerable detection risk can be set. Auditors must keep within this risk level by conducting appropriate fieldwork.

An audit by internal audit can only be initiated as part of the regular annual audit planning or in response to a duly approved request. This ensures that audits cannot be conducted arbitrarily. Even if internal audit acts in response to a self request, the request will only be accepted after a critical review by the internal audit management in cooperation with the responsible board member. The audit request is always assessed with risk exposure in mind. For this reason, each requested audit is subjected to a risk assessment.

18.3.8 Audit Findings and Conclusions

Audit findings should be formally documented and should include the process area audited, the objective of the process, the control objective, the results of the test of that control, and a recommendation in the case of a control deficiency. An audit finding form serves the purpose of documenting both control strengths and weaknesses and can be used to review the control issue with the responsible IT manager to agree on corrective action. Analysis of the results of tests and interviews should occur as soon as possible after they are completed. Too often, the bulk of audit analysis occurs when the auditor sits down and writes the audit report. Timely analysis enables the auditor to determine the causes and exposures of findings early in the audit. Conclusions are auditors' opinions, based on documented evidence, that determine whether an audit subject area meets the audit objective. All conclusions must be based on factual data obtained and documented by the auditor as a result of audit activity. Recommendations are formal statements that describe a course of action that should be implemented to restore or provide accuracy, efficiency, or adequate control of audit subjects.

18.3.9 Audit Reports

Formal communication issued by the audit department describing the results of the audit is called an audit report. Audit report should include the audit scope and objectives, a description of the audit subject, a narrative of the audit work activity performed, conclusions, findings, and recommendations. To be effective, audit reports must be timely, credible, readable, and have a constructive tone. All types of audit reports in the entire organization must be prepared consistently, correctly, and free of influence by external parties. Internal auditors frequently have access to information which may be considered sensitive from a commercial, political, or security point of view. The internal audit department and its personnel must exercise due professional care to ensure that such information is properly safeguarded and thus should establish procedures and controls to assure the physical security of working papers.

18.3.10 Working Papers

Working papers are the formal collection of pertinent writings, documents, flowcharts, correspondence, results of observations, plans for tests, results of tests, the audit plan, minutes of meetings, computerized records, data files or application results, and evaluations that document the auditor activity for the entire audit period. It is normal to treat working papers, communications with audited entities and draft reports as confidential documents until recognized and established procedures for their release have been followed.

Information technology auditors in large corporations can make a major contribution to the dynamic computer system controls by insisting on a policy of comprehensive testing for all new systems and all changes to existing systems. Information technology auditors test the systems through audit functions.

18.4 Controls

The increase in the usage of information technology in business operations has increased the associated potential for loss in varied forms, as procedures once formed and established manually are entirely automated. Conventional systems and procedures relied on manual checks and verifications to ensure the accuracy and completeness of data and records. Manual controls are replaced by computer application control function with the expansion of information systems in the organization. Applications that anticipate exceptions, which were previously handled manually on an ad hoc basis, are required. Without such control routines, incomplete or incorrect transactions can be processed unnoticed.

The information needs of different levels of management and the information processing capabilities have both become more complex and have merged. As more business functions are computerised, business activities and management have become dependent on electronic data processing and the internal controls that ensure accuracy and completeness. Traditional controls of manual systems have become outmoded with their audit methods, tools, and techniques, as a result of changes in the structure and form of computer application systems.

An information protection program should be a part of any organization's overall asset protection program. This program is not established to meet security needs or audit requirements. It is a business process that provides management with the processes needed to perform the fiduciary responsibility. Management is charged with a trust to ensure that adequate controls are in place to protect the assets of the enterprise. An information management program that includes policies, standards, and procedures will allow management to demonstrate a standard of care.

Controls are either user actions or system-programmed routines designed to mitigate risks and achieve established organizational objectives. Controls are designed and implemented to reduce the risk of error in financial data or other data to an acceptable level and to safeguard organizational assets. There are a number of possible options for the implementation of controls, such as automated, system-based controls, manual controls or computer-dependent processes, in order to achieve the desired outcome. The grouping of a series of controls associated with a particular business process or topic is referred to as a control framework. For example, interfaces can be considered as a means of providing input into a system. Controls over interfaces should be carefully reviewed because of the volume of transactions and the automated methods used for interfaces. Input controls are meant to minimize risks associated with data input into the system. They ensure the authenticity, accuracy, completeness, and timeliness of data entered into an application. Processing controls ensure the accuracy, completeness, and timeliness of data during either batch or online processing. Logging error activity, reporting open errors, and recording error correction as a part of the transaction's audit trial can ensure transaction completeness. Output controls ensure the integrity of output and the correct and timely distribution of output produced. Functional testing and acceptance testing is a key to application controls and they ensure that the application fulfils the agreed-upon functional expectations of the users, meets established usability criteria, and satisfies performance guidelines prior to being implemented into production.

Presence of controls in a computerized system is significant from the audit point of view as these systems may allow duplication of input or processing, conceal or make some of the processes invisible, and in some of the auditee organizations where the computer systems are operated by outside contractors employing their own standards and controls, making these systems vulnerable to remote and unauthorised access. Apart from this, the significance of controls lies in the following possibilities.

- Data loss due to file damage, data corruption (manipulation), fire, burglary, power failure (or fluctuations), viruses, etc.
- Error in software can cause manifold damage as one transaction in a computer system may affect data everywhere.
- Computer abuse, such as fraud, vengeance, negligent use, etc., is a great potential danger.
- Absence of audit trails makes it difficult for an auditor to ensure efficient and effective functioning of a computerized system.

An effective control system provides reasonable, but not absolute assurance for the safeguarding of assets, the reliability of financial information, and the compliance with laws and regulations. Reasonable assurance is a concept that acknowledges that control systems should be developed and implemented to provide management with the appropriate balance between risk of a certain business practice and the level of control required to ensure business objectives are met. The cost of a control should not exceed the benefit to be derived from it. The degree of control employed is a matter of good business judgment. Therefore, control procedures need to be developed so that they decrease risk to a level where management can accept the exposure to that risk.

18.4.1 Internal Controls

The control environment is the control consciousness of an organization. It is the atmosphere in which people conduct their activities and carry out their control responsibilities. An effective control environment is an environment where competent people understand their responsibilities, the limits to their authority, and are knowledgeable, mindful, and committed to doing what is right and doing it the right way. They are committed to following an organization's policies and procedures and its ethical and behavioral standards. The control environment encompasses technical competence and ethical commitment. It is an intangible factor that is essential to effective internal control.

Internal control is recognized as a vital feature of modern management, with the extent and effectiveness of internal controls being just as important to the auditor. The use of IT-based systems, no matter how small, large, or sophisticated, does not in any way reduce the need for the auditor to evaluate the reliability of that system in determining the scope and character of the audit.

Internal controls will vary from enterprise to enterprise. They need to be tailored to the relevant industry (or industries) that the organization operates within; they are also typically unique for each enterprise. They are determined by its business activities and processes as well as its financial controls. They are closely related to the IT systems and databases that the enterprise uses for financial and other reporting.

A business entity or activity adopts internal controls to safeguard its assets, check the accuracy and reliability of its accounting data, promote operational efficiency, and encourage adherence to prescribed managerial policies. Internal control is a process designed to provide reasonable assurance regarding the achievement of objectives in the following categories.

- Effectiveness and efficiency of operations
- Reliability of financial reporting
- Compliance with applicable laws and regulations

Internal control keeps an organization on course toward its objectives and the achievement of its mission and minimizes surprises along the way. Internal control promotes effectiveness and efficiency of operations, reduces the risk of asset loss, and helps to ensure compliance with laws and regulations. Internal control also ensures the reliability of financial reporting.

Internal control consists of the following five interrelated components.

- Control (or operating) environment
- Risk assessment
- Control activities
- Information and communication
- Monitoring

All five internal control components must be present to conclude that internal control is effective. The evaluation of internal control is therefore probably more vital in the audit of an information system than in a manual system. It is generally accepted that the auditor is limited to determining whether the system is such that a relatively large number of people must be involved in order to circumvent the internal controls. A strategic map is a picture of the business. A strategic map that is developed and tailored to an enterprise enables senior managers, as well as middle managers, expert business staff, and IT staff to see the data, activities, and processes. From the strategic map and underlying strategic model, the Governance Analysis Framework becomes visible.

18.4.2 IT Controls

In business and accounting, IT controls are specific activities performed by persons or systems designed to ensure that business objectives are met. Information technology controls are specific IT processes designed to support a business process. These can be categorized as either general controls or application controls.

18.4.3 General Controls

Information technology general controls represent the foundation of the IT control structure. General controls are those controls that are pervasive to all systems components, processes, and data for a given organization or systems environment. General controls refer to the overall controls an entity has over its entire IS function. These controls affect all applications processed by the IS department and are often referred to as integrity controls because of their pervasive effect. Such controls are designed to ensure that the procedures programmed into a computer system are appropriately implemented, maintained, and operated and that only authorized changes are made to programs and data. The purpose of general controls is to establish a framework of overall control over the IS activities and to provide a reasonable level of assurance that the overall objectives of internal control are achieved.

18.4.4 Application Controls

Application controls are those controls that are appropriate for individual accounting subsystems, such as payroll or accounts payable. They relate to the processing of individual applications and help ensure that the transactions occurred, are authorized, and are completely and accurately recorded, processed, and reported. Application controls refer to controls that are specific to individual accounting applications, that is, they relate to and are unique to, particular accounting applications or functions, for example, debtors, creditors, payroll, or inventory. The purpose of application controls is to establish specific control procedures over the accounting applications in order to provide reasonable assurance that transactions are authorized and recorded, and are processed completely, accurately, and on a timely basis.

18.4.5 Management Controls

Management controls are those functions that govern the operation and use of the system. Operational controls are the implementation of those management ideas. Technical controls are the technological specifics that give meaning to the operational controls.

18.4.6 IT Controls Practices

Many IT organizations view IT controls as an externally imposed regulatory cost and burden on already stretched IT resources. Executives can leverage the compliance and controls mandate to set the tone at the top. This can be done by following documented process and procedures in key functional areas, which is a strategy for performance improvement. The requirement of needing IT controls to meet various regulatory requirements offers a unique opportunity to make organizational and cultural changes so that everyone in the organization understands they are expected to follow documented process and procedures.

Information technology controls consist of practices designed to maintain the integrity and availability of information processing functions, networks, and associated application systems. These controls apply to business application processing in computer centres by ensuring complete and accurate processing. These controls ensure that correct data files are processed, processing diagnostics and errors are noted and resolved, applications and functions are processed according to established schedules, file backups are taken at appropriate intervals, recovery procedures for processing failures are established, software development and change control procedures are consistently applied, and actions of computer operators and system administrators are reviewed. Additionally, these controls ensure that physical security and environmental measures are taken to reduce the risk of sabotage, vandalism, and destruction of networks and computer processing centres.

Audit trail provides the records of all the events that take place within an application system. These trails are in-built in a well-designed system and application software and do not disappear as was considered by many auditors in earlier days.

Relying so heavily on computer-based controls, auditors of all kinds will need to have at least a basic understanding of the various information technology or general controls, such as systems implementation and maintenance, and computer operations. Auditors will do much of their auditing on a continuous basis from workstations. Using an auditor's workstation on the system, auditors will monitor the flow of transactions and use techniques such as interactive test facilities. From an audit perspective, auditors will have to understand the system development life cycle to evaluate whether appropriate controls and quality are built into the system, maintenance controls to understand whether changes are made properly, computer operations for access and recoverability, and program and data file security for validity and integrity. Auditors must have a detailed understanding of a customized software packages and how it is implemented in the given environment. Auditors must view management audit reports in terms of audit objectives: completeness, accuracy, existence, and valuation.

18.4.7 Nature of Controls

Each organization should identify the events and circumstances whose occurrence could result in a loss to the organization. These are called exposures. Controls are those acts which the organization should implement to minimize the exposures. There are four types of controls, which are as follows.

- Preventive controls that keep something from happening. They prevent the cause of exposure from occurring or at least minimize the probability of unlawful event taking place. For example, policy, separation of duty, and authorization processes are all preventive controls. This also includes security controls at various levels such as hardware, software, application software, database, network, etc.
- Detective-analytical controls that monitor the activity and the processes to determine if the preventive controls have failed or if something is out of compliance. When a cause of exposure has occurred, detective controls report its existence in an effort to arrest the damage further or minimize the extent of the damage. For example, change monitoring and verification are detective controls.
- Corrective controls restore the situation back to the expected state. They are designed to recover from a loss situation. For example, if a system crashes due to a failed change, reloading all applications from the last known good image to bring the system back online serves as a corrective control. Business continuity planning is a corrective control. Without

corrective controls in place, a bank will face the risk of loss of business and other losses due to its inability to recover essential IT based services, information, and other resources after a disaster has taken place.

The combination of the three types of controls creates a system of checks and balances to help ensure that the processes, people, and technology operate within the prescribed bounds.

The use of a computer to record financial transactions changes dramatically the way in which financial information is processed and stored which in turn may affect the organization and the procedures employed to achieve adequate internal control. Though the objectives of internal control will not be affected by the presence of an information system, the control procedures adopted by an entity will be affected by the characteristics of the information systems in existence. The control procedures adopted to achieve the objectives of internal control in an IS environment will include both manual and computer procedures. These control procedures may be classified as either general controls or application controls and should ensure that an entity's transactions are completely and accurately processed and recorded in accordance with the management's authorization and that assets are safeguarded.

18.4.8 Assessing System Reliability

Weak or ineffective internal controls, such as inadequate record keeping, external audit, or loan review, has caused operational losses in many organizations, large IT establishments (banks) and has contributed to the failure of others. Some of these cases involved insider fraud that could have been prevented or discovered through effective control mechanisms before the fraud resulted in loss.

18.4.9 Controls and Classifications

Control requirements vary depending on the industry and type of business, but some of the most prevalent compliance requirements today are the Sarbanes-Oxley Act and the Payment Card Industry's Data Security Standard (PCI DSS). The Sarbanes-Oxley Act (SOX), named after Senator Paul Sarbanes and Representative Michael Oxley, came into force in July 2002. It is built over the principles of integrity, reliability, and accountability. It was created to ensure that financial records were complete and accurate (integrity), that the information was reliable, and that management would be held accountable.

In the past five or six years we have seen heavy emphasis on financial reporting controls and improving the quality and reliability of financial reporting. And it looks like the next five or six years will continue to add emphasis on 'protecting' our data (and not just financial data). So while we are assessing and improving financial application controls, we will see increasing emphasis on access controls, general IT controls, controls over how we manage and audit changes to systems and infrastructure, and the controls for how we authorize and authenticate changes to sensitive data and database systems.

Control is the limit placed on the ability of an individual, a group of individuals, organizations, or another state to have partial or full access to the data contained on a database. Partial data access is the inability to do any of the following.

- View all of the data entered and stored by the system
- Append data
- Edit data
- Copy data
- Distribute/share the data by any means

Controls should be used to limit access in a manner consistent with any confidentiality requirements and protect the data from unauthorised changes. Of greatest importance is the protection of primary data from accidental corruption. The master copy of data must always be 'write protected'. Although control and security are important, they should not hinder legitimate access.

18.5 Knowledge Audit and Evaluation

There were many statements gleamed from the knowledge management (KM) works and writings, including a proliferation of definitions that sometimes disagreed with each other. Knowledge is often an important source of competitive advantage and it is essential to protect it. Knowledge management is rapidly being introduced to technical organizations and is becoming a key element of successful enterprises. It has a strong potential to become foundational in solving an enterprise's problems, enhancing innovation, and providing a basis for integrating technology, organization, leadership, and learning. The new business environment demands foresight, conversion, innovation, and adaptation in contrast to the traditional emphasis on optimization [Stankosky 2005]. The new business environment demands knowledge growth and maturity that is relevant, applicable, and value-added. In short, it must be able to solve enterprise-wide problems to be effective and valuable. Knowledge creation should be the foundation of a company's human resource strategy. It is critical that knowledge be interpreted, deciphered, analysed, and applied in terms of relevance to the knowledge worker.

Knowledge falls into two categories: explicit knowledge such as copyright or information codified in handbooks, systems or procedures; and tacit knowledge that is retained by individuals, including learning, experience, observation, deduction, and informally acquired knowledge. Knowledge gaps make an organization more vulnerable to competition. The downsizing strategy that many firms have followed has highlighted the dangers of getting rid of people with expertise and experience in the pursuit of short-term cost savings [Kourdi 2003].

The mobility of knowledge workers and the increased value of information and knowledge have significantly raised the importance of knowledge assets and their proper usage for higher competitiveness and growth. Evaluation is an in-depth examination of program or policy success, relevance, and cost-effectiveness.

A knowledge audit (K-audit) is a systematic examination and evaluation of organizational knowledge assets and is usually recommended in industries as an important first step prior to the launching of any knowledge management program. A K-audit (an assessment of the way knowledge processes meet an organization's knowledge goals) is to understand the processes that constitute the activities of a knowledge worker, and see how well they address the 'knowledge goals' of the organization.

A K-audit is designed to uncover the breadth, depth, and location of an organization's knowledge, and it has the following three core components.

- Defining what knowledge assets exist, especially information or skills that would be difficult or expensive to replace.
- Locating those assets; that is, discovering who keeps or 'owns' them.
- Classifying them and assessing how they relate to other assets. In this way, opportunities can be found in other parts of the organization.

A K-audit, however, refers to the auditing process of creating, acquiring, retaining, distributing, transferring, sharing, and re-using the institutional knowledge of an organization, i.e., it refers to how the knowledge is managed in the organization. Common K-audit tools include site observation, interviews, questionnaires, focus group, and workshop. Generally speaking, a K-audit could be divided into four parts: background study, data collection, data analysis, and data evaluation.

The output of K-audit will deliver the following knowledge inventory benefits.

- Identification of core knowledge assets and flows—who creates, who uses
- Identification of gaps in information and knowledge needed to manage the business effectively
- Areas of information policy and ownership that need improving
- Opportunities to reduce information handling costs
- Opportunities to improve coordination and access to commonly needed information
- A clearer understanding of the contribution of knowledge to business results.

The K-audit, if properly carried out, contributes to building a KM strategy based on extended knowledge of the company status, its internal and external environment, and thus enable the organization to take appropriate decisions to overcome existing gaps and possible drawbacks.

Summary

As we have seen throughout this chapter, changes in technology have been so fast and so dramatic that their effects on the auditor are truly significant. The audit function, whether internal or external, is a part of the corporate environment. It is a process to objectively validate, verify, and substantiate a process, activity, function, system, subsystem, or project within a company. Auditors have a unique set of skills and abilities that allows them to evaluate varied issues and environments. There may be elimination of some of the traditional controls, such as batch totaling, and more reliance on computer matching, to ensure completeness. Authorization will be directed more than ever by the computer, with extremely heavy reliance on access controls for approval to proceed. Traditional paper that allowed a historical audit will disappear, and auditors will have to adopt a forward-looking prospective way to audit.

This chapter presents a simple, logical, common sense way of looking at internal control and risk management that integrates them conceptually and so opens the door to integrating them in practical ways too. The literature on risk and control is littered with diagrams of cubes, circles, triangles, pyramids, trees, and so on, but of course none of these exist physically. They are attempts to make sense of patterns of human behaviour and computer processing spread out over time and space, and mixed with other behaviour so that they are hard to see.

Key Terms

Audit risk It is the risk that an auditor may arrive at the wrong conclusions and opinions of the work that they have undertaken.

Audit scope It refers to the activities covered by an internal audit. Audit scope often includes: audit objectives, nature and extent of auditing procedures performed, time period audited, related non-audit activities that delineate the boundaries of the audit.

Audit working papers These record the information obtained, the analyses made, and the conclusions reached during an audit. Audit working papers support the bases for the findings and recommendations to be reported.

Audit universe It is an inventory of audit areas that is compiled and maintained to identify areas for audit during the audit planning process.

Business risk Business risk is a concept used by auditors and managers to express concerns about the probable material effects of an uncertain environment on achieving established objective.

Code of ethics This is a code that promotes an ethical culture in the global profession of internal auditing. A code of ethics is necessary and appropriate for the profession of internal auditing, founded as it is on the trust placed in its objective assurance about risk, control, and governance.

Control Any action taken by the management, the board, and other parties to enhance risk management and increase the likelihood that established objectives and goals will be achieved is called control. Management plans, organizes, and directs the performance of sufficient actions to provide reasonable assurance that objectives and goals will be achieved.

Control environment The attitude and actions of the board and management regarding the significance of control within the organization is called control environment. It provides the discipline and structure for the achievement of the primary objectives of the system of internal control.

Control framework It is a recognized system of control categories that covers all internal controls expected in an organization.

Control processes The policies, procedures, and activities that are a part of a control framework, designed to ensure that risks are contained within the risk tolerances established by the risk management process.

Detection risk The probability that an incorrect audit conclusion will be drawn from the results of the examination or that the audit work will fail to detect any serious errors is called detection risk.

Effect Effect is the risk or exposure the auditee, organization and/or others encounter because the condition is not the same as the criteria.

Effective control It is present when the management directs systems in such a manner as to provide reasonable assurance that the organization's objectives and goals will be achieved.

External auditors It refers to those audit professionals who perform independent annual audits of an organization's controls.

Inherent risk There are risks that an account or class of transactions contains material misstatements irrespective of the effects of the controls.

Planning risk It is the risk that the planning process is flawed. In risk assessment, it is the risk that the assessment process is inappropriate or improperly implemented.

Risk analysis The assessment of risk, the management of risk, and the process of communicating about risks is called risk analysis. A systematic use of available information to determine how often specified events may occur and the magnitude of the consequences.

Risk classification It is a part of the risk assessment process that categorizes risks, typically into high, medium, low, and intermediate values.

REVIEW QUESTIONS

18.1 What is audit?
18.2 What is the need for audit?
18.3 What are the skills needed to audit information systems? Are they technical or non-technical?
18.4 Why is IT control and auditability so important in today's virtual environment?
18.5 What are management policies and procedures and why are they so important to the audit process?
18.6 Who are internal auditors? What are their roles and responsibilities?
18.7 Who are external auditors? What are their roles and responsibilities? Provide and discuss two examples of cases with external auditors and their roles.
18.8 What is audit planning?
18.9 Why is audit planning required?
18.10 What are controls and why is it required for an organization?
18.11 What are internal controls?
18.12 What are the benefits of introducing internal controls in an organization?
18.13 Discuss the need for knowledge audit.

Projects

18.1 Visit the websites of four external audit organizations: two private and two government sites. Provide a summary of their roles, function, and responsibilities.

18.2 Visit the nearest IT organization and understand the IT controls deployed within the organization. Prepare a report on it.

18.3 Conduct an mock knowledge audit drill for an IT organization and prepare a report.

18.4 ABC unlimited decides to trade in lighting products that are unique in concept and that appeal to a broad audience. ABC recently developed a strategic road map to align its IT with its overall business strategy with the help of a local consulting firm. ABC needed to align its IT infrastructure, processes, and applications with its strategic goals. The company knew that to compete more effectively, it would have to improve its customer focus and supply chain efficiency and support these areas with transparent IT solutions, compliant with the company's strategic IT vision. The company's main goals in undertaking a transformation of its IT infrastructure and processes were to support the creation of a comprehensive business, achieve profitable growth, reduce costs and improve customer focus and supply chain efficiency. With this clear vision of where it needed to go, ABC sought a consulting partner with expertise in the trading industry to develop the business case for implementing new IT infrastructure and processes, including recommendations for new major IT application installations and integration across web integrated functional areas.

The consultancy firm teamed with ABC to deliver the company's IT strategy plan, including the business case for required investments. The team used the consultant's proprietary methodology to evaluate ABC's strategic IT processes. The resulting road map aligns the company's IT strategy with its larger business goals and addresses the business requirements and issues. The actual implementations of recommended IT solutions will be completed during the next two years, delivering a solid return on investment (ROI) once the implementation is completed. The most important part of the solution was the implementation of an enterprise resource planning (ERP) system. The common ERP system is the key to ABC's cost reductions and profitable growth through the integration of production, supply, and customer service. It is expected that through this ERP implementation, a better fusion between IT and business will be achieved, enabling a more efficient supply chain and improved logistics for purchasing and distribution. Plan an annual IT audit plan for ABC.

REFERENCES

Braun, Robert L. and Harold E. Davis 2003, 'Computer-assisted Audit Tools and Techniques: Analysis and Perspectives,' *Managerial Auditing Journal*.

Cangemi, Michael P. and Tommie 2003, *Managing the Audit Function: A Corporate Audit Department Procedures Guide*, John Wiley.

Champlain, Jack J. 2003, *Auditing Information Systems Second Edition*, John Wiley.

Coderree, David 2009, *Internal Audit Efficiency through Automation*, John Wiley.

Gallegos, Frederick, et al. 2004, *Information Technology Control And Audit*, Auerbach.

Kagermann, Henning, et al. 2008, *Internal Audit Handbook Management with the SAP®-Audit Roadmap*, Springer.

Kourdi, Jeremy 2003, 'Business Strategy A Guide to Effective Decision-Making,' *The Economist*.

Leitch, Matthew 2008, *Intelligent Internal Control and Risk Management*, Gower.

Musaji, Yusufali F. 2002, *Integrated Auditing of ERP Systems*, John Wiley.

Stankosky, Michael 2005, *Creating the Discipline of Knowledge Management*, Elsevier.

Taylor, Hugh 2006, *The Joy of SOX*, John Wiley.

Tricker, Ray 2005, *ISO 9001:2000 Audit Procedures*, Elsevier.

PART V

IT Applications—Business Systems

- **Chapter 19** Governance
- **Chapter 20** Connected World and E-commerce
- **Chapter 21** Information Systems and Business Systems

CHAPTER

19 Governance

Risk is a choice rather than a fate. The actions we dare to take, which depend on how free we are to make choices, are what the story of risk is all about. And that story helps define what it means to be a human being.
−Peter L. Bernstein

Learning Objectives

After reading this chapter, you should be able to understand:

- the need for organizational governance
- the differences between the various governance mechanisms
- IT governance and its applications to an organization

19.1 Introduction

Massive changes in technology are clearly part of the cause of the current industrial revolution and its associated excess capacity. Both within and across industries, technological developments have had far-reaching impact. The changes in computer technology, including miniaturization, have not only revamped the computer industry, but also redefined the capabilities of countless other industries. Fibre optics and other telecommunications technologies are bringing about vast increases in worldwide capacity and functionality. The vast improvements in telecommunications, including computer networks, electronic mail, teleconferencing, and facsimile transmission are changing the workplace in major ways that affect the manner in which people work and interact. It is far less valuable for people to be in the same geographical location to work together effectively, and this is encouraging smaller, more efficient, entrepreneurial organizing units that cooperate through technology, creating virtual organizations.

Organizations have always tried to adapt their organizational structure to suit their organizational strategy. Virtual organizations are a set of organizations that rely on multiparty cooperative relationships between people across structural, temporal, and geographic boundaries. Flexibility is brought about in part by reconfigurable networks of computer-based communications that allow organizations to coordinate their activities, and in part by a management philosophy based on collaboration and innovation. Virtual organizations tap talented specialists, avoid many of the regulatory costs imposed on permanent structures, and bypass the inefficient work rules and high wages imposed by unions. In doing so, they increase efficiency and thereby contribute to the organization. These changes in managing and organizing principles have contributed significantly to the productivity of the world's capital stock and economized on the use of labour and raw materials. There is a need to bring in effective governance to manage such virtual organizations. By nature, organizations abhor control systems, and ineffective governance is a major part of the problem with internal control mechanisms. This chapter is an effort to bring the pertinent facts and fictions together with an intent to inform the information technology (IT) auditor, the role to be played in the corporate or enterprise IT governance.

19.2 Governance

The concept of governance is not new. It is as old as human civilization. Don McLean points out that the word governance is derived from the Latin word *gubernare*, which refers to the action of steering a ship. This etymology suggests a broader definition for governance. One important implication of this broader view is that governance includes multiple tools and mechanisms. Simply put, governance is the process of decision-making and the process by which decisions are implemented. Development agencies, international organizations, and academic institutions define governance in different ways. Governance is the system of values, policies, and institutions by which a society manages its economic, political, and social affairs through interactions within and among the state, civil society, and private sector. It is the way a society organizes itself to make and implement decisions achieving mutual understanding, agreement, and action. It comprises the mechanisms and processes for citizens and groups to articulate their interests, mediate their differences, and exercise their legal rights and obligations. It is the rules, institutions, and practices that set limits and provide incentives for individuals, organizations, and firms. Governance, including its social, political and economic dimensions, operates at every level of human enterprise, be it the household, village, municipality, nation, region or globe (refer to UNDP Strategy Note on Governance for Human Development, 2004).

Governance of organizations in a dynamic and complex business environment is something which needs discussion. It is seen that the traditional organizational structure is crumbling under the weight of ever increasing regulations that drive greater accountability and transparency. Smart companies are on the forefront of building new and improved control structures, including IT controls that support and enhance this new compliance environment.

Governance model and the system of governance trace their roots back to the definitions of governance and risk. The governance model is a framework and processes, including the activities, tools, and methodologies that may be described, documented, taught, and put in place. The system of governance is the active involvement of the board, executive management, and indeed the entire enterprise in the dynamic, real-time operation of the governance model. Unfortunately, as with any system found in nature, the system of governance is invisible.

19.3 Corporate Governance

Corporate governance became a dominant business topic in the wake of the spate of corporate scandals of 2002. Good corporate governance is important to professional investors. Corporate governance deals with the ways in which suppliers of finance to corporations assure themselves of getting a return on their investment.

Corporate governance is defined as a system of structuring, operating, and controlling a company with a view to achieve long-term strategic goals to satisfy shareholders, creditors, employees, customers, and suppliers, and complying with the legal and regulatory requirements, apart from meeting environmental and local community needs. The SEBI Committee on Corporate Governance defines corporate governance as the acceptance by management of the inalienable rights of shareholders as the true owners of the corporation and of the management's own role as trustees on behalf of the shareholders.

> **Time to Start Thinking Differently about Business**
>
> So troubled were we by companies playing fast and loose with accounting, data management, and other practices that we sicced the politicians on them. The result is a host of new regulations and lawsuits that heap billions of dollars in costs onto thousands of companies in the name of weeding out billions of dollars of wrongdoing among a handful of them. Any chance we are overreacting or focusing on the wrong places?
>
> No way, say proponents of the crackdown, who point to the recent spate of CEO firings and departures as evidence that regulations like Sarbanes-Oxley are thrusting accountability back into the executive suite. If not for new rules that require boards to appoint more independent directors and act transparently, they say, the financial and ethical offenses that forced management changes at Boeing, Fannie Mae, Hewlett-Packard, and other companies would have continued unchecked.

Corporate governance is the system within an organization that protects the interests of its diverse stakeholder groups. The best approaches recognize that stakeholders are more than shareholders, and include customers, employees, suppliers, retirees, communities, lenders and other creditors. Good corporate governance is evolving from command-and-control models to a more proactive and continuous process that assesses, sources, measures, and manages risks across the enterprise. Effective corporate governance instills a culture of sound business practices and ethics; an understanding of company risks and how to manage them.

Creative accounting is a practice which has emerged in a big way. It refers to accounting practices that may or may not follow the letter of the rules of standard accounting practices but certainly deviate from the spirit of those rules. Enron has made everyone from politicians to comedians aware of the potentially disastrous results of creative accounting practices. Enron is not an isolated situation. It is merely the most extreme example of problematic financial reporting leading to creative accounting.

Corporate governance is the system by which organizations are directed and controlled. The business dependency on IT has resulted in the fact that corporate governance issues can no longer be solved without considering IT. Corporate governance should therefore drive and set enterprise governance of IT. The organization, management process, and information subsystem is an integral element not only of the governance model but of the extended enterprise as a complete system. Corporate governance is, by no means, an easy task, and probably many times more difficult with a system based on millions and millions of paper documents. The MCA-21 project, so-called by the Ministry of Corporate Affairs, to reflect India's corporate governance goals for the twenty first century, has begun to address the complex issue. The MCA-21 project rolled out the nearly paperless system across the country, starting with Coimbatore in Tamil Nadu. Today, almost 6 lakh companies in the country make their filings online.

19.4 IT Governance

Information and the technology that supports it represent a company's most valuable assets. What is even more important is that today's competitive and rapidly changing business environment requires increased quality, functionality, and ease of use from organizations' IT systems. It is clear that most enterprises rely on IT for their competitive advantage and cannot afford to devote anything less to it than, say, financial supervision or general corporate governance. Therefore, the time has come for company board members to create committees that proactively take charge of IT governance. As leading global businesses increasingly recognize the imperative for strong IT governance, some boards are stepping up to adopt a much stronger oversight role, and leading institutions are proposing those organizations as role models for the rest of the business world. Organizations need to satisfy the quality, fiduciary, and security requirements for their information, as for all assets. Managements must optimize the use of available resources, including data, application systems, technology, facilities and people. To discharge these responsibilities, as well as to achieve its objectives, management must understand the status of its own IT systems and decide what security and control they should provide.

Despite the significant role IT plays in business, most boards of directors/boards have remained largely in the dark when it comes to IT strategy and governance. With the passage of the Sarbanes-Oxley Act in the USA in 2002, and an ever-increasing corporate focus on ensuring prudent returns on technology investments, the notion of IT governance became a major issue in all applicable domains. Although the term IT governance is a relatively new addition, significant previous work is reported on IT decisions rights and IT loci of control, notions that are synonymous with the current understanding of IT governance. Management specialists dealing with the IT governance, look at the increasing importance of IT governance within an organization than before.

Governance of IT is about defining and embedding processes and structures in the organizations that enable both business and IT people to execute their responsibilities in creating value from IT-enabled business investments. The IT Governance Institute further defines IT governance as the management process which ensures delivery of the expected benefits of IT in a controlled way to enhance the long-term success of an enterprise.

Information technology governance is recognized as an integral part of enterprise governance. It consists of the leadership and organizational structures and processes that ensure that an organization's IT sustains and extends the organization's strategies and objectives. It is defined as the framework for leadership, organizational structures and business processes, standards and compliance to these standards, which ensures that the organization's information systems (IS) support and enable the achievement of its strategies and objectives. This implies that IT governance impacts all layers in the organization, from operational management to senior and executive management, and finally the board of directors. It also suggests that IT governance is fundamentally different from IT management.

Information technology governance, in simple terms, can be said to be a method for chief information officers (CIOs) to manage IT strategy and execution by enabling a consolidated view of key governance functions such as project, demand, resource, risk and performance management. Chief information offices of organizations in a single line of business, tend to focus on the business and IT needs that are pertinent to a single corporate entity. Group CIOs should ideally use corporate IT governance practices and put a system in place that will design and assign decision rights. An accountability framework to encourage desirable behaviour in the use of IT is recommended. However, when IT governance is extended to cover all the aspects of the corporate functions, then governance of IT becomes an integral part of corporate governance. It addresses the definition and implementation of processes, structures, and relational mechanisms in the organization that enable both business and IT people to execute their responsibilities in support of business/IT alignment and the creation of business value from IT-enabled business investments.

> As an example of its growing importance, the International Organization for Standardization (ISO) released a new worldwide ISO standard defined as corporate governance of IT (ISO/IEC 38500:2008) in 2008. In this standard, ISO puts forward the following six principles for the governance of IT.
>
> - **Principle 1:** *Responsibility:* Individuals and groups within the organization understand and accept their responsibilities in respect of both supply of, and demand for IT. Those with responsibility for actions also have the authority to perform those actions.
> - **Principle 2:** *Strategy:* The organization's business strategy takes into account the current and future capabilities of IT; the strategic plans for IT satisfy the current and ongoing needs of the organization's business strategy.
> - **Principle 3:** *Acquisition:* Information technology acquisitions are made for valid reasons, on the basis of appropriate and ongoing analysis, with clear and transparent decision-making. There is appropriate balance between benefits, opportunities, costs, and risks, in both the short term and the long term.
> - **Principle 4:** *Performance:* Information technology is fit for supporting the organization, providing the services, levels of service, and service quality required to meet current and future business requirements.
> - **Principle 5:** *Conformance:* Information technology complies with all mandatory legislations and regulations. Policies and practices are clearly defined, implemented, and enforced.
> - **Principle 6:** *Human Behaviour* Information technology policies, practices, and decisions demonstrate respect for human behaviour, including the current and evolving needs of all the people in the process.
>
> Adapted from ISO/IEC 38500:2008 – Corporate Governance of Information Technology.

Although the above definitions differ in some aspects, they focus on the same issues such as achieving the link between business and IT and the primary responsibility of the board.

With higher volumes of failures resulting in poor organizational performance, combined with increasing attention on corporate governance and a growing realization that IT portfolios are complex, one would expect more enterprises to implement frameworks for IT governance. This is required to ensure tight focus on their IT processes, people, and priorities on achieving business goals.

Clearly, well-managed information technology is a business enabler. All directors, executives, and managers, at every level in any organization of any size, need to understand how to ensure that their investments in information and information technology enable the business.

Every deployment of IT brings with it immediate risks to the organization, and therefore every director or executive who deploys or every manager who makes any use of IT, needs to understand these risks and the steps that should be taken to counter them.

Furthermore, IT governance integrates and institutionalizes good practices to ensure that the enterprise's IT supports the business objectives. Information technology governance enables the enterprise to take full advantage of its information, thereby maximizing benefits, capitalizing on opportunities, and gaining competitive advantage. Ultimately, IT governance concerns can be framed by the following two overarching goals.

- The ability of IT to deliver value to the business, which is driven by the strategic alignment of IT with business
- The mitigation of IT risks, which is driven by embedding accountability into the enterprise

Within these two larger goals, five domains (focus areas) of IT governance are identified, three of which are drivers and two are outcomes (ITGI, 2003, p. 19). The drivers include IT strategic alignment, IT resource management, and IT performance management. outcomes include IT risk management and IT value delivery.

19.5 Operational Risk and Governance

The prevailing definition of operational risk (op risk) comes from the Basel Committee on Banking Supervision. It states that operational risk is the risk of loss from inadequate internal processes or failed internal control. These processes may regard people, tools, methods, procedures, or systems. Operational risk also stems from external events that are not always under the control of our organization. Therefore, operational risk control must take full account of the likelihood that the evolving organizational structure will resemble an orchestra with 300 professionals and one conductor, as Peter Drucker suggests. To account for operational risks that go beyond fraud and other well-known cases, many financial institutions have their own definitions, which largely complement that of Basel Committee.

Key indicators for operational risk include: outstanding risk claims; number of errors, by channel, frequency of other incidents; impact of each class of incidents in economic terms; legal issues connected to operational risk; level and sophistication of staff training; staff turnover; and the way in which jobs are organized and supported including IT supports. A good way to look at operational risk events is that they are the results of ill-defined, inadequate, and failed internal processes or external impacts overwhelming internal defences.

The Basel Committee has defined an appropriate operational risk management environment in the following terms. Operational risk strategy must reflect the institution's tolerance for risk; the board should be responsible for approving the basic structure of managing operational risk; and senior management must have the responsibility for developing operational risk. Aptly, the Basel Committee on Banking Supervision underlines the fact that internal control must enable senior management to monitor the effectiveness of all operational risk checks and balances. Operational risk presents a complex picture, and the Basel Committee has given guidelines for its identification.

19.6 Organizational Framework—Value Creation

Until the last decade, established organizations usually had sufficient time to adjust to all but the most catastrophic changes in business circumstances. If, for example, a company's sales increased or decreased unexpectedly, the change invariably took place at a rate that was, in corporate terms, easily controllable and not usually seen as a reason for undue concern. In many cases, it was just a matter of adjusting the number of employees and the long-established organization's considerable expertise in handling problems of this type.

During the 1990s all this began to change. The competitive pressure became so strong that the continued existence of everything and everyone in business was soon being challenged. Technology developments changed some processes, made others obsolete and effectively moved some from one function to another, for example, from finance to IT. The management gurus began to argue that the time had come for major changes. They felt that it is time to knock down the hierarchical walls and create a flatter, more competitive management structure by getting rid of middle management. Over the last few years, management structures have got notably flatter in many organizations, with many of the redundant middle managers being used in new, often technical specialist roles that reflect the organizations' changing circumstances. One has to accept that there is both a threat and an opportunity continuously present in every function in every organization. If the function is managed well the organization will have a competitive advantage; if managed badly the organization will be in trouble.

Generally, IT has a reputation that it is a bottomless cost sink run by people who cannot explain what they are doing, cannot deliver what they promise, seem to feel contempt for their customers, and appear to have no understanding of business concerns. The dependency on IT becomes even more imperative in our knowledge-based economy, where organizations are using technology in managing, developing, and communicating intangible assets, such as information and knowledge. Corporate success can be attained when information and knowledge, provided and sustained by technology, is secure, accurate, reliable, and provided to the right person, at the right time, at the right place. This major IT dependency implies a huge vulnerability that is inherently present in IT environments.

> **IT Governance Solutions: Major Hollywood Studio**
>
> A major Hollywood studio with internal MIS Clients worldwide and over 300 client territories required details at the individual Client/Territory level. The client billing included all labour, materials and equipment, and accounts payable expenses. It identified costs by the systems owned by the client and by the type of work performed (activity based management system) such as large and medium enhancement projects, small enhancements, and incidents. All work originated in Remedy, where it is given an identifying ticket number and associated to a system.
>
> Niku Project Management was customized to add additional project-level information passed from Remedy such as system name and ID, and activity-based management task name. In addition, task-level information was customized to add the same information for tasks pushed by Remedy that are not projects—small enhancements and incidents are tasks pushed from Remedy to Niku as tasks within annual plans and all client billing information is contained in the task custom fields. To implement the actual client billing process, the Niku Financial Module was configured and populated with master data including systems, clients/territories, client/territory ownership percentage of systems, labour rate matrix, material rate matrix, and material resources. As a part of this IT governance program, Remedy Help Desk used with new modules added for change management and asset management and customizations to reflect the business needs of the client billing system and the way work is managed in Remedy.
>
> Source: http://www.xinify.com/downloads/it_governance.pdf

Information technology has the potential not only to support existing business strategies, but also to shape new strategies. In this sense, IT becomes not only a success factor for survival and prosperity, but also an opportunity to differentiate and to achieve competitive advantage. Information technology management is focused on the effective and efficient internal supply of IT services and products and the management of present IT operations. Information technology governance is much broader and concentrates on performing and transforming IT to meet present and future demands of the business and the customers. It should be an enabler to compete in the dynamic business world. Competitiveness will normally involve some mixture of quality, continuous service improvement, speed of performance, and cost reduction. The exact mix will depend upon the nature of the organization.

According to the IT Governance Institute (ITGI, 2003), fundamentally, IT governance is concerned about two things: IT's delivery of value to the business and mitigation of IT risks. The first is driven by strategic alignment of IT with the business. The second is driven by embedding accountability into the enterprise. Both need to be measured adequately. This leads to the following five main focus areas for IT governance, all driven by stakeholder value.

- Value delivery
- Strategic alignment
- Resource management
- Risk management
- Performance management

To achieve effective governance, executives require that controls be implemented by operational managers within a defined control framework for all IT processes.

The purpose of IT governance is to direct IT endeavours, to ensure that IT's performance meets the following objectives.

- Alignment of IT with the enterprise and realization of the promised benefits
- Use of IT to enable the enterprise by exploiting opportunities and maximizing benefits
- Responsible use of IT resources
- Appropriate management of IT-related risks

A governance and control framework needs to serve a variety of internal and external stakeholders, each of whom has specific needs. These are as follows.

- Stakeholders within the enterprise who have an interest in generating value from IT investments:
  - Those who make investment decisions
  - Those who decide about requirements
  - Those who use IT services
- Internal and external stakeholders who provide IT services:
  - Those who manage the IT organization and processes
  - Those who develop capabilities
  - Those who operate the services
- Internal and external stakeholders who have a control/risk responsibility:
  - Those with security, privacy and/or risk responsibilities
  - Those performing compliance functions
  - Those requiring or providing assurance services

If a company is going to run IT like a business, it must understand how businesses are run. Some of IT's problems can be traced to rigid functional approaches, leading to silos. Business gurus have been addressing this issue since the 1980s, and the most compelling framework has been the value chain, developed by Michael Porter in his landmark book *Competitive Advantage*.

The term 'process framework' is generally used to characterize comprehensive and systematic representations of a major business area by focusing on its activities. There is typically an attempt made to distinguish between what the organization is doing and who is doing it.

Process management provides a methodology for managing this white space between the boxes on the organization chart. Process maturity, in general, is often assessed with a staged framework such as the Capability Maturity Model. Maturity implies definition, adherence, repeatability, measurability, and continuous improvement; it does not imply sophistication of technical platforms or depth of staff expertise.

At a very broad level, organizations can approach governance on an ad hoc basis and create their own frameworks, or they can adopt standards that have been developed and perfected through the combined experience of hundreds of organizations and people. Information technology governance helps ensure achievement of critical success factors by efficiently and effectively deploying secure, reliable information, and applied technology.

19.7 Internet Governance

Governance of the Internet is multifaceted, complex, and far from transparent. While Internet governance is a phrase increasingly seen in literature, its precise meaning remains somewhat diffused. Discussions on Internet governance have been taking place for several years. To a large extent, the global debates on the subject grew in volume due to the technology boom of the late 1990s and the heavy involvement and interest of the information and communication technology (ICT) private sector in the process.

Governance is ensured through a control structure with defined responsibilities. Internet, however, is not owned by an individual or a firm. There exists no central authority on the Internet. Instead, there exists a multitude of actors, institutions and bodies, exerting control or authority in a variety of ways, and at multiple levels. This does not necessarily imply anarchy. These participants in general have formal and well-defined roles, and they address specific tasks or responsibilities.

The core of the information society does not lie in a single nodal point but is spread at its edges in the form of its users, the people. The power of the Internet is that everyone can create content.

One of the working definitions of Internet governance is that it is the development and application by governments, the private sector, and civil society, in their respective roles, of shared principles, norms, rules, decision-making procedures and programmes that shape the evolution and use of the Internet. Internet governance embraces issues concerned not just with the infrastructure for transmitting data but also the information content of the transmitted data.

Internet governance encompasses a range of issues and actors, and takes place at many layers. Throughout the network, there exist problems that need solutions, and, more importantly, potential that can be unleashed by better governance. At a macroscopic level, the infrastructure layer can be considered the foundational layer of the Internet. It includes the copper and optical fibre cables and radio waves that carry data around the world and into users' homes. It is upon this layer that the other layers of communication are built. Governance of the infrastructure layer is, therefore, critical to maintaining the seamlessness and viability of the entire network.

Given the infrastructure layer's importance, it makes sense that a wide range of issues requiring governance can be located at this level. Since traditional universal access regulation involves fixed-line telephony, national and international telecommunications regulators are usually the most actively involved in governance for universal access.

Next-generation technologies also require governance to ensure that they are deployed in a manner that is harmonious with pre-existing (legacy) systems. Such coordination is essential at every layer of the network, but it is especially critical at the infrastructure layer. If new means of transmitting information cannot communicate with older systems, then that defeats the very purpose of deploying new systems. Standards are among the most important issues addressed by Internet governance at any layer. As noted, the Internet is only able to function seamlessly over different networks, operating systems, browsers, and devices because it sits on a bedrock of commonly agreed-upon technical standards.

19.8 Governance of Internal IT Processes

In today's regulated environment, shareholders have become more demanding and are paying more attention to the governance and compliance strategies of an enterprise. There are various regulatory compliance requirements today that are mandated by the Organisation for Economic Co-operation and Development's (OECD) Principles of Corporate Governance, Basel II, Sarbanes–Oxley, and respective Stock Exchange guidelines. The rationale behind such regulations is to ensure a verifiable process to manage corporate risks and instill a corporate environment of respect for all stakeholders.

Organizations are required to provide an assurance to the accuracy and integrity of both financial reports and core business processes. Therefore, IT controls have become integral to the effective governance of the modern enterprise. Corporate IT groups have recognized the inherent value of corporate and IT governance leading to the birth of the notion of business and IT alignment.

19.8.1 Modern Governance of IT

The ITGI proposes that information security governance should be considered a part of IT governance, and that the board of directors are informed about information security, set direction to drive policy and strategy, provide resources to security efforts, assign management responsibilities, set priorities, support changes required, define cultural values related to risk assessment, obtain assurance from internal or external auditors, and insist that security investments be made measurable and reported on for program effectiveness.

Additionally, the ITGI suggests that the management should identify and document security policies with business input. The management should ensure that roles and responsibilities are defined and clearly understood, threats and vulnerabilities are identified, security infrastructures are implemented, control frameworks are implemented as approved by the governing body. The management should also ensure that timely implementation of priorities are carried out, monitoring of breaches are in-built in the systems, periodic reviews and tests are conducted, awareness education is viewed as critical and delivered, and that security is built into the system's development life cycle.

Multiple frameworks have been created to support auditing of implemented security controls. These resources are valuable for assisting in the design of a security program, as they define the necessary controls for providing secure IS. In recent years, increased attention has been devoted to internal control by auditors, managers, accountants, and legislators. Five recently issued documents are the result of continuing efforts to define, assess, report on, and improve internal control. They are as follows.

- Committee of Sponsoring Organizations of the Treadway Commission's Internal Control—Integrated Framework (COSO)
- Information Systems Audit and Control Foundation's COBIT (Control Objectives for Information and Related Technology)
- Institute of Internal Auditors Research Foundation's Systems Auditability and Control (SAC)
- American Institute of Certified Public Accountants' Consideration of the Internal Control Structure in a Financial Statement Audit (SAS 55), as amended by Consideration of Internal Control in a Financial Statement Audit
- An Amendment to SAS 55 (SAS 78)

19.8.2 COSO

The COSO was formed in 1985. It began when five private-sector organizations that were concerned about the apparent increasing frequency of fraudulent financial reporting, came together to sponsor the National Commission on Fraudulent Financial Reporting, more commonly called the Treadway Commission after its chairman, James C. Treadway, Jr, a former SEC commissioner. The sponsoring organizations were:

- American Accounting Association (AAA)
- American Institute of Certified Public Accountants (AICPA)

- Financial Executives Institute (now Financial Executives International, (FEI)
- Institute of Internal Auditors (IIA)
- National Association of Accountants (now known as the Institute of Management Accountants, IMA)

The COSO report defines internal control as a process, effected by an entity's board of directors, management, and other personnel, designed to provide reasonable assurance regarding the achievement of objectives in the following categories, namely effectiveness and efficiency of operations, reliability of financial reporting, compliance with applicable laws and regulations. The COSO identifies five areas of internal control necessary to meet financial reporting and disclosure objectives. These areas include:

- control environment,
- risk assessment,
- control activities,
- information and communication, and
- monitoring.

The report offers guidance for public reporting on internal control and provides materials that management, auditors, and others can use to evaluate an internal control system. The two major goals of the report are to:

- establish a common definition of internal control that serves many different parties and
- provide a standard against which organizations can assess their control systems and determine how to improve them.

The COSO internal control model has been adopted as a framework by some organizations working towards Sarbanes-Oxley Section 404 compliance.

19.8.3 COBIT

Control Objectives for Information and Related Technologies (COBIT) is an IT-focused governance and control framework created by the ITGI and Information Systems Audit and Control Association (ISACA). Developed as an open standard, COBIT is being increasingly adopted globally as the governance and control model for implementing and demonstrating effective IT governance. The first, second, third, and fourth editions of COBIT were published in 1994, 1998, 2000, and 2005 respectively. It is based on established frameworks such as the Software Engineering Institute's Capability Maturity Model, ISO 9000, ITIL, and ISO 17799 (standard security framework, now ISO 27001). The primary motivation for providing this framework was to enable the development of clear policy and good practices for IT control throughout industry worldwide. About 13 of the 34 high-level control objectives are derived directly from the Information Technology Infrastructure Library(ITIL) service support and service delivery areas.

> **The Manta Group**
>
> A boutique management consulting firm, The Manta Group, has found information technology (IT) governance to be a strong differentiating factor for its clients. The current global, networked business environment now demands that IT improve organizational processes via well-defined controls and metrics. The Manta Group uses Control Objectives for Information and Related Technology (COBIT) to help clients improve their processes and achieve alignment with business goals through relevant and practical controls and metrics. After extensive research and review, as well as hands-on expertise in the field, the firm has found COBIT to be an internationally accepted governance framework to provide a complete and concise model for governing and attaining value from investments in IT.
>
> The Manta Group and its clients have found COBIT to be the most effective governance framework due to its coverage footprint as well its result/outcome-oriented approach, where every task and action is measured by a specific contribution to a goal. The firm has successfully established COBIT as the IT governance framework for its clients that have a demand for IT governance.
>
> *Source*: http://www.isaca.org/Template.cfm?Section=Case_Studies3&Template=/ContentManagement/ContentDisplay.cfm&ContentID=28371

The COBIT framework provides a tool for the business process owner who facilitates the discharge of this responsibility. The framework continues with a set of 34 high-level control objectives, one for each of the IT processes, grouped into the following four domains.

- Planning and organization
- Acquisition and implementation
- Delivery and support
- Monitoring

This structure covers all aspects of information and the technology that supports it. By addressing these 34 high-level control objectives, the business process owner can ensure that an adequate control system is provided for the IT environment.

The COBIT framework allows management to benchmark the security and control practices of IT environments, allows users of IT services to be assured that adequate security and control exists, and auditors to substantiate their opinions on internal control and to advise on IT security and control matters. Control Objectives for Information and Related Technology is intended to be used at the highest level of IT governance. It provides an overall governance framework based on a high-level process model of a generic nature that makes it applicable to most organizations. Processes and standards that cover specific areas in more detail, such as ITIL and ISO 27001, can be mapped to the COBIT framework to create a hierarchy of guidance materials.

19.8.4 SAC

The Systems Auditability and Control (SAC) Report defines a system of internal control as a set of processes, functions, activities, subsystems, and people who are grouped together or consciously segregated to ensure the effective achievement of objectives and goals. It defines the system of internal control, describes its components, provides several classifications of controls, describes control objectives and risks, and defines the internal auditor's role. The report provides guidance on using, managing, and protecting information technology resources and discusses the effects of end-user computing, telecommunications, and emerging technologies. The report emphasizes the role and impact of computerized IS on the system of internal controls. It stresses on the need to assess risks, to weigh costs and benefits, and to build controls into systems rather than add them after implementation. The SAC Report provides the following five classification schemes for internal controls in information systems.

- Preventive, detective, and corrective
- Discretionary and non-discretionary
- Voluntary and mandated
- Manual and automated
- Application and general controls

These schemes focus on when the control is applied, whether the control can be bypassed, who imposes the need for the control, how the control is implemented, and where in the software the control is implemented.

The SASs 55 (1988b) and 78 (1995) provide guidance to external auditors regarding the impact of internal control on planning and performing an audit of an organization's financial statements. The SAS 78 replaces the definition of the internal control structure in SAS 55 with that of internal control in the COSO report, except that SAS 78 emphasizes the reliability of financial reporting objective by placing it first. The SAS 78 defines internal control as a process, effected by an entity's board of directors, management, and other personnel, designed to provide reasonable assurance regarding the achievement of objectives in the following categories: reliability of financial reporting effectiveness and efficiency of operations, and compliance with applicable laws and regulations.

19.8.5 IT Control Dependencies

A comparison of the five documents reveals that each builds on the contributions of the previous documents. Control Objectives for Information and Related Technology incorporates both COSO and SAS as a part of its source documents. It takes its definition of control from COSO and its definition of IT control objectives from SAC. The COSO uses the internal control concepts in both SAS 55 and SAC, and SAS 78 amends SAS 55 to reflect the contributions to internal control concepts made by COSO.

Systems Auditability and Control embodies the internal control concepts developed in SAS 55, In particular, SAS 78 responds to the call for a reconciliation of the internal control concepts presented in the COSO report and SAS 55. Control Objectives for Information and Related Technology and SAC examine control procedures relative to an entity's automated information system; COSO and SASs 55/78 discuss the control procedures and activities used throughout an entity.

19.8.6 Benefits of IT Governance

Information technology goals vary considerably across organizations. They may be relatively modest. For example, eliminating inaccuracies and inefficiencies in administrative processes will emerge as a benefit of IT governance. Information technology governance may be central to a company's strategy, namely supporting a seamless global supply chain, flawless customer service, or leading-edge research and development. Information technology spending can be designed to meet immediate needs and allowed for an array of future benefits only if IT and business goals are clearly defined.

Given the uncertain returns on IT spending, many executives wonder whether they are spending too much—or perhaps even too little. But in successful companies, senior managers approach the question very differently.

First, they determine the strategic role that IT will play in the organization, and only then do they establish a company-wide funding level that will enable technology to fulfil that objective.

It is seen that companies of a few hundred people have a couple of IT projects underway. Clearly, not all of them are equally important. But we find that senior managers are often reluctant to step in and choose between the projects that will have a significant impact on the company's success and those that provide some benefits, but are not essential. The failure of senior managers to choose a manageable set of IT priorities can also lead to disaster. One need only remember Hershey Foods' infamous decision in 1999 to implement several major systems simultaneously, including customer relationship management (CRM), enterprise resource management (ERP), and supply chain management, which ultimately resulted in the company's inability to deliver.

19.9 E-governance Framework

As the information revolution gradually permeates developing countries, more and more governments are embracing e-governance as a tool for enhancing interaction with citizens, increasing productivity in the delivery of government services, and improving transparency and accountability. Electronic governance uses Internet and communication technologies to automate governance in innovative ways, so that it becomes more efficient, more cost-effective, and empowers the human race even more.

Achieving success in e-governance requires active partnerships between government, citizens, and the private sector. The e-governance process needs continuous input and feedback from the customers, the citizens, resident, businesses, and officials who use electronic public services. Their voices and ideas are essential to making e-governance work. It is important to have a national strategy for e-government. The strategy should specify target groups and needs and identify priorities. It should also be part of an overall national ICT strategy that addresses affordability and accessibility.

---

e-Perolehan

e-Perolehan is the secure e-procurement service of the government of Malaysia. e-Perolehan is financed through a build-operate-transfer (BOT) scheme involving Commerce Dot Com Sdn. Bhd., which is financing the project. Suppliers can host

Contd

> their products and prices online free of charge, reducing their overhead costs, while government departments can easily access the pricing information online. Commerce Dot Com Sdn. Bhd receives a transaction fee charged on each completed sale.
>
> Significantly, the government's commitment to e-Perolehan was coupled with the establishment of a network of procurement telecentres nationwide to enable smaller-sized suppliers to trade online with the government. The telecentres, located in all state and district capitals, help non-IT savvy suppliers submit registration applications and provide catalogue details.
>
> *Source*: http://home.eperolehan.com.my/en/default.aspx, downloaded on 2 June 2009

E-government initiatives should be monitored and measured in order to replicate successes and achieve citizen trust, and avoid wasting resources on projects that are not serving genuine needs. The e-government vision has to be connected to local culture and local administration.

19.9.1 Definition of E-governance

E-governance is the use of ICT by mobilizing government resources and utilizing the internal information resources by the government employees with the help of citizens' acceptability to the changes taking place to provide better services to them. It can also be defined as the application of electric means in

- the interaction between government and citizens and government and businesses and
- internal government operations to simplify and improve democratic, government, and business aspects of the government.

E-government projects often require substantial financial resources and interdisciplinary skills to plan, install, and operate such systems and services effectively. As one analysis has pointed out, it becomes difficult at times for governments to take on the projects completely with their own resources. This promotes governments to get into partnerships to leverage the strengths and resources of its partners. These collaborations or partnerships can be built up with the private sector as well as the other stakeholders in the process of e-government including non-government organizations (NGOs).

19.9.2 E-governance Initiatives of India

The Government of India (GoI) has taken major initiatives to accelerate the development and implementation of e-governance and to create right environment for introducing G2G, G2B, G2E, and G2C services within the country. The National Policy on Open Standards for e-governance provides a set of guidelines for the uniform and reliable implementation of e-governance solutions. The policy is applicable to all systems used for e-governance. All new e-government infrastructure systems and government to public systems will conform to the standards based on the open standards policy. E-governance promotion is based on the following two important planks.

- Reduce delay and inconveniences through technology interventions including the use of modern tools, techniques, and instruments of e-governance
- Promote knowledge sharing to realize continuous improvement in the quality of governance

The government of Karnataka has set up, through the public-private partnership (PPP) model, a network of 800 telecentres at village hobli level. It has also got set up 177 back offices at the taluka level. The project was christened as Nemmadi and is a Rs 30-crore public-private initiative. It is an extension of the celebrated Project Bhoomi that computerized 2 crore rural land records in 2001. These telecentres would deliver a range of G2C and B2C services at the citizen's doorsteps. This initiative was actually started in 2004 with a pilot project at Mandya, in a town called Maddur to determine its efficacy across villages. The initiative on e-governance was designed to create efficient and smart virtual offices of the state government in all the villages. It was initiated to provide copies of land records and 38 other citizen-centric services of the revenue department in a convenient and efficient manner through 800 village telecentres across rural Karnataka. The initiative also mandated to scale-up the operations to cover all other G2C services of all the departments and to enhance the accountability, transparency, and responsiveness of the government to citizen's needs. The initiative also wanted to enable government departments and agencies to focus on their core functions and responsibilities by freeing them from the routine operations, such as issuing of certificates, land records, collection of utility bills of citizens, and thereby enhancing the overall productivity of the administrative machinery.

The Government of India has approved the National E-governance Plan (NeGP) in May 2006 with the vision 'make all government services accessible to the common man in his locality, throughout common service delivery outlets and ensure efficiency, transparency and reliability of such services at affordable costs to realize the basic needs of the common man'.

For example, the Karnataka government's treasury department wanted to remedy check-mate inefficiency, data inaccuracy, and the misappropriation of funds at the department level. Inaugurated in July 2003, Khajane aimed to remove basic systemic deficiencies by harnessing high network availability and accessibility, and creating pinpoint accuracy by replicating all existing treasury data into master files. The project began as a pilot in January 2002 across four sites: Tumkur, Bangalore Urban, Gubbi, and Shiggaon aimed at e-enabling the interaction between government departments/officials and the concerned treasury which manages the allocated funds and resources.

The NeGP currently consists of 27 mission mode projects (MMPs) and eight support components to be implemented at the central, state, and local government levels. These include projects such as income tax, customs and excise and passports at the central level, land records, agriculture and e-district at the state level and panchayats and municipalities at the local level. There are also a number of integrated MMPs such as e-procurement, service delivery gateway and EDI which are integrated MMPs, where delivery of services envisaged in the project entail coordinated implementation across multiple Departments of Government.

For example, Indian Farmers Fertilizer Co-operative (IFFCO) is among the largest manufacturers of nitrogenous and phosphatic fertilizers in the world. It had a manual procurement system but it was plagued by problems of inefficiency. Vendors from distant locations were finding it difficult to contact IFFCO, collect bid documents, and submit quotes on time. The IFFCO planned for an e-procurement system partnering with Microsoft under 'Enterprise-GO' program. It saved on cost of printing, dispatch, human resource, and filing, while vendors saved on cost of travelling, preparation of bid, and dispatch.

19.9.3 State-wide Area Network

Government has approved the scheme for establishing state-wide area networks (SWANs) in 29 states and six union territories (UTs) across the country at a total cost of Rs 3,334 crore. This scheme envisages establishment of an intra-government network with a minimum of 2 Mbps connectivity from the state headquarters to block headquarters through district headquarters. The SWAN project provides the connectivity to facilitate the rolling out of citizen-centric services under various MMPs under NeGP. The scheme has the following two implementation options.

- PPP Model: The first option is to call for a bid from private entities, which would set up and operate the SWAN in the state for a period of five years. The payments to the successful bidder would begin only after the infrastructure has been set up. The contract signed with the private party provides for strict service level agreements (SLAs) and there is a provision of penalties to be imposed on the party should it fail to meet those service levels.
- National Information Center (NIC) Model: In the second model, the state has the option for going directly to the NIC. The NIC would set up and maintain the SWAN for the state for a period of five years.

19.9.4 State Data Centres (SDCs)

State data centres (SDCs) are proposed to be established across 29 states and six UTs in the country along with disaster recovery (DR), in order to provide shared, secured, and managed infrastructure for consolidating and securely hosting state-level data and applications. The SDC would provide better operations and management control and minimize overall cost of data management, IT management, deployment, etc. The SDCs would ordinarily be located at the state headquarters and help the state government, state line ministries and departments in providing central repository (database consolidation), application consolidation, state intranet/Internet portal, state messaging infrastructure, remote management, business continuity site, etc. needed for their G2G, G2C, and G2B services. The various MMPs, both at the central level, state level, and also the integrated services of the NeGP are expected to use SDCs to deliver their services. The SDC scheme was approved by the government in January 2008, at a total cost of Rs 1,633 crore. The Department of Information Technology (DIT) has sanctioned SDC proposals of 23 states at a total cost of 1,077 crore.

19.9.5 E-governance PPP Projects

The last couple of years have seen e-governance drop roots in India. Information technology enables the delivery of government services as it caters to a large base of people across different segments and geographical locations. The effective use of IT services in government administration can greatly enhance existing efficiencies, drive down communication costs, and increase transparency in the functioning of various departments.

For e-governance initiatives, the three Ps: public-private partnership (PPP) are a must. The PPP model of development focuses on collaboration between the public and the private sector. It recognizes the importance of the private sector in reaching development goals by promoting business, creating income, providing jobs, as well as developing a sense of corporate social responsibility. The public and the private sector recognize overlaps of their goals, see the opportunities for cooperation, and work side-by-side in mutual projects. Lately, the government has started using the PPP model for implementing e-governance projects. The PPP model is nothing new, and has frequently been used by the government to deploy various other types of projects, such as those related to infrastructure development. A PPP project combines the skills and synergies of the government and the private sector to deploy projects successfully and on time.

The PPP model applies well when delivery of services are basic and permanent, setting up of infrastructure is for steady return on investment (ROI) in the long term when the information is not so sensitive and where the government wants to hold the control for various reasons. In general, a good rule of thumb for financing e-governance projects can be as follows.

- Highly sensitive information as well as commercially non-viable portals to be operated by the government only (and paid for by the government)
- Information and conveniences for the ordinary citizen where there is a financial or other implication (such as land records, passport information, etc.) to be operated by the private sector but content to be solely the prerogative of the government
- Ordinary services and information to be managed by the private sector and financed by the private sector, under the periodic supervision of the government

19.9.6 E-governance BOOT Projects

In this model, the asset created by the private party is transferred to the government after a pre-specified time, and such models are adopted where the technology is time-tested and assets are expected to outlast the concession period.

> ### E-Seva E-Government Initiative of Andra Pradesh
>
> A frequently cited example of an e-government public-private partnership is e-Seva, an innovative project between the government of India's state of Andhra Pradesh. E-Seva project is one of the most popular and best known examples of e-government projects in India in the recent times. The project first started in the state of Andhra Pradesh and has attained national and international fame as the most innovative way of providing integrated services to the general people. E-Seva provides more than one hundred services, ranging from the payment of utility bills to the registration of motor vehicles. In response to access barriers, e-Seva was launched with 43 service centres in the city of Hyderabad, later expanded to 213 towns, and most recently has been extending into rural areas. As a measure of success, e-Seva completes over 1.6 million transactions per month in the city of Hyderabad alone. In this case, Tata Consulting, under the 'Build-Own-Operate-Transfer, (BOOT)' model, built the e-Seva portal and runs the service, charging normal fees for the various government services and keeping part of the revenue.
>
> Source: http://unpan1.un.org/intradoc/groups/public/documents/APCITY/UNPAN005029.pdf, downloaded on 2 June 2009.

19.9.7 Benefits of E-governance

The benefits to be expected from e-government through e-governance can be categorized into the following three major categories.

- Improved delivery of public services, in terms of availability, ease of use, and cost savings to the government, to businesses and to individuals
- Improved transparency, accountability and democracy and reduced opportunities for corruption
- Broader economic and societal gains

SUMMARY

This chapter records and interprets some important models and practices in IT governance and strategic alignment domain. Information technology governance is defined in the chapter and its relationship with corporate governance and IT management is clarified. This chapter introduces current and prior IT governance literature across key focus areas being strategic alignment of business and IT systems, delivery of value from IT systems, risk management of IT systems, management of IT resources and measurement of the performance of IT systems. The technological complexity of computer science provides both benefits and difficulties for IT auditing. The IT auditor is allowed a liberty about the reliability of certain hardware and software components in the information systems. But if this knowledge is abused, it may become extremely difficult for the IT auditor to identify and detect the abuse.

The need and importance of IT audit changes in the corporate governance scenario and the induction of latest information and communication technology hardware and software systems have made the audit and control of information systems almost mandatory. As leading global businesses increasingly recognize the imperative for strong IT governance, some boards are stepping up to adopt a much stronger oversight role, and leading institutions are proposing those organizations as role models for the rest of the business world. The IT Assurance Guide provides detailed material to the assurance professional helping him/her to test control design, test the outcome of control objectives, and test the impact of control weaknesses.

Control Objectives for Information and Related Technology (COBIT) is a powerful framework to implement enterprise governance of IT. However, COBIT also provides in-depth support to execute IT assurance assignments. To support this, COBIT's IT Assurance Guide offers two elements, a detailed roadmap showing how typical IT assurance activities can be planned (IT assurance planning), scoped (IT assurance scoping), and executed (IT assurance execution) based on COBIT and a detailed guidance in the IT assurance execution domain providing specific steps to test the control design, test the outcome of the control objectives, and test the impact of control weaknesses, based on COBIT.

Key Terms

Accountability It is the requirement that officials answer to stakeholders on the disposal of their powers and duties, act on criticisms or requirements made of them and accept (some) responsibility for failure, incompetence, or deceit.

Effectiveness It is the capacity to realize organizational or individual objectives. Effectiveness requires competence; sensitivity and responsiveness to specific, concrete, human concerns; and the ability to articulate these concerns, formulate goals to address them and develop and implement strategies to realize these goals

Governance It is the exercise of political, economic, and administrative authority in the management of a country's affairs at all levels. Governance is a neutral concept comprising the complex mechanisms, processes, relationships, and institutions through which citizens and groups articulate their interests, exercise their rights and obligations and mediate their differences.

Organization It is a social group with a structure designed to achieve collective goals. Organizations provide the basis for purposeful collective action.

Transparency It means sharing information and acting in an open manner. Transparency allows stakeholders to gather information that may be critical to uncovering abuses and defending their interests. Transparent systems have clear procedures for public decision-making and open channels of communication between stakeholders and officials, and make a wide range of information accessible.

REVIEW QUESTIONS

19.1 Discuss the difference between IT audit and IT assurance.

19.2 Explain the relationship between the control objectives and the control practices and illustrate with an example.

19.3 Explain what is meant by application controls and why these are not specific IT management controls.

19.4 Explain the difference between accountability and responsibility and illustrate with IT examples.

19.5 Explain the relationship between enterprise governance of IT and business/IT alignment.

19.6 Discuss how business/IT alignment can be measured and determine which is the most practical approach.

19.7 What are the key enterprise governance of IT practices to enable business/IT alignment?

19.8 Explain and illustrate value-based scoping for IT assurance.

19.9 Explain and illustrate risk-based scoping for IT assurance.

19.10 In reporting on control weaknesses, the assurance professional should focus on business risk issues. Explain and illustrate.

19.11 Define what an IT governance awareness campaign is and explain why it is an important IT governance mechanism.

Projects

19.1 Visit the nearest firm which is dependent on IT and prepare a report on the IT governance as practised by the organization.

19.2 ABC unlimited decides to trade in lighting products that are unique in concept and that appeal to a broad audience. ABC recently developed a strategic road map to align its IT with its overall business strategy with the help of a local consulting firm. ABC needed to align its IT infrastructure, processes and applications with its strategic goals. The company knew that to compete more effectively, it would have to improve its customer focus and supply chain efficiency and support these areas with transparent IT solutions, compliant with the company's strategic IT vision. The company's main goals in undertaking a transformation of its IT infrastructure and processes were to support the creation of a comprehensive business, achieve profitable growth, reduce costs and improve customer focus and supply chain efficiency.

With this clear vision of where it needed to go, ABC sought a consulting partner with expertise in the trading industry to develop the business case for implementing new IT infrastructure and processes, including recommendations for new major IT application installations and integration across web integrated functional areas.

The consultancy firm teamed with ABC to deliver the company's IT strategy plan, including the business case for required investments. The team used the consultant's proprietary methodology to evaluate ABC's strategic IT processes. The resulting road map aligns the company's IT strategy with its larger business goals and addresses the business requirements and issues. The actual implementations of recommended IT solutions will be completed during the next 2 years, delivering a solid return on investment (ROI) once the implementation is completed. The most important part of the solution was the implementation of an enterprise ERP system. The common ERP system is the key to ABC's cost reductions and profitable growth through the integration of production, supply and customer service. It is expected that through this ERP implementation, a better fusion between IT and business will be achieved, enabling a more efficient supply chain and improved logistics for purchasing and distribution.

Identify the COBIT processes and control objectives relevant to handle the process of defining and implementing the IT strategy defined in the case.

The solution was to bring in an ERP package. Identify appropriate control objectives for designing an audit plan, and justify your selection of the relevant control objectives.

19.3 Refer to the case available in Chapter 1. Call it the resilience gap. The world is becoming turbulent faster than organizations are becoming resilient. The evidence is all around us. Big companies are failing more frequently. Of the 20 largest US bankruptcies in the past two decades, ten occurred in the last two years. Corporate earnings are more erratic. Prepare an IT governance framework for deployment for ABC Unlimited to help ABC with a good corporate governance framework.

REFERENCES

Bernstein, Peter L. 1996, *Against the Gods: The Remarkable Story of Risk*, John Wiley Sons Inc.

Betz, Charles T. 2007, *Architecture and Patterns for IT Service Management, Resource Planning, and Governance*, Elsevier.

Bevir, Mark 2007, *Encyclopedia of Governance*, Vol 1, Sage.

Bygrave, Lee A. and Jon Bing (Eds) 2009, *Internet Governance Infrastructure and Institutions*, Oxford University Press.

Calder, Alan and Steve Watkins 2008, *International IT Governance*, Kogan Page.

Cater-Steel, Aileen 2009, *Information Technology Governance and Service Management: Frameworks and Adaptations*, Information Science Reference.

De Haes, S. and Van Grembergen, W. 2008a, Practices in IT Governance and Business/IT Alignment, Information Systems Control Journal, vol. 2.

De Haes, S., and Van Grembergen, W. 2008b, Analysing the Relationship between IT Governance and Business/IT Alignment maturity, in Proceedings of the 41st Hawaii International Conference on System Sciences (HICSS).

Maizlish, Bryan and Robert Handler 2005, *IT Portfolio Management Step-by-Step*, John Wiley.

McIntyre-Mills, Janet 2006, *Systemic Governance and Accountability*, V-3, Springer.

Nicolai, J. Foss and Snejina Michailova 2009, *Knowledge Governance Processes and Perspectives*, Oxford University Press.

Shah, John S. 2003, *Corporate Governance and Risk*, John Wiley.

Van Grembergen, Wim and Steven De Haes 2008, *Implementing Information Technology, Governance: Models, Practices, and Cases*, IGI Publishing.

Van Grembergen, Wim and Steven De Haes 2009, *Enterprise Governance of Information Technology, Achieving Strategic Alignment and Value*, Springer.

CHAPTER

20 Connected World and E-commerce

The dissemination of information is one of the cornerstones of modern civilization.
–John F. Budd

Learning Objectives

After reading this chapter, you should be able to understand:

- the concepts of e-commerce
- the various types of e-commerce
- and differentiate value creation in e-business

20.1 Websites and E-business

The unprecedented progress of computers and communications technology in the last few decades has increased the information intensity of most activities in value chains. A company is perceived as a series or chain of activities. The best-known value configuration create value through efficient production of goods and services based on a variety of resources. Primary activities in the value chain include inbound logistics, production, outbound logistics, marketing and sales, and service. Support activities include infrastructure, human resources, technology development, and procurement. That we live today in an information economy is a frequently encountered assertion that few people would have any disagreement with. In developed economies today, information has come to play an important role in almost every walk of life. The rapid development and utilization of Internet and associated technologies have increased the speed of information transformation among countries in the world that economizes the cost and time to transmit the information of the international trade transactions. Undoubtedly, the emergence of new communications and information technologies has significantly reduced the existing physical, social, and political barriers that may limit business firms to exchange and trade at a global scale.

From a business perspective, globalization is the process of internationalization of business practices and existence to reach a global market with almost no limitations. Globalization occurs when business companies decide to expand their business practices and processes beyond their national or local boundaries and take part in the global economy. As a result, organizations will present themselves in foreign markets whether physically and/or virtually to target and reach global customers and compete internationally. Hence, globalization is the name that has been given to the social, economic, and political processes that have produced the characteristic conditions of contemporary existence. This perception has made it possible to begin to imagine the world as a single, global space linked by a wide array of technological, economic, social, and cultural forces. For example, today consumers can make more informed decisions in their purchasing activities. Producers, on the other hand, can now decide more easily on what to produce, how to produce, and for whom to produce. The Internet and the web system form the basic infrastructure of many e-commerce systems.

In a web-based e-commerce system, the web browser is the client interface and the web server and application server are the main parts of the service system. Refer to the earlier chapters on Internet and web system.

20.2 E-business

Internet has introduced new forms of business interactions such as electronic business (e-business) and electronic commerce (e-commerce). The term commerce is defined by some as describing transactions conducted between business partners. E-business refers to a broader definition of e-commerce, including servicing customers, collaborating with business partners, and conducting electronic transactions within an organization. In the last couple of years, e-business technologies have undergone generations of transitions from static web pages, to interactive media, and to dynamic commerce. Early adopters of e-business primarily used static websites for broadcasting business information. The next generation websites began to support information requests and customer profiles, moving beyond the one directional information flow of the initial websites. The subsequent generation e-business technologies transformed relatively primitive websites into the world of e-commerce, transferring the Internet into a new sales channel. The new technologies at the heart of e-business open up limitless possibilities not just to consider the re-engineering of existing processes but also to design, develop, and deploy fundamentally new ways of conceiving and executing business processes.

E-business is transforming the solutions available to customers in almost every industry, that is, the breadth of solutions and how the solutions are obtained and experienced. In short, e-business offers the platform for new forms of marketplace strategy models. The focus of e-business is on information integration with the aim at transforming business processes across organizations and throughout the entire value chain including customers, suppliers, distributors, and even competitors, and extending the Internet to the physical world.

20.3 E-commerce

The advent of e-commerce has brought about a revolution in business processes and redefined the notion of a customer. E-commerce is driving the new economy and the Internet is the primary facilitator. The most important technology and business innovations that enabled this revolution are the extension of the Internet to the Web and the electronic transaction of business known as e-commerce. Refer to the earlier chapters on Internet for a better understanding on Internet and enabling languages.

E-commerce is usually associated with buying and selling over the Internet or conducting any transaction involving the transfer of ownership or rights to use goods or services through a computer-mediated network. From a communications perspective, e-commerce is the delivery of goods, services, information, or payments over computer networks or by any other electronic means. From a business process perspective, e-commerce is the application of technology towards the automation of business transactions and workflow. From a service perspective, e-commerce is a tool that addresses the desire of firms, consumers, and management to cut service costs while improving the quality of goods and increasing the speed of service delivery. From an online perspective, e-commerce provides the capability of buying and selling products and information on the Internet and other online services. From a collaboration perspective, e-commerce is the framework for inter- and intra-organizational collaboration. From a community perspective, e-commerce provides a gathering place for community members, to learn, transact, and collaborate.

There are many start-up companies that are using the Internet to communicate, advertise, provide information via search engines, and assist individuals and companies in setting up attractive websites. As we are in the new millennium, it is found that cyber shopping has become the order of the day. Consumers find it easier to cyber shop than to take time from their schedules to visit malls. One can shop any time depending on the availability of time. An electronic shop is a web-based software system that offers goods and services, generates bids/offers, accepts orders, and handles delivery and modes of payment. A visitor to the electronic shop can find out about the products and services offered by it. Intending to buy, he/she communicates a minimum amount of data about himself/herself and establishes a user profile, alongwith payment and delivery arrangements.

An important precondition to the success of cyber shopping malls is the construction of appropriate customer interfaces. The customer interface is defined as the user interface of cyber shopping mall systems, through which customers interact to search for the target items and to purchase the identified items. Designing a customer interface involves activities such as content design, structure design, navigation design, and graphic design. Some cyber shopping malls use hierarchies to organize individual pages. For example, the top-level page depicts the overall structure, the second-level pages describe product categories (such as home electronics), the third-level pages describe individual stores, and finally the bottom-level pages describe individual products.

One company that has struck gold is the classic Amazon.com, which has become an all-familiar website. Established in 1995 by Jeff Bezos, Amazon.com (www.amazon.com) is one of the most well-known e-commerce sites in general and online bookseller in particular. It is a typical example of business to consumer (B2C) e-commerce in which a business sells already manufactured products to the consumers directly on the Internet. Indeed,

Amazon is selling information but in the form of tangible books that one orders, pays for by credit card, and receives from a book warehouse. The value chain has been established so that inventory is identified and accumulated, and elements, such as inventory location, inventory renewal, physical order delivery, and customer billing, are provided to complete the e-commerce cycle.

Due to technological changes, the information factor has become more significant than the production factor. Many companies and organizations have moved their business processes onto the Web and realized customer relationships with the help of electronic means of information and communication, leading to the term 'electronic business'. The goal for companies that use information and communication technologies is to procure and analyse information on market participants as well as on existing and potential customers and thus develop and sell products and services that are promising. While doing so, web-based information systems naturally become a strategic tool. For example, Amazon.com runs with a web front-end, hiding the business complexities to the user front. It allows the development of the market and the behaviour of market participants to be studied and interpreted. Information systems (IS) support the design and production of goods and services. The physical marketplace, concerned with existing physical raw materials, products, and resources, is still present. Here, physical value-creating processes are required in order to procure, develop, and distribute material goods. However, due to the development of the Internet, the physical marketplace is now being supplemented. As a result, one can now speak of a digital marketspace where digital products and services can be developed and sold inside networks. Thousands of new companies have created new marketplaces and new opportunities worldwide. The most visible impact to the average consumer is in the explosion of digital content availability and the plethora of new e-tail sites to purchase everything from books to airline tickets to groceries.

In simple terms, e-commerce is the online transaction of business, featuring linked computer systems of the vendor, host, and buyer. E-commerce includes electronic trading, electronic messaging, electronic data interchange (EDI), electronic mail (e-mail), electronic catalogues, Internet, intranet and extranet services. It also pertains to any form of business transaction in which the parties interact electronically, rather than by physical exchanges or direct physical contact.

For example, banking in rural areas was more or less thrust on government-run nationalized banks, while private banks focused on more immediately-profitable urban centres. ICICI Bank wanted to step in and reinvent the landscape of commodity-based financing and built the e-commodity based Financing (eCBF). The eCBF is a web-based system that can be accessed by locally appointed warehouse management agents (WMAs), who interact with farmers on ICICI Bank's behalf. Since they operate from remote locations, WMAs can only access eCBF on narrow-band Internet—sometimes using CDMA-based mobiles as a medium. The rural business was launched with skeletal capabilities, and as more processes began to emerge, the bank's IT team, in conjunction with the rural agri-business team, identified new requirements and best ways to add relevant controls. e-Commodity based Financing also helps ICICI Bank provide financial support to corporates in the same way it helps farmers.

20.4 Business on the Net

A business model can be defined as the method by which a firm builds and uses its resources to offer its customers better value than its competitors use. It reflects the company's way of competing, whether it concerns being unique or being the most cost-efficient company in the industry. Launching a business on the Internet requires careful planning, understanding the target customer base, and choosing the right products and services to offer. This first planning step involves strategic questions such as the following.

- Who will buy the product?
- How familiar are you with the Internet?
- Are you planning to be a short-termer or a long-termer?
- Who are your competitors?
- How good will your product(s) look?
- How will you present your product offerings?
- How will you manage and process transactions?
- How will the product be shipped?
- How will you handle unexpected change?
- How will you get and use feedback?

The business planning and standardizing phase is followed by the hardware, software, security, and set-up phase, the design phase, the marketing phase, the fulfilment phase, and finally the maintenance and enhancement phase. The other action points in launching a business on the Internet includes resolving the software and hardware issues, especially with respect to linking to the Internet service provider (ISP) that will put the business on the Internet. For a fee, the ISP gives the new firm a software package, user name, password, and access phone number.

Next comes the site readiness which should generate repeat customers. Assuming the buyer has gone through the ordering process, how can the experience end on a good note? The delivery of the product is critical. The system should include a tracking system to let the shopper know when and who received the product. Customer service contributes a great deal to creating customer loyalty. In addition to being enjoyable, the shopping experience should be risk-free for the firm and the firm's customer. This means implementing powerful security measures for the website and the servers to protect them and the transactions from hackers.

20.4.1 Software Development

Retailers spend a tremendous amount of time and energy testing in-store merchandizing strategies to garner maximum attention and interest, building unique promotional units, offering eye-level shelf space, and displaying engaging end-caps and kiosks, especially during the holidays. E-commerce sites, nowadays, expose variable degrees of sophistication, functionality, and complexity. Most e-commerce sites offer lists of available products, usually organized in categories. For each product, a brief description, the price, and possibly an image are made available to e-customers; more information items may be included depending on the e-commerce domain.

E-commerce sites provide a different challenge to the developers and the investors. A website is the interface through which employees and customers interact with the organization. In that sense, it is analogous to a brick and mortar store. High website usability is, therefore, akin to a user-friendly and pleasant store environment and influences the website traffic.

Software development is an inherently risky endeavour. Complex systems are difficult to deliver without significant defects, even when adequate time and resources are available. E-commerce systems are even more challenging because they are subject to additional pressures that increase the risks associated with software delivery. It is important to explore the effects of design parameters, such as network delays, navigation strategies, and layout and information content, on the usability of websites and web-based systems.

A great number of e-commerce systems adopt the tiered and multitier approach to software architecture; the simplest variation integrates a web browser, a web application server, and a back-end database. If we think of these subsystems as residing in different architectural layers, then the boundary between each layer is an appropriate place for business logic and security measures such as firewalls and virtual private network routing. In terms of e-commerce applications, client-side programming is generally used for processing a sale transaction, providing information on businesses, and updating information in the back-end server systems. Client-side programming in carried out in different ways, which include using hypertext markup language (HTML), JavaScript, Java Applets, and ActiveX controls. Furthermore, one could also use plugins, which are applications of different sorts that are embedded in a web page for performing special functions. A very important factor in client-side programming is the downloading time as mentioned above. This is the time required to download a web page and its associated elements from the server side to the client-side over the Internet. Unlike company-internal software development, which proceeds according to schedules set within the organization, e-commerce systems are often developed under the pressures of Internet time. Here the expectation is that new technologies and solutions are rolled out every three to six months, under the credo 'evolve rapidly, or die'.

A modern e-commerce system brings together content and functionality on several levels. For example, client-side user interface involves the identification of abstract user interface objects and their interaction with other abstract user interface objects, as well as other objects in the system such as domain objects or data management objects. The abstract user interface objects specify what the user interface does, not how the user interacts with the system to perform tasks. The logical design involves, among other things, characterization of the flow of interaction. The flow of interaction requires an understanding of the possible sequences involved in the interaction between the application and the software. This is frequently embedded in the user's perception of how the tasks should be carried out. It could be considered as a delegation/monitoring/control paradigm of the task to the machine.

E-commerce applications are characteristic of session traffic and remember interactions. The flow of interaction is maintained with the help of sessions. A session is a sequence of individual requests of different types made by a single customer during a single visit to an e-commerce site. During a session, a customer can issue consecutive requests of various functions such as browse, search, select, add to shopping cart, register, and pay. To attract and capture a web consumer, a number of design and engineering techniques can be employed. In the case of a web application, identifying the locale and preferences of the customer is required in order to provide content that is tailored to their language and culture.

In the traditional approach to e-commerce, the Web uses the identified locale and customer preferences to present and express information in a personalized manner. Personalization and usability enable customers to feel almost as if they are interacting directly with representatives of the business during website visits. Attracting and capturing a customer are the peak challenges of e-commerce websites.

When considering the organization of electronic products and services, the point for the time being is to find a suitable form of customer–vendor cooperation with the help of a business model. Such forms of cooperation vary from open marketplaces with negotiable goods and value realized over tightly organized hierarchical networks to self-organized and loosely-coupled communities. Most of the complexity in e-commerce development is associated with the integration and testing of the various components in the software architecture.

Server-side applications run on a web server or a dedicated application server. Some of the popular server-side programming methods include the use of Java servelets, common gateway interface (CGI), and active server pages. A servlet is a small piece of server-side application, which can be viewed as the server-side analog of an applet. In a typical servlet application, a servlet-enabled web server receives an HTTP request from the client. The CGI programming allows a web client to pass data to a server-side application. Some of the prominent CGI languages include Perl, Apple Script, Unix Shell Scripting, and TCL. Active server page (ASP) is a scripting technique that runs on web servers rather than web clients. The server-side code written in ASP can be embedded in the HTML document, which allows one to insert it into web pages even though it is executed on the server.

E-commerce applications, ranging from B2C applications such as e-shopping to business to business (B2B) applications such as virtual market place, require one to connect to and access information from the back-end database system. For example, Java Database Connectivity (JDBC) is an application program interface (API) specification that provides a set of interfaces and classes to perform database-related operations.

20.4.2 Payment Systems

An e-commerce system becomes functional when payment functions are incorporated into the system. In the physical world, there are four main types of payment methods: cash, credit card, cheque, and credit/debit (funds transfer). Each of these payment methods has its own unique characteristics. To build a complete e-commerce system, we also need to implement these payment methods in cyber space. Secure electronic transaction (SET) is specially developed to provide secure credit card payment over the Internet. It is now widely supported by major credit card companies including Visa and Mastercard. It aims at satisfying the following security requirements in the context of credit card payment.

- Confidentiality: Sensitive messages are encrypted so that they can be kept.
- Integrity: Nearly all messages are digitally signed to ensure content integrity.
- Authentication: Authentication is performed through a public key infrastructure.

The authentication system of SET is based on the X.509 digital certificate framework and permits merchants, cardholders, and acquirers to verify the identities of each other by exchanging digital certificates. This is carried out by the SET protocol in four phases: initiation, purchase, authorization, and capture. Another Internet payment method is by using smart cards. Conventional credit cards and bank cards can be regarded as the first generation smart cards. Such smart cards are governed by the ISO 7816 standards, which are as follows.

- ISO 7816-1: Defines the physical characteristics
- ISO 7816-2: Defines the dimensions and location of the physical contact
- ISO 7816-3: Defines the electrical signals
- ISO 7816-4: Defines the file system and communication protocol
- ISO 7816-5: Mainly defines the numbering system

20.4.3 Value Creation

The value chain concept was developed and popularized in 1985 by Michael Porter, in *Competitive Advantage*, a seminal work on the implementation of competitive strategy to achieve superior business performance. Value chain describes the full range of activities which are required to bring a product or service from conception, through the different phases of production, delivery to final consumers, and final disposal after use.

The rising integration of world markets through trade has brought with it a disintegration of multinational firms, since companies are finding it advantageous to outsource an increasing share of their non-core manufacturing and service activities both domestically and abroad. Thus, electronic markets are built on value chains. Market participants change the intermediation in electronic markets by offering services in lieu of particular value-creating stages. With this sort of intermediation, the value chains are broken up by third-party providers. Lead firms increase complexity when they place new demands on the value chain, such as when they seek just-in-time supply and when they increase product differentiation. This enables companies to concentrate on core competencies and to delegate less important activities to collaborating partners. By streamlining the processes that generate the goods and services that customers value, fewer resources need to be expended, and the margin between customer value and the cost of delivery increases, improving a firm's profit margin [Feller 2006]. New value can be created in e-business by the ways in which transactions are enabled.

The term 'value' refers to the total value created in e-business transactions, regardless of whether it is the firm, the customer, or any other participant in the transaction who appropriates that value. Value chains are mapped and analysed using value chain analysis (VCA). Dell, for example, buys supplies from component manufacturers, hires workers, and accesses capital through the stock market. These inputs are transformed into a personal computer (PC), which is sold to a Dell customer. In the process, value is created whenever the the PC user's willingness to pay exceeds the opportunity cost of the resources used to provide this offering.

When the Municipal Corporation of Delhi (MCD) engineering department introduced an e-tendering facility, it marked a departure from the traditional, paper-based system of tendering. Figures suggest that the e-tendering process has achieved significant time- and resource-efficiencies. The e-tendering system itself has helped MCD take huge strides in the preliminary stages of procurement. For instance, over 30,000 tenders have been transacted through the e-tendering system by MCD, making it one of the world's highest volumes in numbers by any government organization. The e-tendering system has enabled MCD to post and amend tender documents, and view and compare bids online. This has substantially reduced administrative costs and eliminated the difficulties associated with paperwork.

20.4.4 Types of E-commerce

The emergence of e-commerce has created fundamental changes to the way business is conducted. These changes are altering the way in which every enterprise acquires wealth and creates shareholder value. The myriad of powerful computing and communications technology enabling e-commerce, allow organizations to streamline their business processes, enhance customer service, and offer digital products and services. There are a number of major areas in which information technology (IT) has impacted market structure, company conduct, and the nature of competition. Apart from some of these major issues dealing with operational costs, transactions, and structure, there are a number of other management issues that arise. Some of the common e-business models include the following.

Business to Consumer (B2C) As the name suggests, it is the model involving businesses and consumers. This is the most common e-commerce segment. In this model, online businesses sell to individual consumers. The basic concept behind this type is that the online retailers and marketers can sell their products to the online consumer by using clear data which is made available via various online marketing tools.

Business to Business (B2B) It is the largest form of e-commerce involving business. In this form, the buyers and sellers are both business entities and do not involve an individual consumer. It is like the manufacturer supplying goods to the retailer or wholesaler. Eor example, Dell sells computers and other related accessories online but it is does not manufacture all those products.

Consumer to Consumer (C2C) It facilitates the online transaction of goods or services between two people. Though there is no visible intermediary involved the parties cannot carry out the transactions without the platform which is provided by the online market maker such as eBay.

Peer to Peer (P2P) It is more than that just an e-commerce model. It is a technology in itself which helps people to directly share computer files and computer resources without having to go through a central web server. To use this, both sides need to install the required software so that they can communicate on a common platform.

M-commerce It refers to the use of mobile devices for conducting the transactions. The mobile device holders can contact each other and can conduct the business. Even the web design and development companies optimize the websites to be viewed correctly on mobile devices.

There are other types of e-commerce business models, such as business to employee (B2E), government to business (G2B), and government to citizen (G2C), but in essence they are similar to the above mentioned types. Moreover, it is not necessary that these models are dedicatedly followed in all the online business types. It may be that a business is using all the models or only one of them or some of them as per its needs.

20.5 Digital Markets

Markets play a central role in the economy and facilitate the exchange of information, goods, services, and payments. Markets have three main functions: matching buyers to sellers; facilitating the exchange of information, goods, services, and payments associated with a market transaction; and providing an institutional infrastructure, such as a legal and regulatory framework which enables the efficient functioning of the market. Internet-based electronic marketplaces leverage IT to perform these functions. The implementation and use of IT is enabling e-commerce and markets, which are characterized by multichannel transactions and relational processes which may span different activities of the value chain. Digital markets are the new mutants of organized market places. Digital markets have evolved into interconnected one-stop shops, providing specialized services with affiliate firms.

A digital marketplace is a virtual place, where buyers and sellers meet in order to openly negotiate over the goods offered and their prices. These websites facilitate the exchange of digital and material goods and services, as providers and buyers haggle among themselves over the price. Different service providers can offer and sell products and services on such websites.

Electronic business models have evolved from basic electronic procurement and e-commerce into more complex electronic market ecosystems. These digital market environments are characterized by rapid exchange of information within a virtual network of customers and suppliers working and evolving together to create and recreate value-added processes. Digital marketplaces compete with established, conventional marketplaces for patronage and trading volume. Business models for e-business can be characterized according to positioning, customer benefits, products and services, choice of business web, and profit model. Consumers have more possibilities to search and discover products on the Internet. Much-improved search engines (Google,

Yahoo!), comparison shopping (Buy.com, Kelkoo, PriceGrabber.com, Shopping.com), and discovery mechanisms such as collaborative filtering (used, for example, by Amazon.com and Netflix) enable consumers to hone in on products they could not find a decade ago in the aisles of shopping malls. Indeed, digital markets are now operating in a variety of industries, including industrial metals, chemicals, energy supplies, food and grain, construction, automotive and so on.

With publication costs decreasing, even small vendors at remote locations publicize the existence of their merchandize through context-based advertisement programs such as AdSense from Google. Establishing the sources of profit with which the company refinances itself constitutes a central element of the business model.

The 'Long Tail' article by Chris Anderson in the October 2004 *Wired* magazine describes the niche strategy of businesses, such as Amazon.com or Netflix, that sell a large number of unique items, each in relatively small quantities. The emergence of long-tail business is shaping the digital market. The rise of long-tail business will be a phenomenon that advertisers, marketers, and product designers must pay attention to. Instead of catering to a mass audience, advertisers will target much narrower, dedicated audiences with their messages. The ability to profile the increasingly fragmented channels and understand the rising market transparencies will become more important.

20.5.1 Infomediary

An infomediary is a neutral entity, a third-party provider of unbiased information; it does not promote or try to sell specific products in preference over other products. It could be a website that gathers and organizes large amounts of data and acts as an intermediary between those who want the information and those who supply the information. The role of the intermediary can be divided into two parts: putting potential trading partners in contact and providing each side of the trading relation with information about the other side. In the first function, clients of the intermediary, buyers and seller, value the diversity and breadth of offering of the intermediary. This can be because the higher the number of sellers present at the intermediary, the higher is the probability that one of them offers a low price, or because the higher the number of buyers coming to the intermediary, the higher is the probability that one will value the supplier's good highly.

For example, infomediaries, such as Autobytel.com and BizRate.com, offer consumers a place to gather information about specific products and companies before they make purchasing decisions. The other variant of infomediary collects the personal information from the buyers and markets that data to businesses. For example, BBH Infomediary provides a single connectivity point between financial service companies and their external service providers and internal systems. Infomediary business has powerful network-effect economies of scope. The more customer data the infomediary has, the more it can serve the needs of the customer base, and the more valuable its marketing services are to vendors. An infomediary service will command great customer loyalty, because switching costs are high. For example, Google is valued by consumers because it has a superior search algorithm that allows them to find information and products rapidly without fear that the search result may have been slanted toward one producer or another.

20.5.2 E-auctions

An electronic auction (e-auction) is a means of carrying out purchasing negotiations via the Internet. E-auctions are efficient mechanisms to allocate resources in e-commerce. In this context, online auctions, institutions where goods are traded on the Internet by the process of bidding and allocating through competition, are among the most widely studied and employed means of interaction. Auctions on the Web are often focused on certain products and customer segments and they supplement the traditional methods of selling.

Auction providers aim to open up a new sales channel in the hope of producing advertising effects. Auction scenarios consist of two clearly distinct components: protocols, and strategies. The protocol is set by the marketplace owner before execution and is publicly known to all the participants. In contrast, the strategy is determined by each individual participant and is typically private. Nevertheless, protocols and strategies are inextricably linked because the effectiveness of a strategy is very much determined by the protocol. Thus, a strategy that is effective for one protocol may perform very poorly or may even be invalid for other protocols.

An Internet auction house supports the seller with regard to an optimal price strategy, takes over marketing, and carries out credit assessment and payment handling. In Internet auctions, market participants need to be fully aware that they are dealing with legally binding transactions with quality standards that are customary to the industry. The operators of auctions and auction portals are intermediaries who increase the volume of transactions using innovative services. For example, eBay started at the end of the 1990s, acting as a low-key collectors of eBay exchange and junk market, but it has since developed into an important electronic marketplace. eBay is a neutral third party that provides customers (buyers) and content providers (sellers) with a platform for negotiations and transactions. The central value of eBay lies in the offer of a trust-building platform for the exchange of information, the opportunity for dynamic price discovery, and the impetus for service completion.

20.5.3 Agents

Agent technology is part of the science of artificial intelligence (AI). An agent, according to the *Concise Oxford English Dictionary*, is one who or that which exerts power or produces an effect. We all use agents in our daily life for a variety of tasks ranging from the mundane and boring ones to the highly complex ones. Computer programs can be used to carry out tasks that have been delegated to them by a human user and act in this respect as an agent, albeit not a human one. They are, therefore, commonly called software agents or intelligent agents. Some programs are called agents simply because they can be scheduled in advance to perform tasks on a remote machine. By using software agents, we relinquish part of the direct control we have and substitute it with indirect control through the agent.

A software agent is an autonomous entity capable of performing actions and interactions typically based on notions of beliefs and goals. Software agent is a kind of an umbrella term for software programs that display, to some extent, attributes commonly associated with agency. This includes attributes such as autonomy, authority, and reactivity. In software environments agents are mostly used for:

- solving the technical problems of distributed computing and
- overcoming the limitations of user interface approaches.

Providing software agents with a sense of their surroundings and the ability to react to changes in their environment is one of the great challenges of artificial-intelligence research. To this end, different agent architectures and multi-agent systems are being developed that enable agents to operate effectively in a given environment, thus providing room for indirect management. In order to meet their design objectives, agents must be able to operate without the direct intervention of humans and should be in control of their own actions and internal state

Agent systems can consist just one such agent or a collection of agents performing different tasks based on individual or common goals. Due to the individualistic characteristics described above, an agent system can collectively draw on further advantages, including mobility, dynamic sizing, and complex cooperation through negotiation. Software agents perceive the environment in which they operate either through sensory input or by using an internal model of their surroundings. The software agent uses the information regarding its surroundings as the basis for its decisions. An agent can reach a decision through a reactive process, a deliberative process, or a combination of both. The agent executes its decision using its actuators. Reactive agents are a class of agents that do not possess internal, symbolic models of their environments and do not use complex symbolic reasoning to accomplish their goals; instead they act in a stimulus-response manner to the present state of the environment in which they are embedded. With sophisticated communication, agents can interact to cooperatively achieve global tasks and goals. But this coordination needs more than sufficient shared semantics. It also requires planning and scheduling techniques to govern the order and partition of tasks. Mobile agents are agents that can move across different systems to perform specific tasks. Due to their mobile function, it is expected that mobile agents will play an increasingly important role in the future e-commerce system. For example, IBM's Aglet is widely used for developing mobile-agent-based e-commerce applications because it is lightweight and simple to use. Basically, the activities of Aglets are event-driven. That means that when a certain event occurs as detected by some 'listeners,' the specific actions for that event will be executed.

Agents can perform a variety of different roles. These are as follows.

- Monitoring auctions in order to keep the user informed of the latest progress of various auctions
- Analyzing the market situation and history in order to predict probable trends
- Deciding when, how many, and how much to bid in order to get the best deals

An agent in continuous double auctions (CDA) is a software package that can be viewed as a delegate of a user to achieve a good performance, which usually means a good profit. A CDA is a marketplace where there are agents selling goods (sellers) and agents buying goods (buyers). The sellers and buyers in one CDA market trade single-type (homogeneous) goods. An ask is the price submitted by a seller to sell a unit of goods. Similarly, a bid is the amount submitted by a buyer to buy a unit of goods. Sellers and buyers can submit their asks and bids at any time during a CDA. Although all the bidding strategies obey the basic protocol of CDAs, many differences exist in each specific rule and factor considered in the strategies.

To evaluate the performance of an agent in these auctions in a meaningful way, specific laboratorial market environments should be designed, which simulate markets in real life. Normally, two kinds of trading environments need to be considered. The simple and conventional kind is the static market environment in which the same traders are required to join every round and not allowed to leave the market freely. The more complex kind is a dynamic market environment. If an agent demonstrates a good performance in both kinds of laboratorial markets, the overall performance of the agent is demonstrated to be good.

20.5.4 Digital Supermarkets

Digital supermarkets are available on the Web. Digital supermarket selects suitable products and services from different producers, decides on the appropriate market segments, set prices, and supervise the fulfilment of the transaction. They buy products and services according to their discretion, and to a great extent set the purchase prices. They then determine the selling prices and discounts for the assortment of goods. They also control the sale and distribution of the goods. Amazon.com is an example. The products sold by this supermarket are, to a large extent, standardized. They are easily catalogued and described and visualized electronically in varying detail. The digital supermarket can be operated to a great extent by intelligent software agents. Simple agents advise the buyers and look for and valuate the desired products. Intelligent agents help the customers clarify their desires and select an attractive combination from the variety of offers.

An electronic distribution network transfers material products, intangible products, and services from the producer to the user. This model acts as a transportation company, electricity provider, financial service, courier service and postal service, communications network operator, or logistics company. The distribution network connects the producer of products and services with the purchaser or customer. It can, therefore, consist of a physical or digital network and distribution system.

In real-world electronic markets, the market participants do not always appear to form clearly definable web business models. Business websites often exhibit characteristics of several types of web business models.

ICT and Farming Community

Guided by the principle that an informed farmer is an empowered farmer, the Madhya Pradesh farmer welfare and agriculture development department took the initiative of bringing the benefits of ICT to the farming community in Madhya Pradesh. The project included development of a web-based Application Software Monitoring and Management Information Decision Support System (MIDSS). The project included setting up the requisite infrastructure, undertaking training of the department staff, and creating backlog of data entry of previous years data. The web-based portal offers three types of services: government to consumer, government to business, and government to government. The government to government segment, provides static information on different aspects of agriculture such as package practices of different crops, farm management practices including crop management technology, water and soil conservation technology, pesticides disease management technology, etc.

'Kisan Call Center' was established in September 2008 with a toll free number where in farmers could dial in and get information from subject-matter experts on agriculture related queries. The call centre currently has around 30 SMEs on horticulture and veterinarians.

Source: http://www.mpkrishi.org/

20.5.5 E-commerce and E-procurement

E-procurement systems are usually integrated with corporate enterprise systems and organizational intranets. They typically consist of two parts. One part resides on the top of the company's intranet behind its firewall, where employees can search and place order for desired supplies. The purchase orders, after they have been approved and consolidated, are sent out to a third party, usually a neutral electronic marketplace. This is where the second part of the e-procurement system resides. At the electronic marketplace, these orders are transformed into various formats according to different protocols so that they can be received and processed by different suppliers. E-procurement refers to all of the connective processes between companies and suppliers that are enabled by electronic communication networks. Enterprise resource planning (ERP) systems are used for the information technical support of procurement in companies. Different models of e-procurement are present in the marketplace. They have been developed according to who controls the marketplace. In the sell-side market model, the supplier provides the purchase software and an electronic catalogue. It also provides the entire business logic for the procurement process, including the product catalogue in an information system, for example, Dell which supplies computers and peripheral devices. In the buy-side market model, the buyer must run and maintain the appropriate software, together with extracts from the product catalogue. The supplier is only responsible for the content management, and regularly transmits changes in the product catalogue.

In the electronic marketplace for e-procurement, the required software solutions and the catalogues are operated by a third-party provider. The third-party provider can uniformly display and describe the products with its software solution. An example is the auction platform Fastparts (http://www.fastparts.com), which deals with standardized electronic components.

20.5.6 E-commerce and E-contracting

Contracts are very much a part of living in a society. In the broadest sense, a contract is simply an agreement that defines a relationship between one or more parties. A commercial contract, in simplest terms, is an agreement made by two or more parties for the purpose of transacting business. Because a contractual relationship is made between two or more parties who have potentially adverse interests, the contract terms are usually supplemented and restricted by laws that serve to protect the parties and to define specific relationships between them in the event that provisions are indefinite, ambiguous, or even missing. When one party enters into a commercial contract with an unfamiliar and distant party across a country border, a contract takes on added significance. The creation of an international contract is a more complex process than the formation of a contract between parties from the same country and culture. In a cross-border transaction, the parties usually do not meet face-to-face, they have different societal values and practices, and the laws to which they are subject are imposed by different governments with distinct legal systems. These factors can easily lead to misunderstandings, and therefore the contracting parties should define their mutual understanding in contractual, and preferably written, terms.

For contracts made between parties within the same country, missing or indefinite terms may be filled in by local laws or practices. The rationale is that the parties are likely intended to follow the local laws and practices with which they were familiar. If the parties are from different countries, their intentions cannot be so easily implied because they herald from different legal systems and no doubt utilize dissimilar business practices. For this reason, it is essential that international contracts spell out in definite terms the various rights and obligations of each party.

Contract terms tend to differ depending on whether the parties intend to establish a long-term relationship. In a one-time transaction, the terms of the sale are established and in the absence of a major problem, the parties rarely renegotiate. If the parties intend to work together for a long time, they may still set up a contract for a single transaction, but generally the contract will allow the parties to alter terms as necessary to make the performance of the contract profitable for both. Alternatively, the parties may enter into an agreement that establishes the basic parameters of all transactions that will subsequently occur between them.

In electronic markets, business transactions can be carried out without the business partners ever coming into physical contact with one another. There is a need for a suitable software system to record the mutual agreement between the market participants over the exchange of goods or services, to facilitate the signing of contracts in a legally binding manner, and to archive for monitoring purposes.

Before market participants enter into electronic negotiations, the participants would like to verify the identity of the opposite party. This is done through sites or institutions which issue legal identity certificates for individuals. These so-called certification authorities or certification sites guarantee that the market participant is in fact the person he/she claims to be. This is done by a certification authority when it issues an electronic certificate based on identification documents from the market participant. The certificate issued by a certification authority for the market participant is limited and can be brought into the contract negotiation process.

E-contracting supports the electronic negotiation process by creating standards for the production of a legal document with a digital signature. When an electronic negotiation process is successful it results in an electronic document called electronic contract. This document legally binds the contracting parties to their tasks and responsibilities.

A digital signature or electronic signature is a procedure that guarantees the authenticity of a document. A certification site (certification authority or trust centre) is an institution which certifies the allocation of public signatures to real people. It is responsible for the production, issuance, and administration of certificates. In the process, it must reliably identify the applicant.

Secure electronic transaction is considered one of the most secure protocols in e-payment. It is based on a public key cryptosystem. The SET protocol was developed by some of the large credit card companies.

20.5.7 E-retailing

Electronic retailing (e-retailing) essentially consists of the sale of goods and services. Traditional retailing essentially involves selling to a final customer through a physical outlet or through direct physical communication. This normally involves a fairly extensive chain starting from a manufacturer to a wholesaler and then to the retailer who through a physical outlet has direct contact with the final customer. However, e-retailing permits the customer to shop anywhere around the globe without being restricted to his local vicinity. Different models of e-retailing are being practised.

20.5.8 E-commerce and E-marketing

The Internet is continuing to grow rapidly and seamlessly across borders and into an online world already inhabited by over a billion customers. Given its scale and the benefits it offers to these customers and businesses, it is a big part of the future of all businesses. Marketing has pretty much been around forever in one form or another. The methods of marketing have changed and improved, and we have become a lot more efficient at telling our stories and getting our marketing messages out there. Electronic marketing (E-marketing) is the product of the meeting between modern communication technologies and the age-old marketing principles that humans have always applied. E-marketing, also often referred to as online marketing, assumes that the classical rules of marketing hold. Very simply put, e-marketing refers to the application of marketing principles and techniques via electronic media and more specifically the Internet.

E-marketing is a subset of e-business that utilizes electronic medium to perform marketing activities and achieve desired marketing objectives for an organization. Internet marketing, interactive marketing, and mobile marketing are all forms of e-marketing. The well-known AIDA (attention, interest, desire, action) formula or modified forms of it are still valid: attention must be attracted to a website, interest aroused, desire created, and sales actions triggered.

Internet marketing allows the marketer to reach consumers in a wide range of ways and enables them to offer a wide range of products and services. E-marketing includes, among other things, information management, public relations, customer service and sales. With the range of new technologies becoming available all the time, this scope can only grow.

For example, Google has built a billion dollar business simply by charging for mouse clicks, some costing up to 50 dollars. Perhaps it is even weirder when we consider that in future most of the business dealings will be wireless. For example, Cisco systems announced many years ago that they will no longer do business with suppliers who cannot take orders over the Web. The Web is effective for the following tasks and return on investment (ROI) models.

- Brand building
- Lead generation
- Online sales (e-commerce)
- Customer support
- Market research
- Word of mouth, word of Web, buzz marketing
- Content services
- Web publishing

E-marketing is the use of Internet as a media to market and promote your business globally. In today's world, the most cost-effective method of global business promotion is e-marketing. While the fundamental principles of traditional marketing still apply to e-marketing, the power of e-marketing stems from some of its unique advantages. Internet advertising has recaptured the imagination of marketers, who see an enormous potential to raise the profile of their brands through vehicles such as paid search and online video. Internet (digital) marketing is the practice of promoting products and services using digital distribution channels to reach consumers in a timely, relevant, personal, and cost-effective manner. The digital world has developed faster than the tools needed to measure it. Critical contributors to the broadening mission of marketers include the Internet and evolving distribution models, which are profoundly changing the way consumers research and buy products. In addition, third parties such as bloggers and the creators of user-generated media are having a greater influence on corporate reputations. In categories as diverse as electronics, financial services, and health care, consumers increasingly ignore push marketing, preferring instead to use the Internet to research products and decide which ones to buy.

The proliferation of distribution touch points and the more rapid growth of the low and high ends of the market at the expense of the middle are forcing marketers to take low-cost, time-saving, 'facts-only' sales approaches and, at the same time, higher-value, more service-oriented approaches, often through alternative distribution channels.

Websites receive Internet traffic or online visit of customers under two broad categories. One kind of traffic is the organic traffic and the other kind of traffic is the paid website traffic. Organic traffic, as the name implies, is traffic that comes to your website naturally and without being driven there by a specific marketing campaign. In essence, Website visitors are there because they found the site and thought it had something they wanted. And like anything organic, organic traffic is not there instantly; it takes time and nurturing to grow into something healthy and with longevity.

Paid website marketing has the advantage of driving traffic immediately to your website. This is great for launching a site, or for a special promotion. Popular paid options include (but are not limited to):

- newspaper magazine and TV ads,
- purchasing of banner ads on other websites,
- launching a search engine marketing (SEM) campaign, and
- distributions of mass e-mails and press releases.

A secondary benefit of paid website marketing is that when done properly, it can help lay the seeds of organically generated traffic.

20.6 Electronic Data Interchange

Paper-driven processes are being re-engineered to capture the benefits of doing business electronically. Businesses are implementing electronic commerce to meet the needs of an increasingly competitive world. Data Interchange Standards Association (DISA), the standards body advances the foundation of electronic trade and commerce by supporting and promoting standards used for business-to-business (B2B) data exchange. Providing administrative and technical support to the Accredited Standards Committee (ASC) X12, DISA helps individuals and organizations improve business processes, reduce costs, increase productivity, and take advantage of new opportunities. Electronic Data Interchange (EDI) is the computer-to-computer exchange of business data in standard formats. The National Institute of Standards and Technology in a 1996 publication defines Electronic Data Interchange as "the computer-to-computer interchange of strictly formatted messages that represent documents other than monetary instruments. The EDI can be formally defined as

'The transfer of structured data, by agreed message standards, from one computer system to another without human intervention'. The development of Electronic Data Interchange and Electronic Fund Transfer (EFT) can be said to be the beginnings of the digital exchange of accounting information among trading partners. Electronic Data Interchange and EFT both involve exchange of data electronically and sound very similar to e-commerce.

In EDI, information is organized according to a specified format set by both parties, allowing a 'hands-off' computer transaction that requires no human intervention or rekeying on either end. All information contained in an EDI transaction set is, for the most part, the same as on a conventionally printed document.

20.6.1 Need for EDI

Electronic Data Interchange is a way of business life. It is based on the principle of trust and contractual obligations. The EDI documents generally contain the same information that would normally be found in a paper document used for the same organizational function. Organizations that send or receive documents between each other are referred to as 'trading partners' in EDI terminology. The trading partners agree on the specific information to be transmitted and how it should be used. Once Evidence Act and other laws of the land recognize EDI transactions and provide for the same by fast settlement of disputes, it should be possible to do away with requirements for paper documentation, i.e., there would be no necessity to submit invoices, packing list, B/L, etc. in paper.

20.6.2 EDI Standards

The EDI standards were designed to be independent of communication and software technologies. The EDI standards are developed and maintained by the Accredited Standards Committee (ASC) X12. The standards are designed to work across industry and company boundaries. The Accredited Standards Committee (ASC) X12 develops EDI standards and related documents for national and global markets.

The EDI, if properly implemented, can streamline supply chain management, reduce labour costs and errors, increase processing speed, and accelerate cash flows. The primary problems with EDI are as follows.

- The formats are highly structured and sometimes proprietary.

- The structure of the data format limits the amount of information in the EDI messages.
- Specialized software that is expensive to install and maintain is required.
- It offers few financial benefits to suppliers.

20.7 Electronic Data Security

The modern economy and the future wealth and prosperity of industry and commerce rely increasingly on the exchange of data and information, in electronic form, between business partners. The speed and reliability of the information exchanged coupled with the spread in the distributed use and application of IT are increasingly affecting the competitiveness of businesses and international trade. Electronic information exchanged in this way is growing in volume because of the increasing number of business partners that may be involved and the numerous documents that need to be exchanged.

Data security is important because there is interesting information inside a company for outsiders to make money with. So the risks through the years of computing became larger, because more people get and use a computer for all kind of aims. Organizations do not want any risks, because they have an impact on their profit. Also there would be a chance that some other organization earns moment with a product that they have made. So organizations want no risks, but that is impossible, so they want to minimize the risks. A risk depends on the impact, how great the chance would be that the risks are happening, and corresponding damage. (Refer to the sections on risk management elsewhere in this book)

In general, moving data off premises increases the number of potential security risks, and appropriate precautions must be taken. Furthermore, although the name 'cloud computing' gives the impression that the computing and storage resources are being delivered from a celestial location, the fact is that the data is physically located in a particular country and is subject to local rules and regulations. For example, in the US, the US Patriot Act allows the government to demand access to the data stored on any computer; if the data is being hosted by a third party, the data is to be handed over without the knowledge or permission of the company or person using the hosting service. Since most cloud computing vendors give the customer little control over where the data is stored (e.g., Amazon S3 only allows a customer to choose between US and European Union (EU) data storage options), the customer has little choice but to assume the worst and that unless the data is encrypted using a key not located at the host, the data may be accessed by a third party without the customer's knowledge. Large cloud computing providers with data centres spread throughout the world have the ability to provide high levels of fault tolerance by replicating data across large geographic distances.

One must assume that EDI may be used across a wide-ranging messaging continuum covering different types of network services and various value-added application platforms [Humphreys 2004]. This range of communications provision will reflect a need for different levels and types of security to protect these EDI messages. The EDI components chain and the emerging EDI enabling technologies to support this chain are migrating the proprietary/direct-link type of offering to the Open-EDI approach based on international standards. The EDI security appears at several interrelated stages of system technology. The basic security objectives that may need to be met at each stage are those of authentication and integrity, non-repudiation, access control, availability, audit, and accountability. These objectives will need to be satisfied by both logical and legal controls and procedures, which are supported by a range of technologies, tools, and standards. Protection in an EDI messaging environment is essentially concerned with the non-repudiable submission, delivery, and receipt of messages in a way that preserves the integrity, confidentiality, and availability of the messages being communicated. The current messaging standards provide the means of applying security mechanisms to meet different types of security objectives and levels of security.

SUMMARY

This chapter introduced the process of the organizational transformation to globalization and e-business environment along with related concepts, stages, and activities. E-commerce refers to a wide range of online business activities for products and services. It also pertains to any form of business transaction in which the parties interact electronically, rather than by physical exchanges or direct physical contact.

While some use e-commerce and e-business interchangeably, they are distinct concepts. In e-commerce, information and communications technology (ICT) is used in inter-business or inter-organizational transactions and in business-to-consumer (B2C) transactions.

E-commerce does not refer merely to a firm putting up a website for the purpose of selling goods to buyers over the Internet. For e-commerce to be a competitive alternative to traditional commercial transactions and for a firm to maximize the benefits of e-commerce, a number of technical as well as enabling issues have to be considered. Although e-commerce developers are under pressure to produce working systems quickly, the use of an incremental prototyping model helps to improve the quality of delivered systems by uncovering defects and performance problems earlier in the development cycle.

Key Terms

Electronic cheque Customers pay for merchandise by writing an electronic cheque that is transmitted electronically by e-mail, fax, or phone. The 'cheque' is a message that contains all of the information that is found on an ordinary cheque, but it is signed digitally, or indorsed. The digital signature is encoded by encrypting with the customer's secret key. Upon receipt, the merchant or 'payee' may further indorse by encoding with a private key. When the cheque is processed, the resulting message is encoded with the bank's secret key, thus providing proof of payment.

Merchant account A merchant account is a relationship between a business (i.e., a merchant) and a merchant bank which allows the retailer or merchant to accept credit card payments from customers.

Micro transactions or micro payments Micro transactions are transactions of tiny amounts—a few cents or a few dollars, typically made in order to download or access graphics, games, and information.

Prototyping The creation of an initial version of a software system, often with limited functionality, to support problem analysis, requirements definition, interface design, etc.

Review Questions

20.1 What is e-business? Differentiate it with e-commerce.
20.2 What are the types of e-commerce?
20.3 Discuss the need for e-commerce security.
20.4 Prepare a report on your understanding of EDI as practised today.
20.5 Discuss digital markets and their implications on information economy.
20.6 Discuss eauctions.

Projects

20.1 Refer to the case in Chapter 1. Prepare a draft proposal to enable ABC Unlimited to move towards retail ecommerce practices.

REFERENCES

Bean, James 2003, *Engineering Global E-Commerce Sites*, Morgan Kaufmann Publishers.

Feller, Andrew, Dan Shunk, and Tom Callarman 2006, *Value Chains Versus Supply Chains*, BPTrends.

Humphreys, Ted 2004, *Electronic Data Interchange (EDI) Messaging Security*, Information Security.

In Lee 2009, *Emergent Strategies for E-Business Processes, Services, and Implications: Advancing Corporate Frameworks*, Information Science Reference.

Khosrow-Pour, Mehdi 2006, *Encyclopedia of E-Commerce, E-Government, and Mobile Commerce*, Idea Group.

Kotler, Philip 2003, *Marketing Insights from A to Z, 80 Concepts Every Manager Needs To Know*, John Wiley.

Salazar, Angel J, and Steve Sawyer 2007, *Handbook of Information Technology in Organizations and Electronic Markets*, World Scientific.

CHAPTER

21 Information Systems and Business Systems

Observe always that everything is the result of change, and get used to thinking there is nothing Nature loves so well as to change existing forms and to make new ones like them.
–Marcus Aurelius
Roman Emperor

Learning Objectives

After reading this chapter, you should be able to understand:

- the nature of information systems domain and the need for information systems
- the need for ERP systems
- the differences between the ERP systems
- the differences between the commercial ERP systems
- and differentiate between CRM systems
- the need for supply chain systems

21.1 Information Systems

Corporations today rely on Information Technology (IT) to realize their business value. Business value is just one output of the collection of processes through which businesses today try to maximize the age-old equation of profit equals revenue minus expenses. It is a multidimensional output and different observers apply different weights to different dimensions at different times.

Information technolog has become a matter of serious concern for managements today. The IT department has evolved from a narrowly focused data processing element of the accounting department to a function that supports and, in many cases, drives nearly every area of a company. The principal driver behind this remarkable, rapid creation of a vibrant, sophisticated, and enormous industry is the quest for business productivity improvement.

The proliferation of computers and the subsequent mechanization has introduced dramatic changes to business process and has introduced the concept of information systems (IS). As computer technology has advanced, industries world over have become increasingly dependent on computerized information systems to carry out their operations and to process, maintain, and report essential information. The concept of multinational sales was not a viable option until modern transportation and communication systems were developed to a level that made dealing with large distances possible.

The goal of a company is to increase the wealth of its shareholders and benefit its employees. A primary method of doing this is by maintaining and increasing its ability to grow. One way to continue to grow is to expand into new and emerging markets, which entails branching out into the international marketplace. For example, a multinational manufacturing company that does business in a number of countries it has a substantial commitment of its resources in international business; it engages in international production in a number of countries, and it has a worldwide perspective in its management. By branching out into multiple countries, the company continues to grow and reduces the risks it faces if conditions in one country changes drastically. This strategy requires a good communication mechanism at the the back-end like an enterprise resource planning (ERP) system. For example, today Robert Bosch GmbH differs from many global companies not only in its diversity and breadth, but also in its corporate structure and mission. Bosch employs approximately 238,000 people, more than half of whom are employed in facilities outside of Germany. Bosch's global sales are generated in nearly 250 subsidiaries and associated companies in more than 50 countries. Bosch operates roughly 260 production sites worldwide, of which nearly 200 are located outside Germany.

The concept of an information system began to emerge around the year 1960, and although it may be considered a well-established concept it is difficult to define it precisely. Information systems are of vital importance to many organizations across a wide range of sectors of the economy. An information system is a system designed to support operations, management, and decision-making in an organization. It is a system that facilitates communication among its users (Figure 2.1). Typically, these information systems are taking care of some specific part of the business. Information systems must, however, be able to communicate with each other, because strategic decisions usually require information from several functional areas within an organization.

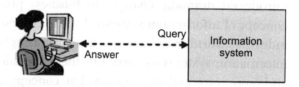

FIGURE 21.1 Information System

Information systems are strategic to the extent that they are used to support or enable different elements of an organization's business strategy. This fluent alignment of information systems to serve business is one of the biggest concerns among managers. The business data is managed through the information systems and stored in the databases. Organizations need to be able to process this data and use the information efficiently in order to succeed.

There is a common agreement that the landscape for business and information technology has been changing in the last decade. The widespread use of information technology and the connectivity granted by the Internet provide new methods for interconnecting organizations and customers, enabling transaction costs across organization boundaries to be driven down and making the traditional departmental or hierarchical structure less attractive either from a strategic or economical point of view.

Information systems are changing the industry structure and in doing so has altered the rules of competition. The growing competitive pressure is forcing organizations to rethink continuously the ways in which they do business and the type of business that they do, leading to a different way of managing. Competitiveness favours those who spot new trends and act on them. For example, British Petroleum deployed an information system strategy and transformed from a typical public sector enterprise to a virtual team network and emerged as an efficient oil enterprise globally. Proctor & Gamble adopted a business strategy driven information system and was able to reduce channel inventory with improved supply to logistics services.

21.1.1 Process

There are many different types of work people do across the world. In all the cases one tangible/intangible thing is produced or modified. This is called a case. Each case involves a process being performed. A process is a structured, measured set of activities designed to produce a specified output for a particular customer or market. It implies a strong emphasis on how work is done within an organization. A process consists of a number of tasks that need to be carried out and a set of conditions that determine the order of the tasks. A process can also be called a procedure. A task is a logical unit of work that is carried out as a single whole by one resource. A resource is the generic name for a person, machine, or group of persons or machines that can perform specific tasks. Two or more tasks that must be performed in a strict order are called a sequence. There are also tasks that can be performed in parallel. A task can also be defined as a process that cannot be subdivided any further, that is, it is an atomic process. A task is, therefore, an atomic process for the person defining or ordering it, but for the person carrying it out it is often a non-atomic one. Processes are divided into three categories: primary, secondary, and tertiary. Primary processes are those that produce the company's products or services. They, therefore, are known also as production processes. They deal with cases for the customer. As a rule, they are the processes that generate income for the company, and are clearly customer-oriented. Secondary processes are those that support the primary ones. They, therefore, are also known as support processes. Tertiary processes are the managerial processes that direct and coordinate the primary and secondary processes.

A business process is a series of steps designed to produce a product or a service. A business process is a collection of related structural activities that produce a specific outcome for a particular customer. It is a set of logically related business activities that combine to deliver something of a value to a customer. A business process not only processes the value for the customer but also represents real-time integration of the processes of a company with those of its suppliers, business partners, and customers.

Processes may be classified according to two dimensions: degree of mediation and degree of collaboration. The degree of mediation refers to the sequential flow of input and output among the participants' functions in a business process. A process at a high degree of mediation involves a large number of intermediate steps, performed in various functions that contribute indirectly to the process outcome. A process at a low degree of mediation has several functions that contribute directly to the process outcome without the mediation of sequential steps. The degree of collaboration dimension is related to the degree of collaboration between functions through information exchange.

The lifecycle of a case is defined by a process. Because each case has a finite lifetime with a clear beginning and end, it is important that the process also conforms with this. Processes are generally identified in terms of beginning and endpoints, interfaces, and organization units involved, particularly the customer unit. So each process also has a beginning and an end, which respectively mark the appearance and completion of a case. High impact processes will have process owners. Processes are currently invisible and unnamed because people think about the individual departments more often than the process with which all of them are involved. So companies that are currently used to talking in terms of departments such as marketing and manufacturing must switch to giving names to the processes that they have such that they express the beginning and end states. Business processes represent a new approach to coordination across the firm. A core business process usually creates value by the capabilities it gives the company for competitiveness. A limited number of such core business processes can be identified in any company, and enhancing those processes can lead to business improvement. For example, supply chain processes in general span the areas of demand planning, order fulfilment (sales), distribution, production, and procurement. For example, part of the structure that keeps ITC's gigantic IT machine moving is that each business unit (BU) creates a multifunction planning taskforce—which includes IT managers—to undertake an annual business planning exercise. The yearly huddle introduces a comprehensive IT plan and roadmap as part of the division's overall business plan.

The designer of an executable process must be aware of both business and technical perspectives. From the business perspective, it is important that a process (i.e., business process) is modelled to closely follow the business activities, events, and message exchanges.

21.1.2 Process Reengineering

Reengineering is the fundamental rethinking and radical redesign of business processes to achieve dramatic improvements in critical, contemporary measures of performance such as cost, quality, service, and speed. The concept of reengineering traces its origins back to management theories developed as early as the nineteenth century Davenport93. The purpose of reengineering is to 'make all business processes the best-in-class'. Frederick Taylor suggested in the 1880s that managers use process reengineering methods to discover the best processes for performing work, and that these processes be reengineered to optimize productivity.

The key question for organizations is how the reengineering significantly improves the cost, quality, service, and speed simultaneously. To accomplish this, companies develop new processes to produce the results important to customers. They look for ways to become more flexible and responsive. When the environment is fairly stable, work has to be divided into simple, repetitive tasks for a largely unskilled, uneducated workforce to create efficiencies of scale. Layers of supervision and controls are required to link these simple tasks together connecting people who perform complex, multi-disciplinary tasks by a general understanding and agreement on vision and processes. These phases are collectively called the reengineering process and it will allow the organizations to grow at a rapid rate. Hence, reengineering is a holistic solution for companies which require radical redesigning for quantum improvement in its performance.

21.1.3 Business Process Reengineering

Business modelling is already a widespread term. The ability to utilize advanced technology for modelling, analysis, and simulation of various aspects of ever-changing businesses has made a significant contribution to the way businesses are planned and operated these days. It is important for effective business decision-making to have clear and concise modelling that allows the extraction of critical values from business processes and specifies the rules to be enforced accurately.

Business process reengineering (BPR) is the popular term for optimization of organizational processes and structures following the introduction of new information technologies into an organization [Boudreau 1996]. Business Process Reengineering evolved from the experiences of the US-based companies in the 1980s. These companies radically changed their processes by applying modern information technology innovatively, in pursuit of change management and improved performance. It is one of the fundamental steps undertaken prior to ERP implementation. Business process reengineering provides scope for problem identification and also solutions to implement the successful business operations. There are many new elements in BPR such as extensive use of IT and new perspectives on organizational structure. It is the analysis and redesign of workflow within and between enterprises. It is a methodology that promotes change and introduces new processes and new styles of working. So certain elements will be required to make change possible. These elements are known as enablers and may be defined as elements that act as vehicles for processes to change. Business Process Reengineering is not a one-time solution, rather it is a continuous process of screening and reengineering/modification to bring excellence in the service delivery to stakeholders.

The business case for a process framework approach has been well stated by management theorists for years. Businesses are about activities, and optimizing these activities is a primary challenge. The creation and use of process frameworks to understand a large and complex set of activities is currently the standard industry practice. Framing IT activities in terms of business process leads to increased credibility with business stakeholders. Metrics-based process management means the use of management concepts familiar to business partners—leading and lagging indicators such as critical success factors and key performance indicators and so forth.

21.1.4 Classical Business

A business is a complex system. It has a specific purpose or goal. A business consists of a vertical organization of departments with defined functions for each department. Some of the functions overlap with other departments. All the functions of the business interact to achieve the goal. A model of the business system describes the goals and internal structures.

The following entities are generally identified as a part of the business system.

Resources They are objects that are used in the business. They are consumed, produced, transformed, or used by the business process. A resource can be a physical resource. This is an entity that occupies a volume of space. Examples include raw materials, parts, products, etc. A resource can be an abstract resource, an information object, or human beings.

Processes Processes can be classified into the following along two dimensions.

- Degree of mediation: This refers to the sequential flow of input and output among the participant's functions in a business process. A high degree of mediation involves a large number of intermediate steps, performed in various functions that contribute indirectly to the process outcome. A low degree of mediation implies the direct contribution of several functions to the process outcome without the mediation of sequential steps.
- Degree of collaboration: This defines the degree of collaboration through information exchange.

Goals Goals describe the desired state of a business process. They motivate activities leading to state changes along the desired direction. A goal can be broken down into sub-goals. This is called goal modelling.

Rules A business rule is a statement that can control a business process. Business rules define constraints conditions and policies to guide the business process. Rules ensure that the business is run according to predefined external or internal restrictions or goals. The rules control the business. There are three types of business rules. These are:

- Derivations
- Constraints
- Existence

21.1.5 Enterprise Application Integration

Many of the operational systems in use today, say, in manufacturing, include all of the processes from managing inventory and the shop floor to providing strategic information for planning a business's future. This includes making informed decisions on product lines which are likely to succeed and which should be discontinued, based on trend analyses and other information these systems can provide.

Enterprise application integration is an emerging generation of integration software that addresses more effectively the need to integrate both intra- and inter-organizational systems. In doing so, it securely incorporates functionality from disparate applications. As companies strengthen their relationships and collaborate at an inter-organizational level, the chain itself gains more links and therefore, increases management and coordination efforts. An enterprise is no longer viewed as a single corporation, it is a loose collection of trading partners that can contract with manufacturers, logistics companies, and distribution organizations. Therefore, a comprehensive integration of business processes and both intra- and inter-organizational applications is needed to support long-term co-ordination, survival, and growth. Such integration increases the automation of business processes and significantly reduces manual tasks, redundancy of data, and functionality. It combines traditional integration technologies (e.g., database-oriented middleware) with new enterprise application integration (EAI) technologies (e.g., adapters, message brokers) to support the efficient incorporation of information systems. Thus, application integration results in supporting data, objects, and processes incorporation as well as custom applications, packaged systems, and e-business solutions integration.

Businesses see tighter integration with customers, suppliers, and partners as a way to alter the marketplace to their advantage and to realize the same benefits of traditional process synchronization, streamlining and integration, albeit applied to processes that extend beyond company boundaries. The web technology utilized has global reach by default and, furthermore, the security implications of opening up internal systems to third parties are global. Businesses want to exploit improvement in international transportation and relaxation of trade regulations to regionalize or globalize selected business processes, either in the interests of internal efficiency or better meeting demands of global customers or partners.

Teamlease Switches to ERP

Teamlease services is a Rs 700 crore staffing company with over 650 clients. It operates from 19 geographical locations in India and provides a range of temporary and permanent manpower solutions from establishing a co-employment relationship with clients to undertaking turnkey and recruitment mandates for permanent fulfillment. In order to meet the current demands of business, Teamlease moved into an ERP solution to improve its ability to handle over 20,000 charts of accounts, 2,000 invoices and over 1,00,000 payments. The move to ERP enabled Teamlease with improved reporting functionalities across multiple departments in zero downtime, resulting in a more scalable transaction management system. The switch enhanced integration between payroll and financial systems giving Teamlease a holistic view of their business.

Source: http://www.cio.in/case-study/teamlease-switches-erp-counter-tally-woes

---

To support efficient and flexible supply chain organizations need to integrate their applications on enterprise and cross-enterprise level. Through the integration of internal and external business activities, companies are continuing to find ways to further improve their efficiency and streamline key business processes. To accomplish this, organizations have not only coordinated functional staff and business processes but have sought to integrate enterprise applications into systems that can link members of the supply chain together via data, information, and knowledge.

First generation EAI focused on data integration and stressed on converting data between disparate formats. Enterprises looked for integrating packages at the application level. Due to the popularity of packaged applications, most of the enterprises started investing huge revenue on buying and customizing these packages as they wanted to implement the best industry practices and processes in their organization.

The dispersed application packages were to be integrated. The various levels of integration are as described below.

21.1.6 Data-level Integration

Enterprise application integration technologies that provide data-level integration services are designed to help decoupling systems by allowing each application to access its own data store. This approach mainly focuses on converting data between disparate formats highlighting low-level aspects such as the byte format of data, character encoding, etc. But these integrations lead to redundant data storages across the organization. It was also a huge overhead for the system to keep the data stores in sync if changes are frequent.

21.1.7 Function-level Integration

Function-level integration involves one application programmatically invoking the functionality of the other without using an intermediate software. Here the applications are modified to exchange and understand each other's messages. This makes the applications tightly coupled.

Here the integration is not flexible to manage changes from either side of the applications. Any change may lead to recoding, retesting, and redeployment of the applications, which involve a lot of cost overhead.

21.1.8 Brokered Integration

Application are integrated through an intermediate application called broker. The applications are permitted to exchange messages between them through the broker. It is the broker's responsibility to explicitly transform incoming messages and route them to the destination applications. It basically focuses on the connection and routing of messages between the applications. It manages the state of data flow but does not manage the business activity flow.

It was not too late before the enterprises realized that these packages created silos of information and business rules across the organizations, which needed to be integrated.

A business process can be defined as a set of interrelated tasks linked through a number of decision activities. Business processes have starting points and ending points and they are repeatable. Moreover, business processes encapsulate the knowledge of operations and services provided by an organization.

Today's market environment feature customer control as the primary force. In today's context, customers are assuming control. Customers control the sales by telling the suppliers what they want, when they want, and how much they are willing to pay. Customers have power over the salesman due to easy access to more information.

21.2 Workflow Automation

Workflow is usually regarded as the computerized facilitation or automation of a business process, in whole or in part. It consists of a coordinated set of activities that are executed to achieve a predefined goal. Workflow management aims at supporting the routing of activities (i.e., the flow of work) in an organization such that the work is efficiently done at the right time by the right person with the right software tool. It focuses on the structure of work processes, and not on the content of individual tasks. Individual tasks are supported by specific application programs. Workflow management links persons to applications in order to accomplish the required tasks.

21.2.1 Workflow Management System

A workflow system is an information system based on a workflow management system (WFMS) that supports a specific set of business processes through the execution of a process specification. A process specification (or workflow schema) describes a type of process that can be interpreted as a template for the execution of concrete workflow instances. A workflow engine interprets the workflow description, which is represented in a workflow programming (or modelling) language.

Modelling processes in a real-life situation is often done in a less formal language. People tend to understand informal models easily, and even if models are not executable, they can help a great deal when discussing process definitions. However, at some point in time, these models usually have to be translated into a specification that can be executed by an information system. This translation is usually done by computer scientists, which explains the fact that researchers in that area have been trying to formalize informal models for many years now. A data model is a map for organizational data that includes entities and relationships that are important to the business. Managing information is important in an organization for cost-effective development and operation of information systems. Poorly implemented systems frequently result from a lack of efficient data management. Developing an understanding of the relationships between data is critical in managing the data resource.

Workflows are typically case-based, that is, every piece of work is executed for a specific case. Workflow modelling is a prerequisite for planning workflow execution, and for analysing, training, executing, modifying, and archiving workflows. In order to cope with the complexity of workflow models, it is useful to look at them from different perspectives. The workflow model specifies all aspects of a workflow that are relevant for workflow execution. Each workflow schema defines a workflow type. For each workflow type, many workflow instances may be created. A workflow schema contains at least information about tasks that are to be executed, relationships between tasks, and conditions that enable tasks, and resources and resource management rules. Each workflow type has a life cycle that starts with the recognition of a need for a new workflow type in an organization. The new workflow type is planned, its boundaries are clarified, and its schema is modelled and evaluated. For each workflow type, an event that initiates the workflow execution and an event that defines the end of the workflow execution are defined. The workflow execution ends in a predefined termination state, representing either successful termination or termination in an exceptional manner.

When we talk of the supply chain, we refer to the network of processes, resources, and enterprises that provides for the supply of goods from raw materials through processing and production to the end consumer as a workflow. For example, logistics is concerned with the organization, coordination, and control of the flow of goods through the supply chain. Most enterprises are concerned with managing both their inbound and outbound supply chains. The former comprise the flow of raw materials, components, and products, which they purchase to manufacture and supply products to their customers and to support their business operations. The latter is the flow of their products from source through a distribution system to their end customers.

21.2.2 Value Chain

Value propositions define the relationship between what a supplier offers and what a customer purchases, by identifying how the supplier fulfils the customer's needs across different customer roles. It describes the benefits and therefore the value a customer or a value partner gains from the business. Value proposition can be defined as the description of the value that a product, service, or process will provide a customer with. A value process defines exactly how the value was processed by one actor that was required by the other to fulfil the demand. A value process includes the process of anything that contributes as part of the satisfaction of the customer, whereas a business process means the processing of each value unit which is the detailed total number of activities and resources required to deliver a specific value. A unit of value is created and supplied by one or more business process units. The business process is required for the value process. A value process cannot deliver a value without a business process and a business process does not know what to process without a value process.

There are two elements in the value process: value addition activities and value creation activities. Value addition activities are activities that are required to add some value to the final value indirectly, whereas value creation activities are the activities that are required to create some value to be added to the final value directly. A set of such value activities are performed to form the value process.

To better understand the activities through which a firm develops a competitive advantage and creates shareholder value, it is useful to divide the business system into a series of value-generating activities referred to as the value chain (Figure 21.2). Understanding the fundamental processes a product goes through, from raw material to manufacturing, to inventory, and then to the ultimate buyer was clearly delineated in Porter's concept of competitive advantage. Porter also recognized the importance of the supporting functions such as accounting and human resources. Certainly, the internal business processes must be connected and integrated in an expeditious fashion. Computer systems must track the processes and provide management with the data and information needed to optimize the value chain, ensuring that the right combination of products and services reaches the customer. Many companies have indeed automated the value chain over the years. However, the dramatic changes in technology, such as the introduction of the client/server concept, data warehousing, and data mining, and the Internet with its inherent capabilities in being able to communicate not only within the organization but outside as well, have changed the playing field. Probably the most significant change has been the opportunity to connect and extend the value chain to include outside vendors, business partners, and customers. This presents an entirely new dimension, but it may well be the key to competitive advantage, keeping in mind that competitive advantage is the ultimate objective of value chain automation.

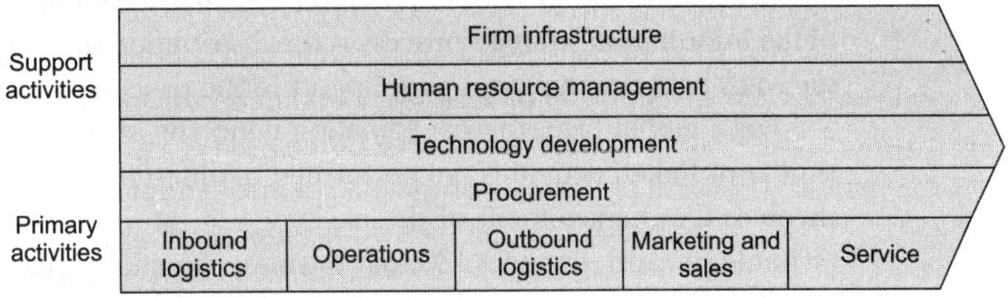

Support activities	Firm infrastructure				
	Human resource management				
	Technology development				
	Procurement				
Primary activities	Inbound logistics	Operations	Outbound logistics	Marketing and sales	Service

FIGURE 21.2 Porter's Value Chain

Managing inventories that one cannot see or does not own may seem a challenging task, but that is where much of the action lies. Examples include Wal-Mart, which allows its suppliers to directly stock Wal-Mart stores and Federal Express, which allows customers to link to its internal system for tracking deliveries.

21.3 Strategic Information Systems

The advent of databases and more sophisticated and powerful mainframe computers gave rise to the idea of developing corporate databases, in order to supply management with information about the business. These database-related developments also required data processing professionals who specialized in organizing and managing data. The idea of a corporate database that is accurate and up to date with all the pertinent data from the production systems and is attractive. It is worth noting that well-organized and well-managed businesses always had 'systems' for business control. In this sense, management information systems always existed and the notion of having such systems in an automated form was quite natural, given the advances of computing technology that were taking place at the time.

In general, the concept of strategy relates to corporate strategy which is the strategy that guides the corporation or enterprise as a whole. Business units within large organizations have business strategies related to their specific product-market situation. The notion of functional strategies such as marketing strategy, manufacturing strategy, personnel strategy, financial strategy, and information strategy, is derived from the corporate or business strategy and Information strategy began to attract interest at the beginning of the 1970s, and many terms have been used since then to address the alignment of information systems and business strategy. The core of the information strategy process is defined on the one hand by methodologies and tools and on the other hand by participants and their roles [Rivard et al. 2004]. An important determinant of the information strategy process is the distribution of the responsibility and the roles between the main participants in the process.

For a global firm, the coordination concerns involve an analysis of how similar or linked activities are performed in different countries. Coordination involves the management of the exchange of information, goods, expertise, technology, and finances. Many business functions play a role in such coordination—logistics, order fulfilment, financial, etc. Coordination involves sharing and use, by different facilities, of information about the activities within the firm's value chain. The innovations in IT in the past two decades have greatly reduced coordination costs by reducing both the time and cost of communicating information. Market and product innovation often involves coordination and partnership across a diverse set of organizational and geographically dispersed entities.

21.3.1 Cycle Time—Lead Time Measurements

Cycle time or lead time is the end-to-end delay in a business process. Cycle time is the time between receipt of the order and delivery of the product. For supply chains, the business processes of interest are the supply chain process and the order-to-delivery process. Correspondingly, we need to consider two types of lead times: supply chain lead time and order-to-delivery lead time (refer to section 21.7 on supply chain)

The order-to-delivery lead time is the time elapsed between the placement of order by a customer and the delivery of products to the customer. If the items are in stock, then it would be equal to the distribution lead time and order management time. If the items are made to order, then this would be the sum of supplier lead time, manufacturing lead time, distribution lead time, and order management time.

Work in progress (WIP) cost represents the cost that occurs due to parts waiting in queues. If the lot size is too big, the holding cost and the WIP cost will be high and the setup cost will be low. However, if the lot size is too small, the holding cost and the WIP cost will be low and the setup cost will be high. Moreover, changing the size of a lot will have an effect on the WIP and the lead-time of other lots. Hence, changing the size of a lot will have an effect on the cost function of other lots also. The WIP and lead-time of a particular lot size are dependent on factors such as the size of remaining lots, machine failures, the variability of processing time, scrap rate, and setup time.

21.4 Enterprise Resource Planning

A company's competitive edge often lies in its ability to make deals, to implement non-standard responses to customer needs, and to innovate in a host of ways. A key component of managing these fast-changing environments is information technology. Companies around the world are taking advantage of IT to radically alter how they conduct business. Starting in the late 1980s and the beginning of the 1990s, new software systems known in the industry as enterprise resource planning or ERP emerged to assist industries in their daily operations. These new systems surfaced in the market targeting mainly large complex business organizations. They were designed to integrate and optimize processes across the entire enterprise. Enterprise resource planning systems are configurable information systems packages that integrate information and information-based processes within and across functional areas in an organization [Ptak 2004]. ERP Enterprise resource planning system has the potential to integrate the processes and functions of the company to provide a comprehensive view of the entire organization. An ERP system has the potential to integrate all process and functions of a company and to present a comprehensive picture of the organization.

Companies have to clearly know what ERP is before thinking of implementing them. Enterprise resource planning systems are comprehensive software packages for business management, encompassing modules supporting functional areas such as planning, manufacturing, sales, marketing, distribution, accounting, financial, human resource management, project management, inventory management, service and maintenance, transportation, e-business and other functional areas [Davenport 1998]. The acquisition of ERP software is not without its challenges. At a cost of several hundreds of thousands or even millions of dollars, the acquisition of ERP software is a high-expenditure activity that consumes a significant portion of an organization's capital budgets. It also brings to light the need for finding the best means for acquiring this type of software so that the right choice can be made. Because of the implementation and the risk of it going awry, ERP implementations are said to be the single business initiative most likely to go wrong.

The ERP system is an information system that integrates business processes with the aim of creating value and reducing costs by making the right information available to the right people at the right time to help them make good decisions in managing resources productively and proactively. They are configurable information systems packages that integrate information and information-based processes within and across functional areas in an organization. They provide seamless integration of all the information flowing through the company by using a single database. For example, MindTree's ERP implementation Project mPower, has given the organization almost real-time visibility into the profitability of all its projects and how every resource is being used. mPower has also improved MindTree's ability to measure utilization more accurately—an important metric in the service industry.

Enterprise resource planning systems are based on the concept of identifying and implementing a set of best practices and procedures, resulting in tools that can be employed by different functions of a corporation to attain total organizational excellence through integration. These complex, expensive, powerful, proprietary systems are off-the-shelf solutions requiring consultants to tailor and implement them based on the company's requirements. Among the most important attributes of ERP systems are their abilities to automate and integrate an organization's business processes, share common data and practices across the entire enterprise; and produce and access information in a real-time environment. These systems have drawn much attention of many researchers and practitioners in the last two decades. These large automated cross-functional systems were designed to bring about improved operational efficiency and effectiveness through integrating, streamlining- and improving fundamental back-office business processes.

Enterprise resource planning covers the techniques and concepts employed for the integrated management of business as a whole from the viewpoint of the effective use of management resources, to improve the efficiency of an enterprize. It attempts to integrate all departments and functions across a company onto a single computer system that can serve all those different departments' particular needs. An ERP software integrates the information used by an organization's many different functions and departments into a unified computing system. That means that instead of using isolated departmental databases to manage information, such as employee records, customer data, purchase orders, and inventory, everyone in the enterprise relies on the same database. This allows employees in different departments to look at the same information. Thus ERP systems are comprehensive. Using the same software company wide, eliminates most of the data integration problems and facilitates oversight of procedures across all company functions. That integrated approach can have a tremendous payback if companies install the software correctly. Enterprise resource planning replaces the old stand-alone computer systems in finance, human resource (HR), manufacturing, and the warehouse with a single unified software program divided into software modules that roughly approximate the old stand- alone systems. The ERP software of most vendors is flexible enough so that you can install some modules without buying the whole package. Standard ERP systems use a single logical database shared by all ERP modules to provide a common view of an organization's data. In the early 1990s, this was a farsighted way to overcome inconsistency and fragmentation; today it seems monolithic and inflexible. But a number of technologies, collectively known as middleware, are eliminating the requirement that all ERP modules share the same database. Middleware allows application components to communicate through standardized messages, which simplify the coupling between systems. The ERP system has a master data which contains information about customers, suppliers and products and other organizational entities such as sales sites, and warehouses. It has a transactional data which contains the processed business information in the normal flow of the business such as sales quotes and orders shipments, purchase orders and invoices. It also has a historical data which is the results of business events retained for analysis and audit purposes.

21.4.1 Need for ERP systems

Most organizations across the world have realized that in a rapidly changing environment, it is impossible to create and maintain a custom designed software package which will cater to all their requirements and also be completely up to date. Realizing the requirement of user organizations some of the leading software companies have designed ERP software which offer an integrated software solution to all the functions of an organization. A successful ERP system makes it possible to develop and implement a variety of flexible supply chain options that can create significant cost and value advantages. It provides a way to increase management control and to increase IT infrastructure capacity and business flexibility and to reduce IT costs.

21.4.2 Evolution of ERP systems

The origin of ERP can be traced back to materials requirement planning (MRP). While the concept of MRP was understood conceptually and discussed in the 1960s, it was not practical for commercial use. It was the availability of computing power (processing capability and storage capacity) that made commercial use of MRP possible and practical. The term ERP systems did not appear until the early 1990s. These systems evolved from material requirements planning, manufacturing resource planning (MRP II), computer integrated manufacturing (CIM), and other functional systems responsible for the automation of business transactions in the areas of accounting and human resources. Due to the purported benefits of ERP systems, many companies consider them as essential information systems infrastructure to be competitive in today's business world and provide a foundation for future growth.

21.4.3 Materials Requirement Planning

The focus of manufacturing systems in the 1960's was on inventory control. Companies could afford to keep lots of just-in-case inventory on hand to satisfy customer demand and still stay competitive. Consequently, techniques of the day focused on the most efficient way to manage large volumes of inventory. In the 1970s, it became increasingly clear that companies could no longer afford the luxury of maintaining large quantities of inventory. This led to the introduction of material requirements planning systems. Materials planning practices were used to improve efficiencies of the manufacturing process at the plant and manufactured goods levels. Material requirements planning systems were centred on the flow of physical resources for manufacturing. The MRP or set of orders is broken up into production bills which are then broken up into the list of materials needed to manufacture all the products included in that bill. It used a bill of materials (BOM), a list of what materials were available to the production process, in combination with a production forecast. Manufacturing resource planning uses the BOM to split up the production of the finished product into smaller units, while engineer-to-order production divides the overall production process into work packages, which are specified in a work breakdown structure (WBS).

A WBS is a hierarchical model of the tasks that need to be carried out in a project and is the basis for the organization and coordination of the project. It contains the work, the time, and the costs that are associated with every task. A provisional WBS is created for the preparatory planning stage. This is done one specific product at a time. The result includes forecasts for future material needs and materials purchase timing. The capacity to handle those materials was disregarded to a large extent. It was a monotonous, and steady way to plan. The planning done by the manufacturers did not involve the consumers they served. The processes were aimed internally at the enterprise. But the planning and the processes were able to at least reduce the costs of production. Material requirements planning was only concerned with materials and has nothing to do with the money needed to acquire them. It assumes that a separate system takes care of that.

However, the demand for consumer goods frequently outran the supply, but the way that the manufacturers dealt with it was to extend the delivery time. Customer expectation was low, customer tolerance was high, and so the volatility level was also quite low. People were willing to wait for the branded goods that they wanted. In the 1980s, companies began to take advantage of the increased power and affordability of available technology and were able to couple the movement of inventory with the coincident financial activity. Manufacturing resources planning systems evolved to incorporate the financial accounting system and the financial management system along with the manufacturing and materials management systems. In order to support the product driven corporate ecosystem, MRP added sales forecasts to calculate the timing of material flows into the production process, based on the expected numbers of product sales at specific times.

Capacity planning and determining on-hand inventory, especially given retailer needs, made the ability to plan the use of materials, given this consumer demand, all that much better. It was a new way to create more efficient production and reduce costs of production and to improve the timeliness of the good's creation. These systems integrated budgeting and financial accounting into the manufacturing planning process. It would be easy to conclude that MRP sees the organization from a manufacturing-centred perspective. Areas such as product design, information warehousing, materials planning, capacity planning, communication systems, human resources, finance, and project management were included in the plan.

21.4.4 Types of Manufacturing

Discrete manufacturing starts when a production order is created and processed. A production order is created either manually or when a planned order that was created in the production and procurement planning process is converted. A production order is a request to the production department to produce or provide products or services at a specific time and in a specific quantity. It specifies the work centre and material components that are to be used for production. The creation of a production order automatically creates reservations for the required material components. Purchase requisitions are created for externally procured material components and services, and capacity requirements are created for the work centres at which the order will be executed.

Repetitive manufacturing is characterized by the interval-based and quantity-based creation and processing of production plans. With repetitive manufacturing, a certain quantity of a stable product is produced over a certain period of time. Repetitive manufacturing is suitable for a variety of industries, such as branded items, electronics, semiconductors, and packaging. Repetitive manufacturing also can be used for pure make-to-stock production. Production in this case has no direct connection to a sales order.

Process manufacturing is characterized by batch-oriented and recipe-oriented production of products or co-products in the process industry. Process manufacturing is used mainly in the following industries: chemicals, pharmaceuticals, food and luxury foods, and process-based electronics. Process manufacturing starts when a process order is created and processed in accordance with a master recipe. A production order is created either manually or when a planned order that was created in the production planning process is converted. A production order is a request to the production department to produce or provide products or services at a specific time and in a specific quantity. It specifies the resource and material components that are to be used for production.

The next major shift during the late 1980s and the early 1990s was that the time to market was getting increasingly shorter. The first to market with a product made the most long-term profit. Lead times expected by the market continued to shorten and customers were no longer satisfied with the service level that was considered world class only a few years earlier. Customers were demanding to have their products delivered when, where, and how they wanted them. Companies began to develop and embrace the philosophies of just in time (JIT) and supplier partnerships as a way to remain competitive. It was no longer possible to remain competitive by focusing only within the four walls. The JIT crusade brought the importance of reliable suppliers to focus.

21.4.5 Order Processing

Order processing is the administrative process of receiving orders from customers, and ensuring that the right goods are delivered on time and to the right place. The customers may be either end-consumers or companies operating as the next process in the distributing network. It frequently represents the interface between, on the one hand, the marketing and selling functions of a company, which are concerned with promoting the products and obtaining the orders, and, on the other hand, the distribution function, which is responsible for delivering the product to the customers. It can be seen as the first stage of the distribution process, and its main purpose is to record all customer orders received by the company to ensure their correct delivery to the customers and to see that a timely and accurate invoice is raised and despatched. To the extent that it captures and holds data on customer orders received, the order processing system is frequently used to provide important information on patterns of both customer buying and product demand and usage for the purpose of forecasting and business planning.

The first step on receipt of an order is order entry, recording the details of the order. This may first allocate a reference number to the order for internal identification tracking and recording the date the order was received. Then the customer's name and, frequently, an internal customer reference number are recorded together with the customer's internal purchase order reference, if there is one, as this is likely to be the means by which the customer identifies the order. The address to which the goods are to be delivered and the address to which the invoice should be sent also need to be recorded at this stage. In many cases, if the order is from a regular customer, much of this information will already be held on the company's records and simply attached to the order record as it is entered.

Should product availability not be sufficient to meet customer requirements, then in discussion with the customer there may be some rescheduling necessary. The instructions for manufacturing, if required, and for picking, marshalling and packing are, however, frequently generated by the order processing system. The progress of these activities is monitored by the order processing system so that the order status is known and can be reported if necessary, both internally and to the customer. Once the product is ready for despatch, the invoice can be raised and sent to the customer.

21.4.6 Current ERP Systems

Enterprise resource planning originates from the large packaged application software that have been widespread since the 1970s. Material requirements planning software supported material handling. The name ERP could have been derived from MRP and its variants.

The ERP systems are commercial packages that are bought from software vendors and therefore they do not meet the needs of an organization. It is argued that ERP packages meet about 70 per cent of an organization's needs. Hence, an organization has the following options. It can change its business process flow and conform to the package. It can customize the package to make it suit the needs of the organization. It can choose to ignore the remaining 30 per cent which is not conforming.

The planning functionality of an ERP, essentially, consists of the following.

- Check the plan in the database before making or allowing any transaction that affects a given resource (like an order from a client)
- In the case that it is available, assign or use the required resource and immediately update its planned availability
- Create a planning requisition if the resource is not available
- Report on several management levels how close by the plan is being followed
- Make the necessary changes to adjust the existing plan

The planning functionality of an ERP is not limited, as it used to be, to a single department. This is advantageous as long as the enterprise is capable of doing that type of planning. In classical strategic planning, employees or representatives from all organizational ranks are commonly co-opted into the process. The decision stage leads to a decision to invest resources in acquiring and assimilating a new system. When deciding to allocate resources for the purchase of a global ERP system, a cost assessment including the costs of implementation processes will be made by representatives of all operational departments. At this point, it is important to nominate the project leader who will manage the implementation process and achieve the organizational improvements that the new system was designed to accomplish.

An ERP software integrates management information and processes, such as financial, manufacturing, distribution, and human resources, for the purpose of enabling enterprise-wide management of resources.

21.4.7 ERP Implementation—Integration

ERP systems have become a great integration challenge, as it is they are possible to reuse common services through common applications, and share business processes and services. Despite ERP systems being introduced as integrated suites, they have failed to achieve application integration and supply chain integration [Chung and Snyder 2000]. Chung and Snyder [2000] claim that ERP systems support generic processes and best practices with organizations attempting to parameterize ERP packages to better support their business processes and strategy. However, customization is a difficult task that causes serious integration problems as ERP systems are complex, non-flexible, and often not designed to collaborate with other autonomous applications. The implementation of an enterprise resource planning system in a company can have different degrees of complexity. Chen01. The complexity of ERP implementation is exacerbated when ERP systems are globally implemented since the integrative nature of this technology calls for the crossing of national or regional, in addition to organizational boundaries Holland99. Since countries, regions, and organizations differ in their absorptive capacity and their societal and organizational culture, implementation of ERP systems in global enterprises requires a detailed examination of potential gaps or inconsistencies in the interaction between new technologies, end-users, and organizations in different countries and/or regions. The implementation strategy generally consist of three major parts, implementation orientation, scope of standardization, and implementation procedure. Referring to its business orientation, an organization may take one of the three implementation orientations to plan the implementation of a global ERP system in its several sites, namely global rollout, local implementation, or mixed-mode implementation.

A *global rollout* implies that there will be a creation of a global template which will be implemented in each site without any customization. This template could also be the vendor's off-the-shelf solution.

Local implementation is the direct opposite to global rollout orientation. Every site is enabled to customize any kind of ERP system to the company's demands. Therefore, the idea of decentralization is realized within this approach.

Mixed-mode implementation dilutes the global rollout and local implementation orientations. On one hand, data must be harmonized to generate global economies of scale. On the other hand, there will be a high degree of freedom to choose an ERP system according to local demands which fits the individual site. These two aspects result most likely in standardization of interfaces among sites, while hosting and administration of the systems remains the responsibility of the individual sites.

According to Shang and Seddon, the following five-dimensional framework, which is built on a large body of previous research into IT benefits, has been organized around operational efficiency and managerial and strategic effectiveness, as the outlook of strategic managers are too broad to identify casual links between enterprise system investment and benefit realization, and those of operational managers are too narrow to consider all relevant organizational goals.

Operational benefits include cost reduction, cycle time reduction, productivity improvement, quality improvement, and customer service improvement.

Managerial benefits include better resource management, improved decision making, and planning and performance improvement.

Strategic benefits include support business growth, support business alliance, building business innovation, building cost leadership, generating product differentiation and enabling worldwide expansion.

Infrastructural benefits include IT cost reduction, and increased IT infrastructure stability.

Organizational benefits include facilitating business learning, empowerment, and increased employee morale and satisfaction.

21.4.8 Popular ERP systems

Major ERP vendors attempting to help organizations reduce their organizational disintegration costs include SAP (www.sap.com), Baan (www.baan.com), Oracle (www.oracle.com), J D Edwards (www.jdedwards.com). Microsoft (www.microsoft.com), and others.

SAP

Enterprise resource planning (ERP) systems like SAP R/3 are business software tools that help managing important parts of business, including product planning, parts purchasing, maintaining inventories, interacting with vendors, customer service, and tracking orders. Systems, Applications and Products in Data Processing (SAP AG) was started by five former IBM engineers in Germany in 1972 for producing integrated business application software for the manufacturing enterprise. Its first ERP product, R/2, was launched in 1979 using a mainframe-based centralized database that was then redesigned as client/server software R/3 in 1992. The SAP R/3 has been designed for the client-server environment and can be used to illustrate both client-server and browser-server environments. The SAP R/3 is an ERP system. SAP R/3 is based on a commercial relational database system which is used to store data. The SAP stores master data, transaction data and metadata that describe the database structure at the database server. Microsoft and SAP are cooperating in a 10-year agreement to integrate Microsoft's .Net with SAP's NetWeaver Platform. The stated goals are to make it easier for Microsoft Office users to connect to SAP applications and to make Visual Studio the development tool of choice for SAP connections.

The SAP R/3 is based on a three-tier client/server-architecture with the following layers.

- The presentation layer: It provides a graphical user interface (GUI) usually running on personal computers (PCs) that are connected with the application servers via a local (LAN) or a wide area network (WAN).
- The application layer: It comprises the business administration know-how of the system. It processes predefined and user-defined application programs such as online transaction processing (OLTP) and the implementation of decision support queries. Application servers are usually connected via a LAN with the database server.
- The database layer: It is implemented on top of a commercial database product that stores all the data of the system.

Application servers can run on various operating systems such as Uniplexed Information and Computing System (UNIX), its variations, Virtual Memory System (VMS) and Windows New Technology (NT), and its successors. Business logic runs on application servers, and these servers also perform basic services.

The Advanced Business Application Programming (ABAP) language can be used to define and manipulate such stored data. The ABAP/4 is a Fourth Generation Language (4GL) whose origins can be found in report/application generator languages. It is a language that is constantly evolving with recent releases incorporating object-oriented capabilities (ABAP objects). The robustness of the language is evident in the wide range of functionality and high performance capabilities within the R/3 system, allowing applications to process huge amounts of customer data. The ABAP/4 is an interpreted language, which makes it very easy to integrate the new ABAP/4 application programs into the system. Like ordinary data, all the ABAP/4 application programs are managed by the R/3 data dictionary and the program code is stored in the SAP database.

A business application programming interface (BAPI) is a precisely defined interface providing access to processes and data in business applications such as R/3. A BAPI is defined as application programming interface (API) methods of SAP objects present in business object repository (BOR). Business object resporitory is an object oriented repository in the R/3 system. Business objects define:

(i) logical structures of the business such as plants, warehouses, storage locations,

(ii) business documents such as purchase requisition, quotation, purchase order, and

(iii) resources such as materials, equipment, vendors.

A business object is considered as a set of entity types that share a common external interface. A business object usually consists of a source entity type and all those entity types that are hierarchically dependent on it. The internal structure of a business object is described by means of the Structured Entity–Relationship Model (SERM), a data modelling language anchored on four concepts: entity types, attributes, relationship categorie, and specialization categories. Entity types serve to model items needed to carry out a business process. Attributes represent properties the business users wish to attach to an entity type.

Event-driven Process Chains

The SAP business requirement documents specify business scenarios that are represented in terms of event-driven process chains (EPC), based on modelling notation and accepted by SAP as a process modelling standard for R/3 implementation projects.

An EPC consists of three main elements. Combined, these elements define the flow of a business process as a chain of events. The elements used are the following.

- Functions, which are the basic building blocks. A function corresponds to an activity (task, process step) which needs to be executed. A function is drawn as a box with rounded corners. The functions (processes) represent time consuming business transactions that use, update, or create business objects. Within a company, a process describes what an organizational unit does. In the SAP R/3 system, a process means a transaction.

- Events, which describe the situation before and/or after a function is executed. The events refer to points in time and are defined as data items that should be available at the start and at the end of process' execution. Events trigger or drive the process flows that follow. Functions are linked by events. An event may correspond to the position of one function and act as a precondition of another function. Events are drawn as hexagons.
- Connectors, which can be used to connect functions and events. The connectors define alternative or parallel workflows through the process chains. This way, the flow of control is specified. There are three types of connectors, namely (and), (xor), and (or). Connectors are drawn as circles, showing the type in the centre of the circle.

SAP uses the term logical unit of work (LUW) for transactions (Figure 21.3). Basically, an SAP LUW has the same ACID properties as database system transactions. An SAP LUW can span several dialog steps and an SAP LUW is either executed completely or not at all (i.e., atomicity). To synchronize LUWs, SAP implemented its own locking scheme which is managed by a (centralized) enqueue server which runs in one of the application servers(Figure 21.4). Basically, SAP also implemented its own monitor consisting of a message handler and request queue in every application server. A transaction screen is considered the atom of the interaction, but still they can be complex. Moreover, the number of transactions available is determined by the functionality required by the ERP adapter during ERP implementation, and the consequence is complex menu structures.

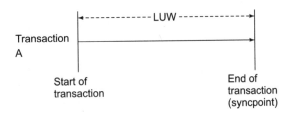

FIGURE 21.3 Logical Unit of Work

The SAP Internet Transaction Server (ITS) is a web server. The primary function of SAP ITS is to establish communication between the Internet and the SAP R/3 system, a difficult task given the technical differences. The SAP ITS is a cornerstone for the mySAP.com service that provides a multitude of web functionalities. The presentation layer contains SAP GUI, which is an interface operated by users to access data, launch applications, and display data. This GUI provides a consistent look and feel across different platforms.

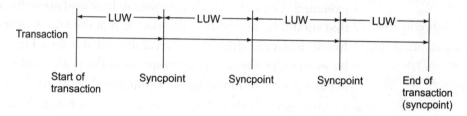

FIGURE 21.4 Transaction

SAP has institutionalized systematic reuse approaches and developed and successfully maintained the infrastructure of processes, people, and tools for customers to reuse. SAP has made the reuse practice an integral part of the R/3. For example, the Kuoni Travel Group was looking for a more centralized enterprise IT infrastructure. Kuoni India's IT function became part of Kuoni's global initiative to introduce a common working environment across all its locations. The project called Future Business in IT Architecture (FITA), started with three key focus areas: trimming application portfolios, standardizing IT infrastructure and setting up a homogenous functional organization. They rolled out SAP to take care of most of the back-end functions of the business on a captive platform called Shared Services Center. On the application portfolio front, Kuoni India adopted a three-tier distribution network. Various travel products are sold by Kuoni branches, its franchisees, and agents across the country.

Accounting module maps business transactions in accordance with their financial value and is responsible for planning, controlling, and monitoring the value flow within the enterprise. It is subdivided into financial accounting and managerial accounting, in accordance with the addressee group. Managerial accounting consists of cost accounting and activity accounting, and its purpose is to provide the decision-makers in the enterprise with quantitative information. Financial accounting is structured in accordance with statutory regulations; enterprises use it to comply with requirements for disclosure with regard to external parties, in particular tax authorities and investors. The main components that support the tasks of accounting are financial accounting (FI), investment management (IM), and controlling (CO). These modules are further subdivided into corresponding sub-modules. Human resources management (HRM) module is divided into the areas of personnel planning and development, and personnel administration and payroll. Personnel planning and development supports the strategic utilization of staff by providing functionality that enables the enterprise to systematically and qualitatively manage its staff. Personnel administration and payroll comprises all administrative and operational human resources activities. Logistics in the business context structures the flow of materials, information, and production from the supplier through production to the customer.

Microsoft Dynamics

Microsoft Dynamics is a scalable and global enterprise resource planning solution that can be tailored to support the exact needs of your manufacturing business, while delivering a low total cost of ownership (TCO). The solution has broad and advanced functionality that supports the entire business process in a company. Microsoft Dynamics delivers this through its flexibility, ease of use, connectivity, rapid development toolkit, unique architecture and advanced functionality covering manufacturing, supply chain management (SCM), customer relationship management (CRM), financial management, project management, HRM, business analysis, and e-business. Microsoft Dynamics gives executives direct insight into change through its business analysis functionality including the Microsoft Dynamics Balanced Scorecard and Microsoft Dynamics Business Process Management modules.

21.4.9 Business Benefits

Enterprise resource planning provides two major benefits that do not exist in non-integrated departmental systems, namely a unified enterprise view of the business that encompasses all functions and departments; and an enterprise database where all business transactions are entered, recorded, processed, monitored, and reported. This unified view increases the requirement for, and the extent of, interdepartmental cooperation and coordination. But it enables companies to achieve their objectives of increased communication and responsiveness to all stakeholders.

It is generally a misleading perception that implementing an ERP system will improve an organization's functionalities overnight. The high expectation of achieving all-round cost savings and service improvements is very much dependent on how good the chosen ERP system fits to the organizational functionalities and how well the tailoring and configuration process of the system matched with the business culture, strategy, and structure of the organization. Overall, an ERP system is expected to improve both the backbone and front-end functions simultaneously. Organizations choose and deploy ERP systems for many tangible and intangible benefits and strategic reasons. In many cases, the calculation of return on investment (ROI) is weighted against the many intangible and strategic benefits.

21.5 Change Management

Any organization faces many potential barriers to adaptability, some specific to itself. Companies tend to be organized as hierarchies, with the most experienced and successful people on top. When there is a change in the company's business and the IT system's environment, these individuals may have difficulty in recognizing the change and may draw too heavily on their earlier experiences. Managing change is tough, but part of the problem is that there is little agreement on what factors most influence transformation initiatives. Ask five executives to name the one factor critical for the success of these programs, and the one would probably get five different answers. That is because each manager looks at an initiative from his or her viewpoint and, based on personal experience, focuses on different success factors. The experts, too, offer different perspectives. In recent years, many change management gurus have focused on soft issues such as culture, leadership, and motivation. Such elements are important for success, but managing these aspects alone is not sufficient to implement transformation projects. Soft factors do not directly influence the outcomes of many change programs. So what is required is a strategic planning. The process of strategic planning starts with identifying the impetus for changes in the company's business and IT systems, and their expected strategic and operational benefits. This process helps people understand the need for change; it sparks their interest in it and promotes their commitment to the process. Strategic planning associated with ERP implementation relates to process design, process performance measurement, and continuous process improvement, also known as business process reengineering.

Strategic planning is carried out through a clear understanding of the business and change implications. Expertise and creative thinking are an individual's raw materials, his or her natural resources. From the context of change management, creative thinking refers to how people approach problems and solutions and their capacity to put existing ideas together in new combinations. But a third factor, motivation, determines what people will actually do. Managers can influence all three components of creativity, namely expertise, creative-thinking skills, and motivation. But the fact is that the first two are more difficult and time-consuming to influence than motivation.

Reasons for implementing ERP systems, implementation management approaches and factors critical to successful implementations have been presented and discussed in literature [Bingi et al. 1999]. Bancroft et al. [1998] were among the first who stressed that ERP implementation is a change management process. Companies have not always carefully assessed all the possible implications of their decisions to implement ERP systems. Any change initiative must be to assess the level of agreement in the organization along two critical dimensions. The first dimension is the extent to which people agree on what they want, the results they seek from their participation in the enterprise and the trade-offs they are willing to make in order to achieve those results. Employees at Microsoft for instance, have historically been united around a common goal—to dominate the desktop. The second dimension is the extent to which people agree on cause and effect—which actions will lead to the desired outcome.

21.6 Customer Relationship Management (CRM)

Customer relationship management (CRM) technology evolution has kept pace with the IT community at large. Hence proprietary vendor architectures have been largely overhauled to accommodate open standards (e.g., XML, WSDL, SOAP) and service-based component architectures. Customer relationship management has its roots in relationship marketing which supports the proposition that a firm can boost its profitability by establishing long-term relationships with its customers. The term 'the relationship economy' resulted from a broader recognition of the role that knowledge and technology have in catalysing economic growth. Knowledge has always been central to economic development when impacted by the introduction of new technology. Extrapolation of relationship management had a logical path towards CRM. The mental model of CRM looks like this—customers are so well-understood that customer life cycle (engage/transact/fulfil/service) treatments can be differentiated based on the customer's strategic value to the organization. Customer relationship management is defined as a business strategy aimed at gaining long-term competitive advantage by optimally delivering customer value and extracting business value simultaneously. Customer relationship management integrates customer-centric efforts such as marketing campaigns, call centres, help desks, sales force automation and customer analytics [Lindstrand et al. 2006]. An organization manages its customers according to customer profitability indicators in addition to other business metrics. Yet, the new role of customer segment manager is added to work with product and brand management. The firm derives its value in the form of cost savings since it is less expensive to retain existing customers than to expend time and energy on constantly acquiring new customers.

The model implies that customer segmentation strategies evolve beyond the use of demographics and psycho-graphics to include cross-channel interaction and buying behaviours, interaction preferences, predictive behaviour modelling, and the segment's strategic value. Value is not just about financial value; it is also about how well a particular customer type supports the organization's business strategy. When it comes to CRM solutions, one size does not fit all. Technology investments must be balanced with business process re-engineering and customer life cycle implementation. Organizations must be aware that upgrading will be critical, CRM will move from being niche to being mainstream, and there is an end in sight! Customer relationship management has the ability to provide competitive differential.

The needs of large enterprises vary considerably from smaller organizations and even from each other. Large enterprises have complex, global customer relationships with unique, industry-specific requirements. They also often have disparate, poorly-integrated information and autonomous departments with stand-alone or home-grown CRM solutions. A large enterprises should seek CRM solutions from a vendor that offers comprehensive functionality, flexible deployment, and integration options, and deep industry domain expertise. The CRM systems are highly important for most businesses today.

Customer relationship management evolving definition is now significantly more complex and more enigmatic than it was at any time in the last few years. There are countless attempts to come up with a definition, as widely disparate as the agendas of those presenting them. However, there are some universally recognized practices, concepts, and features that can be described by the CRM industry.

Customer relationship management has a history, an extensive body of experiences, and a significant corpus of literature on its failures, successes, and, even its definition. The most detailed definition has been given by Christopher,

Payne, and Ballantyne (2002)—CRM is a strategic approach to improving shareholder value through the development of appropriate relationships with key customers and customer segments. Customer relationship management unites the potential of IT and relationship marketing strategies to deliver profitable, long-term relationships, Importantly, CRM provides enhanced opportunities to use data and information, both to understand customers and implement relationship marketing strategies better. This requires a cross-functional integration of people, operations and marketing capabilities enabled through information technology and applications. Perception has shifted traditionally and CRM had been seen as a technology initiative, though there has always been a politically correct insistence that it is not a technology. That is characterized by the CRM software vendor created mantra, 'it is not a technology, it is a system.'

21.6.1 Need for CRM Systems

Despite substantial investments in customer relationship management applications, there is a lack of research demonstrating the benefits of such investments. A primary motivation for a firm to implement CRM applications is to track customer behaviour to gain insight into customer loyalty, customer tastes, and evolving needs. By organizing and using this information, firms can design and develop better products and services. Loyalty, in day-to-day life, implies an unselfish belief in institutions, or unswerving fidelity in marriage, or emotional commitment to friends. Loyalty also suggests monogamy—one choice above all others. However, customers, loyalty towards vendors is not like that. There is not a customer alive who will consider using one shop for every need. When retailers look at winning and keeping the loyalty of their customers they are looking to achieve a little extra goodwill, a slight margin of preference, an incremental shift in buying behaviour. This can add up to a massive contribution to the financial success of the business.

KPMG calls simple Wal-Mart style discounting as purge loyalty. KPMG identifies some other ways in which loyalty strategies can work. These are as follows.

- 'Pure' loyalty means strengthening the existing bond between the customer and the retailer, so the retailer can find out what the customer wants, and give that customer more of it.

- 'Pull' loyalty means attracting customers by augmenting a retail offer, so customers will find that buying one product means they get an offer on another, linked product.
- 'Push' loyalty means creating a scheme to encourage us to use a way of shopping that we would not have done before, pushing customers through new channels, or trying to create new types of behaviour.

Customer relationship management applications facilitate organizational learning about customers by enabling firms to analyse purchase behaviour across transactions through different channels and customer touchpoints. Glazer (1991) provides examples of how FedEx and American Airlines used their investments in IT systems at the customer interface to gain valuable customer knowledge. More recently, firms have invested in an integrated set of tools and functionalities offered by leading software vendors to gather and store customer knowledge. Firms with greater deployment of CRM applications are in a better position to leverage their stock of accumulated knowledge and experience into customer support processes. Customer relationship management encompasses far more than the challenges of marketing. It is useful to see CRM strategy operating in two halves, namely structural CRM, the means by which a company joins up its operations to deliver better service to customers and better value to the business; and active CRM, the means by which it exploits that structural investment to drive sales, reduce costs, and improve the customer experience. For example, Yes Bank managed cutomer relations via Excel sheets, which lacked interactive features and wasn't the smartest way to run the bank's sales force. Yes Bank move to Collaborative CRM (YCCRM) changed all that with banking heavily on employee collaboration.

21.6.2 Evolution of CRM systems

Over the past 10 years, CRM systems have evolved from departmental client/server systems to semi-integrated systems. These systems usually addressed the needs of isolated departments such as the marketing department, sales and service, technical support, or the call centre. These systems were limited in their scope and focused inwardly on solving immediate problems of one area within a company. They were often developed by various vendors at different times, with no attempt to integrate information between systems. Soon the focus shifted to customer relationships, with an emphasis on supporting integration of all customer information for a complete 360-degree view of the customer and all customer interactions. What has characterized CRM' move into the mainstream business thinking is the integration of its principles into the business ecosystem as it transformed. It has gone from a reactive product-driven corporate ecosystem to a real-time proactive customer ecosystem, though it is not fully transformed to this state yet.

The way businesses have been traditionally organized, along functional and product lines, may be insufficient to take full advantage of the apparent and latent opportunities in measuring customer activity [Newell 2003]. Many companies are seeking to shift the central focus of corporate activity away from products and on to customers or at the very least to learn new ways of managing customer-facing activities. To effect this change, businesses will need to build out new, more robust measurement systems, replacing or standing alongside existing product oriented measurement systems. Designing and managing these measurement systems and the CRM technologies around them requires new combinations of skills and roles, for which many companies have not planned.

21.6.3 Process Flow

A CRM system is to some extent a groupware application for managing your business. Groupware is a term used to describe computer software designed to help a group of people work together cooperatively. Fundamentally, a CRM system captures information about accounts and contacts. 'Accounts' in the above context the complete set of other firms an organization does business with such as partners, suppliers, and customers. The CRM system also keeps track of new business leads and once qualified, converts these leads into opportunities and relates these new opportunities to the accounts, contacts, and your own employees in charge of selling to those accounts.

21.6.4 Popular CRM Systems

SugarCRM is both a company and an Open Source project (Figure 21.5). SugarCRM the company was created as a commercial open source company and funded by Silicon Valley venture capital firms. The open-source product is called Sugar Open Source, and the commercial products are called Sugar Pro and Sugar Enterprise. The SugarCRM Open Source project has its official home at http://www.sugarforge.org/. SugarCRM is representative of the best CRM systems available in the market for the manner in which the systems are used or navigated. Various key elements of the screen layout overleaf have numbered highlights, where 1 is Navigation Tabs, 2 is Navigation Shortcuts

Box, 3 is Last Viewed, 4 is Search Box, 5 is User Management Links, 6 is Quick New Item Box, and 7 is Main Screen Body.

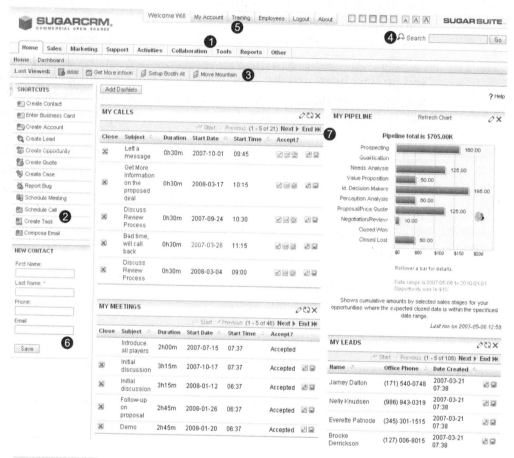

FIGURE 21.5 SugarCRM—Home Screen

That Microsoft Dynamics CRM is arguably one of the easiest-to-use CRM applications on the market should come as no surprise to anyone. Combined with the need for most businesses to have a CRM solution and the market dominance of Windows (and, specifically, Outlook), this gives Microsoft a significant edge on integration and ease of use. Microsoft introduced Business Contact Manager (BCM) with Office 2003 as a tool for the small business owner to manage contacts and accounts in ways similar to a fullblown CRM system. However, the area where BCM really comes up short is its inability to integrate workflow for the automation of routine and/or necessary tasks.

Microsoft announced the release of Microsoft Dynamics CRM 4.0 in January 2008.

21.6.5 Business Benefits

Customer relationship management's apparent benefit was that it seemed to be able to automate much of the 'customer-facing' activity of a company. For firms evaluating CRM applications, it is important to understand the conditions under which deployment of those applications contribute to improved customer knowledge and customer satisfaction. Since CRM measurement systems can be used to understand past and future customer behaviour, the ability for companies to convert that knowledge into business results can be a significant form of competitive advantage. Knowledge about how a company interacts with its customers is specific to the company's brand and its customers and therefore is proprietary to that company. Bharadwaj [2000] notes the advantages of gathering customer knowledge from customer encounters and disseminating this knowledge to employees for cross-selling and forecasting product demand. Bolton, Kannan, and Bramlett [2000] provide empirical evidence that IT-enabled loyalty programs enable firms to gain valuable customer knowledge about customers' purchase behaviour. Jayachandran et al. [2004] show that customer knowledge processes enhance the speed and effectiveness of a firm's customer response [Mithas et al. 2005]. Firms with greater supply chain integration are more likely to benefit from CRM applications in terms of customer knowledge and customer satisfaction. The importance of customer knowledge as a mediator for customer satisfaction suggests that in addition to implementing CRM, managers should also ensure that customer knowledge is disseminated across customer touchpoints in order to benefit in terms of customer satisfaction.

Companies implement CRM measurement very differently based on their internal decision-making styles. As companies make decisions about customer strategies, they look to customer measurement to help influence specific decision-makers or the decision-making process or validate initial ideas about how to manage customer relationships.

21.7 Supply Chain and Integrated Supply Chain Management

To support efficient and flexible supply chains organizations need to integrate their applications on enterprise and cross-enterprise level. Through the integration of internal and external business activities, companies are continuing to find ways to further improve their efficiency and streamline key business processes. To accomplish this, organizations have not only coordinated functional staff and business processes but have sought to integrate enterprise applications into systems that can link members of the supply chain together via data, information and knowledge.

The term 'supply chain management' or SCM is gaining currency, implying that there is something different about managing the supply chain. Suppl chain management is a set of approaches utilized to efficiently integrate suppliers, manufacturers, warehouses, and stores, so that merchandize is produced and distributed at the right quantities, to the right locations, and at the right time, in order to minimize system wide costs while satisfying service-level requirements.

21.7.1 Supply Chain

A supply chain is the system of organizations, people, technology, activities, information, and resources involved in moving a product or service from supplier to customer. It is the set of value adding activities that connects a firm's suppliers to the firm's customers. It is a network of supplier, manufacturing, assembly, distribution, and logistics facilities that perform the functions of procurement of materials, transformation of these materials into intermediate and finished products, and the distribution of these finished products to customers. A supply chain (SC) includes all the participants and processes involved in the satisfaction of customer demand: transportation, storages, retailers, wholesalers, distributors, and factories. A large number of participants, a variety of relations and processes, dynamics, the uncertainty and stochastics in material and information flow, and numerous managerial positions prove that supply chains should be considered as a complex system in which coordination is one of the key elements of management.

It is understood that business-to-business (B2B) supply chains are composed of discrete activities with each supply chain member holding one leg of the giant chain. Supply chain activities transform natural resources, raw materials- and components into a finished product that is delivered to the end customer.

21.7.2 Supply Chain Systems

In sophisticated supply chain systems, used products may re-enter the supply chain at any point where residual value is recyclable. The deming flow diagram describes a business process as a continuous process connected on one end with the supplier and on the other end with the customer. Three types of flows occur throughout the supply chain: (1) product, (2) information, and (3) funds. These flows travel both upstream and downstream within the supply chain. One way of classifying supply chain management analysis is to divide the area into the strategic components like supply chain design (SCD), and supply chain operations or execution (SCE). Supply chain design is the process of determining the supply chain infrastructure that will be used to satisfy customer demands. Supply chain execution is the process of determining solutions to more tactical and operational issues such as local inventory polices and deployment, manufacturing and service schedules, transportation plans, etc.

Supply chain management revolves around efficient integration of suppliers, manufacturers, warehouses and stores, and hence it encompasses the firm's activities at many levels, from the strategic level through the tactical, to the operational level. These levels are as follows.

- The strategic level deals with decisions that have a long lasting effect on the firm. This includes decisions regarding the number, location, and capacity of warehouses and manufacturing plants, and the flow of material through the logistics network.
- The tactical level includes decisions which are typically updated anywhere between once every week, once every month, or once every year. These include purchasing and production decisions, inventory policies, and transportation strategies including the frequency with which customers are visited.
- The operational level refers to day-to-day decisions such as scheduling, lead-time quotations, routing, and truck loading.

Porter introduced the value chain concept as a systematic way of examining all the activities a firm performs and how they interact to provide competitive advantage. This giant chain is composed of strategically relevant activities that create value for a firm's buyers. Agility of the supply chain has become a focus of organizations faced with today's highly discontinuous business environment characterized by rapid technological change, global competition, and demanding customers. Many believe that agility is fundamental to maintaining a leadership position in the global market.

21.7.3 Integrated Supply Chain Management

Integrated supply chain management involves designing, managing, and integrating a company's own supply chain with that of its suppliers and customers. Integrated supply chain comprises all activities associated with the flow and transformation of products from the raw material stage through delivery to the end customer.

Market leaders, such as Wal-Mart and Dell, understand that the supply chain can be a strategic differentiator. They constantly search for new ways to add value and push the boundaries of performance. And they keep refining their supply chains so they stay one step ahead of the competition. The advantage of the supply chain approach is that the processes are examined from the point of view how they contribute to the targets of the supply chain management. Therefore, the integration between different logistical functions, for instance sales planning and production planning, is stronger within the focus of the supply chain management approach. Supply chain networks have gained prominence in the last decade. Valid reasons for their growing importance include the following.

- Global dispersion of manufacturing and distribution facilities
- Demand for customized products for local markets
- Competitive pressures
- Rapid advances in information technologies in the form of electric data interface (EDI), Internet technologies, electronic commerce, etc.

Know your Firm–Wal-Mart

Wal-Mart is the largest retailer in the world. With its new Wal-Mart Supercenters, the regular Wal-Mart discount stores, and grocery stores, the company has become the second largest grocer in the United States behind Kroger. Its megastores average approximately 180,000 square feet and are located mainly in the South and Midwest. The company's Sam's Club stores operate in 48 states. In 1998, the company operated 1,921 discount stores and 441 Supercenters. Wal-Mart also has stores in South and Central America, Asia, and Europe.

- 1962: Sam Walton opens the first Wal-Mart store in Rogers, Arkansas
- 1969: The company incorporates as Wal-Mart Stores, Inc.
- 1976: *Forbes* magazine ranks Wal-Mart number one on their list of discount and variety stores

Contd

- 1985: Wal-Mart launches the 'Made in the USA' program to encourage buying domestically produced products
- 1987: Wal-Mart acquires 18 Supersave Wholesale Clubs and renames them Sam's Club
- 1994: A joint venture is formed to open stores in Hong Kong and China
- 1996: Fiji Photo purchases rights to Wal-Mart's photo finishing labs

Source: www.walmart.com

Ideally a supply chain management project starts with a business case to define the targets and quantify the benefits of the project and is followed by a high-level process design. The high-level process design defines which processes are in the scope, whether they are local or a global and is used to define the according roles and responsibilities. Depending on the impact of the organizational changes, change management gains increasing importance to support the acceptance of the new processes and thus indirectly of the new planning system. Supply chains exist in both service and manufacturing organizations, although the complexity of the chain may vary significantly from industry to industry and firm to firm. Depending on the type of product or service and the sequence of steps in the supply chain process, supply chains can be categorized into various structures such as pipelined structure; late customization, divergent structure, and convergent structure. In the pipeline structure, the product goes through a series of production/assembly stages as in mass production or continuous manufacturing. In the customization structure, the initial stages will produce standard items which are assembled and customized to specific requirements either in local plants or in the distribution process. Personal computers, integrated circuit (IC) chips, disk drives, laser printers, electronic gadgets, etc. fall in this category. Here the product variety is obtained in late stages of manufacturing/assembly. Inventories are maintained at the sub-assembly level and customization facilities or plants will assemble these or configure these rapidly into customer desired products. In the diverging structure, customization starts in early production phases. A wide variety of finished products are produced with a limited number of raw materials or components. Examples of such supply chains include electro-mechanical systems such as motors, textiles, metal fabrications, and chemicals.

> **Celerant Launches RFID For POS**
>
> Celerant Technology announced the release of RFID for POS, launched at the NRF Show held at the Jacob Javits Centre recently. This cutting-edge technology allows for faster checkouts and more efficient inventory tracking.
>
> 'RFID seems to have real-life applications in the warehousing environment. However, there is not much RFID to speak of at the POS or retail store back-office,' stated Ian Goldman, President CEO, Celerant. 'For RFID to work in real life, it must be incorporated into the entire product lifecycle (receiving, transfers, sales, returns and, etc.). Once that is done, retailers can enjoy a POS transaction as simple as a customer dropping 10 garments on a counter, or a store-to-store transfer essentially as simple with an inventory clerk.'
>
> In addition to reduced transaction time at the POS, RFID will also allow for more precise inventory tracking, increased theft control and more efficient transfers.

A supply chain management has sub-modules for procurement of materials, transformation of the materials into products and distribution of products to customers. Successful supply chain management allows an enterprise to anticipate demand and deliver the right product to the right place at the right time at the lowest possible cost to satisfy its customers.

21.7.4 Bullwhip Effect

One of the most influential articles to be published in the *Harvard Business Review* was Theodore Levitt's Marketing myopia. Levitt focused on the capacity of technological change to undermine the competitive foundations of an industry, but other types of shifts in the wider business environment can also cause them to crumble. Supply chain coordination functions well as long as all stages of the chain take actions that together increase total supply chain profits. Each participant (phase) of the chain should maintain its actions in a good relation to other participants and the supply chain in general and make decisions beneficial to the whole chain. If the coordination is weak or does not exist at all, a conflict of objectives appears among different participants, who try to maximize personal profits. Besides, all the relevant information for some reason can be unreachable to chain participants, or the information can get deformed in non-linear activities of some parts of chain which leads to irregular comprehension. All these lead to the so-called bullwhip effect resulting from information disorder within a supply chain.

The bullwhip effect increases the level of inventory leading to warehouse space shortage, all of which leads to an increase in holding or carrying costs of storage services. It prolongs the lead time and demands more efficient transportation to satisfy the increased demand, which leads to a high transportation cost. It also increases labor costs decreasing the level of product availability, which can lead to deficiency of retail inventory.

21.7.5 Logistics

Logistics as a discipline began within the military and the origin of the word logistics comes from *loger*, the ancient French term for a soldier's barracks building or quarters. Logistics became a topic for discussion in the business world in the 1960s and 1970s and began its rise to prominence in the 1980s. It was truly in the 1990s however that Logistics began to garner the appreciation it deserved. Logistics and supply chain management has come to the boardrooms and lots of discussions on implementation and team formations are taking place in corporate. As believed in the industry 90 per cent of all costs lies in supply chain and to effectively implement world class SCM practices in other words means 'saving money'.

Effective logistics is the key to compete in today's market. It is a process, a supply pipeline, which stretches from your vendor into your customer. This is true whether you compete in the US, India, Europe, or internationally. Competitors, customers, and suppliers are present worldwide. Logistics can contribute to a competitive advantage if it is viewed as a comprehensive process with a single objective to make the final product more competitive in the eyes of the customer.

The supply chain process lead time is the time spent by the supply chain to convert the raw materials into final products plus the time needed to reach the products to the customer. It thus includes supplier lead time, manufacturing lead time, distribution lead time, and the logistics lead time for transport of raw materials from suppliers to plants and for transport of semi-finished/finished products in and out of intermediate storage points.

Lead time in supply chains is dominated by the interface delays due to the interfaces between suppliers and manufacturing plants; between plants and warehouses, between distributors and retailers, etc. Lead-time compression is an extremely important topic because of time based competition and the correlation of lead time with inventory levels, costs, and customer service levels.

21.7.6 Supply Chain Process visibility

Supply chain complexity has multiplied exponentially in all markets, but nowhere is the mix of design, manufacturing, and assembly more intricate and delicately balanced than in the electronics industry. With global outsourcing splintering manufacturing processes into ever-smaller specialties, partners in an electronics supply chain require unprecedented levels of collaboration and communication. Supply chain integration refers to the extent to which a firm shares relevant information about its customers with its supply chain partners. Supply chain integration ensures that products and services offered by various organizational units and suppliers are coordinated to provide a better customer experience. For example, Fisher, Raman, and McClelland (2000) [Fisher et al., 2000] note that IT-enabled data accuracy is critical for efficient forecasting and to design agile supply chain management processes.

> **Flextronics–Supply Chain Visibility**
>
> At Flextronics, the world's leading electronics manufacturing services (EMS) provider, headquartered in Singapore, achieving a better understanding of volatile fluctuations in customer demand has contributed to its success. The 14.5 billion global company, which contracts with as many as 2,500 suppliers in 32 countries, coordinates design, engineering, manufacturing, and logistics operations across its network for leading original equipment manufacturers (OEMs). Flextronics worked with RiverOne Inc., an Irvine (Calif.) provider of supply-chain management software for manufacturers, to develop a new tool: FLEXPASS, a clear, integrated technology platform that enables Flextronics to collaborate quickly and easily with its supply chain partners. FLEXPASS links RiverOne technology with Flextronics' enterprise resource planning systems. The result is an environment where suppliers can view demand changes, adjust production plans, and share information about delivery dates and volumes.

21.8 Business Analytics and Knowledge Management

The field of business analytics has improved and provides business users with better insights, particularly from operational data stored in transactional systems. Business intelligence is an architecture and a collection of integrated operational as well as decision-support applications and databases that provide the business community easy access to business data. It is the analytical process of reasoning, forecasting and measuring business actions and processes based on extracted patterns in collected business data and business plans.

The business analytics software market comprises performance management (PM) tools and applications and data warehouse (DW) platform software. Software is used to access, transform, store, analyse, model, deliver, and track information to enable fact-based decision-making and extend accountability by providing all decision-makers with the right information, at the right time, using the right technology. Corporations have started capitalizing on the techniques of business analytics to achieve new breakthroughs in process performance. Business analytics is broad enough to include capabilities and solutions that benefit a variety of disciplines.

As Davenport and Prusak (1998) explain, transforming information into knowledge involves making comparisons, thinking about consequences and connections, and engaging in conversations with others. Strategy is a type of knowledge and is itself an outcome of knowledge processing. There are two knowledge processes: knowledge production, the process an organization executes that produces new general knowledge and other knowledge whose creation is non-routine; and knowledge integration, the process that presents this new knowledge to individuals and groups comprising the organization. Knowledge management is the set of processes that seeks to change the organization's present pattern of knowledge. The applications of IT in business is well positioned with business analytics and knowledge management.

21.9 Management Information Systems

A management information system (MIS) is a system or process that provides the information necessary to manage an organization effectively. The MIS and the information it generates are generally considered essential components of prudent and reasonable business decisions. By its very nature, management information is designed to meet the unique needs of individual institutions.

As a result, MIS requirements will vary depending on the size and complexity of the operations. For example, systems suitable for community, sized institutions will not necessarily be adequate for larger institutions. However, basic information needs or requirements are similar in all financial institutions regardless of the size. The complexity of the operations and/or activities, together with institution size, point to the need for MIS of varying degrees of complexity to support the decision-making processes.

21.10 Geographic Information System

A Geographic Information System (GIS) is simply using geography and technology to help people make decisions and allow them to better understand our world. It is a collection of computer hardware, software, and geographic data for capturing, storing, updating, manipulating, analysing, and displaying all forms of geographically referenced information. The GIS takes layers of information for viewing and analysis. These layers are represented as points, lines, polygons, and annotation. Due to the fact that each layer has attribute data associated to it through a database, the map can be 'asked' questions by the user and the results displayed for review and decision-making.

Summary

Business process reengineering is a methodology by which important improvements are obtained, although it requires big changes in organization and work style. This involves the need to change or even increase working styles, job functions, needed knowledge, and organization values. In this way, reengineering requires long-time dedication, resources, and effort. These are made easier by using elements called enablers. The value process and the business process are the major elements of business modelling.

Order processing is the interface between sales and marketing on the one hand, and distribution and manufacturing on the other. Order processing comprises order entry, checking customer credit worthiness and stock availability, order confirmation, order progressing and reporting, stock allocation and sometimes invoicing. Although no two companies are the same, standard tools and techniques are available. The concept of these tools and techniques is similar to the varied tools that are available in a toolbox. Each tool and its possible application must be well understood before the best selection can be made for the job at hand. Different situations and industries mandate the use of different tools. The pitfalls to avoid and secrets for a successful implementation are identified to best utilize this critical enterprise communications tool to increase overall enterprise profitability and growth now and in the future.

It is important to recognize that customer relationship management targets customer life cycle management, consisting of the customer life cycle, its underlying business processes, and the enabling technology ecosystem. Ultimately, organizations will transform from product-centricity to customer-centricity.

Key Terms

Content It is the information conveyed by a website, such as company information, product information, news releases, order history, and user preferences.

CRM It is the business process that captures and stores information about the customer, including preferences, past and present orders, business relationships, etc.

ERP It is the global management of business processes such as procurement, administration, project management, sales force automation, product development, and order fulfillment.

ERP software ERP software is a program that unites all the operations of an enterprise into one control. The significance of this program is that it makes use of one device to maneuver and operate the entire organization, its vital and routine activities.

Review Questions

21.1 What is an Information System?
21.2 Discuss MRP and its implications in manufacturing.
21.3 What do you understand by ERP and what is the need for ERP for an organization?
21.4 Differentiate between cycle-time and lead-time?
21.5 What is CRM and what is the need for CRM
21.6 What is supply chain? Differentiate between a supply chain and integrated supply chain management.

Projects

21.1 ABC unlimited is a retail chain establishment with multiple outlets. As you are aware that you have helped the organization move their data centre to a new headquarters building several miles away. Refer to details of the case in Chapter 1.

Two months ago, ABC Unlimited approached the consulting firm 'Retail Automation Consulting Services' to facilitate ABC in the selection and installation of a new enterprise resource planning system for their centralized operations, with several retail outlets spanning the continent from Canada to Mexico. Retail Automation Consulting Services was represented by Apostle. CEO David had approached CEO Rathod, to help ABC Unlimited to acquire and implement a suitable ERP solution. As Apostle was heading the ERP consulting line of business at Retail Automation Consulting, Apostle took over this exercise.

Apostle and the retail manager were sitting at opposite ends of a chipped Formica table. Through the window Apostle could sense the New York wind gusting; inside the thin walls of the small break room vibrated with the muffled din of heavy equipment. Coffee steamed in Styrofoam cups, and scattered about the table lay newspapers and year-old magazines. The retail manager swirled the coffee in his cup, then he looked Apostle straight in the eye and said 'Get the hell out of here'. Disagreement and negotiation can be messy and uncomfortable, but they are also a healthy means of constructive change. Every department or functional unit within an enterprise may have its own point of view, with objectives that, while not directly in opposition with the others, create subtle conflict.

Apostle was wondering on what to do. As Apostle did recall, ABC has emerged successful and has outlets across the world. There was a statement that there is no night time for ABC since ABC is always open across the world. This spread had created the disparity in data reporting which has created the need for ERP. Although many talented individuals representing years of specialized industry experience worked at these retail outlets, each individual rarely communicated with his or her peers at the other sites. This was partly due to their geographic separation, but the most significant cause was more subtle and difficult to overcome. Each of these retail outlets had been an independent profit centre managed by able group of people. Each location enjoyed a proud heritage, where local managers and employees maintained their own customs and business practices. Executives hoped that a new ERP system would be the catalyst to bring these disparate sites together, sharing ideas to develop enterprise wide best practices. By marshalling their considerable design and engineering talent in collaboration with their customers, they would develop a coordinated supply chain enabling them to better service national accounts, thus establishing a competitive advantage in what was a relatively unsophisticated and localized niche industry.

A couple of regions were non-committal on the move to ERP. As the incidents were probed, it was discovered that the retail managers were particularly concerned that a new ERP system would push more work than they could handle, while at the same time requiring unnecessary data capture activity. As a consultant and a change management leader, Apostle coordinated with ABC Unlimited, to work out on a plan on ERP solutions. ABC feels that an ERP solution is required to get a comprehensive and real time view of business across all its outlets. In fact, the very thought of moving to ERP has excited all the employees at the HQ and at the outlets. ERP was implemented successfully over a period of 15 months.

Discuss the technical and management plans which will enable ABC to carry out the migration to ERP with no detectable disruptions. As you are aware, there are open source and commercial ERP software systems available for the customers. Prepare a report on the features of an open source ERP software available on the Internet and advice the organization on the selection of a suitable ERP system.

21.2 ABC unlimited is a retail chain establishment with multiple outlets. As you are aware you have helped the organization move their data centre to a new headquarters building several miles away. Refer to details of the case in Chapter 1.

ABC Unlimited, is working out on a plan to go for CRM solutions to understand customer behaviour. They in fact strongly believe that a suitable CRM solution will help them understand the customers more than anything else. However, they are confused on the variety of CRM solutions available in the market. They are bogged down with the exhaustive amount of literature flooded by the various vendors.

They feel that an CRM solution is required to get a comprehensive and real time view of business across all its outlets. In fact the very thought of moving to CRM has excited all the employees at the HQ and at the outlets.

Discuss the technical and management plans which will enable ABC to carry out the migration to CRM with no detectable disruptions. As you are aware there are open source and commercial ERP software systems available for the customers. Prepare a report on the features of an open source CRM software available in the Internet and advice the organization on the selection of a suitable CRM system.

REFERENCES

Abramowicz, Witold and C. Heinrich 2007, *Technologies for Business Information Systems*, Springer, May.

Bancroft, N., H. Seip, and A. Sprengel 1998, *Implementing SAP R/3–How to Introduce a Large System into a Large Organization*, second edition, Manning Publications Co.

Basili, Victor R. 2004, *Process Improvement in Practice: A Handbook for IT Companies*, Kulver.

Bharadwaj, A. 2000, *A Resource Based Perspective on Information Technology Capability and Firm Performance: An Empirical Investigation*, MIS Quarterly, 24 (1), 169–96.

Bingi, P., M.K. Sharma, and J.K. Godla 1999, *Critical issues affecting an ERP implementation*, Information Systems Management 16 (3), pp. 7–14.

Bolton, Ruth N., P.K. Kannan, and Matthew D. Bramlett 2000, *Implications of Loyalty Program Membership and Service Experiences for Customer Retention and Value*, Journal of the Academy of Marketing Science, 28 (1), 95–108.

Boudreau, M. and D. Robey 1996 'Coping with Contradictions in Business Process Re-engineering,' *Information Technology and People*, 9, pp. 40–57.

Chen, I.J. 2001, 'Planning for ERP Systems: Analysis and Future Trend,' Business Process Management Journal 7 (5), pp. 374–386.

Chorafas, Dimitris N. 2001, *Integrating ERP, CRM, Supply Chain Management, and Smart Materials*, CRC Press

Chung, S. and Snyder, C. 2000, 'ERP Adoption: A Technological Evolution Approach,' International Journal of Agile Management Systems, 2(1): 24-32.

Davenport, T.H. 1993, *Process Innovation: Reengineering Work Through Information Technology*, HBS Press, Boston, MA.

Dickersbach Jörg Thomas 2009, *Supply Chain Management with APO Structures, Modelling Approaches*, Third Edition, Springer.

Fisher, Marshall L., Ananth Raman, and Anna Sheen McClelland 2000, *Rocket Science Retailing Is Almost Here: Are You Ready?* HBR, 78 (July–August), 115–24.

Glazer, Rashi 1991, 'Marketing in an Information Intensive Environment: Strategic Implications of Knowledge as an Asset,' Journal of Marketing, 55 (October), 1–19.

Greenberg, Paul, *CRM at the Speed of Light: Essential Customer Strategies for the 21st Century*, Third Edition, McGraw Hill.

Harrison, Terry P., Hau L. Lee, and John J. Neale 2005, *The Practice of Supply Chain Management*, Springer.

Holland, C., B. Light 1999, 'A Critical Success Factors Model for ERP implementation,' IEEE Software, May/June, 30–36.

Hossain, Liaquat, Jon David Patrick, and M.A. Rashid 2002, *Enterprise Resource Planning: Global Opportunities and Challenges*, Idea Group Publishing.

Jacot, Allen, Michael Jacot, and John Stern 2009, *JD Edwards EnterpriseOne: The Complete Reference*, McGraw Hills.

Jayachandran, Satish, Kelly Hewett, and Peter Kaufman 2004 *Customer Response Capability in a Sense-and-Respond Era: The Role of Customer Knowledge Process*, Journal of the Academy of Marketing Science, 32 (Summer), 219–33.

Lindstrand, Angelika, Jan Johanson, and Dharma Deo Sharma 2006, *Managing Customer Relationships on the Internet*, Elsevier.

Mithas, Sunil, M.S. Krishnan, and Claes Fornell, 2005, 'Why Do Customer Relationship Management Applications Affect Customer Satisfaction?' Journal of Marketing Vol. 69, 201–209

Murthy, Christopher 2005, *Competitive Intelligence Gathering, Analysing and Putting it to Work*, Gower.

Newell Frederick 2003, *Why CRM Doesn't Work: How to Win by Letting Customers Manage the Relationship*, Bloomberg.

Oba, Michiko 2001, Multiple Type Workflow Model for Enterprise Application Integration, Proceedings of the 34th Hawaii International Conference on System Sciences.

Ptak, Carol A. 2004, *ERP Tools, Techniques, and Applications for Integrating the Supply Chain*, CRC Press.

Sirkin, Harold L., Perry Keenan, and Alan Jackson 2005, *The Hard Side of Change Management*, HBR, Oct.

Stapleton, Gregg and Catherine J. Rezak 2004, Change Management Underpins a Successful ERP Implementation at Marathon Oil Journal of Organizational Excellence.

Suzanne, Rivard et al., 2004, *Information Technology and Organizational Transformation*, Elsevier.

Zrimsek, Brian, and Derek Prior 2003, ERP *Systems Implementation: Drivers of Post-Implementation Success*, Gartner.

Index

Access control 298
Access control list 298, 490
Access point 215
Access point device 215
ACID 353
    atomicity 353
    consistency 353
    durability 353
    isolation 353
Ad hoc 215
American National Standards Institute 170
ARPA 409
Assemblers 88
Assembly language 113
Audit 560
    knowledge 560

Backdoor 298
Bastion host 298
Biometric systems 286
Bots 269
Browser 373
    FireFox 374
    Internet Explorer 373
Bullwhip effects 670
Byte 124

Cable 165
    attenuation 165
    connectors 188
    crosstalk 165
    impedance 165
    noise 165
    resistance 165
    STP 176
    twisted-pair 173
    UTP 174
    wall plates 190
Cabling 186
    backbone 187
    horizontal 186
    pathways 187
    structured cabling system 168
    testing and certification 191
Capability Maturity Model 580
CCTV 286
Change management 465, 658

atomic change 465
isolated change 465
Checksum 124
Chief executive officer 4
Chief financial officer 6
Chief information officer 6, 122
Chief technology officer 5
Circuit switched network 420
Cloud computing 75
    Amazon Elastic Compute Cloud 80
    cloud platforms 78
    Cumulus 79
    Eucalyptus 79
    Google App Engine 80
    hybrid clouds 78
    Nimbus 79
    private cloud 78
    public clouds 78
    Windows Azure 80
Cluster 73
    active/active 74
    active/passive 74
    Beowulf 72
    clustering 72
    commodity 73
    failback 74
    failover 73
    high availability 73
CMS 365
    Alfresco 365
    Documentum 365
    Mambo 365
    MCMS 367
    need 362
    portal 362
COBIT 34 458
Computing 35, 76
    cloud 76
    green 35
    grid 36 74
Computing infrastructure 11
Configuration item 465
Consolidation 37
    server 37
    storage 38
Content management solution 362
Cookies 379

session cookie 379
CRM 659
    business benefits 665
    process flow 663
    SugarCRM 663
CSMA/CD 129
Cycle time 642

DARPA 409
Database 341
    ACID properties 352
    file 342
    object 342
    transaction 352
Data centre 35
Data files 223
    non-proprietary 223
    proprietary 223
Data management 340
Data mining 359, 360
    CRISP 360
    SEMMA 360
Data storage 220
DBMS 342
    DB2 351
    Microsoft SQL Server 351
Demilitarized zone 278, 298
Department of Defense 126
Dreamweaver 381
Dense wavelength division multiplexing 172

Enterprise application integration 636
    brokered integration 637
    data-level integration 637
    function-level integration 637
Electronic data interchange 622
Email 268
    spam 269
Enterprise content management 361
EPC Global 290
Enterprise resource planning 643
    business object 654
    evolution 646
    integration 651
    local implementation 651
    Microsoft Dynamics 657

Index

mixed mode implementation 651
Ethernet 16, 128, 166
   fast 129
   gigabit 129
Event 464
Extranet 439

Fibre 179
   cladding 179
   core 179
   graded index 184
   multimode 184
   sheath 180
   single mode 183
   step-index 184
   types 183
Fibre channel 229
Fibre-optic cable 177
Fibre optics 177
   electronic bottleneck 181
Firewall 274, 300
   defined 276
Firm 53, 56
   Adobe 312
   America Online 378
   AT&T Corporation 181
   Cisco 142
   Dell 56
   Fortinet 279
   Gemini Communications Ltd 287
   ICICI 604
   IFFCO 592
   Intel 53
   ITC 632
   MindTree 644
   Oracle 349
   Robert Bosch 629
   Teamlease 636
   Wal-Mart 668
Four A framework 6
Front Page 381

Gartner 9
   five C model 9
Gartner Group 5
GIS 673
Globalization 600
Governance 33
Grid 74
   data 75
   services 74

HTML 376
   hyperlinks 377
HTTP 379
   stateless 379

ICT 121
IDS 282
Incident management 464
Information technology 4, 454
Information warfare 300
Infrastructure 4, 74
Integrated supply chain 668
Intermediary 610
Internet 407
   Internet Activities Board 408
   Internet Engineering Planning Group 409
   Internet Engineering Steering Group 409
   Internet Engineering Task Force 408
   Internet Research Task Force 409
   Society 408
Internet protocol 416
   addressing mechanism 416
   functions 416
   limitations 417
   V4 address system 417
   V6 address system 418
Internet service providers 123, 409
Internet services 440
Intrusion detection 300
International Organization for Standards 124
ISP 134
ITIL 458

Local area network 16
Load balancers 13
Log 297
   activity logging 298
Logistics 671
Lotus notes 326, 330
LUN 222

Metropolitan area network 17
Management 26
   configuration 462
   data 26
   inventory-asset 462
   remote infrastructure 28

SLA 29
Mobile ad hoc network 19
Manufacturing 648
   discrete 648
   just in time 649
   process 648
   repetitive 648
Metric 459
   hierarchical 459
Microsoft Corporation 300, 303
MIS 673
Modelling 638
MRP 646
   MRPII 646
MS Office 303
   font 305, 309
   formatting 306, 309
   keyboard shortcuts 308, 311
   MS Excel 309, 313
   MS Word 301, 304
   Pivot tables 318, 321
   PowerPoint 321, 324
   styles 323, 326
   text alignment 307, 310
   View 324, 327
   Word templates 304, 307

Network 18
   ad hoc 18, 215
   EPON 21
   GPON 22
   LAN 16
   MAN 17
   mobile 20
   optical 20
   passive optical 20, 21
   PSTN 420
   telecommunication 420
   WAN 17
   wireless network 18
Networking infrastructure 11
Network interface card 129

Open Office 330
   Calc 327, 331
   Draw 327, 331
   Impress 327, 331
   macros 328, 331
   Writer 327, 331
Open source initiative 106
Operating system 90, 302
   multiprogramming 90
   virtual memory 90

Index

Windows 98 300–304
Order processing 649
Open system interconnection
   123–124
   application layer 126
   data link layer 125, 414
   network layer 125, 413
   physical layer 125, 414
   presentation layer 126
   reference model 412
   session layer 126, 413
   transport layer 125

Picture archiving and communication system 25
Porter's value chain 640
Process 631
   business 456, 631
   business process reengineering 633
   degree of collaboration 635
   degree of mediation 635
   reengineering 633
Processer
   access time 50
   CISC 49
   fetching 49
   RISC 49
Protocol 123, 331, 334, 411
   ARP 413
   FTP 441
   HTML 376
   ICMP 413
   Internet message access protocol 331, 334
   IP 127, 416
   MPLS 420
   post office protocol 331, 334
   storage 227
   storage-iSCSI 228
   TCP 128
   TCP/IP 411, 415
   Telnet 441
   UDP 128

RAID 11
RFID 289
   active tags 298
   EPC Global 289
ROI 437
Router 12

SAP 652
   ABAP 653
   BAPI 654
   BOR 654
   Internet transaction server 655
   logical unit of work 655
   service access points 125
Structured cabling system 168
SCSI 222
Security 283
   biometric systems 286
   bots 269
   cameras 285
   DoS 268
   Firewall 273
   insider threat 269
   RFID 288
   social engineering 272
   spam 268
   worm 267
Service level agreements 460
Services 439
   FTP 441
   service coupling 445
   telnet 441
   web 443
SGML 375
   XML 380
Storage management initiative 15
Storage Networking Industry Association 15
Software 105
   FOSS 105
   license 107
   open source 105
SOX 560
Storage 222
   DVD 225
   holographic 226
   solid state 225
Storage area networks 229
Storage infrastructure 14–15
Storage media 222
   types 223
Shielded twisted pair 11
Structured cabling 11
Supply chain 639, 665
Supply chain systems 666
Switch 13, 136
Switches 20
   optical 20
Systems 22

access control 488
expert 23
geographic information 23
health information 25
management information 23

TCP-IP 126
Telecommunications Industry Association 170
Topology 131
Transaction 355
   external 355
   flat 355
   internal 355
   lock 356
   transaction processing monitor 356
Transaction processing 352
   application 354
   system 352
Trunks 421

Unshielded twisted pair 11

Video walls 287
Virtualization 36 65
   hypervisors 67
   Servers 71
VoIP 421
   soft phone 423
Vulnerability 263

Wide area network 17
Wavelength division multiplexing 20
Weblog 437
Web servers 371
   Apache 372
Windows
   Longhorn 103
   Vista 103
   Windows Server 103
   Windows XP 102
Work breakdown structure 646
Workflow 638
   Workflow management system 638
Work in progress 642
Worm 300
WWW 440
   global village 446
   search engine 441
   surfing 446